College Algebra

Henry Burchard Fine

AMS CHELSEA PUBLISHING
American Mathematical Society • Providence, Rhode Island

The Publisher wishes to thank Paul Kirchner for his help
in bringing this book back into print.

2000 *Mathematics Subject Classification.* Primary 12–01, 40–01.

For additional information and updates on this book, visit
www.ams.org/bookpages/chel-354

Library of Congress Cataloging-in-Publication Data

Fine, Henry B. (Henry Burchard), 1858–1928.
College algebra / Henry Burchard Fine.
p. cm.
Originally published: Boston : Ginn and C., 1905.
Includes index.
ISBN 0-8218-3863-6 (acid-free paper)
1. Algebra. I. Title.

QA154.F49 2005
512—dc22 2005048172

Reprinted by the American Mathematical Society, 2005.
Printed in the United States of America.

∞ The paper used in this book is acid-free and falls within the guidelines
established to ensure permanence and durability.
Visit the AMS home page at `http://www.ams.org/`

10 9 8 7 6 5 4 3 2 1 10 09 08 07 06 05

PREFACE

In this book I have endeavored to develop the theory of the algebraic processes in as elementary and informal a manner as possible, but connectedly and rigorously, and to present the processes themselves in the form best adapted to the purposes of practical reckoning.

The book is meant to contain everything relating to algebra that a student is likely to need during his school and college course, and the effort has been made to arrange this varied material in an order which will properly exhibit the logical interdependence of its related parts.

It has seemed to me best to divide the book into two parts, a preliminary part devoted to the number system of algebra and a principal part devoted to algebra itself.

I have based my discussion of number on the notion of cardinal number and the notion of order as exhibited in the first instance in the natural scale $1, 2, 3, \cdots$. There are considerations of a theoretical nature in favor of this procedure into which I need not enter here. But experience has convinced me that from a pedagogical point of view also this method is the best. The meaning of the ordinal definition of an irrational number, for example, can be made clear even to a young student, whereas any other real definition of such a number is too abstract to be always correctly understood by advanced students.

My discussion of number may be thought unnecessarily elaborate. But in dealing with questions of this fundamental character a writer cannot with a good conscience omit points which properly belong to his discussion, or fail to give proofs of statements which require demonstration. I hope the details of the discussion will interest the more thoughtful class of students; but all that the general student need be asked to learn from it is the ordinal character of the real numbers and of the relations of equality and inequality among them, and that for all numbers, real and complex, the fundamental operations admit of definitions which conform to the commutative, associative, and distributive laws.

In the second or main part of the book I begin by observing that in algebra, where numbers are represented by letters, the laws just mentioned are essentially the definitions of the fundamental operations. These algebraic definitions are stated in detail, and from them the entire theory of the algebraic processes and the practical rules of reckoning are subsequently derived deductively.

I shall not attempt to describe this part of the book minutely. It will be found to differ in essential features from the text-books in general use. I have carefully refrained from departing from accepted methods merely for the sake of novelty. But I have not hesitated to depart from these methods when this seemed to me necessary in order to secure logical consistency, or when I saw an opportunity to simplify a matter of theory or practice. I have given little space to special devices either in the text or in the exercises. On the other hand, I have constantly sought to assist the student to really master the general methods of the science.

Thus, instead of relegating to the latter part of the book the method of undetermined coefficients, the principal method of investigation in analysis, I have introduced it very early and have subsequently employed it wherever this could be done to advantage. This has naturally affected the arrangement of topics. In particular I have considered partial fractions in the chapter on fractions. They belong there logically, and when adequately treated, supply the best practice in elementary reckoning that algebra affords.

Again, I have laid great stress upon the division transformation and its consequences, and in connection with it have introduced the powerful method of synthetic division.

The earlier chapters on equations will be found to contain a pretty full discussion of the reasoning on which the solution of equations depends, a more systematic treatment than is customary of systems of equations which can be solved by aid of the quadratic, and a somewhat elaborate consideration of the graphs of equations of the first and second degrees in two variables.

The binomial theorem for positive integral exponents is treated as a special case of continued multiplication, experience having convinced me that no other method serves so well to convey to the student the meaning of this important theorem. I have introduced practice in the use of the general binomial theorem in the chapter on fractional exponents, but have deferred the proof of the theorem itself, together with all that relates to the subject of infinite series, until near the end of the book.

In the chapters on the theory of equations and determinants there will be found proofs of the fundamental theorems regarding symmetric functions of the roots of an equation and a discussion of the more important properties of resultants. These subjects do not belong in an elementary course in algebra, but the college student who continues his mathematical studies will need them. The like is to be said of the chapters on infinite series and of the chapter on properties of continuous functions with which the book ends.

The ideas which underlie the first part of the book are those of Rowan Hamilton, Grassmann, Helmholtz, Dedekind, and Georg Cantor. But I do not know that any one hitherto has developed the doctrine of ordinal number from just the point of view I have taken, and in the same detail.

In preparing the algebra itself I have profited by suggestions from many books on the subject. I wish in particular to acknowledge my indebtedness to the treatises of Chrystal.

The book has been several years in preparation. Every year since 1898 the publishers have done me the courtesy to issue for the use of the freshmen at Princeton a pamphlet containing what at the time seemed to me the most satisfactory treatment of the more important parts of algebra. With the assistance of my colleagues, Mr. Eisenhart and Mr. Gillespie, I endeavored after each new trial to select what had proved good and to discard what had proved unsatisfactory. As a consequence, much of the book has been rewritten a number of times. No doubt subsequent experience will bring to light many further possibilities of improvement; but I have hopes that as the book stands it will serve to show that algebra is not only more intelligible to the student, but also more interesting and stimulating, when due consideration is given to the reasoning on which its processes depend.

HENRY B. FINE

PRINCETON UNIVERSITY
June, 1905

CONTENTS

PART FIRST — NUMBERS

PART SECOND — ALGEBRA

A COLLEGE ALGEBRA

PART FIRST—NUMBERS

I. THE NATURAL NUMBERS—COUNTING, ADDITION, AND MULTIPLICATION

GROUPS OF THINGS AND THEIR CARDINAL NUMBERS

Groups of things. In our daily experience things present themselves to our attention not only singly but associated in groups or assemblages. **1**

The fingers of a hand, a herd of cattle, the angular points of a polygon are examples of such groups of things.

We think of certain things as constituting a *group*, when we distinguish them from other things not individually but as a whole, and so make them collectively *a single object* of our attention.

For convenience, let us call the things which constitute a group the *elements* of the group.

Equivalent groups. One-to-one correspondence. The two groups of letters ABC and DEF are so related that we can combine *all* their elements in pairs by matching elements of the one with elements of the other, one element with one element. Thus, we may match A with D, B with E, and C with F. **2**

Whenever it is possible to match all the elements of two groups in this manner, we shall say that the groups are *equivalent;* and the process of matching elements we shall call bringing the groups into a one-to-one relation, or a relation of *one-to-one correspondence.*

3 **Theorem.** *If two groups are equivalent to the same third group, they are equivalent to one another.*

For, by hypothesis, we can bring each of the two groups into one-to-one correspondence with the third group. But the two groups will then be in one-to-one correspondence with each other, if we regard as mates every two of their elements which we have matched with the same element of the third group.

4 **Cardinal number.** We may think of all possible groups of things as distributed into classes of equivalent groups, any two given groups belonging to the same class or to different classes, according as it is, or is not, possible to bring them into one-to-one correspondence.

Thus, the groups of letters $ABCD$ and $EFGH$ belong to the same class, the groups $ABCD$ and EFG to different classes.

The property which is common to all groups of one class, and which distinguishes the groups of one class from those of another class, is the *number of things* in a group, or its *cardinal number*. In other words,

The number of things in a group, or its cardinal number, is that property which is common to the group itself and every group which may be brought into one-to-one correspondence with it.

Or we may say: "The cardinal number of a group of things is that property of the group which remains unchanged if we rearrange the things within the group, or replace them one by one by other things"; or again, "it is that property of the group which is independent of the *character* of the things themselves and of their *arrangement* within the group."

For rearranging the things or replacing them, one by one, by other things will merely transform the group into an equivalent group, § 2. And a property which remains unchanged during all such changes in the group must be independent of the character of the things and of their arrangement.

Part. We say that a first group is a *part* of a second group when the elements of the first are some, but not all, of the elements of the second. **5**

Thus, the group ABC is a part of the group $ABCD$.

From this definition it immediately follows that

If the first of three groups be a part of the second, and the second a part of the third, then the first is also a part of the third. **6**

Finite and infinite groups. We say that a group or assemblage is *finite* when it is equivalent to no one of its parts; *infinite* when it *is* equivalent to certain of its parts.* **7**

Thus, the group ABC is *finite;* for it cannot be brought into one-to-one correspondence with BC, or with any other of its parts.

But any never-ending sequence of marks or symbols, the never-ending sequence of numerals 1, 2, 3, 4, $\cdots$, for example, is an *infinite* assemblage.

We can, for instance, set up a one-to-one relation between the entire assemblage 1, 2, 3, 4, $\cdots$ and that part of it which begins at 2, namely,

between $\qquad 1, 2, 3, 4, 5, \cdots \qquad$ (a)

and $\qquad 2, 3, 4, 5, 6, \cdots, \qquad$ (b)

by matching 1 in (a) with 2 in (b), 2 in (a) with 3 in (b), and so on, — there being for every numeral that we may choose to name in (a) a corresponding numeral in (b).

Hence the assemblage (a) is equivalent to its part (b). Therefore (a) is infinite.

Less and greater cardinal numbers. Let M and N denote any two finite groups. It must be the case that **8**

1. M and N are equivalent,

or 2. M is equivalent to a part of N,

or 3. N is equivalent to a part of M.

* Of course we cannot actually take account of all the elements, one by one, of an infinite group — or assemblage, as it is more often called. We regard such an assemblage as *defined* when a law has been stated which enables us to say of every *given* thing whether it belongs to the assemblage or not.

In the first case we say that M and N have the same cardinal number, § 4, or *equal* cardinal numbers; in the second case, that the cardinal number of M is *less than* that of N; in the third, that the cardinal number of M is *greater than* that of N.

Thus, if M is the group of letters abc, and N the group $defg$, then M is equivalent to a part of N, to the part def, for example.

Hence the cardinal number of M is less than that of N, and the cardinal number of N is greater than that of M.

9 **Note.** It follows from the definition of finite group, § 7, that there is no ambiguity about the relations "equal," "greater," and "less" as here defined.

Thus, the definition does not make it possible for the cardinal number of M to be at the same time equal to and less than that of N, since this would mean that M is equivalent to N and also to a part of N, therefore that N is equivalent to one of its parts, § 3, and therefore, finally, that N is infinite, § 7.

10 **Corollary.** *If the first of three cardinal numbers be less than the second, and the second less than the third, then the first is also less than the third.*

For if M, N, P denote any groups of things of which these are the cardinals, M is equivalent to a part of N, and N to a part of P; therefore M is equivalent to a part of P, §§ 3, 6.

11 **The system of cardinal numbers.** By starting with a group which contains but a single element and repeatedly "adding" one new thing, we are led to the following list of the cardinal numbers:

1. The cardinal number of a "group" like |, which contains but a single element.

2. The cardinal number of a group like ||, obtained by adding a single element to a group of the first kind.

3. The cardinal number of a group like |||, obtained by adding a single element to a group of the second kind.

4. And so on, without end.

We name these successive cardinals "one," "two," "three," ⋯ and represent them by the signs 1, 2, 3, ⋯.

12 **Observations on this system.** Calling the cardinal number of any finite group a *finite cardinal,* we make the following observations regarding the list of cardinals which has just been described.

First. Every cardinal contained in this list is finite.

For the group I is finite, since it has no part to which to be equivalent, § 7; and each subsequent group is finite, because a group obtained by adding a single thing to a finite group is itself finite.* Thus, II is finite because I is; III is finite because II is; and so on.

Second. Every finite cardinal is contained in the list.

For, by definition, every finite cardinal is the cardinal number of some finite group, as *M*. But we can construct a group of marks III ⋯ I equivalent to any given finite group *M*, by making one mark for each object in *M*. And this group of marks must have a last mark, and therefore be included in the list of § 11, since otherwise it would be never-ending and therefore itself, and with it *M*, be infinite, § 7.

Third. No two of these cardinals are equal.

This follows from the definition in § 8. For, as just shown, all of the groups I, II, III, ⋯ are finite; and it is true of every two of them that one is equivalent to a part of the other.

* We may prove this as follows (G. Cantor, *Math. Ann.*, Vol. 46, p. 490):

If M *denote a finite group, and* e *a single thing, the group* M e, *obtained by adding* e *to* M, *is also finite.*

For let $G \equiv H$ denote that the groups G and H are equivalent.

If Me is not finite, it must be equivalent to some one of its parts, § 7.

Let P denote this part, so that $Me \equiv P$.

(1) Suppose that P does not contain e.

Let f denote the element of P which is matched with e in Me, and represent the rest of P by P_1.

Then since $Me \equiv P_1 f$ and $e \equiv f$, we have $M \equiv P_1$.

But this is impossible, since M is finite and P_1 is a part of M, § 7.

(2) Suppose that P does contain e.

It cannot be that e in P is matched with e in Me, for then the rest of P, which is a part of M, would be equivalent to M.

But suppose that e in P is matched with some other element, as g in Me, and that e in Me is matched with f in P.

If $Me \equiv P$ be true on this hypothesis, it must also be true if we recombine the elements e, f, g so as to match e in P with e in Me, and f in P with g in Me. But, as just shown, we should then have a part of P equivalent to M. Hence this hypothesis also is impossible

THE NATURAL SCALE. EQUATIONS AND INEQUALITIES

13 **The natural numbers.** We call the signs 1, 2, 3, ⋯ — or their names "one," "two," "three," ⋯ positive integers or *natural numbers.* Hence

A natural number is a sign or symbol for a cardinal number.

14 **The natural scale.** Arranging these numbers in an order corresponding to that already given the cardinals which they represent, § 11, we have the never-ending sequence of signs

$$1, 2, 3, 4, 5, \cdots,$$

or "one," "two," "three," "four," "five," ⋯, which we call the *natural scale,* or the scale of the natural numbers.

15 *Each sign in the scale indicates the number of the signs in that part of the scale which it terminates.*

Thus, 4 indicates the number of the signs 1, 2, 3, 4. For the number of signs 1, 2, 3, 4 is the same as the number of groups I, II, III, IIII, and this, in turn, is the same as the number of marks in the last group, IIII, § 8. And so in general.

16 **The ordinal character of the scale.** The natural scale, by itself considered, is merely an assemblage of *different signs* in which there is a first sign, namely 1; to this a definite next following sign, namely 2; to this, in turn, a definite next following sign, namely 3; and so on without end.

In other words, the natural scale is merely an assemblage of different signs which follow one another in *a definite and known order,* and having a first but no last sign.

Regarded from this point of view, the natural numbers themselves are merely *marks of order,* namely of the order in which they occur — with respect to time — when the scale is recited.

17 It is evident that the scale, in common with all other assemblages whose elements as given us are arranged in a definite and known order, has the following properties:

1. We may say of any two of its elements that the one "precedes" and the other "follows," and these words "precedé" and "follow" have the same meaning when applied to any one pair of the elements as when applied to any other pair.

2. If any two of the elements be given, we can always determine *which* precedes and which follows.

3. If a, b, and c denote any three of the elements such that a precedes b, and b precedes c, then a precedes c.

An assemblage may already possess these properties when given us, or we may have imposed them on it by some rule of arrangement of our own choosing. In either case we call the assemblage an **ordinal system.**

Instances of the first kind are (1) the natural scale itself; (2) a sequence of events in time; (3) a row of points ranged from left to right along a horizontal line. An instance of the second kind is a group of men arranged according to the alphabetic order of their names.

18 An assemblage may also have "coincident" elements. Thus, in a group of events two or more may be simultaneous.

We call such an assemblage *ordinal* when the relations 1, 2, 3 hold good among its *non-coincident* elements — it being true of the coincident elements that

4. If a coincides with b, and b with c, then a coincides with c.

5. If a coincides with b, and b precedes c, then a precedes c.

19 It is by their *relative order in the scale* that the natural numbers indicate relations of greater and less among the cardinal numbers.

For of any two given cardinals that one is greater whose natural number occurs later in the scale.

And the relation: "if the first of three cardinals be less than the second, and the second less than the third, then the first is less than the third," is represented in the scale by the relation: "if a precede b, and b precede c, then a precedes c."

In fact, we seldom employ any other method than this for comparing cardinals. We do not compare the cardinal numbers of groups of things directly, by the method of § 8. On the contrary, we represent them by the appropriate natural numbers, and infer which are greater and which less from the relative order in which these natural numbers occur in the scale. The process causes us no conscious effort of thought, for the scale is so vividly impressed on our minds that, when any two of the natural numbers are mentioned, we instantly recognize which precedes and which follows. Thus, if we are told of two cities, A and B, that the population of A is 120,000, and that of B, 125,000, we immediately conclude that B has the greater number of inhabitants, because we know that 125,000 occurs later in the scale than 120,000 does.

20 **Numerical equations and inequalities.** In what follows, the word "number" will mean natural number, § 13; and the letters a, b, c will denote *any* such numbers.

21 When we wish to indicate that a and b denote the same number, or "coincide" in the natural scale, we employ the **equation**

$$a = b, \text{ read "} a \text{ equals } b\text{."}$$

22 But when we wish to indicate that a precedes and b follows in the natural scale, we employ one of the **inequalities**

$$a < b, \text{ read "} a \text{ is less than } b\text{";}$$

$$b > a, \text{ read "} b \text{ is greater than } a\text{."}$$

23 Of course, strictly speaking, these words, "equal," "less," and "greater," refer not to the signs a and b themselves, but to the cardinals which they represent. Thus, the phrase, "a is less than b," is merely an abbreviation for, "the cardinal which a represents is less than the cardinal which b represents."

But *all* that the inequality $a < b$ means *for the signs* a *and* b *themselves* is that a precedes b in the scale.

24 **Rules of equality and inequality.** From §§ 17, 18 and these definitions, §§ 21, 22, it immediately follows that

1. If $a = b$ and $b = c$, then $a = c$.
2. If $a < b$ and $b < c$, then $a < c$.
3. If $a = b$ and $b < c$, then $a < c$.

COUNTING

Arithmetic is primarily concerned with the ordinal relations existing among the natural numbers, and with certain operations by which these numbers may be combined. **25**

The operations of arithmetic have their origin in *counting*.

Counting. To discover what the cardinal number of a given group of objects is, we *count* the group. **26**

The process is a very familiar one. We label one of the objects "one," another "two," and so on, until there are no objects left — being careful to use these verbal signs "one," "two," ⋯, without omissions, in the order of their occurrence in the scale, but selecting the objects themselves in any order that suits our whim or convenience; and the sign or label with which the process ends is what we seek, — the *name* of the cardinal number of the group itself. For owing to the ordinal character of the scale, this last sign indicates how many signs have been used all told, § 15, and therefore how many objects there are in the group, § 8.

Thus, the *process* of counting may be described as bringing the group counted into one-to-one correspondence, § 2, with a part of the natural scale — namely, the part which begins at "one" and ends with the last number used in the count.

Observe that the natural numbers serve a double purpose in counting: (1) We use a certain group of them as mere counters in carrying out the process, and (2) we employ the last one so used to record the result of the count.

We have intimated that it is immaterial in what order we select the objects themselves. This may be proved as follows:

Theorem. *The result of counting a finite group of objects is the same, whatever the order in which we select the objects.* **27**

Suppose, for example, that the result of counting a certain group were 99 when the objects are selected in one order, P, but 97 when they are selected in another order, Q.

The group which consists of the first 97 objects in the order P would then be equivalent to the entire group in the order Q, for, by hypothesis, both have been matched with the first 97 numbers of the natural scale, § 3.

But this is impossible, since it would make a part of the group equivalent to the whole; whereas the group is, by hypothesis, a finite group, § 7.

28 **Another definition of cardinal number.** We may make the theorem just demonstrated the basis of a definition of the cardinal number of a *finite* group, namely:

The cardinal number of a finite group of things is that property of the group because of which we shall arrive at the same natural number in whatever order we count the group.

This is the definition of cardinal number to which we are naturally led if we choose to make the natural scale, defined as in § 16, our *starting point* in the discussion of number.

ADDITION

29 **Definition of addition.** To *add* 3 to 5 is to find what number occupies the third place after 5 in the natural scale.

We may find this number, 8, by counting *three* numbers forward in the scale, beginning at 6, thus: 6, 7, 8.

We indicate the operation by the sign $+$, read "plus," writing $5 + 3 = 8$.

And in general, to *add* b to a is to find what number occupies the bth place after a in the natural scale.

Since there is no last sign in the scale, this number may always be found. We call it the *sum* of a and b and represent it in terms of a and b by the expression $a + b$.

30 **Note.** The process of finding $a + b$ by counting forward in the scale corresponds step for step to that of adding to a group of a things the elements of a group of b things, one at a time. Hence (1) the result of the latter process is a group of $a + b$ things, § 8, and (2) if a and b denote finite cardinals, so also does $a + b$. See footnote, p. 5.

Since $a + 1$, $a + 2$, and so on, denote the 1st, 2d, and so on, numbers after a, the sequence $a + 1, a + 2, \cdots$ denotes all that portion of the scale which follows a. **31**

Hence any given number after a may be expressed in the form $a + d$, where d denotes a definite natural number.

The process. To add large numbers by counting would be very laborious. We therefore memorize sums of the smaller numbers (addition tables) and from these derive sums of the larger numbers by applying the so-called "laws" of addition explained in the following sections. **32**

The laws of addition. Addition is a "commutative" and an "associative" operation; that is, it conforms to the following two laws: **33**

The commutative law. $a + b = b + a$, **34**

The result of adding b *to* a *is the same as that of adding* a *to* b.

The associative law. $a + (b + c) = (a + b) + c$, **35**

The result of first adding c *to* b *and then adding the sum so obtained to* a, *is the same as that of first adding* b *to* a *and then adding* c *to the sum so obtained.*

Note. In practice, we replace the expression $(a + b) + c$ by $a + b + c$, our understanding being that the expression $a + b + c + \cdots$ represents the result of adding b to a, c to the sum so obtained, and so on. **36**

Proofs of these laws. We may prove these laws as follows. **37**

First. The *commutative law:* $a + b = b + a$.

Thus, the sums $3 + 2$ and $2 + 3$ are equal.

For $3 + 2$ represents the number found by first counting off *three* numbers, and after that *two* numbers, on the natural scale. Thus,

The group counted	1, 2, 3, 4, 5,	(a)
the counters,	1, 2, 3, 1, 2.	(b)

But as there is a one-to-one relation between the groups of signs (a) and (b), and every one-to-one relation is reciprocal, § 2, we may interchange the rôles of (a) and (b); that is, if we make (b) the group counted, (a) will represent the group of counters.

Hence finding $3 + 2$ is equivalent to counting the group of signs

$$1, 2, 3, \quad 1, 2. \tag{b}$$

In like manner, finding $2 + 3$ is equivalent to counting the group

$$1, 2, \quad 1, 2, 3. \tag{c}$$

But as (b) and (c) consist of the same signs and differ only in the manner in which these signs are arranged, the results of counting them are the same, § 27; that is,

$$3 + 2 = 2 + 3.$$

Similarly for any two natural numbers, a and b.

Second. The *associative law:* $a + (b + c) = (a + b) + c$.

For in counting to the bth sign after a, namely to $a + b$, and then to the cth sign after this, namely to $(a + b) + c$, we count $b + c$ signs all told, and hence arrive at the $(b + c)$th sign after a, namely at $a + (b + c)$.

The notion of cardinal number is involved in the proofs just given. But addition may be defined and its laws established independently of this notion, as is shown in the footnote below.*

* The Italian mathematician Peano has defined the system of natural numbers without using the notion of cardinal number, by a set of "postulates" which we may state as follows — where "number" means "natural number."

1. The sign 1 is a number.
2. To each number a there is a next following number — call it $a+$.
3. This number $a+$ is never 1. 4. If $a+ = b+$, then $a = b$.
5. Every given number a is present in the sequence $1, 1+, (1+)+, \cdots$.

The numerals $2, 3, \cdots$ are defined thus: $2 = 1+$, $3 = 2+$, $\cdots$.

The *sum* $a + b$ is to mean the number determined (because of 5) by the series of formulas $a + 1 = a+$, $a + 2 = (a + 1)+$, $\cdots$.

The series of formulas just written is equivalent to the single formula

$$6. \quad a + (b + 1) = (a + b) + 1.$$

From 6, by "mathematical induction," we may derive the laws of addition:

$$7. \quad a + (b + c) = (a + b) + c. \qquad 8. \quad a + b = b + a.$$

First. If 7 is true when $c = k$, it is also true when $c = k + 1$. For, by 6 and 7,

$$a + [b + (k + 1)] = a + [(b + k) + 1] = [a + (b + k)] + 1$$
$$= [(a + b) + k] + 1 = (a + b) + (k + 1).$$

But, by 6, 7 is true when $c = 1$.

Hence 7 is true when $c = 2$, $\therefore$ when $c = 3$, $\therefore \cdots$ when $c =$ any number, by 5.

Second. We first prove 8 for the particular case: 8'. $a + 1 = 1 + a$.

If 8' is true for $a = k$, then $(k + 1) + 1 = (1 + k) + 1 = 1 + (k + 1)$, by 6.

Hence if 8' is true for $a = k$, it is also true for $a = k + 1$.

Hence, since 8' is true for $a = 1$, it is true for $a = 2$, $\therefore$ for $a = 3$, $\cdots$.

Finally, if 8 be true for $b = k$, it is true for $b = k + 1$. For, by 7 and 8',

$$a + (k + 1) = (a + k) + 1 = 1 + (a + k)$$
$$= 1 + (k + a) = (1 + k) + a = (k + 1) + a.$$

Hence, since 8 is true (by 8') when $b = 1$, it is true for $b = 2$, $\therefore$ for $b = 3$, $\therefore \cdots$.

See Stolz and Gmeiner, *Theoretische Arithmetik*, pp. 13 ff., and the references to Peano there given; also Huntington in *Bulletin of the American Mathematical Society*, Vol. IX, p. 40. H. Grassmann (*Lehrbuch der Arithmetik*) was the first to derive 7 and 8 from 6.

General theorem regarding sums. By making repeated application of these laws, §§ 34, 35, it can be shown that **38**

The sum of any finite number of numbers will be the same, whatever the order in which we arrange them, or whatever the manner in which we group them, when adding them.

Thus, $a + b + c + d = a + c + b + d.$

For $a + b + c + d = a + (b + c) + d$ § 35

$= a + (c + b) + d$ § 34

$= a + c + b + d.$ § 35

Rules of equality and inequality for sums. First. From the definition of sum, § 29, and the rules of § 24, it follows that **39**

1. If $a = b$, then $a + c = b + c$.
2. If $a < b$, then $a + c < b + c$.
3. If $a > b$, then $a + c > b + c$.

Here 1 is obvious, since if $a = b$, then a and b denote the same number. We may prove 3 as follows, and 2 similarly.
If $a > b$, let $a = b + d$, § 31.
Then $a + c = (b + d) + c = (b + c) + d$, §§ 34, 35, $\therefore > b + c$.

Second. From 1, 2, 3 it follows conversely that

4. If $a + c = b + c$, then $a = b$.
5. If $a + c < b + c$, then $a < b$.
6. If $a + c > b + c$, then $a > b$.

Thus, if $a + c = b + c$, then $a = b$.
For otherwise we must have either $a < b$ and therefore $a + c < b + c$ (by 2), or else $a > b$ and therefore $a + c > b + c$ (by 3).

Third. It also follows from 1, 2, 3 that

7. If $a = b$, and $c = d$, then $a + c = b + d$.
8. If $a < b$, and $c < d$, then $a + c < b + d$.
9. If $a > b$, and $c > d$, then $a + c > b + d$.

Thus, if $a = b$, then $a + c = b + c$, and if $c = d$, then $b + c = b + d$. Hence $a + c = b + d$.

MULTIPLICATION

40 **Definition of multiplication.** To *multiply* a by b is to find the sum of b numbers, each of which is a.

We call this sum the *product* of a by b and express it in terms of a and b by $a \times b$, or $a \cdot b$, or simply ab.

Hence, by definition,

41 $$ab = a + a \cdots \text{to } b \text{ terms.}$$

42 We also call a the *multiplicand*, b the *multiplier*, and a and b the *factors* of ab.

43 **The process.** To find products by repeated addition would be very laborious. We therefore memorize products of the smaller numbers (multiplication tables), and from these derive products of the larger numbers by aid of the laws of addition and the laws of multiplication explained in the following sections.

44 **The laws of multiplication.** Multiplication, like addition, is a commutative and an associative operation, and it is "distributive" with respect to addition; that is, it conforms to the following three laws:

45 **The commutative law.** $ab = ba$,

The result of multiplying a *by* b *is the same as that of multiplying* b *by* a.

Thus, $2 \cdot 3 = 6$ and $3 \cdot 2 = 6$.

46 **The associative law.** $a(bc) = (ab)c$,

The result of multiplying a *by the product* bc *is the same as that of multiplying the product* ab *by* c.

Thus, $2(3 \cdot 4) = 2 \cdot 12 = 24$; and $(2 \cdot 3)4 = 6 \cdot 4 = 24$.
In practice we write abc instead of $(ab)c$. Compare § 36.

47 **The distributive law.** $a(b + c) = ab + ac$,

The result of first adding b *and* c, *and then multiplying* a *by the sum so obtained, is the same as that of first multiplying* a *by* b *and* a by c, *and then adding the products so obtained.*

Thus, $3(4+5) = 3 \cdot 9 = 27$; and $3 \cdot 4 + 3 \cdot 5 = 12 + 15 = 27$.

Proofs of these laws. We may prove these laws as follows: **48**

First. The *distributive law:* $ab + ac = a(b+c)$. (1)

For $ab + ac = (a + a + \cdots \text{ to } b \text{ terms}) + (a + a + \cdots \text{ to } c \text{ terms})$ § 41
$= a + a + a + \cdots \text{ to } (b+c) \text{ terms} = a(b+c)$. §§ 35, 41

Hence $$a(b + c + \cdots) = ab + ac + \cdots. \qquad (2)$$

Thus, $a(b+c+d) = a(b+c) + ad = ab + ac + ad$. by (1) and § 35

We also have $$ac + bc = (a+b)c. \qquad (3)$$

For $ac + bc = (a + a + \cdots \text{ to } c \text{ terms}) + (b + b + \cdots \text{ to } c \text{ terms})$
$= (a+b) + (a+b) + \cdots \text{ to } c \text{ terms} = (a+b)c$. § 38

Second. The *commutative law:* $ab = ba$.

$ab = (1 + 1 + \cdots \text{ to } a \text{ terms})b$
$= 1 \cdot b + 1 \cdot b + \cdots \text{ to } a \text{ terms}$ by (3)
$= b + b + \cdots \text{ to } a \text{ terms} = ba$. § 41

Third. The *associative law:* $(ab)c = a(bc)$.

$(ab)c = ab + ab + \cdots \text{ to } c \text{ terms}$ § 41
$= a(b + b + \cdots \text{ to } c \text{ terms}) = a(bc)$. by (2) and § 41

General theorem regarding products. These laws can be extended to products of any finite number of factors. Thus, **49**

The product of any finite number of factors is independent of the order in which the factors are multiplied together.

Rules of equality and inequality for products. These are: **50**

1. If $a = b$, then $ac = bc$.
2. If $a < b$, then $ac < bc$.
3. If $a > b$, then $ac > bc$.
4. If $ac = bc$, then $a = b$.
5. If $ac < bc$, then $a < b$.
6. If $ac > bc$, then $a > b$.

Here 1 is obvious, since if $a = b$, then a and b denote the same number. We may prove 3 as follows, and 2 similarly.

If $a > b$, let $a = b + d$. Then $ac = (b + d)c = bc + dc, \therefore > bc$.

The rules 4, 5, 6 are the converses of 1, 2, 3 and follow from them by the reasoning used in § 39.

From 1, 2, 3, by the reasoning employed in § 39, it follows that

$$\text{If } a = b \text{ and } c = d, \text{ then } ac = bd.$$

$$\text{If } a < b \text{ and } c < d, \text{ then } ac < bd.$$

$$\text{If } a > b \text{ and } c > d, \text{ then } ac > bd.$$

II. SUBTRACTION AND THE NEGATIVE

THE COMPLETE SCALE

51 **Subtraction.** To subtract 3 from 5 is to find what number occupies the 3d place *before* 5 in the natural scale.

We find this number, 2, by counting three numbers *backward* in the scale, beginning at 4, thus: 4, 3, 2.

We indicate the operation by the sign, –, read "minus," writing $5 - 3 = 2$.

And, in general, to *subtract* b from a is to find what number occupies the bth place before a.

We call this number the *remainder* obtained by subtracting b from a, and represent it in terms of a and b by the expression $a - b$. We also call a the *minuend* and b the *subtrahend.*

52 **Addition and subtraction inverse operations.** Clearly the third number before 5 is also the number from which 5 can be obtained by *adding* 3.

And, in general, we may describe the remainder $a - b$ either as the bth number before a, or as the number from which a can be obtained by adding b, that is, as the number which is defined by the equation

53 $$(a - b) + b = a.$$

Again, since saying that 7 occupies the 3d place after 4 is equivalent to saying that 4 occupies the 3d place before 7, we have $4 + 3 - 3 = 4$. And so, in general,

$$(a + b) - b = a. \qquad 54$$

Since $a + b - b = a$, § 54, subtraction undoes addition; and since $a - b + b = a$, § 53, addition undoes subtraction. We therefore say that addition and subtraction are *inverse* operations. 55

The complete scale. The natural scale does not fully meet the requirements of subtraction; for this scale has a *first* number, 1, and we cannot count backward beyond that number. 56

Thus, on the natural scale it is impossible to subtract 4 from 2.

But there are important advantages in being able to count backward as freely as forward. And since the natural scale is itself merely a system of signs arranged in a definite order, there is no reason why we should not extend it backward by placing a new ordinal system of signs before it.

We therefore invent successively the signs: 0, which we place before 1; -1, which we place before 0; -2, which we place before -1; and so on.

In this manner we create the *complete scale*

$$\cdots, -5, -4, -3, -2, -1, 0, 1, 2, 3, 4, 5, \cdots,$$

which has neither a first nor a last sign or "number," and on which it is therefore possible to count backward, as well as forward, to any extent whatsoever.

Observe the *symmetry* of this scale with respect to the sign 0. As 3 is the third sign *after* 0, so -3 is the third sign *before* 0; and so in general. 57

Meaning of the new numbers. One of these new signs, namely 0, may be said to have a cardinal meaning. Thus, counting backward from 3 corresponds to the operation of removing the elements of any group of 3 things, one at a time. This operation may be continued until *all* the elements have 58 been removed, and we may call 0 the sign of the cardinal number of the resulting "group" of no elements. We therefore often regard 0 as one of the natural numbers.

But $-1, -2, -3, \cdots$ have no cardinal meaning whatsoever.

On the other hand, all these new signs have the same *ordinal* character as the natural numbers. Every one of them occupies a definite position in an ordinal system which includes the natural numbers also. And we may consider it *defined* by this position precisely as we may consider each natural number defined by its position in the scale. We regard this as a sufficient reason for calling the signs $-1, -2, -3 \cdots$ *numbers.*

59 **Positive and negative.** To distinguish the new numbers -1, $-2, -3, \cdots$ as a class from the old, we call them *negative,* the old *positive.*

The numbers of both kinds, and 0, are called *integers* to distinguish them from other numbers to be considered later.

60 **Algebraic equality and inequality.** Let a, b, c denote any numbers of the complete scale. According as a precedes, coincides with, or follows b, we write $a < b$, $a = b$, or $a > b$.

61 Since by definition the complete scale is an ordinal system, § 17, the rules of § 24 apply to it also; thus,

$$\text{If } a < b \text{ and } b < c, \text{ then } a < c.$$

62 When $a < b$, that is, when a precedes b in the complete scale, it is customary to say that a is *algebraically less* than b, or that b is *algebraically greater* than a.

Observe that the words "less" and "greater," as thus used, mean "precede" and "follow," in the complete scale — this and nothing more. Thus, "-20 is less than -18" means merely "-20 precedes -18."

63 **Absolute or numerical values.** We call 3 the *numerical value* of -3 or its *absolute value,* and use the symbol $|-3|$ to represent it, writing $|-3| = 3$. Similarly for any negative number.

The numerical value of a positive number, or 0, is the number itself. Thus $|3| = 3$.

Numerical equality and inequality. Furthermore we say of any two numbers of the complete scale, as $\mathbf{a}$ and $\mathbf{b}$, that $\mathbf{a}$ is *numerically* less than, equal to, or greater than $\mathbf{b}$, according as $|\mathbf{a}| <, =,$ or $> |\mathbf{b}|$. **64**

Thus, while -3 is *algebraically* less than 2, it is *numerically* greater than 2, and while -7 is algebraically less than -3, it is numerically greater than -3.

OPERATIONS WITH NEGATIVE NUMBERS

New operations. We also invent operations by which the negative numbers and 0 may be combined with one another and with the natural numbers, as the latter are themselves combined by addition, multiplication, and subtraction. **65**

We call these operations by the same names, and indicate them in the same way, as the operations with natural numbers to which they correspond.

Employing $\mathbf{a}$, as in § 60, to denote *any* number of the complete scale, but a and b to denote natural numbers only, we may define these new operations as follows:

Definitions of addition and subtraction. These are: **66**

1. $\mathbf{a} + b$ is to mean the bth number after $\mathbf{a}$.
2. $\mathbf{a} - b$ is to mean the bth number before $\mathbf{a}$.
3. $\mathbf{a} + 0$ and $\mathbf{a} - 0$ are to mean the same number as $\mathbf{a}$.
4. $\mathbf{a} + (-b)$ is to mean the same number as $\mathbf{a} - b$.
5. $\mathbf{a} - (-b)$ is to mean the same number as $\mathbf{a} + b$.

In other words, adding a *positive* number b to *any* number $\mathbf{a}$ is to mean, as heretofore, counting b places forward in the scale; subtracting it, counting b places backward: while adding and subtracting a *negative* number are to be equivalent respectively to subtracting and adding the corresponding positive number.

Thus, by 1, $-3+2=-1$, since -1 is the 2d number after -3.

by 2, $2-5=-3$, since -3 is the 5th number before 2.

by 4, $-5+(-2)=-5-2=-7$ (by 2).

by 5, $-6-(-2)=-6+2=-4$ (by 1).

67 **Definition of multiplication.** This is:

1. $0\cdot a$ and $a\cdot 0$ are to mean 0.
2. $a(-b)$ and $(-a)b$ are to mean $-ab$.
3. $(-a)(-b)$ is to mean ab.

In other words, a product of two factors, neither of which is 0, is to be *positive* or *negative* according as the factors have the *same* or *opposite signs.* And in every case the numerical value of the product is to be the product of the numerical values of the factors.

Thus, by 2, $3\times -2=-6$, and $-3\times 2=-6$.

by 3, $-3\times -2=6$.

68 **The origin and significance of these definitions.** Observe that the statements of §§ 66 and 67 are neither assumptions nor theorems requiring demonstration, but what we have called them — *definitions of new operations.*

Thus, it would be absurd to attempt to *prove* that $2(-3)=-2\cdot 3$ with nothing to start from except the definition of multiplication of natural numbers, § 40, for the obvious reason that -3 is not a natural number. The phrase "2 taken -3 times" is meaningless.

But why should such operations be invented? To make the negative numbers as serviceable as possible in our study of relations among numbers themselves and among things in the world about us.

The new operations have not been invented arbitrarily; on the contrary, they are the natural extensions of the old operations to the new numbers.

In dealing with the natural numbers, we first defined addition as a *process* — counting forward — and then showed that the results of this process have two properties *which are independent of the values of the numbers added*, namely:

1. $a + b = b + a$. 2. $a + (b + c) = (a + b) + c$.

Similarly we proved that products possess the three general properties:

3. $ab = ba$. 4. $a(bc) = (ab)c$. 5. $a(b + c) = ab + ac$.

When we employ *letters* to denote numbers, these properties 1–5 become to all intents and purposes our *working definitions* of addition and multiplication; for, of course, we cannot then actually carry out the processes of counting forward, and so on.

Clearly if corresponding operations with the new numbers are to be serviceable, these "definitions" 1–5 must apply to them also. And §§ 66, 67 merely state the solution of the problem:

To make such an extension of the meanings of addition, multiplication, and subtraction that sums and products of any numbers of the complete scale may have the properties 1–5, and that subtraction may continue to be the inverse of addition.

Thus, (1) when we define adding a *positive* number b to $\mathbf{a}$ as counting forward, and subtracting it as counting backward, we are merely repeating the old definitions of addition and subtraction.

(2) From this definition of addition it follows that $-b + b = 0$.

But if the commutative law $a + b = b + a$ is to hold good, we must have $-b + b = b + (-b)$, and therefore $b + (-b) = 0$; or, since $b - b = 0$, we must have $b + (-b) = b - b$.

This suggests the definition $\mathbf{a} + (-b) = \mathbf{a} - b$.

(3) If our new addition and subtraction are, like the old, to be *inverse* operations, we must also have, as in § 66, 5, $\mathbf{a} - (-b) = \mathbf{a} + b$.

(4) Again, to retain the old connection between addition and multiplication, § 41, we must have, as in § 67, 2,

$$(-a)b = -a + (-a) + \cdots \text{ to } b \text{ terms}$$
$$= -a - a - \cdots \text{ to } b \text{ terms} = -ab.$$

(5) If the commutative law $ab = ba$ is to hold good, we must also have $a(-b) = (-b)a = -ba = -ab$, as in § 67, 2.

(6) Similarly, $0 + 0 + \cdots$ to a terms $= 0$, and this fact together with our wish to conform to the law $ab = ba$ leads to the definitions of § 67, 1, namely, $0 \cdot \mathbf{a} = 0$ and $\mathbf{a} \cdot 0 = 0$.

(7) Finally, it follows from (6) that $(-a)(-b+b) = -a \cdot 0 = 0$.

But if the distributive law is to hold good, we must also have $(-a)(-b+b) = (-a)(-b) + (-a)b = (-a)(-b) - ab$, by (4).

We therefore have $(-a)(-b) - ab = 0$. And since also $ab - ab = 0$, we are thus led to define $(-a)(-b)$ as ab, as in § 67, 3.

69 **The operations just defined conform to the commutative, associative, and distributive laws.** It remains to prove that the new operations are in complete agreement with the laws which suggested them.

To begin with, we have

$$\mathbf{a} + (b + c) = \mathbf{a} + b + c, \tag{1}$$

$$\mathbf{a} - (b + c) = \mathbf{a} - b - c, \tag{2}$$

$$\mathbf{a} + b - b = \mathbf{a} - b + b = \mathbf{a}, \tag{3}$$

as follows from the definitions of addition and subtraction as counting forward and backward, by the reasoning in §§ 37, 52.

I. The commutative law, $\mathbf{a} + \mathbf{b} = \mathbf{b} + \mathbf{a}$.

First, $-a + b = b + (-a)$.

For if $a > b$, let $a = d + b$. §§ 31, 34

Then $-a + b = -(d + b) + b$

$= -d - b + b = -d$; by (2) and (3)

and $b + (-a) = b - (b + d)$, § 66, 4

$= b - b - d = -d$. by (2)

Proceed in a similar manner when $b > a$.

Second, $-a + (-b) = -b + (-a)$.

For $-a + (-b) = -(a + b) = -(b + a) = -b + (-a)$, by (2) and § 66, 4.

II. The associative law, $\mathbf{a} + (\mathbf{b} + \mathbf{c}) = (\mathbf{a} + \mathbf{b}) + \mathbf{c}$.

First, $\mathbf{a} + [b + (-c)] = \mathbf{a} + b + (-c)$.

For if $b > c$, let $b = d + c$. §§ 31, 34

Then $\mathbf{a} + [b + (-c)] = \mathbf{a} + [d + c + (-c)] = \mathbf{a} + d,$

and $\mathbf{a} + b + (-c) = \mathbf{a} + d + c + (-c) = \mathbf{a} + d.$ by (3) and § 66, 4

Proceed in a similar manner when $c > b$.

Second, $\mathbf{a} + [(-b) + c] = \mathbf{a} + (-b) + c.$

This follows from I and the case just considered.

Third, $\mathbf{a} + [-b + (-c)] = \mathbf{a} + (-b) + (-c).$

This follows from (2) and § 66, 4, since $-b + (-c) = -(b + c)$.

III. **The commutative law, ab = ba.**

First, $(-a)b = b(-a)$.

For $$(-a)b = -ab = -ba = b(-a).$$ § 45; § 67, 2

Second, $(-a)(-b) = (-b)(-a)$.

For $$(-a)(-b) = ab = ba = (-b)(-a).$$ § 45; § 67, 3

IV. **The associative law, a (bc) = (ab) c.**

First, $(-a)[(-b)(-c)] = [(-a)(-b)](-c)$.

For $$(-a)[(-b)(-c)] = (-a) \cdot bc = -abc,$$ § 46; § 67, 2, 3

and $$[(-a)(-b)](-c) = ab \cdot (-c) = -abc.$$ § 67, 2, 3

Second, the other cases may be proved in the same way.

V. **The distributive law, a (b + c) = ab + ac.**

First, $a[b + (-c)] = ab + a(-c)$.

For $$[b + (-c)]a = [b + (-c)] + [b + (-c)] + \cdots \text{ to } a \text{ terms}$$
$$= b + b + \cdots \text{ to } a \text{ terms} + (-c) + (-c) + \cdots \text{ to } a \text{ terms}$$
$$= ba + (-c)a.$$ § 41; § 67, 2; II and III

Hence $$a[b + (-c)] = ab + a(-c)$$ by III

Second, from this case the others readily follow.

Thus, $$(-a)[b + (-c)] = -a[b + (-c)]$$
$$= -[ab + a(-c)] = (-a)b + (-a)(-c).$$

70 **The general result.** As has already been observed, § 68, in literal arithmetic or algebra, the laws $a + b = b + a$, and so on, are equivalent to *definitions* of addition and multiplication, even when the letters a, b, c denote natural numbers. And we have now shown that these definitions apply to all numbers of the complete scale.

By means of these laws we may change the form of a literal expression without affecting its value, whatever numbers of the complete scale the letters involved in the expression may denote.

Thus, whether a, b, c, d denote positive or negative integers, we have

$$(a + b)(c + d) = (a + b)c + (a + b)d$$
$$= ac + bc + ad + bd.$$

71 **Rules of equality and inequality for sums.** We may prove by reasoning similar to that in § 39 that

According as $\quad \mathbf{a} <, =, \text{ or } > \mathbf{b},$

so is $\quad \mathbf{a} + \mathbf{c} <, =, \text{ or } > \mathbf{b} + \mathbf{c};$

and conversely.

Hence it is true for positive and negative numbers alike that

72 *An equation remains an equation and the sense of an inequality remains unchanged when the same number is added to both sides, or is subtracted from both.*

73 **Rules of equality and inequality for products.** Observe that changing the signs of any two numbers $\mathbf{a}$ and $\mathbf{b}$ reverses the order in which they occur in the complete scale, § 57.

Thus, we have $-3 < -2$, but $3 > 2$; $-5 < 2$, but $5 > -2$.

From this fact and the reasoning of § 50 it follows that

According as $\quad \mathbf{a} <, =, \text{ or } > \mathbf{b},$

so is $\quad \mathbf{a}c <, =, \text{ or } > \mathbf{b}c,$

but $\quad \mathbf{a}(-c) >, =, \text{ or } < \mathbf{b}(-c);$

and conversely. Hence

Multiplying both sides of an equation by the same number, positive or negative, leaves it an equation. **74**

Multiplying both sides of an inequality by the same positive number leaves its sense unchanged.

But multiplying both sides of an inequality by the same negative number changes its sense, from $<$ to $>$, or vice versa.

From the first of these rules and the definition of multiplication by 0, namely $a \cdot 0 = 0$, we derive the following important theorem:

1. If $a = b$, then $ac = bc$. **75**

2. If $ac = bc$, then $a = b$, *unless* $c = 0$.

The exceptional case under 2 should be carefully observed.

Thus, from the true equation, $2 \cdot 0 = 3 \cdot 0$, of course it does not follow that $2 = 3$.

Zero products. *If a product be* 0, *one of its factors must be* 0. **76**

Thus, if $ab = 0$, either $a = 0$ or $b = 0$.

For, since $0 \cdot b$ is also equal to 0,

we have $ab = 0 \cdot b$,

and therefore $a = 0$, *unless* $b = 0$. § 75

Numerical values of products. The numerical value of a product of two or more factors is the product of the numerical values of the factors. **77**

Thus, $|(-2)(-3)(-4)| = |-24| = 24$; and $|-2| \cdot |-3| \cdot |-4| = 24$.

Numerical values of sums. The numerical value of a sum of two numbers is the *sum* of their numerical values when the numbers are of *like* sign, but the numerical *difference* of these values when the numbers are of *contrary* sign. **78**

Thus, $|-3 + (-5)| = |-8| = 8$; and $|-3| + |-5| = 3 + 5 = 8$.

But $|2 + (-5)| = |-3| = 3$; and $|-5| - 2 = 3$.

THE USE OF INTEGRAL NUMBERS IN MEASUREMENT

79 **Measurement.** We use numbers not only to record the results of *counting* groups of distinct things, but also to indicate the results of *measuring* magnitudes, such as portions of time, straight lines, surfaces, and so on.

80 We *measure* a magnitude by comparing it with some particular magnitude of the same kind, chosen as a *unit* of measure.

81 If the magnitude contains the unit a certain number of times exactly, we call this number its *measure.*

In particular, we call the measure of a line segment the *length* of the segment.

Thus, we may measure a line segment by finding how many times we can lay some chosen unit segment, say a foot rule, along it.

If we find that it contains the foot rule exactly three times, we say that it is *three feet long*, or that its length — that is, its measure — is 3.

82 The usefulness of the natural numbers in measurement is due to the fact that, *by their relative* **positions** *in the natural scale*, they indicate *the relative* **sizes** *of the magnitudes whose measures they are.*

83 **Application of the negative numbers to measurement.** We often have occasion to make measurements in opposite "directions" from some fixed "point of reference."

Thus, we measure time in years *before* and *after* the birth of Christ, longitude in degrees *west* and *east* of Greenwich or Washington, temperature in degrees *below* and *above* zero.

We may then distinguish measurements made in the one direction from those made in the other by the simple device of representing the one by *positive* numbers, the other by *negative* numbers.

84 Thus, consider the following figure:

$$\begin{array}{cccccccccc} \cdots -4 & -3 & -2 & -1 & 0 & 1 & 2 & 3 & 4 \cdots \\ \hline \cdots\ P_{-4} & P_{-3} & P_{-2} & P_{-1} & O & P_1 & P_2 & P_3 & P_4 \cdots \end{array}$$

Here the fixed point of reference, or *origin*, is O, the unit is OP_1, and the points $P_2, P_3, \cdots, P_{-1}, P_{-2}, \cdots$ are such that $OP_1 = P_1P_2 = P_2P_3 = \cdots = P_{-1}O = P_{-2}P_{-1} = \cdots$.

Above these points we have written in their proper order the numbers of the complete scale, so that 0 comes over O.

The *distance* of each point P from O, — that is, the length of the segment OP, — is then indicated by the *numerical value* of the number written above it; and the *direction* of P from O is indicated by the *sign* of that number.

Thus, -3 over P_{-3} indicates that P_{-3} is distant 3 units to the left of O.

Moreover, the *order* in which the points occur on the line is indicated by the order in which the corresponding numbers occur in the scale.

85 **Points used to picture numbers.** Inasmuch as there is a one-to-one relation, § 2, between the system of points $\cdots, P_{-2}, P_{-1}, O, P_1, P_2, \cdots$ and the system of numbers $\cdots, -2, -1, 0, 1, 2, \cdots$, *either* system may be used to represent the other. In what follows we shall frequently use the points to picture the numbers.

III. DIVISION AND FRACTIONS

DIVISION REPEATED SUBTRACTION

86 **The two kinds of division.** There are *two* operations to which the name *division* is applied in arithmetic and algebra. The one may be described as *repeated subtraction*, the other as the *inverse of multiplication*. There is a case in which the two coincide. We call this the case of *exact division*.

87 **Division repeated subtraction.** To divide 7 by 3 in the first of these senses is to answer the two questions:

1. What multiple of 3 must we subtract from 7 to obtain a remainder which is less than 3?

2. What is this remainder?

We may find the answer to both questions by repeatedly subtracting 3. Thus, since $7 - 3 = 4$ and $4 - 3 = 1$, we must subtract 3 *twice*, or, what comes to the same thing, we must subtract 3×2. And the remainder is 1.

This kind of division, then, is equivalent to *repeated subtraction.* Its relation to subtraction is like that of multiplication to addition.

Observe that the four numbers 7, 3, 2, 1 are connected by the equation

$$7 = 3 \cdot 2 + 1.$$

And so in general, if a and b are any two natural numbers, to *divide a* by b, in the sense now under consideration, is to find two natural numbers, q and r, one of which may be 0, such that

88 $$a = bq + r \text{ and } r < b.$$

89 We call a the *dividend*, b the *divisor*, q the *quotient*, and r the *remainder*.

90 **Note.** When a and b are given, two numbers q and r satisfying § 88 may always be found.

Thus, if $a < b$, we have $q = 0$ and $r = a$.

If $a \geqq b$, it follows from §§ 31, 35 that we can continue the sum $b + b + \cdots$ until it either equals a or will become greater than a if we add another b. And if q denote the number of terms in this sum, we shall have, § 41, either $a = bq$, or $a = bq + r$, where $r < b$.

Again, when a and b are given, but *one pair* of numbers q and r satisfying § 88 exists.

For were there a second such pair, say q', r', we should have

$$bq + r = bq' + r', \text{ and therefore } b(q - q') = r' - r.$$

But this is impossible, since $r' - r$ would be numerically less than b, but $b(q - q')$ not numerically less than b.

91 **Exact division.** If the dividend a is a multiple of the divisor b, as when $a = 12$ and $b = 3$, the remainder r is 0. We then say that a is *exactly divisible* by b. In this case the equation of § 88 reduces to $a = bq$, or

92 $$qb = a.$$

93 Hence when a is exactly divisible by b, we may also define the quotient q as *the number which, multiplied by* b, *will produce* a.

94 In this case, furthermore, we may indicate the division thus, $a \div b$, and express the quotient q in terms of a and b by one of the symbols $\frac{a}{b}$ or a/b, writing $q = \frac{a}{b}$ as well as $qb = a$.

THEOREMS AND FORMULAS RESPECTING EXACT DIVISION

95 **Theorem 1.** *Exact division and multiplication are inverse operations; that is,*

$$a \div b \times b = a, \quad \textit{and} \quad a \times b \div b = a.$$

These equations follow from the definitions in § 93 and § 87 respectively.

96 **Theorem 2.** *When division is exact, multiplying dividend and divisor by the same number leaves the quotient unchanged.*

For if $\quad a = qb,$ then $am = q \cdot bm.$ §§ 50, 46

That is, if $\quad q = \frac{a}{b},$ then $q = \frac{am}{bm}.$ § 94

97 **Theorem 3.** *Exact division, like multiplication, is distributive with respect to addition and subtraction; that is,*

$$\frac{a}{c} + \frac{b}{c} = \frac{a+b}{c}, \quad \textit{and} \quad \frac{a}{c} - \frac{b}{c} = \frac{a-b}{c}.$$

For if $\quad a = qc,$ and $b = q'c,$

we have $\quad a + b = qc + q'c = (q + q')c.$ §§ 39, 47

Hence $\quad \frac{a+b}{c} = q + q' = \frac{a}{c} + \frac{b}{c}.$ § 94

And similarly for subtraction.

Thus, $\quad \frac{18}{3} + \frac{9}{3} = 6 + 3 = 9;$ and $\frac{18+9}{3} = \frac{27}{3} = 9.$

98 **Formulas for adding and subtracting quotients.** These are

$$\frac{a}{b} + \frac{c}{d} = \frac{ad + bc}{bd}; \quad \frac{a}{b} - \frac{c}{d} = \frac{ad - bc}{bd}.$$

For $$\frac{a}{b}+\frac{c}{d}=\frac{ad}{bd}+\frac{bc}{bd}=\frac{ad+bc}{bd}.$$ §§ 96, 97

And similarly for subtraction.

Thus, $\frac{18}{3}+\frac{10}{5}=6+2=8$; and $\frac{18\cdot 5+10\cdot 3}{3\cdot 5}=\frac{120}{15}=8.$

99 **Formula for multiplying quotients.** This is

$$\frac{a}{b}\cdot\frac{c}{d}=\frac{ac}{bd}.$$

For if $a=qb$, and $c=q'd$, we have $ac=qq'\cdot bd$. §§ 50, 45, 46

Hence $$\frac{a}{b}\cdot\frac{c}{d}=q\cdot q'=\frac{ac}{bd}.$$ § 94

Thus, $$\frac{15}{3}\cdot\frac{6}{2}=5\cdot 3=15;\text{ and }\frac{15\cdot 6}{3\cdot 2}=\frac{90}{6}=15.$$

100 **Formula for dividing one quotient by another, when this division is exact.** This is

$$\frac{a}{b}\div\frac{c}{d}=\frac{a}{b}\cdot\frac{d}{c}=\frac{ad}{bc}.$$

For if $a=qb$, $c=q'd$, and also $q=q''q'$,

we have $$\frac{a}{b}\div\frac{c}{d}=q\div q'=q'',$$ § 94

and $$\frac{ad}{bc}=\frac{qbd}{bq'd}=\frac{q}{q'}=q''.$$ §§ 96, 94

Thus, $$\frac{24}{6}\div\frac{10}{5}=4\div 2=2;\text{ and }\frac{24\cdot 5}{6\cdot 10}=\frac{120}{60}=2.$$

101 **Exact division for negative numbers.** Evidently the definition of quotient given in § 93 has a meaning for negative numbers also, whenever the numerical value of the dividend is exactly divisible by that of the divisor. Expressing these quotients as in § 94, we have the following theorem:

102 **Theorem 4.** *If* a *is exactly divisible by* b, *then*

$$\frac{-a}{b}=-\frac{a}{b};\qquad \frac{a}{-b}=-\frac{a}{b};\qquad \frac{-a}{-b}=\frac{a}{b}.$$

For if $a = qb$, we have $\quad -a = (-q)b.$ §§ 73, 67

Hence $\quad \dfrac{-a}{b} = -q = -\dfrac{a}{b}.$ § 94

And similarly in the other cases.

Zero in relation to exact division. 1. On the other hand, the definition of quotient given in § 93 is *meaningless* when the *divisor* is 0. **103**

For $q \times 0 = 0$, no matter what number q may denote. Hence (1) *every* number is one which multiplied by 0 gives 0; and (2) there is *no* number which multiplied by 0 will give a.

In other words, according to the definition of § 93 and § 94, the symbol $0/0$ would denote *every* number, and $a/0$ *no* number whatsoever. NB

2. But when the *dividend* is 0, and the divisor some number b which is *not* 0, the definition of § 93 *has* a meaning. In fact, the quotient denoted by $0/b$ is 0.

For, according to § 94, $0/b$ should denote the number which multiplied by b gives 0; and 0 is that number (and the only one), since $0 \cdot b = 0$.

FRACTIONS. DIVISION THE INVERSE OF MULTIPLICATION

The second kind of division mentioned in § 86 is the generalization of exact division defined as in § 93. It requires that *fractions* be introduced into the number system. We seek an *ordinal* definition of these new numbers, like that given the negative numbers in § 56. One is suggested by the following theorem, in which a, b, c, d denote natural numbers.

Theorem 5. *When* a *is exactly divisible by* b, *and* c *by* d, *the quotients* a/b *and* c/d *occur in the natural scale in the same relative order as the products* ad *and* bc*; that is,* **104**

$$a/b <, =, \text{ or } > c/d, \text{ according as } ad <, =, \text{ or } > bc.$$

1. For if $\quad \dfrac{a}{b} = \dfrac{c}{d}$, then $\dfrac{a}{b}bd = \dfrac{c}{d}db.$ § 50

But $\quad \dfrac{a}{b}b = a$, and $\dfrac{c}{d}d = c.$ §§ 93, 94

Hence $\quad ad = bc.$

And we can show in a similar manner that

If $a/b<c/d$, then $ad<bc$; and if $a/b>c/d$, then $ad>bc$.

2. But from all this it follows, conversely, that

$$\text{If } ad=bc, \text{ then } a/b=c/d.$$

For otherwise we should have either (1) $a/b<c/d$, and therefore $ad<bc$, or (2) $a/b>c/d$, and therefore $ad>bc$.

And we can show in the same way that

$$\text{If } ad<bc, \text{ then } a/b<c/d; \text{ and if } ad>bc, \text{ then } a/b>c/d.$$

105 **Enlarging the ordinal number system.** But the relative order of ad and bc in the scale is known, whether the values assigned a, b, c, d be such as make a divisible by b, and c by d, or not.

Therefore, take *any* two natural numbers, a and b, of which b is not 0, and with them form the expression $\frac{a}{b}$, or a/b.

If a is exactly divisible by b, let a/b denote, as heretofore, the natural number which is the quotient of a by b; but if not, regard a/b for the moment merely as a new symbol, read "a over b," whose relation to division is yet to be given, § 122.

Then give to *all* such symbols a/b, c/d, and so on, the property of order already possessed by those which denote natural numbers, by supposing them arranged in accordance with the rule: a/b shall precede, coincide with, or follow c/d, according as ad precedes, coincides with, or follows bc.

Or, employing the signs $<$, $=$, $>$, as heretofore, to mean "precede," "coincide with," "follow," —

106 *Let* $a/b <, =,$ *or* $> c/d$, *according as* $ad <, =,$ *or* $> bc$.

Thus, $4/5$ is to precede $7/8$, that is, $4/5<7/8$, since $4 \cdot 8<7 \cdot 5$. Again, $2/3$ is to lie between 0 and 1, or $0<2/3<1$. For $0/1<2/3$, since $0 \cdot 3<2 \cdot 1$; and $2/3<1/1$, since $2 \cdot 1<1 \cdot 3$.

107 To such of the symbols a/b as denote natural numbers this rule assigns their proper places in the scale itself; while to the rest it assigns places between consecutive numbers of the scale.

Note. To find the place thus given any *particular* symbol a/b with respect to the numbers of the scale, we have only to reduce a to the form $a = bq + r$, where $r < b$, § 88. Then if $r = 0$, so that $a = bq$, our rule makes a/b coincide with q. But if r is not 0, our rule places a/b between q and $q + 1$. **108**

The entire assemblage of symbols a/b thus defined and arranged — like the natural scale which forms part of it — is an *ordinal* system. **109**

For it has all the properties of an ordinal system which were enumerated in §§ 17, 18.

Thus, if $a/b < c/d$, and $c/d < e/f$, then $a/b < e/f$.

For if $a/b < c/d$, and $c/d < e/f$,

we have $ad < bc$, and $cf < ed$. § 106

Multiplying the sides of the first of these inequalities by the corresponding sides of the second, we have

$adcf < bced$. § 50

Hence $af < be$, § 50

and therefore $a/b < e/f$. § 106

Fractions. When a/b does not denote a natural number, we call it a *fraction;* and we call a its *numerator,* b its *denominator,* and both a and b its *terms.* Hence **110**

A fraction is a symbol of the form a / b, *defined by its position in an ordinal system which includes the natural numbers.*

From an ordinal point of view, therefore, we are justified in calling fractions *numbers.**

* The rule of § 106 may also be used to define symbols of the form $1/0, 2/0$, and so on, *ordinally.*

Thus, by the rule, $1/0$ will follow every number a/b whose denominator b is not 0. For $1/0 > a/b$, since $1 \cdot b > a \cdot 0$.

Again, $1/0$, $2/0$, and so on, will occupy the same place in our ordinal system. For $1/0 = 2/0$, since $1 \cdot 0 = 2 \cdot 0$.

But the rule will give no definite position to the symbol $0/0$. For whatever the values of a and b, we should have $0/0 = a/b$, since $0 \cdot b = a \cdot 0$.

111 **Negative fractions.** We also form fractions whose numerators, denominators, or both, are *negative* integers, as $\frac{-a}{b}$, $\frac{a}{-b}$, $\frac{-a}{-b}$, defining them ordinally as follows:

1. $\frac{-a}{b} = -\frac{a}{b}$; $\frac{a}{-b} = -\frac{a}{b}$; $\frac{-a}{-b} = \frac{a}{b}$.

2. *Every negative fraction shall precede* 0.

3. *Negative fractions shall be arranged with respect to one another (and negative integers) in accordance with the rule:*

$$-\frac{a}{b} <, =, \text{ or } > -\frac{c}{d}, \text{ according as } -ad <, =, \text{ or } > -bc.$$

112 **The system of rational numbers.** To distinguish integers and fractions alike from other numbers which we have yet to consider, we call them *rational numbers.* And we call the system which consists of all these numbers the *rational system.*

This system possesses an important property which does not belong to its part, the integral system, namely:

113 *The rational system is* **dense**; *that is, between every two unequal rational numbers there are other rational numbers.*

For let $\frac{a}{b}$ and $\frac{c}{d}$ be any two fractions, such that $\frac{a}{b} < \frac{c}{d}$. We can prove as follows that the fraction $\frac{bc + ad}{2bd}$ lies between $\frac{a}{b}$ and $\frac{c}{d}$.

Since $\frac{a}{b} < \frac{c}{d}$, we have $ad < bc$. § 106

1. If we add ad to both sides of $ad < bc$, we have, by §§ 39, 50, 106,

$$2ad < bc + ad, \qquad \therefore a(2bd) < b(bc + ad), \qquad \therefore \frac{a}{b} < \frac{bc + ad}{2bd}.$$

2. If we add bc to both sides of $ad < bc$, we have similarly

$$bc + ad < 2bc, \qquad \therefore (bc + ad)d < c(2bd), \qquad \therefore \frac{bc + ad}{2bd} < \frac{c}{d}.$$

Thus, between $\frac{3}{4}$ and $\frac{5}{6}$ we have $\frac{4 \cdot 5 + 3 \cdot 6}{2 \cdot 4 \cdot 6} = \frac{38}{48} = \frac{19}{24}$.

Hence, when speaking of rationals, one must carefully avoid such expressions as the "*next* number greater or less" than a given number; for *no such number exists.* To each integer there is such a next *integer*, but between any rational and a *rational* assigned as the next, there are always other rationals. **114**

Operations with fractions. In what follows let a, b, c, d denote any given integers, positive or negative. **115**

In §§ 98–102 we proved that, when a/b and c/d denote integers, we have

1. $\frac{a}{b} + \frac{c}{d} = \frac{ad + bc}{bd}$. 2. $\frac{a}{b} - \frac{c}{d} = \frac{ad - bc}{bd}$.

3. $\frac{a}{b} \cdot \frac{c}{d} = \frac{ac}{bc}$. 4. $\frac{a}{b} \div \frac{c}{d} = \frac{ad}{bc}$, when $\frac{ad}{bc}$ is an integer.

But the second member of each of the equations 1, 2, 3, 4 has a meaning even when a/b and c/d are not integers. Each of them is a definite fraction of the kind defined in §§ 110, 111.

Hence 1, 2, 3, 4 at once suggest an extension of the meanings of addition, subtraction, multiplication, and division which will make these operations applicable to fractions, namely:

The *sum* of two fractions a/b and c/d is to mean the fraction $(ad + bc)/bd$. **116**

The *difference* obtained by subtracting the fraction c/d from the fraction a/b is to mean the fraction $(ad - bc)/bd$. **117**

The *product* of two fractions a/b and c/d is to mean the fraction ac/bd. **118**

The *quotient* resulting from dividing the fraction a/b by the fraction c/d is to mean the fraction ad/bc. **119**

Observe that these *definitions* are equivalent to the *rules* for reckoning with fractions given in elementary arithmetic.

The commutative, associative, and distributive laws control these generalized operations. **120**

Thus, $$\frac{a}{b} \cdot \frac{c}{d} = \frac{ac}{bd} = \frac{ca}{db} = \frac{c}{d} \cdot \frac{a}{b}.$$ §§ 118, 69

121 **The rules of equality and inequality,** §§ 71, 73, also hold good for these operations.

Thus, if $\frac{a}{b}\cdot\frac{e}{f}=\frac{c}{d}\cdot\frac{e}{f}$, then $\frac{a}{b}=\frac{c}{d}$.

For if $\frac{a}{b}\cdot\frac{e}{f}=\frac{c}{d}\cdot\frac{e}{f}$, then $aedf = cebf$. §§ 118, 106, 111

Hence $ad = cb$, and therefore $\frac{a}{b}=\frac{c}{d}$. §§ 73, 106, 111

122 **Definition of a fraction as a quotient.** The fraction a/b may now be described as *the number which multiplied by* b *will produce* a, that is, as the number which is defined by the equation

123
$$\frac{a}{b}\cdot b = a.$$

For $\frac{a}{b}\cdot b=\frac{a}{b}\cdot\frac{b}{1}=\frac{ab}{1\cdot b}=\frac{a}{1}\cdot\frac{b}{b}=a.$ §§ 106, 111, 118

124 **Division the inverse of multiplication.** From §§ 118, 119, it follows that

$$\frac{a}{b}\div\frac{c}{d}\times\frac{c}{d}=\frac{a}{b} \text{ and } \frac{a}{b}\times\frac{c}{d}\div\frac{c}{d}=\frac{a}{b};$$

in other words, that multiplication and division, as defined in §§ 118, 119, are *inverse operations.* Compare § 55.

For, by §§ 118, 119 and §§ 106, 111, we have

$$\frac{a}{b}\div\frac{c}{d}\times\frac{c}{d}=\frac{ad}{bc}\cdot\frac{c}{d}=\frac{adc}{bcd}=\frac{a}{b}\cdot\frac{dc}{cd}=\frac{a}{b};\ \frac{a}{b}\times\frac{c}{d}\div\frac{c}{d}=\frac{ac}{bd}\div\frac{c}{d}=\frac{acd}{bdc}=\frac{a}{b}\cdot\frac{cd}{dc}=\frac{a}{b}.$$

Hence we may describe the kind of division now before us as *the inverse of multiplication* and say

125 *To divide* a/b *by* c/d *is to find a number which multiplied by* c/d *will produce* a/b.

By introducing fractions into our number system, we have made it possible always to find such a number, *except when the divisor* c/d *is* 0.

This is the usual meaning of division in arithmetic and algebra. It is the generalization of exact division, § 93.

Reducing a fraction to its lowest terms. Irreducible fractions. **126** If the numerator and denominator of a fraction have a common factor, we can remove it from both without changing the value of the fraction.

For $\frac{am}{bm} = \frac{a}{b}$, since $am \cdot b = a \cdot bm$, § 106.

When all such common factors have been removed, the fraction is said to be *in its lowest terms*, or to be *irreducible.*

Theorem. *If* a/b *be an irreducible fraction, and* a'/b' *any other fraction which is equal to it, then* a' *and* b' *are equimultiples of* a *and* b *respectively.* **127**

For since $a'/b' = a/b$, and therefore $a'b = ab'$, a is a factor of $a'b$.

But, by hypothesis, a has no factor in common with b. Hence a must be a factor of a', § 492, 1.

We therefore have $a' = ma$, where m is some integer.

But substituting ma for a' in $a'b = ab'$, we have $mab = ab'$, and therefore $b' = mb$, § 50.

Corollary. *If two irreducible fractions are equal, their numerators must be equal, and also their denominators.* **128**

THE USE OF FRACTIONS IN MEASUREMENT

Fractional lengths. The definition of length given in § 81 **129** only applies to such line segments **S** as contain the unit segment **s** a certain number of times exactly.

But even if **S** does not contain **s** exactly, it may still be *commensurable* with **s**; that is, it may contain the *half*, the *third*, or some other *aliquot part* of **s** exactly. In that case we *define* its length as follows:

If a given line segment contains the b*th part of the unit segment* a *times exactly, we say that its* **length** *is the fraction* a/b. **130**

Thus, if **S** contains the 10th part of **s** exactly 7 times, the length of **S** (in terms of **s**) is 7/10.

131 **Note.** Observe that if a/b is the length of **S** in terms of **s** according to this definition, so also is every fraction of the form ma/mb.

For if **S** contains the bth part of **s** exactly a times, it will contain the mbth part of **s** exactly ma times.

132 Fractions are useful in measurement for the same reason that integers are useful: namely, *by their relative positions in the rational system, they indicate the relative sizes of the segments whose lengths they are.*

For if a/b and c/d are the lengths of **S** and **T** in terms of **s**, so also are ad/bd and bc/bd, § 131; that is, the bdth part of **s** is contained in **S** exactly ad times, in **T** exactly bc times.

Hence **S** $<, =,$ or $>$ **T**, according as $ad <, =,$ or $> bc$,

that is, **S** $<, =,$ or $>$ **T**, according as $a/b <, =,$ or $> c/d$. § 106

133 **Note.** It hardly need be said that the definition of *length* here given is equivalent to the definition of *fraction* given in elementary arithmetic, and that greater or lesser fractions are there *defined* as fractions which correspond to greater or lesser line segments or other magnitudes.

134 **Rational numbers pictured by points.** Fractions, as well as integers, may be pictured by points on an indefinite straight line, § 85.

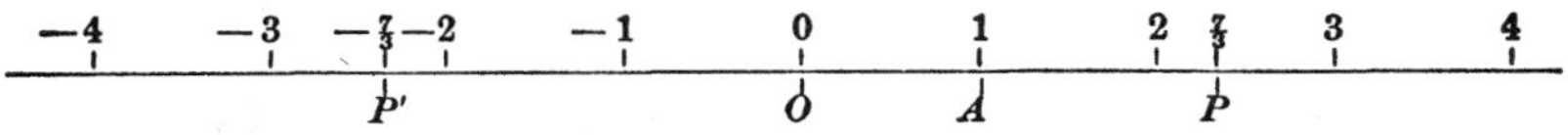

-4 -3 $-\frac{7}{3}$ -2 -1 0 1 2 $\frac{7}{3}$ 3 4

P' O A P

Thus, to construct a point, P, which will picture 7/3 in the same way that A pictures 1, we have only to start at the origin O and lay off the *third* part of the unit OA *seven* times to the right.

P', the corresponding point to the left of O, is the picture of $-7/3$.

We proceed in a similar manner in the case of any given fraction, positive or negative.

135 All such points are arranged along the line in an order corresponding to that of the rationals which they picture. With this in mind we often speak of one rational as lying to the left or right of another rational, or as lying between two other rationals.

IV. IRRATIONAL NUMBERS

PRELIMINARY CONSIDERATIONS

Definitions. The product aa is represented by a^2, read "a square"; the product aaa, by a^3, read "a cube"; the product aaa . . . to n factors, by a^n, read "the nth power of a." **136**

In the symbols a^2, a^3, a^n, the numbers 2, 3, n are called *exponents; a* itself is called the *base.*

Finding a^2 from a is called *squaring* a; finding a^3, *cubing* a; finding a^n, *raising* a *to the* nth *power.*

The operation which consists in raising a given number to a given power is also called *involution.*

Roots and logarithms. If, as we are supposing, a is a rational number, and n a positive integer, a^n is also a rational number. Call this number b; then **137**

$$a^n = b.$$

This equation suggests two new problems:

First. To assign values to n and b, and then find a.

Second. To assign values to a and b, and then find n.

Thus, (1) let $n = 2$ and $b = 9$. The equation then becomes

$$a^2 = 9,$$

and we find that $a = 3$ or -3; for both $3^2 = 9$ and $(-3)^2 = 9$.

Again, (2) let $a = 2$ and $b = 8$. The equation then becomes

$$2^n = 8,$$

and we find that $n = 3$: for $2^3 = 8$.

When $$a^n = b,$$ **138**

1. a is called the nth root of b, and is expressed in terms of n and b by the symbol $\sqrt[n]{b}$, the simpler symbol $\sqrt{b}$, read "square root of b," being used when $n = 2$.

2. n is called the *logarithm of* b *to the base* a, and is expressed in terms of a and b by the symbol $\log_a b$.

Thus, since $3^2 = 9$ and $(-3)^2 = 9$, both 3 and -3 are square roots of 9, and both *may* be written $\sqrt{9}$; but see § 139.

Again, 2 is the logarithm of 9 to the base 3; that is, $2 = \log_3 9$.

139 **Note.** Instead of representing *both* the square roots of 9 by the symbol $\sqrt{9}$, we may represent the positive one, 3, by $\sqrt{9}$, and the negative one, -3, by $-\sqrt{9}$. This is the usual method of representing square roots in elementary algebra, and we shall follow it.

140 **Evolution and finding logarithms.** The operation by which $\sqrt[n]{b}$ is found, when n and b are given, is called *extracting the* n*th root of* b, or *evolution.*

The operation by which $\log_a b$ is found, when a and b are given, is called *finding the logarithm of* b *to the base* a.

Both these operations are *inverses* of involution, §§ 55, 124.

141 **Note.** The reason that involution has two inverses, while addition and multiplication each has but one, will be seen by comparing the three equations

1. $a + b = c$. 2. $ab = c$. 3. $a^b = c$.

Since $a + b = b + a$, and $ab = ba$, the problem: Given c and b in 1 or 2, find a, is of the same kind as the problem: Given c and a, find b.

But since a^b is not equal to b^a, the problem: Given c and b in 3, find a, is wholly different in kind from the problem: Given c and a, find b.

142 **New numbers needed.** We shall subsequently study these new operations in detail; for in algebra they are second in importance to the four fundamental operations only. But the point which now concerns us is this: *They necessitate further extensions of the number system.*

In fact, it is at once evident that $\sqrt[n]{a}$ can denote a rational number in exceptional cases only.

Thus, to cite the simplest of illustrations, neither $\sqrt{-1}$ nor $\sqrt{2}$ can denote a rational number. For

1. Since the square of every rational number is positive, no rational exists whose square is -1. Hence $\sqrt{-1}$ cannot denote a rational number.

2. No rational number exists whose square is 2. For clearly 2 is not the square of any integer, and we can show as follows that it is not the square of any fraction.

Suppose p/q to be a fraction in its lowest terms, such that

$$(p/q)^2 = 2, \text{ or } p^2/q^2 = 2/1.$$

But since p^2/q^2 is in its lowest terms, § 492, 2, it would follow from this, § 128, that $p^2 = 2$, which is impossible, since p is an integer.

Therefore $\sqrt{2}$ cannot denote a rational number.

It can be shown in the same way that if a/b be any fraction in its lowest terms, $\sqrt[n]{a/b}$ cannot denote a rational number, unless both a and b are nth powers of integers.

We are to make good this deficiency in our number system by creating two new classes of numbers: the *irrational numbers*, of which $\sqrt{2}$ is one, and the *imaginary numbers*, of which $\sqrt{-1}$ is one.

We shall treat the irrational numbers in the present chapter and the imaginary numbers in the chapter which follows.

THE ORDINAL DEFINITION OF IRRATIONAL NUMBERS

In the present chapter the letters a, b, c, and so on, will denote any *rational* numbers, whether positive or negative, integral or fractional.

General properties of the rational system. The rational numbers constitute a system which has the following properties: **143**

1. It is an *ordinal* system.

2. It is *dense;* that is, between every two unequal numbers of the system, a and b, there lie still other numbers of the system.

3. The sum, difference, product, and quotient of every two numbers of the system are themselves numbers of the system, the quotient of any number by 0 excepted.

By the definitions which follow, we shall create a more extended system which possesses these same three properties, and which includes the rational system.

Separations of the first kind. 1. The number $\frac{1}{3}$ separates the remaining numbers of the rational system into two classes: **144**

the one class consisting of all rationals which *precede* (are less than) $\frac{1}{3}$, the other of all rationals which *follow* (are greater than) $\frac{1}{3}$. Let us name these two classes of numbers C_1 and C_2 respectively.

C_1 $\frac{1}{3}$ C_2

In the figure, the half line to the left of the point $\frac{1}{3}$ contains the point-pictures of all numbers in the class C_1, and the half line to the right the point-pictures of all numbers in the class C_2, § 134.

From §§ 109, 111, and 113, it immediately follows that

1. Each number in C_1 precedes every number in C_2.
2. There is no last number in C_1, and no first in C_2.

Thus, were there a last number in C_1, there would be numbers between it and $1/3$, § 113, which is impossible since, by hypothesis, all rationals less than $1/3$ are in C_1.

145 2. Instead of thus separating the rational system into the *three* parts C_1, $\frac{1}{3}$, C_2, we may join $\frac{1}{3}$ to C_1, so forming a class C_1' made up of C_1 and $\frac{1}{3}$, and then say:

The number $\frac{1}{3}$ separates the entire rational system into *two* parts, C_1' and C_2, such that:

1. Each number in C_1' precedes every number in C_2.
2. There is a *last* number in C_1', namely $\frac{1}{3}$, but there is no first number in C_2.

146 3. Or we may join $\frac{1}{3}$ to C_2, call the resulting class C_2', and then say:

The number $\frac{1}{3}$ separates the entire rational system into *two* parts, C_1 and C_2', such that:

1. Each number in C_1 precedes every number in C_2'.
2. There is no last number in C_1, but there is a *first* number in C_2', namely $\frac{1}{3}$.

It is evident that each of the rational numbers defines similar separations of the rational system.

147 **Conversely**, if we are able, in any way, to separate the entire rational system into two parts, B_1 and B_2, such that each number in B_1 precedes every number in B_2 and that there is either a *last* number in B_1 or a *first* in B_2, the separation will serve to distinguish this last or first number from all other numbers and, in that sense, to *define* it.

Thus, let us assign the *negative* rationals to B_1 and the *remaining* rationals to B_2. There is then no last number in B_1, but 0 is the first number in B_2. And zero is distinguished from all other numbers when called the first number in B_2, as perfectly as by the symbol 0.

Note. Obviously there cannot be both a last number in B_1 and a first in B_2. For there must then be rationals between these two numbers, § 113, whereas, by hypothesis, *every* rational belongs either to B_1 or to B_2. **148**

Separations of the second kind. But we can also, in various ways, separate the *entire* rational system into a part A_1 in which there is *no last number*, and a part A_2 in which there is *no first number*. **149**

Thus, since no rational exists whose square is 2, § 142, *every* rational is either one whose square is *less than* 2, or one whose square is *greater than* 2.

Let A_2 consist of all *positive* rationals whose squares are *greater* than 2, and let A_1 consist of all the other rational numbers. Then

1. Each number in A_1 precedes every number in A_2.

For let a_1 be any number in A_1, and a_2 any number in A_2.

Evidently $a_1 < a_2$, if a_1 is negative or 0; and if a_1 is positive, $a_1^2 < a_2^2$, and therefore $a_1 < a_2$.

2. There is no last number in A_1, and no first in A_2.

For when any positive rational, a_1, has been assigned whose square is less than 2, we can always find a greater rational whose square is also less than 2, § 183, 2 (3); hence no number can be assigned which is the *last* in A_1. Similarly no rational can be assigned which is the *first* in A_2.

The new number $a = \sqrt{2}$. The relation between the two classes of numbers, A_1 and A_2, is therefore precisely the same as that between the classes C_1 and C_2 in the separation corresponding to $\frac{1}{3}$, which was described in § 144. **150**

But no rational number exists which can be said to correspond to the separation A_1, A_2, or to be defined by it.

For since *every* rational belongs either to A_1 or to A_2, no rational exists which lies between A_1 and A_2, as $\frac{1}{3}$ lies between C_1 and C_2.

And since there is no last number in A_1 and no first in A_2, no rational exists which corresponds to this separation as $\frac{1}{3}$ corresponds to the separation C_1', C_2 of § 145, or to the separation C_1, C_2' of § 146. (Compare § 147.)

Hence this separation A_1, A_2 creates a place for a *new* ordinal number, namely, a number which shall follow all numbers in A_1 and precede all in A_2.

We invent such a number. For the present we may represent it by the letter $\mathbf{a}$; later, when multiplication has been defined for $\mathbf{a}$, we shall find that $\mathbf{a}^2 = 2$, and we can then replace $\mathbf{a}$ by the more significant symbol $\sqrt{2}$, § 182.

151 We then *define* this new number $\mathbf{a}$ *as that number which lies between all positive rationals whose squares are less than* 2 *and all whose squares are greater than* 2.

We may also express this definition by the formula

$$a_1 < \mathbf{a} < a_2$$

where a_1 and a_2 denote any numbers whatsoever in A_1 and A_2 respectively, and, as heretofore, $<$ means "precedes."

152 **Note.** Observe that this definition is of the same kind as the definitions of the negative and fractional numbers given in §§ 56, 110. Like these numbers, $\mathbf{a}$ is a symbol defined by its position in an ordinal system of symbols which includes the natural numbers. It therefore has precisely the same right as they to be called a number.

Our reason for inventing this and similar numbers is also the same as our reason for inventing negative numbers and fractions. They serve a useful purpose in the study of relations among the numbers which we already possess, and among things in the world about us.*

* We may add that there would be no objection *from an ordinal point of view* to our inventing more than one number to correspond to the separation A_1, A_2, say two numbers, $\mathbf{a}$ and $\mathbf{b}$, defined ordinally by the formula $a_1 < \mathbf{a} < \mathbf{b} < a_2$. But there are objections of another kind to our inventing more than one such number. See page 67, footnote (3).

The irrational numbers in general. The real system. The particular separation of the rational system which we have been considering is but one of an infinite number of possible separations of a similar character. **153**

For every such separation we invent a new number, defining it ordinally with respect to the numbers of the rational system precisely as we have defined the number $\mathbf{a} = \sqrt{2}$ in § 151.

To distinguish these new numbers from the rational numbers, we call them *irrational numbers*, or simply *irrationals*.

Again, to distinguish the rational and irrational numbers alike from the imaginary numbers, which we have yet to consider, we call them *real numbers*.

Finally, we call the system which consists of all the rational and irrational numbers the *system of real numbers*, or the *real system*.

Hence, using $\mathbf{a}$ to denote *any* irrational number, we have the following general definition of such a number:

An irrational number, $\mathbf{a}$, *is defined whenever a law is stated which will assign every given rational to one, and but one, of two classes,* A_1, A_2, *such that* (1) *each number in* A_1 *precedes every number in* A_2 *and* (2) *there is no last number in* A_1 *and no first number in* A_2*; the definition of* $\mathbf{a}$ *then being: it is the one number which lies between all numbers in* A_1 *and all in* A_2. **154**

It is here implied that there are numbers in *both* the classes A_1 and A_2; also that A_1 and A_2 together comprise the *entire* rational system.

An irrational number, $\mathbf{a}$, is said to be *negative* or *positive* according as it precedes or follows 0. **155**

The real system is an ordinal system: that is, the numbers which constitute it are arranged in a definite and *known* order, § 17. For the definition of each irrational indicates how it lies with respect to every *rational:* and from the definitions of any two given irrationals we can at once infer how they lie with respect to *one another.* **156**

Thus let **a** and **b** denote any two given irrationals; then

1. If every rational which precedes **a** also precedes **b**, and every rational which follows **a** also follows **b**, the numbers **a** and **b** occupy the same position relative to the numbers of the rational system. By our definition of an irrational number, therefore, § 154, **a** and **b** denote one and the same number. We indicate this by the formula:

$$a = b.$$

2. If among the rationals which follow **a** there are some which precede **b**, then **a** itself must precede **b** (or **b** follow **a**). We indicate this by the formula:

$$a < b \text{ or } b > a.$$

3. If among the rationals which precede **a** there are some which follow **b**, then **a** itself must follow **b** (or **b** precede **a**). We indicate this by the formula:

$$a > b \text{ or } b < a.$$

157 It thus appears that when any two different real numbers are given, we can at once infer which precedes and which follows; also, that we may always draw the following conclusions with respect to three given real numbers, **a**, **b**, **c**:

If $a = b$, and $b = c$, then $a = c$.

If $a < b$, and $b < c$, then $a < c$.

If $a = b$, and $b < c$, then $a < c$.

158 **The real system is dense.** For there are rational numbers not only between any two unequal rationals, § 113, but also between any two unequal irrational numbers, and between any two numbers one of which is rational and the other irrational, § 156.

159 **The real system is continuous.** The real system, therefore, possesses the first and second of the properties of the rational system enumerated in § 143. But it possesses an additional property not belonging to the rational system, namely:

If the entire real system be separated into two parts, R_1 *and* R_2*, such that each number in* R_1 *precedes every number in* R_2*, there is either a last number in* R_1 *or a first in* R_2*, but not both.*

For in separating the *real* system into the parts R_1 and R_2 we separate the *rational* system into two parts, A_1 and A_2, the one part consisting of all the rationals in R_1, the other of all the rationals in R_2.

Every rational belongs either to A_1 or to A_2, and each rational in A_1 precedes every rational in A_2.

Let **a** be the number which the separation A_1, A_2 defines, §§ 147, 154.

Then either **a** is a *rational* — namely the last number in A_1 or the first in A_2, § 147, — or, if there be no last number in A_1 and no first in A_2, **a** is an *irrational* lying between A_1 and A_2, § 154.

1. If **a** is the last number in A_1, it is also the last in R_1; for were there any number in R_1 after **a**, there would be rationals between it and **a**, that is, rationals in A_1 after **a**, which is impossible.

2. Similarly, if **a** is the first number in A_2, it is also the first in R_2.

3. If **a** is irrational, it must, by hypothesis, belong either to R_1 or to R_2. If **a** belongs to R_1, it is the last number in R_1; for were there any number in R_1 after **a**, there would be rationals between it and **a**, § 158, that is, rationals in A_1 after **a**, which is impossible. And, in like manner, if **a** belongs to R_2, it is the first number in R_2.

Finally, there cannot be both a last number in R_1 and a first in R_2, since there would be rationals between these two numbers, § 158, that is, rationals belonging neither to A_1 nor to A_2; which is impossible.

To indicate that the real system is dense and at the same time possesses the property just described, we say that it is *continuous*.

160 **Theorem.** *A real number,* **a**, *either rational or irrational, is defined whenever a law is stated by means of which the entire real system may be separated into two parts,* R_1, R_2, *such that each number in* R_1 *precedes every number in* R_2*; this number,* **a**, *then being either the last number in* R_1 *or the first in* R_2.

This is an immediate consequence of §§ 147, 159.

APPROXIMATE VALUES OF IRRATIONALS

161 Given any irrational number, **a**, defined as in § 154. By the method illustrated below we can find a pair of rationals, the one less and the other greater than **a**, which differ from each other as little as we please. Such rationals are called *approximate values* of **a**.

Let **a** be the irrational, $\sqrt{2}$, which lies between all positive rationals whose squares are less, and all whose squares are greater than 2.

1. We may find between what pair of consecutive *integers* **a** lies by computing the squares of 1, 2, 3, $\cdots$ successively, until we reach one which is greater than 2.

We see at once that $1^2 < 2$ and $2^2 > 2$.

Hence **a** lies between 1 and 2, or $1 < \mathbf{a} < 2$.

2. We may then find between what pair of consecutive *tenths* **a** lies by computing 1.1^2, 1.2^2, $\cdots$ successively, until we reach one which is greater than 2.

We thus obtain $1.4^2 < 2$ and $1.5^2 > 2$; for $1.4^2 = 1.96$, $1.5^2 = 2.25$.

Hence **a** lies between 1.4 and 1.5, or $1.4 < \mathbf{a} < 1.5$.

3. By a similar procedure we find, successively,

$$1.41 < \mathbf{a} < 1.42, \quad 1.414 < \mathbf{a} < 1.415, \quad \text{and so on without end.}$$

4. Let a_1 denote the nth number in the sequence 1.4, 1.41, 1.414, $\cdots$ thus obtained, and a_2 the nth number in the sequence 1.5, 1.42, 1.415, $\cdots$.

Then
$$a_1 < \mathbf{a} < a_2 \text{ and } a_2 - a_1 = 1/10^n,$$

and by choosing n great enough we can make $1/10^n$ less than any positive number, as δ, we may choose to assign, however small.

5. We call 1.4, 1.41, 1.414 the approximate values of $\mathbf{a} = \sqrt{2}$ to the *first, second, third place of decimals;* and so on.

Evidently the process thus illustrated may be applied to *any* given irrational number, **a**; for all that the process requires is a test for determining whether certain rationals are less or greater than **a**, and the definition of **a**, § 154, will always supply such a test. We therefore have the following theorem:

162 *Let* **a** *denote any given irrational number. If any positive number, as* δ, *be assigned, it matters not how small, we can always find two rationals,* a_1, a_2, *such that*

$$\mathrm{a}_1 < \mathbf{a} < \mathrm{a}_2 \text{ and } \mathrm{a}_2 - \mathrm{a}_1 < \delta.$$

Evidently this theorem is true of rationals also.

Thus, if a denote a given rational number, and $a_1 = a - 1/10^n$, $a_2 = a + 1/10^n$, we have $a_1 < a < a_2$, and we can make $a_2 - a_1 = 2/10^n$ as small as we please by choosing n sufficiently great.

ADDITION, SUBTRACTION, MULTIPLICATION, DIVISION

It remains to give the real system the third of the properties of the rational system enumerated in § 143. For this we shall require the following theorem:

163 **Theorem.** *Let* A_1 *and* A_2 *be two classes of rationals such that*

1. *Each number in* A_1 *is less than every number in* A_2,
2. *There is no last number in* A_1 *and no first in* A_2,
3. *For every positive number,* δ, *that may be assigned, it matters not how small, we can find in* A_1 *a number* a_1, *and in* A_2 *a number* a_2, *such that*

$$\mathrm{a}_2 - \mathrm{a}_1 < \delta.$$

We may then conclude that between A_1 *and* A_2 *there lies one number and but one.*

That there is *at least one* such number follows from 1 and 2, by § 154.

That there *cannot be more than one* such number follows from 3.

For suppose that between every a_1 in A_1 and every a_2 in A_2 there were the two *rationals* d and d', as indicated in the figure:

a_1 d d' a_2

Then for *every* a_1, a_2 we should have

$$a_2 > d', \text{ and } -a_1 > -d, \qquad \S\S\ 73,\ 121$$

and therefore

$$a_2 - a_1 > d' - d, \qquad \S\S\ 39,\ 121$$

which is impossible, since it contradicts 3.

Nor can there be two numbers, one or both of which are *irrational*, lying between every a_1 and a_2; for between these two numbers there would be two *rationals* also lying between every a_1 and a_2, § 158, which we have just shown to be impossible.

164 **Note.** This theorem differs from the definition of an irrational number, § 154, in that it is not here a part of the hypothesis that every rational lies either in A_1 or in A_2.

165 **Addition.** Let $\mathbf{a}$ and $\mathbf{b}$ denote any two *given* real numbers, rational or irrational, and let a_1, a_2, b_1, b_2, denote *any rationals whatsoever* such that

$$a_1 < \mathbf{a} < a_2 \text{ and } b_1 < \mathbf{b} < b_2. \qquad (1)$$

Observe that there is no last number of the kind denoted by a_1 or b_1, and no first number of the kind denoted by a_2 or b_2; and that if any positive number, as δ, be assigned, it matters not how small, we can always choose a_1, a_2, and b_1, b_2, § 162, so that both

$$a_2 - a_1 < \delta \text{ and } b_2 - b_1 < \delta. \qquad (2)$$

When both $\mathbf{a}$ and $\mathbf{b}$ are *rational*, say $\mathbf{a} = \alpha$ and $\mathbf{b} = \beta$, we can find their sum, $\alpha + \beta$, by the rule of § 116; and it follows from (1), by § 121, that

$$a_1 + b_1 < \alpha + \beta < a_2 + b_2.$$

Moreover, whether $\mathbf{a}$ and $\mathbf{b}$ are rational or not, it follows from (1), by § 121, that

$$a_1 + b_1 < a_2 + b_2. \qquad (3)$$

These considerations lead us to *define* the sum of $\mathbf{a}$ and $\mathbf{b}$, when one or both are *irrational*, as follows:

166 *The sum of* $\mathbf{a}$ *and* $\mathbf{b}$, *written* $\mathbf{a} + \mathbf{b}$, *is to mean that number which lies between all the numbers* $a_1 + b_1$ *and all the numbers* $a_2 + b_2$. *In other words, it is the number defined by the formula*

$$a_1 + b_1 < \mathbf{a} + \mathbf{b} < a_2 + b_2,$$

where a_1, a_2, b_1, b_2 *denote any rationals whatsoever such that*

$$a_1 < \mathbf{a} < a_2 \text{ and } b_1 < \mathbf{b} < b_2.$$

To justify this definition we must show that there is *one* and *but one* such number $\mathbf{a} + \mathbf{b}$. This follows from § 163; for

1. Each $a_1 + b_1$ is less than every $a_2 + b_2$.
2. There is no last $a_1 + b_1$ and no first $a_2 + b_2$.

Thus, $a_1' + b_1'$ cannot be the last $a_1 + b_1$; for since there is no last a_1 and no last b_1, we can choose a_1 and b_1 so that $a_1 > a_1'$ and $b_1 > b_1'$, and therefore $a_1 + b_1 > a_1' + b_1'$.

3. If any positive rational, δ, be assigned, we can choose a_1, a_2, b_1, b_2, so that

$$a_2 - a_1 < \delta/2 \text{ and } b_2 - b_1 < \delta/2, \qquad \S\ 162$$

and therefore

$$(a_2 + b_2) - (a_1 + b_1) < \delta. \qquad \S\ 121$$

167 **Definition of $-\mathbf{a}$.** Let $\mathbf{a}$, a_1, a_2 have the same meanings as in § 165. Considerations like those in § 165 lead us, when $\mathbf{a}$ is irrational, to define $-\mathbf{a}$ as follows:

168 *The symbol* $-\mathbf{a}$ *is to mean the number defined by the formula*

$$-a_2 < -\mathbf{a} < -a_1,$$

where a_1, a_2 *denote any rationals whatsoever such that*

$$a_1 < \mathbf{a} < a_2.$$

It follows from § 163 that there is *one* and *but one* such number $-\mathbf{a}$; for

1. Each $-a_2$ is less than every $-a_1$, since $a_1 < a_2$. §§ 73, 111
2. There is no last $-a_2$ and no first $-a_1$. Thus, were there a last $-a_2$, there would be a first a_2; but no such number exists.
3. We can always choose a_1, a_2, so that

$$-a_1 - (-a_2) = a_2 - a_1 < \delta. \qquad \S\ 162$$

169 **Subtraction.** The result of *subtracting* $\mathbf{b}$ from $\mathbf{a}$, written $\mathbf{a} - \mathbf{b}$, is to mean the number $\mathbf{a} + (-\mathbf{b})$; that is,

$$\mathbf{a} - \mathbf{b} = \mathbf{a} + (-\mathbf{b}).$$

The meaning of $\mathbf{a} + (-\mathbf{b})$ itself is known from §§ 166, 168.

It follows from §§ 166, 168 that $\mathbf{a} - \mathbf{b}$ may also be defined by the formula

$$a_1 - b_2 < \mathbf{a} - \mathbf{b} < a_2 - b_1,$$

where a_1, a_2, b_1, b_2 denote any rationals whatsoever such that

$$a_1 < \mathbf{a} < a_2 \text{ and } b_1 < \mathbf{b} < b_2.$$

170 **Multiplication, both factors positive.** Let $\mathbf{a}$ and $\mathbf{b}$ be any two given *positive* numbers, and a_1, a_2, b_1, b_2, any *positive* rationals whatsoever such that

$$a_1 < \mathbf{a} < a_2 \text{ and } b_1 < \mathbf{b} < b_2. \quad (1)$$

When $\mathbf{a}$ and $\mathbf{b}$ are rational, say $\mathbf{a} = \alpha$, $\mathbf{b} = \beta$, it follows from (1), by § 121, that

$$a_1 b_1 < \alpha\beta < a_2 b_2,$$

and in every case it follows that

$$a_1 b_1 < a_2 b_2. \quad (2)$$

We are therefore led, when one or both of the numbers $\mathbf{a}$, $\mathbf{b}$ are *irrational,* to define their product thus:

171 *The product of two positive numbers* $\mathbf{a}$ *and* $\mathbf{b}$, *written* $\mathbf{ab}$, *is to mean that number which lies between all the numbers* $\mathrm{a_1 b_1}$ *and all the numbers* $\mathrm{a_2 b_2}$. *In other words,* $\mathbf{ab}$ *is the number defined by the formula*

$$\mathrm{a_1 b_1} < \mathbf{ab} < \mathrm{a_2 b_2},$$

where $\mathrm{a_1, a_2, b_1, b_2}$ *denote any positive rationals whatsoever such that*

$$\mathrm{a_1} < \mathbf{a} < \mathrm{a_2} \textit{ and } \mathrm{b_1} < \mathbf{b} < \mathrm{b_2}.$$

It follows from § 163 that there is *one* and *but one* such number $\mathbf{ab}$; for

1. Each $a_1 b_1$ is less than every $a_2 b_2$.
2. There is no last $a_1 b_1$ and no first $a_2 b_2$. (Compare proof, § 166, 2.)
3. Any positive δ being given, we can choose a_1, a_2, b_1, b_2, so that

$$a_2 b_2 - a_1 b_1 < \delta.$$

For $$a_2 b_2 - a_1 b_1 = a_2 (b_2 - b_1) + b_1 (a_2 - a_1),$$

and we can choose a_1, a_2, b_1, b_2, § 162, so that

$$b_2 - b_1 < \delta / 2 a_2 \text{ and } a_2 - a_1 < \delta / 2 b_1, \quad (1)$$

and therefore $$a_2 (b_2 - b_1) + b_1 (a_2 - a_1) < \delta. \quad (2)$$

We may make such a choice of a_1, a_2, b_1, b_2, as follows:

First take any *particular* number of the kind b_2, as b_2', and then choose a_1, a_2, so that

$$a_2 - a_1 < \delta / 2 b_2'. \quad (3)$$

Next, using the a_2 thus found, choose b_1, b_2, so that

$$b_2 - b_1 < \delta / 2\, a_2, \qquad \text{as in (1).}$$

Since $b_1 < b_2'$ and therefore $\delta / 2\, b_2' < \delta / 2\, b_1$, it follows from (3) that

$$a_2 - a_1 < \delta / 2\, b_1, \qquad \text{as in (1).}$$

Multiplication, one or both factors negative or 0. Let $\mathbf{a}$ and $\mathbf{b}$ denote any two given positive numbers. Then **172**

1. $\mathbf{a}(-\mathbf{b})$ and $(-\mathbf{a})\mathbf{b}$ are to mean $-\mathbf{ab}$.
2. $(-\mathbf{a})(-\mathbf{b})$ is to mean $\mathbf{ab}$.
3. $\mathbf{a} \cdot 0$ and $0 \cdot \mathbf{a}$ are to mean 0.

Definition of $1/\mathbf{a}$. Let $\mathbf{a}$ be any given *positive* number, and a_1, a_2, any positive rationals whatsoever such that **173**

$$a_1 < \mathbf{a} < a_2.$$

Considerations like those of § 165 lead us, when $\mathbf{a}$ is irrational, to define $1/\mathbf{a}$ as follows:

The symbol $1/\mathbf{a}$ *is to mean the number defined by the formula* **174**

$$1/a_2 < 1/\mathbf{a} < 1/a_1,$$

where a_1, a_2 *denote any positive rationals whatsoever such that*

$$a_1 < \mathbf{a} < a_2.$$

It follows from § 163 that there is *one* and *but one* such number $1/\mathbf{a}$; for

1. Each $1/a_2$ is less than every $1/a_1$, § 106.
2. There is no last $1/a_2$ and no first $1/a_1$. (Compare proof, § 168, 2.)
3. Any positive δ being given, we can choose a_1, a_2 so that

$$1/a_1 - 1/a_2 < \delta.$$

For $\quad 1/a_1 - 1/a_2 < \delta$, if $a_2 - a_1 < \delta \cdot a_1 a_2$. §§ 106, 117

But if a_1' denote any *particular* number of the kind a_1, we can choose a_1, a_2 so that $a_1 > a_1'$ and $a_2 - a_1 < \delta\, a_1'^2$, and therefore $< \delta\, a_1 a_2$.

Definition of $1/(-\mathbf{a})$. Let $\mathbf{a}$ denote any given positive number. Then $1/(-\mathbf{a})$ is to mean $-1/\mathbf{a}$. **175**

176 **Division.** The *quotient* of $\mathbf{a}$ by $\mathbf{b}$ ($\mathbf{b}$ not 0) is to mean the number $\mathbf{a} \cdot 1/\mathbf{b}$, that is,

$$\frac{\mathbf{a}}{\mathbf{b}} = \mathbf{a} \cdot \frac{1}{\mathbf{b}}.$$

The meaning of $\mathbf{a} \cdot 1/\mathbf{b}$ itself is known from the preceding definitions.

When $\mathbf{a}$ and $\mathbf{b}$ are positive, it follows from §§ 171, 174 that we may also define $\mathbf{a}/\mathbf{b}$ by the formula

$$a_1/b_2 < \mathbf{a}/\mathbf{b} < a_2/b_1,$$

where a_1, a_2, b_1, b_2 denote any positive rationals whatsoever such that

$$a_1 < \mathbf{a} < a_2 \text{ and } b_1 < \mathbf{b} < b_2.$$

177 **The commutative, associative, and distributive laws.** The operations just defined are extensions of the corresponding operations for rational numbers. Subtraction continues to be the inverse of addition, and multiplication of division. Finally, *addition and multiplication continue to conform to the commutative, associative, and distributive laws.*

Thus, if $\mathbf{a}$, $\mathbf{b}$, and $\mathbf{c}$ are any three positive numbers defined, as in § 170, by the formulas

$$a_1 < \mathbf{a} < a_2, \qquad b_1 < \mathbf{b} < b_2, \qquad c_1 < \mathbf{c} < c_2,$$

we have

$$\mathbf{a}(\mathbf{b} + \mathbf{c}) = \mathbf{ab} + \mathbf{ac}.$$

For by §§ 166, 171, $\mathbf{a}(\mathbf{b} + \mathbf{c})$ and $\mathbf{ab} + \mathbf{ac}$ are defined by the formulas

$$a_1(b_1 + c_1) < \mathbf{a}(\mathbf{b} + \mathbf{c}) < a_2(b_2 + c_2), \quad (1)$$

$$a_1b_1 + a_1c_1 < \mathbf{ab} + \mathbf{ac} < a_2b_2 + a_2c_2. \quad (2)$$

And since $a_1(b_1 + c_1) = a_1b_1 + a_1c_1$ and $a_2(b_2 + c_2) = a_2b_2 + a_2c_2$, § 120, the numbers defined by (1) and (2) are the same.

178 **The rules of equality and inequality.** These also hold good for sums and products as just defined, namely:

According as $\quad \mathbf{a} <, =, \text{ or } > \mathbf{b},$

so is $\quad \mathbf{a} + \mathbf{c} <, =, \text{ or } > \mathbf{b} + \mathbf{c};$

also, $\quad \mathbf{ac} <, =, \text{ or } > \mathbf{bc}, \quad$ if $\mathbf{c} > 0$,

but $\quad \mathbf{ac} >, =, \text{ or } < \mathbf{bc}, \quad$ if $\mathbf{c} < 0$.

Thus, if $\mathbf{a}<\mathbf{b}$, then $\mathbf{a}+\mathbf{c}<\mathbf{b}+\mathbf{c}$.

For let d and $d+\alpha$ be any two rationals between $\mathbf{a}$ and $\mathbf{b}$, and choose c_1 so that $c_1<\mathbf{c}<c_1+\alpha$.

Then, since $\mathbf{a}<d$ and $\mathbf{c}<c_1+\alpha$, we have $\mathbf{a}+\mathbf{c}<d+c_1+\alpha$, (1)
and since $d+\alpha<\mathbf{b}$ and $c_1<\mathbf{c}$, we have $d+\alpha+c_1<\mathbf{b}+\mathbf{c}$, § 166. (2)

But from (1) and (2) it follows, § 157, that $\mathbf{a}+\mathbf{c}<\mathbf{b}+\mathbf{c}$.

The proof that, if $\mathbf{a}<\mathbf{b}$ and $\mathbf{c}>0$, then $\mathbf{ac}<\mathbf{bc}$, is similar.

But in this case we choose c_1 so that $c_1<\mathbf{c}<c_1(1+\alpha/d)$.

179 From these rules it follows, as in § 39, that if $\mathbf{a}<\mathbf{b}$ and $\mathbf{c}<\mathbf{d}$, then $\mathbf{a}+\mathbf{c}<\mathbf{b}+\mathbf{d}$, and so on; also, as in § 50, when $\mathbf{a}, \mathbf{b}, \mathbf{c}, \mathbf{d}$ are positive, that if $\mathbf{a}<\mathbf{b}$ and $\mathbf{c}<\mathbf{d}$, then $\mathbf{ac}<\mathbf{bd}$, and so on.

180 **On approximate values.** 1. Having now defined subtraction for irrational numbers, § 169, we can state the theorem of § 162 as follows:

When any irrational $\mathbf{a}$ *is given, and any positive rational* δ *is assigned, however small, we can always find rationals,* a_1 *and* a_2, *which will differ from* $\mathbf{a}$ *by less than* δ.

For, by § 162, we can find a_1 and a_2 such that $a_1<\mathbf{a}<a_2$ and $a_2-a_1<\delta$.

But from $\mathbf{a}<a_2$ it follows, § 178, that $\mathbf{a}-a_1<a_2-a_1$, and therefore that $\mathbf{a}-a_1<\delta$.

In like manner, since $-\mathbf{a}<-a_1$, we prove that $a_2-\mathbf{a}<\delta$.

Thus, § 161, we have $\sqrt{2}-1.41<.01$ and $1.42-\sqrt{2}<.01$.

We say of such an a_1 or a_2 that it represents $\mathbf{a}$ with an *error* not exceeding δ.

2. In practical reckoning we employ approximate values of irrational numbers more frequently than the numbers themselves. If a_1 and b_1 are approximate values of $\mathbf{a}$ and $\mathbf{b}$ respectively, then a_1+b_1 will be an approximate value of the sum $\mathbf{a}+\mathbf{b}$. But to insure that the error of a_1+b_1 shall not exceed δ, we must ordinarily choose a_1 and b_1 so that their respective errors shall not exceed $\delta/2$. This follows from the proof in § 166. Similar rules for finding approximate values of $\mathbf{a}-\mathbf{b}$, $\mathbf{ab}$, and $\mathbf{a}/\mathbf{b}$ with errors not exceeding δ, may be derived from the proofs in §§ 168, 171, 174.

INVOLUTION AND EVOLUTION

181 **Powers.** In the case of irrational numbers, as in that of rationals, we represent the products **aa**, **aaa**, $\cdots$, by $\mathbf{a}^2$, $\mathbf{a}^3$, $\cdots$.

182 **Roots.** The *m*th *root* of any given positive number **b**, written $\sqrt[m]{\mathbf{b}}$, is to mean that positive number whose *m*th power is **b**; that is, $\sqrt[m]{\mathbf{b}}$ is to denote the positive number which is defined by the formula $(\sqrt[m]{\mathbf{b}})^m = \mathbf{b}$.

To justify this definition, we must show that one, and but one, such number as it implies actually exists. We accomplish this as follows:

183 **Theorem.** *The real system contains the* m*th root of every positive real number* **b**.

1. If **b** is the *m*th power of a rational number, the truth of the theorem is obvious.

Thus, if $\mathbf{b} = 8/27 = (2/3)^3$, then $\sqrt[3]{\mathbf{b}} = 2/3$.

2. If **b** is *not* the *m*th power of a rational, its *m*th root is that real number **a** which lies between all positive rationals, a_1, whose *m*th powers are less than **b** and all positive rationals, a_2, whose *m*th powers are greater than **b**. Compare § 151.

It follows from § 154 that there is one, and but one, such number **a**, since (1) every positive rational is either an a_1 or an a_2, (2) each a_1 is less than every a_2, and (3) there is no last a_1 and no first a_2.

We may prove (3) as follows:

If there be a last a_1, call it p. Then since $p^m < \mathbf{b}$, there are rationals between p^m and **b**. Let one of them be $p^m + \delta$. We have only to show that we can find a rational $q > p$ such that $q^m < p^m + \delta$, or $q^m - p^m < \delta$; for we shall then have $p^m < q^m < \mathbf{b}$, so that p is *not* the last a_1.

But $q^m - p^m = (q-p)(q^{m-1} + q^{m-2}p + \cdots + qp^{m-2} + p^{m-1})$ § 308

$\therefore \quad < (q-p)ma_2'^{\,m-1}$, if a_2' be any *particular* a_2,

$\therefore \quad < \delta$, if $q = p + \delta / ma_2'^{\,m-1}$.

We can show in a similar manner that there is no first a_2.

This established, it may readily be proved that $\mathbf{a} = \sqrt[m]{\mathbf{b}}$.

For, since $a_1 < \mathbf{a} < a_2$, we have $a_1^m < \mathbf{a}^m < a_2^m$. §§ 171, 181

But $\mathbf{b}$ is the only number between every a_1^m and every a_2^m.

Hence $\mathbf{a}^m = \mathbf{b}$, that is, $\mathbf{a} = \sqrt[m]{\mathbf{b}}$.

Rules of equality and inequality. Let a and b denote any positive real numbers, and m any positive integer. Then **184**

$$\text{According as} \quad a <, =, \text{ or } > b,$$

$$\text{so is} \quad a^m <, =, \text{ or } > b^m, \tag{1}$$

$$\text{and} \quad \sqrt[m]{a} <, =, \text{ or } > \sqrt[m]{b}. \tag{2}$$

We may prove (1) by repeated use of § 179.

Thus, if $a < b$, then $a \cdot a < b \cdot b$, that is, $a^2 < b^2$; and so on.

We derive (2) from (1). Thus, if $a = b$, then $\sqrt[m]{a} = \sqrt[m]{b}$; for were $\sqrt[m]{a} <$ or $> \sqrt[m]{b}$, we should have $a <$ or $> b$.

Rules of exponents. Let a and b denote any two real numbers, and m and n any two positive integers. Then **185**

1. $a^m \cdot a^n = a^{m+n}$. 2. $(a^m)^n = a^{mn}$. 3. $(ab)^m = a^m b^m$.

Thus,

$$a^3 \cdot a^2 = aaa \cdot aa = aaaaa = a^5 = a^{3+2} \qquad \text{§ 177}$$

$$(a^2)^3 = a^2 \cdot a^2 \cdot a^2 = a^{2+2+2} = a^{2\cdot 3} \qquad \text{by 1}$$

$$(ab)^3 = ab \cdot ab \cdot ab = aaa \cdot bbb = a^3 \cdot b^3 \qquad \text{§ 177}$$

And similarly for any other positive integral values of m and n.

A theorem regarding roots. Let a and b denote any positive real numbers, and m any positive integer. Then **186**

$$\sqrt[m]{a}\sqrt[m]{b} = \sqrt[m]{ab}.$$

For $(\sqrt[m]{a} \cdot \sqrt[m]{b})^m = (\sqrt[m]{a})^m \cdot (\sqrt[m]{b})^m = ab$ §§ 182, 185, 3

and $(\sqrt[m]{ab})^m = ab$. § 182.

Hence $(\sqrt[m]{a} \cdot \sqrt[m]{b})^m = (\sqrt[m]{ab})^m$

and therefore $\sqrt[m]{a} \cdot \sqrt[m]{b} = \sqrt[m]{ab}$. § 184, (2)

VARIABLES AND LIMITS

187 **Variables.** We say that a never-ending sequence of numbers, such as

$$a_1,\ a_2,\ a_3,\ \cdots,\ a_n,\ \cdots,$$

is *given* or *known*, if the value of every particular term a_n is known, or can be computed, when the index n which shows its position in the sequence is given.

We often have occasion to consider *variables* which are supposed to be running through such given but never-ending sequences of values.

Thus, $\frac{1}{2}, \frac{2}{3}, \frac{3}{4}, \cdots, \frac{n}{n+1}, \cdots$ is such a *given* never-ending sequence, and x is such a variable if we suppose it to be running through this sequence, that is, to be taking successively the values $\frac{1}{2}, \frac{2}{3}, \frac{3}{4}, \cdots$.

188 **Limits.** As x runs through the sequence $\frac{1}{2}, \frac{2}{3}, \frac{3}{4}, \cdots$, it continually approaches the value 1, and in such a manner that if we assign any positive number, as δ, it matters not how small, the difference $1 - x$ will ultimately *become and remain* less than the number so assigned. Thus, after x reaches the 100th term of the sequence, $1 - x$ will remain less than .01.

We express all this by saying that, as x runs through the sequence $\frac{1}{2}, \frac{2}{3}, \frac{3}{4}, \cdots$, it approaches 1 as *limit*. And in general

189 *A variable* x, *which is supposed to be running through a given never-ending sequence of values, is said to approach the number* a *as* **limit**, *if the difference* a − x *will ultimately become and remain numerically less than every positive number* δ *that we may assign.*

Observe that it is not enough that $a - x$ *become* less than δ; it must also *remain* less, if x is to approach a as limit.

Thus, if x run through the sequence $\frac{1}{2}, 0, \frac{2}{3}, 0, \frac{3}{4}, 0, \cdots$, the difference $1 - x$ will *become* less than every δ that we can assign, but it will not *remain* less than this δ, and x will *not* approach 1 as limit.

In particular, $a - x$ may become 0; that is, x may *reach* its limit a.

To indicate that x is approaching the limit **a**, we write either $x \doteq$ **a**, read "x approaches **a** as limit," or lim $x =$ **a**, read "the limit of x is **a**." **190**

Whether a variable x approaches a limit or not depends entirely on the character of the sequence of values through which it is supposed to be running. **191**

Thus, while x approaches a limit when it runs through the sequence $\frac{1}{2}, \frac{3}{4}, \frac{7}{8}, \cdots$, plainly it does *not* approach a limit when it runs through the sequence 1, 2, 3, 4, $\cdots$, or the sequence 1, 2, 1, 2, $\cdots$.

Hence the importance of the following theorems:

Theorem 1. *If the variable* x *continually increases, but, on the other hand, remains always less than some given number* **c**, *it approaches a limit. And this limit is either* **c** *or some number which is less than* **c**. **192**

For by hypothesis there are numbers which x will never exceed. Assign all such numbers to a class R_2, and all other numbers, that is, all numbers which x *will* ultimately exceed, to a class R_1.

We thus obtain a separation of the entire system of real numbers into two parts, R_1, R_2, so related that each number in R_1 is less than every number in R_2.

Obviously there is no last number in R_1. Hence, § 160, there is a first number in R_2. Call this number **a**. As x increases, it will approach **a** as limit.

For however small δ may be, if only positive, **a** $- \delta$ belongs to the class of numbers R_1, which x will ultimately exceed. Hence x will ultimately remain between **a** $- \delta$ and **a**, and therefore differ from **a** by less than δ.

In the same manner it may be demonstrated that

If the variable x *continually decreases, but, on the other hand, remains always greater than some given number* **c**, *it approaches a limit. And this limit is either* **c** *or some number which is greater than* **c**. **193**

194 **Regular sequences.** It is not necessary, however, that x should always increase or always decrease, if it is to approach a limit.

Thus, x is sometimes increasing and sometimes decreasing, as it runs through the sequence $-\frac{1}{2}, \frac{1}{4}, -\frac{1}{8}, \frac{1}{16}, \cdots$; but it approaches 0 as limit.

We shall prove that x will or will not approach a limit, according as the sequence of values $a_1, a_2, \cdots, a_n, \cdots$ through which it runs, has or has not the character described in the following definition:

195 *The sequence* $a_1, a_2, \cdots, a_n, \cdots$ *is said to be* **regular,** *if for every positive test number* δ *that may be assigned a corresponding term* a_k *can be found, which will differ numerically from every subsequent term by less than* δ.

1. Thus, the sequence 1.4, 1.41, 1.414, $\cdots$ (1), § 161, is regular.

For the difference between the first term, 1.4, and every subsequent term is less than $1/10$; that between the second term, 1.41, and every subsequent term is less than $1/10^2$; that between the nth term and every subsequent term is less than $1/10^n$.

Now, however small δ may be, we can give n a value which will make $1/10^n$ smaller still; and if k denote such a value of n, the kth term of 1.4, 1.41, $\cdots$ will differ from every subsequent term by less than δ.

Thus, if we assign the value $1/500000$ to δ, we have $1/10^6 < \delta$, so that the *sixth* term of 1.4, 1.41, $\cdots$ will differ from every subsequent term by less than this value of δ.

2. The following sequences are also regular:

$$\frac{1}{2}, \frac{3}{4}, \frac{7}{8}, \frac{15}{16}, \cdots, \quad (2) \qquad \frac{3}{2}, \frac{5}{4}, \frac{9}{8}, \frac{17}{16}, \cdots, \quad (3)$$

$$-\frac{1}{2}, \frac{1}{4}, -\frac{1}{8}, \frac{1}{16}, \cdots, \quad (4) \qquad 2, 1, 1, 1, \cdots. \quad (5)$$

Observe that in (2) each term is followed by a greater term, in (3) by a lesser term, in (4) sometimes by a greater term, sometimes by a lesser.

We sometimes encounter regular sequences like (5), all of whose terms after a certain one are the same. Evidently a variable which runs through such a sequence will ultimately become constant, that is, will *reach* its limit.

3. The following sequences are *not* regular:

$$1, 2, 3, 4, \cdots, \quad (6) \qquad \frac{1}{2}, \frac{1}{3}, \frac{1}{2}, \frac{1}{3}, \cdots. \quad (7)$$

For in (6) the difference between a term and a subsequent one may always be indefinitely great, and in (7) it may always be $\frac{1}{8}$, and therefore not less than *every* number, $\frac{1}{9}$ for instance, that we can assign.

Formulas for regular sequences. 1. We may indicate the relation between the term a_k and every subsequent term, a_p, by the formula, § 63: **196**

$$|a_p - a_k| < \delta \qquad \text{for every } p > k. \qquad (1)$$

2. Again, since any terms a_p there may be which are $> a_k$ will lie between a_k and $a_k + \delta$, and any which are $< a_k$ will lie between $a_k - \delta$ and a_k, we may also write

$$a_k - \delta < a_p < a_k + \delta \qquad \text{for every } p > k. \qquad (2)$$

3. It follows from (2) that if some of the terms a_p are less, and some are greater than a_k, the difference between two of these terms may exceed δ, but not 2δ.

But we can always find a term, a_l, which corresponds to $\delta/2$ as a_k corresponds to δ. The difference between every two terms after a_l will then be numerically less than $2(\delta/2)$, or δ; that is, the relation between every two of these terms will be that indicated by the formula

$$|a_p - a_q| < \delta \qquad \text{for every } p > q > l. \qquad (3)$$

Theorem 2. *The variable* x *will approach a limit if the sequence of values* $a_1, a_2, \cdots, a_n, \cdots$, *through which it is supposed to run, is a regular sequence.* **197**

For there are numbers to whose right x will ultimately *remain* as it runs through the sequence $a_1, a_2, \cdots, a_n, \cdots$. (1)

Thus, if δ and a_k have the meanings above explained, x will remain to the right of $a_k - \delta$ after it reaches the value a_k, § 196 (2).

Assign all such numbers to a class, R_1, and all other numbers — that is, all numbers to whose right x will *not* remain — to a class, R_2.

We thus obtain a separation of the entire system of real numbers into two parts, R_1, R_2, so related that each number in R_1 is less than every number in R_2. By § 160, a definite number, $\mathbf{a}$, exists at which this separation occurs.

Thus, if the sequence be $-\frac{1}{2}, \frac{1}{4}, -\frac{1}{8}, \frac{1}{16}, \cdots$, the negative rationals constitute R_1, but 0 and the positive rationals, R_2; and $\mathbf{a}$ itself is 0.

As x runs through the sequence (1), it will approach this number $\mathbf{a}$ as limit.

For assign any positive test number, δ, it matters not how small. Since (1) is regular, we can find a term, a_m, § 196 (3), such that

$$|a_p - a_q| < \delta/2 \qquad \text{for every } p > q > m. \tag{2}$$

But since $\mathbf{a} - \delta/2$ belongs to R_1, all the values of x after a certain one will lie to the right of $\mathbf{a} - \delta/2$. And since $\mathbf{a} + \delta/2$ belongs to R_2, among these values there will be some *after* a_m which lie to the left of $\mathbf{a} + \delta/2$; for otherwise $\mathbf{a} + \delta/2$ would belong to R_1, since x would ultimately *remain* to its right.

Thus, if the sequence be $-\frac{1}{2}, \frac{1}{4}, -\frac{1}{8}, \frac{1}{16}, \cdots$, and $\delta = \frac{1}{10}$, *all* values of x after the *fourth*, $\frac{1}{16}$, lie between $\mathbf{a} - \delta/2$ and $\mathbf{a} + \delta/2$, that is, between $-\frac{1}{20}$ and $\frac{1}{20}$.

Let a_q' denote such a value of x. Then

$$\mathbf{a} - \delta/2 < a_q' < \mathbf{a} + \delta/2,$$

or

$$|\mathbf{a} - a_q'| < \delta/2. \tag{3}$$

From (2) and (3), since $q' > m$, it follows, §§ 78, 178, that

$$|\mathbf{a} - a_p| < \delta \qquad \text{for every } p > q'.$$

In other words, after x reaches the value a_q' the difference $\mathbf{a} - x$ remains numerically less than δ.

Therefore x approaches $\mathbf{a}$ as limit, § 189.

198 **Conversely,** *if* x *is approaching a limit,* $\mathbf{a}$, *the sequence of values* $\mathrm{a}_1, \mathrm{a}_2, \cdots, \mathrm{a}_n, \cdots$, *through which it is supposed to run, must be regular.*

For since the difference $\mathbf{a} - x$ will ultimately become and remain numerically less than every assigned positive number, δ, § 189, we can choose a_k so that

$$|\mathbf{a} - a_k| < \delta/2 \text{ and } |\mathbf{a} - a_p| < \delta/2 \qquad \text{for every } p > k;$$

whence

$$|a_p - a_k| < \delta \qquad \text{for every } p > k.$$

Hence the sequence $a_1, a_2, \cdots, a_n, \cdots$ is regular, § 196 (1).

We may combine §§ 197, 198 in the single statement:

199 *The sufficient and necessary condition that a variable approach a limit is that the sequence of values through which it is supposed to run be a regular sequence.*

SOME IMPORTANT THEOREMS REGARDING LIMITS

In the present section a and b will denote any *given* real numbers, and x and y *variables* which are supposed to run through given never-ending sequences of values.

200 **The limit 0.** From the definition of limit, § 189, it immediately follows that

1. If the variable x will ultimately become and remain numerically less than every positive number, δ, that may be assigned, then x approaches 0 as limit; and conversely.

2. If x approaches a as limit, then $a - x$ approaches 0 as limit; and conversely.

Thus x approaches the limit 0, as it runs through the sequence $\frac{1}{2}, \frac{1}{3}, \frac{1}{4}, \cdots$; and $1 - x$ approaches the limit 0, as x runs through the sequence $\frac{1}{2}, \frac{2}{3}, \frac{3}{4}, \cdots$.

A variable whose limit is 0 is called an *infinitesimal*.

201 **Theorem 1.** *If* $\mathrm{x} \doteq 0$ *and* $\mathrm{y} \doteq 0$, *and* A *and* B *remain numerically less than some fixed number*, c, *as* x *and* y *vary, then* $\mathrm{Ax} + \mathrm{By} \doteq 0$.

For assign any positive number, δ, it matters not how small.

Since $x \doteq 0$, x will ultimately remain numerically $< \delta/2\,c$. § 200, 1

Since $y \doteq 0$, y will ultimately remain numerically $< \delta/2\,c$. § 200, 1

Hence $Ax + By$ will ultimately remain numerically $< 2\,c\frac{\delta}{2\,c}$, $\therefore < \delta$, and therefore approaches 0 as limit, § 200, 1.

Thus, if $x \doteq 0$ and $y \doteq 0$, then $(xy - 3)x + 2\,y \doteq 0$.

202 **Note.** This theorem may readily be extended to any *finite* number of variables.

Thus, if $x \doteq 0$, $y \doteq 0$ and $z \doteq 0$, then $Ax + By + Cz \doteq 0$.

203 **Theorem 2.** *The limit of the sum, difference, product, quotient of two variables which approach limits is the sum, difference, product, quotient of these limits: that is, if* x *and* y *approach the limits* a *and* b *respectively, then*

1. $x + y \doteq a + b$.
2. $x - y \doteq a - b$.
3. $xy \doteq ab$.
4. $x/y \doteq a/b$, unless $b = 0$.

For, since $a - x \doteq 0$ and $b - y \doteq 0$, § 200, it follows from § 201 that

$$A(a - x) + B(b - y) \doteq 0. \quad (1)$$

The formulas 1, 2, 3, 4 may be derived from (1). Thus,

1. $$a + b - (x + y) = (a - x) + (b - y) \therefore \doteq 0, \quad \text{by (1)}$$

that is, $$x + y \doteq a + b. \quad \text{§ 200, 2}$$

2. $$a - b - (x - y) = (a - x) - (b - y) \therefore \doteq 0, \quad \text{by (1)}$$

that is, $$x - y \doteq a - b. \quad \text{§ 200, 2}$$

3. $$ab - xy = (a - x)b + (b - y)x \therefore \doteq 0, \quad \text{by (1)}$$

that is, $$xy \doteq ab. \quad \text{§ 200, 2}$$

4. $$\frac{a}{b} - \frac{x}{y} = \left(\frac{a}{b} - \frac{x}{b}\right) + \left(\frac{x}{b} - \frac{x}{y}\right) = (a - x)\frac{1}{b} - (b - y)\frac{x}{by} \therefore \doteq 0, \quad \text{by (1)}$$

that is, $$x/y \doteq a/b. \quad \text{§ 200, 2}$$

204 **Corollary.** *If* $x \doteq a$, *then* $x^n \doteq a^n$.

205 **Theorem 3.** *The limit of the* n*th root of a variable which approaches a limit is the* n*th root of that limit: that is,*

$$\textit{If } x \doteq a, \textit{ then } \sqrt[n]{x} \doteq \sqrt[n]{a}.$$

1. When $a = 0$. Assign any positive number, δ.

Since $x \doteq 0$, x will ultimately remain numerically $< \delta^n$. § 200, 1

Hence $\sqrt[n]{x}$ will ultimately remain numerically $< \delta$. § 184

Therefore $\sqrt[n]{x} \doteq 0$. § 200, 1

2. When a is not 0. It follows from a later section, § 308, that $x - a$ is always exactly divisible by $\sqrt[n]{x} - \sqrt[n]{a}$, and that the quotient Q does not approach the limit 0 when $x \doteq a$.

It therefore follows from § 203 (1), by setting $A = 1/Q$ and $B = 0$, that

$$\sqrt[n]{x} - \sqrt[n]{a} = (x - a)/Q \doteq 0, \text{ that is, } \sqrt[n]{x} \doteq \sqrt[n]{a}. \quad \text{§ 200, 2}$$

RELATION OF THE IRRATIONAL NUMBERS TO MEASUREMENT

206 **Length of a line segment incommensurable with the unit.** If a line segment **S** be *incommensurable* with the unit segment **s**,— that is, if, as when **S** and **s** are diagonal and side of the same square, we can prove that *no* aliquot part of **s**, however small, is contained in **S** exactly — the definition of length given in § 130 does not apply to **S**.

But there is then a definite *irrational* number, **a**, which stands in the following relation to **S**:

The segments which *are* commensurable with the unit **s** fall into two distinct classes, those which are less than **S** and those which are greater than **S**.

The rational numbers which are their lengths, § 130, fall into two corresponding classes, which we may call A_1 and A_2. Every positive rational belongs either to A_1 or to A_2, each number in A_1 precedes every number in A_2, and, finally, there is neither a last number in A_1 nor a first in A_2.*

There is, then, § 154, a definite irrational number, **a**, which lies between all numbers in A_1 and all in A_2. We call this

* For were there a last number in A_1, then among the segments commensurable with **s** and less than **S** there would be a greatest, say **S′**.

But no such segment exists. For according to the Axiom of Archimedes, explained in the following footnote, we could find an aliquot part of **s** which is less than **S** − **S′**; and the sum of **S′** and this part of **s** would be commensurable with **s**, less than **S** and *greater* than **S′**.

number **a** the *length* of **S**. We therefore have the following definition:

207 *The* **length** *of any segment,* **S**, *incommensurable with the unit,* **s**, *is that irrational number,* **a**, *which lies between all rationals which are lengths of segments less than* **S** *and all rationals which are lengths of segments greater than* **S**.

Thus, $\sqrt{2}$ is the length of the diagonal of a square in terms of the side.

208 If the length of **S** in terms of **s** is **a**, we write **S** = **as**, and that whether **a** is rational or irrational.

209 **Real numbers pictured by points.** As in the figure of § 134, take any right line and on it a fixed point O as *origin;* also some convenient unit, **s**, for measuring lengths. And by the *distance* from O of any point P of the line, understand the *length* of the segment OP in terms of **s**, §§ 130, 207.

We choose as the picture of any given number, **a**, *that point,* P, *of the line whose distance from* O *is the numerical value of* **a**, *the point being taken to the right or left of* O, *according as* **a** *is positive or negative.*

If **a** is a rational number, we can actually construct P, § 134. On the contrary, if **a** is irrational, we usually cannot construct P. We then *assume that* P *exists,* in other words, that on the line there is a single point, P, lying between all points which picture rationals less than **a** and all which picture rationals greater than **a**.*

* This is not the place for a discussion of the axioms of geometry; but we may mention the following because of their relation to the subject of measurement now under consideration.

1. **Axiom of Archimedes.** *If* **s** *and* **S** *denote two line segments such that* **s** < **S**, *we can always find an integer*, m, *such that* ms > **S**.

2. **Axiom of continuity.** *If all the points of a right line be separated into two classes,* R_1 *and* R_2, *such that each point in* R_1 *lies to the left of every point in* R_2, *there is either a last point in* R_1 *or a first in* R_2.

(1) The Axiom of Archimedes is involved in the assumption that every line segment can be measured. For the first step in measuring **S** in terms of **s** is to find an integer, m, such that $(m-1)$ **s** < **S** < m**s**.

(2) The axioms 1 and 2 enable us to *prove* the assumption in § 209 that for every given irrational **a** there exists a corresponding point, P.

For **a** separates the rational system into two parts, which we may name B and C respectively. Call the points corresponding to the numbers in each the

Conversely, when P is given, we can find a, at least approximately, by measuring OP and attaching the + or − sign to the result, according as P is to the right or left of O. **210**

Thus, if P is to the right of O, and we can lay s along OP *five* times, the *tenth* part of s along the part left over *seven* times, and the *hundredth* part of s along the part still left over *six* times, then 5.76 will be the value of a to the second place of decimals.

In this manner we set up a relation of *one-to-one correspondence*, § 2, between all the real numbers and all points on the line; and if a and b denote any two real numbers, and P and Q the corresponding points, P will lie to the left or right of Q according as a is less or greater than b. **211**

Thus, if a and b are positive and $a < b$, and if c denote a rational lying between a and b, and R the corresponding point, we have, § 206,

$$OP < OR \text{ and } OR < OQ \text{ and therefore } OP < OQ.$$

B-points and the C-points respectively. We are to show that there is in the line a definite point, P, which separates all the B-points from all the C-points.

First assign the B-points and all intermediate points to a class R_1 and all points to the right of these to a class R_2, and let P denote the point which this separation defines, by 2.

Next assign the C-points and all intermediate points to a class S_2 and all points to their left to a class S_1, and let Q denote the point which this separation defines, by 2.

The points P *and* Q *must coincide.* For if not, let PQ denote the line segment between them. By 1, we can find an integer, m, such that

$$m \cdot PQ > s, \text{ and therefore } PQ > s/m.$$

But this is impossible. For we can select from B a number b and from C a number c such that $c - b < 1/m$. And if L and M be the points corresponding to b and c respectively, we have

$$LM < s/m, \text{ and } PQ < LM, \text{ and therefore } PQ < s/m.$$

It is this one point, P or Q, that corresponds to a according to § 209.

(3) Finally, observe that corresponding to 2 the system of real numbers has the property described in § 160, and corresponding to 1 the property:

If a *and* b *are any two positive real numbers, we can always find an integer,* m, *such that* $mb > a$.

For, by §§ 108, 176, 178, we can choose an integer, m, such that $m > a/b$ and therefore $mb > a$.

The real system would not possess this property — at least not without a sacrifice of some of its other properties — were we to invent more than one irrational for a separation of the rational system of the kind described in § 154.

Thus, if every rational is either an a_1 or an a_2, and $a_1 < b < c < a_2$ for every a_1, a_2, we should have $c - b < a_2 - a_1$, § 178 and proof of § 163.

But however small a positive number, δ, we might assign, we could find no integer, m, so great that $m(c - b) > \delta$.

For it would then follow that $c - b > \delta/m$, which is impossible since $c - b < a_2 - a_1$ and we can choose a_1, a_2 so that $a_2 - a_1 < \delta/m$.

212 **Theorem.** *If the length of* **S** *in terms of* **T** *is* **a**, *and that of* **T** *in terms of* **s** *is* **b**, *then the length of* **S** *in terms of* **s** *is* **ab**.

1. When **a** and **b** are *rational.*

Let $\mathbf{a} = a/b$ and $\mathbf{b} = c/d$, where a, b, c, d denote integers.

Since **S** contains the bth part of **T** a times, § 130, $b\mathbf{S}$ will contain **T** itself a times, that is,

$$b\mathbf{S} = a\mathbf{T}. \tag{1}$$

Similarly
$$d\mathbf{T} = c\mathbf{s}. \tag{2}$$

But from (1) and (2) it readily follows that

$$bd\mathbf{S} = ad\mathbf{T}, \text{ and } ad\mathbf{T} = ac\mathbf{s},$$

and therefore
$$bd\mathbf{S} = ac\mathbf{s}.$$

That is, the length of **S** in terms of **s** is $\dfrac{ac}{bd}$ or $\dfrac{a}{b} \cdot \dfrac{c}{d}$. § 130

2. When **a** and **b**, one or both, are *irrational.*

Let $\mathbf{S}_1$ and $\mathbf{S}_2$ denote any segments commensurable with **T**, such that

$$\mathbf{S}_1 < \mathbf{S} < \mathbf{S}_2,$$

and let a_1, a_2 be the lengths of $\mathbf{S}_1$, $\mathbf{S}_2$ in terms of **T**, so that

$$\mathbf{S}_1 = a_1\mathbf{T} \text{ and } \mathbf{S}_2 = a_2\mathbf{T}, \text{ where } a_1 < \mathbf{a} < a_2. \qquad \text{§ 208}$$

Similarly, let $\mathbf{T}_1$, $\mathbf{T}_2$ denote any segments commensurable with **s**, such that

$$\mathbf{T}_1 < \mathbf{T} < \mathbf{T}_2,$$

and let b_1, b_2 be the lengths of $\mathbf{T}_1$, $\mathbf{T}_2$ in terms of **s**, so that

$$\mathbf{T}_1 = b_1\mathbf{s}, \text{ and } \mathbf{T}_2 = b_2\mathbf{s}, \text{ where } b_1 < \mathbf{b} < b_2.$$

Then since $\mathbf{S}_1 = a_1\mathbf{T}$, and $\mathbf{T} > \mathbf{T}_1$, and $\mathbf{T}_1 = b_1\mathbf{s}$,

we have, by case 1,
$$\mathbf{S}_1 > a_1b_1\mathbf{s}.$$

Similarly
$$\mathbf{S}_2 < a_2b_2\mathbf{s}.$$

Hence
$$a_1b_1\mathbf{s} < \mathbf{S}_1 < \mathbf{S} < \mathbf{S}_2 < a_2b_2\mathbf{s},$$

and therefore
$$a_1b_1\mathbf{s} < \mathbf{S} < a_2b_2\mathbf{s}.$$

We have thus demonstrated that all the numbers a_1b_1 and a_2b_2 are lengths, in terms of **s**, of segments respectively less and greater than **S**. Therefore **ab**, the one number which lies between all the numbers a_1b_1 and a_2b_2, § 171, is the length of **S** itself in terms of **s**, § 207.

213 **Corollary.** *If the lengths of* $\mathbf{S}$ *and* $\mathbf{T}$ *in terms of* $\mathbf{s}$ *are* $\mathbf{a}$ *and* $\mathbf{b}$ *respectively, then the length of* $\mathbf{S}$ *in terms of* $\mathbf{T}$ *is* $\mathbf{a}/\mathbf{b}$.

For let the length of $\mathbf{S}$ in terms of $\mathbf{T}$ be $\mathbf{x}$.

Then since the length of $\mathbf{S}$ in terms of $\mathbf{T}$ is $\mathbf{x}$, and that of $\mathbf{T}$ in terms of $\mathbf{s}$ is $\mathbf{b}$, the length of $\mathbf{S}$ in terms of $\mathbf{s}$ is $\mathbf{xb}$, § 212.

But, by hypothesis, the length of $\mathbf{S}$ in terms of $\mathbf{s}$ is $\mathbf{a}$.

Hence $$\mathbf{xb} = \mathbf{a},$$

and therefore $$\mathbf{x} = \mathbf{a}/\mathbf{b}.$$

214 **The continuous variable.** One of our most familiar intuitions is that of *continuous motion*.

Suppose the point P to be moving continuously from A to B along the line OAB; and let $\mathbf{a}$, $\mathbf{x}$, and $\mathbf{b}$ denote the lengths of OA, OP, and OB respectively, O being the origin.

According to the assumption of § 209, the segment AB contains a point for every number between $\mathbf{a}$ and $\mathbf{b}$, through which, of course, P must pass in its motion from A to B. This leads us to say that as P moves continuously from A to B, $\mathbf{x}$ increases from the value $\mathbf{a}$ to the value $\mathbf{b}$ *through all intermediate values*, or that $\mathbf{x}$ varies *continuously* from $\mathbf{a}$ to $\mathbf{b}$.

Of course it is impossible actually to trace the variation of this $\mathbf{x}$, since to any given one of its values there is no *next* following value. If we attempt to reason about $\mathbf{x}$ mathematically, we must content ourselves with defining it thus: (1) $\mathbf{x}$ may take every *given* value between $\mathbf{a}$ and $\mathbf{b}$, and (2) if p and q denote any given pair of these values, and $p < q$, then $\mathbf{x}$ will take the value p before it takes the value q. We may add that $\mathbf{x}$ is often called a continuous variable when only the first of these properties is attached to it.

215 **Ratio.** Let M and N denote any two magnitudes of the same kind. By the *measure* of M in terms of N, or the *ratio* of M to N, we mean the very same numbers which we have defined as *lengths* in §§ 81, 130, 207, when M and N denote line segments.

Hence the theorems of §§ 212, 213 regarding lengths hold good for the measures or ratios of any magnitudes of the same kind. In particular,

216 *If the measures of* M *and* N *in terms of the same unit are* **a** *and* **b** *respectively, the ratio of* M *to* N *is* **a/b**.

V. THE IMAGINARY AND COMPLEX NUMBERS

PURE IMAGINARIES

217 The real system does not contain the even roots of negative numbers; for the even powers of all real numbers are positive. Thus the real system does not contain the square root of -1.

To meet this difficulty, we invent a new system of signs called *imaginary* or *complex* numbers.

218 The simplest of these new signs is i, called the *unit of imaginaries*. With this unit and the real numbers, a, we form signs like ai, which we then regard as arranged in the order in which their "coefficients," a, occur in the real system. We thus obtain a new continuous ordinal system of "numbers," which we call *pure imaginaries*.

Thus, proceeding as when developing the real system, we may first form the complete *scale of imaginaries*

$$\cdots \quad -3i, \quad -2i, \quad -i, \quad 0, \quad i, \quad 2i, \quad 3i, \quad \cdots,$$

then enlarge this into a *dense* system by introducing imaginaries with fractional coefficients, and finally into a *continuous* system by introducing imaginaries with irrational coefficients.

Here $2i$ is merely the name of one of our new numbers. Its only property is a definite position in the new ordinal system. But when we have defined multiplication, we shall see that $2i$ also represents the product $2 \times i$ or $i \times 2$. Similarly every pure imaginary ai.

In particular we shall define $0 \cdot i$ as 0. Hence we write 0 for $0i$.

Observe that 0 is the *only* number which is common to the real system and the system of pure imaginaries.

For these new numbers we invent operations which we call *addition* and *multiplication*. They are defined by the following equations: **219**

1. $ai + bi = (a + b)i$. 2. $a \cdot bi = bi \cdot a = abi$.

3. $ai \cdot bi = -ab$.

Thus, 3, the product of two pure imaginaries, ai and bi, is to mean the *real* number, $-ab$, obtained by multiplying the coefficients of ai and bi together and changing the sign of the result.

We define *power* as in § 136. Thus, $(ai)^2 = ai \cdot ai$.

The system of pure imaginaries contains the square roots of all negative numbers in the real system, namely: **220**

$$\sqrt{-1} = i \quad \textit{and} \quad \sqrt{-a^2} = ai.$$

For $i^2 = i \cdot i = 1i \cdot 1i = -1$. § 219, 3

Therefore, i is a square root of -1, § 138. We indicate this root by $\sqrt{-1}$, and thus have $i = \sqrt{-1}$.

In like manner, it may be shown that $-i$ is a square root of -1. We indicate this root by $-\sqrt{-1}$.

Similarly, since $(ai)^2 = ai \cdot ai = -a^2$, we have $ai = \sqrt{-a^2}$.

COMPLEX NUMBERS

To secure a number system which will contain the *higher* even roots of negative numbers, we invent *complex numbers.* These are expressions like $a + bi$, formed by connecting a real number, a, with a pure imaginary, bi, by the sign $+$. They are also often called *imaginary numbers.* **221**

Until addition has been defined for complex numbers the expression $a + bi$ is to be regarded as a single symbol and the sign $+$ as merely a part of this symbol.

Since $a = a + 0i$ and $bi = 0 + bi$, real numbers and pure imaginaries are included among the complex numbers. **222**

223 We regard the complex numbers as arranged in rows and columns in such a manner that all numbers $a + bi$ which have the same b lie in the same *row* and are arranged in this row from left to right in the order of their a's; while all numbers which have the same a lie in the same *column* and are arranged in this column from below upward in the order of their b's. And we may consider any particular complex number *defined* by its position in this "two-dimensional ordinal arrangement."

In § 238 we shall explain a method of picturing this arrangement for *all* values of a and b. We may indicate it as follows for *integral* values of a and b.

...	. .	. .	.	. .	. .	...
...	$-2+2i$	$-1+2i$	$2i$	$1+2i$	$2+2i$	...
...	$-2+i$	$-1+i$	i	$1+i$	$2+i$	...
...	-2	-1	0	1	2	...
...	$-2-i$	$-1-i$	$-i$	$1-i$	$2-i$	...
...	$-2-2i$	$-1-2i$	$-2i$	$1-2i$	$2-2i$	...
...	. .	. .	.	. .	. .	...

This arrangement may also be described as an ordinal system, § 17, whose elements are rows (or columns), each of which is itself an ordinal system of signs of the form $a + bi$.

224 **Definition of equality.** Two complex numbers are said to be *equal* when they occupy the same position in the two-dimensional ordinal arrangement just described. Hence,

225 If $a + bi = c + di$, then $a = c$ and $b = d$; and conversely. In particular, if $a + bi = 0$, then $a = 0$ and $b = 0$; and conversely.

Of two *unequal* complex numbers, like $2 + 3i$ and $3 + i$, we cannot say that the one is *less* or *greater* than — that is, precedes or follows — the other, since complex numbers do not form a *simple* ordinal system.

226 **Definitions of addition, subtraction, multiplication.** The *sum*, *difference*, and *product* of two complex numbers $a + bi$, $c + di$, are to mean the complex numbers which form the second members of the following equations:

1. $(a + bi) + (c + di) = (a + c) + (b + d)i.$

2. $(a + bi) - (c + di) = (a - c) + (b - d)i.$

3. $(a + bi)(c + di) = (ac - bd) + (ad + bc)i.$

According to 1 and 2, addition and subtraction are inverse operations. In particular, by 1, $(a + 0i) + (0 + bi) = (a + 0) + (0 + b)i = a + bi$; that is, $a + bi$ is the sum of a and bi, according to the definition 1.

These definitions are in agreement with the commutative, associative, and distributive laws. In fact, we arrive at them by combining these laws with definitions previously given. Thus,

$$\begin{aligned}(a + bi)(c + di) &= (a + bi)c + (a + bi)di \\ &= ac + bi \cdot c + a \cdot di + bi \cdot di \\ &= (ac - bd) + (ad + bc)i, \text{ since } i^2 = -1.\end{aligned}$$

Corollary. *A product vanishes when a factor vanishes.* **227**

For $(a + bi)(0 + 0i) = (a \cdot 0 - b \cdot 0) + (a \cdot 0 + b \cdot 0)i = 0.$

Division. We define the *quotient* of $a + bi$ by $c + di$ as the complex number which multiplied by $c + di$ will give $a + bi$. When $c + di$ is not 0, there is one and but one such number, namely, that in the second member of the equation **228**

$$\frac{a + bi}{c + di} = \frac{ac + bd}{c^2 + d^2} + \frac{bc - ad}{c^2 + d^2} i.$$

But when $c + di$ is 0, no determinate quotient exists.

For the product of the right member of this equation by $c + di$ is $a + bi$, as the reader may easily verify by aid of § 226.

We discover that this number is the quotient as follows:

If a number exists which multiplied by $c + di$ will give $a + bi$, let it be $x + yi$.

Then $$(x + yi)(c + di) = a + bi. \quad (1)$$

or $$(cx - dy) + (dx + cy)i = a + bi. \quad (2)$$

and therefore $cx - dy = a$ and $dx + cy = b$. (3) § 225

Solving this pair of equations for x and y, we obtain

$$x = \frac{ac + bd}{c^2 + d^2}, \quad y = \frac{bc - ad}{c^2 + d^2}, \text{ unless } c^2 + d^2 = 0. \quad (4)$$

Moreover, since (4) is the *only* pair of values of x and y which will satisfy (3), the corresponding number $x + yi$ is the *only* number which multiplied by $c + di$ will give $a + bi$.

It is evident from (4) that when $c^2 + d^2 = 0$ our definition of quotient is meaningless. But if $c^2 + d^2 = 0$, both $c = 0$ and $d = 0$, since otherwise we should have a positive number equal to 0. And if $c = 0$ and $d = 0$, the divisor $c + di$ is 0.

229 The commutative, associative, and distributive laws. The operations just defined evidently include the corresponding operations with real numbers. Like the latter, *they conform to the commutative, associative, and distributive laws.*

Thus, $$(a + a'i)(b + b'i) = ab - a'b' + (ab' + a'b)i, \quad (1)$$

and $$(b + b'i)(a + a'i) = ba - b'a' + (b'a + ba')i. \quad (2)$$

But the second members of (1) and (2) are equal, by § 177.

Hence $$(a + a'i)(b + b'i) = (b + b'i)(a + a'i).$$

And the remaining laws may be established similarly.

230 Rules of equality. Let $\mathbf{a}$, $\mathbf{b}$, $\mathbf{c}$ denote any complex numbers.

1. If $\mathbf{a} = \mathbf{b}$, then $\mathbf{a} + \mathbf{c} = \mathbf{b} + \mathbf{c}$.
2. If $\mathbf{a} + \mathbf{c} = \mathbf{b} + \mathbf{c}$, then $\mathbf{a} = \mathbf{b}$.
3. If $\mathbf{a} = \mathbf{b}$, then $\mathbf{ac} = \mathbf{bc}$.
4. If $\mathbf{ac} = \mathbf{bc}$, then $\mathbf{a} = \mathbf{b}$, unless $\mathbf{c} = 0$.

1. For let $\mathbf{a} = a + a'i$, $\mathbf{b} = b + b'i$, and $\mathbf{c} = c + c'i$.

If $$a + a'i = b + b'i,$$

then $$a = b \quad \text{and} \quad a' = b'. \qquad \text{§ 225}$$

Hence $$a + c = b + c \text{ and } a' + c' = b' + c', \qquad \text{§ 178}$$

and therefore $$(a + c) + (a' + c')i = (b + c) + (b' + c')i, \qquad \text{§ 225}$$

that is, $$\mathbf{a} + \mathbf{c} = \mathbf{b} + \mathbf{c}. \qquad \text{§ 226}$$

2. If $$\mathbf{a} + \mathbf{c} = \mathbf{b} + \mathbf{c},$$

we have $$\mathbf{a} + \mathbf{c} + (-\mathbf{c}) = \mathbf{b} + \mathbf{c} + (-\mathbf{c}), \qquad \text{by 1}$$

and therefore $$\mathbf{a} = \mathbf{b}. \qquad \text{§ 226}$$

3 and 4. The proofs of these rules are similar to those of 1 and 2 respectively.

Corollary. *If a product vanish, one of its factors must vanish.* **231**

This follows from § 230, 4, by the reasoning of § 76.

Absolute value of a complex number. The positive real number $\sqrt{a^2+b^2}$ is called the *absolute* or *numerical* value of $a+bi$ and is represented by $|a+bi|$. Hence, by definition, **232**

$$|a+bi|=\sqrt{a^2+b^2}.$$

Thus, $|2+i|=\sqrt{4+1}=\sqrt{5}$.

When $b=0$, this definition of numerical value reduces to that already given for real numbers, § 63. For a geometrical interpretation of this definition see § 239.

We also say of two complex numbers that the first is *numerically less than, equal to,* or *greater than* the second, according as the absolute value of the first is less than, equal to, or greater than that of the second. **233**

Thus, $2+3i$ is numerically greater than $3+i$.
For $|2+3i|=\sqrt{13}$ and $|3+i|=\sqrt{10}$, and $\sqrt{13}>\sqrt{10}$.

Theorem 1. *The absolute value of a product of two complex numbers is equal to the product of their absolute values.* **234**

Let the numbers be $\mathbf{a}=a+a'i$ and $\mathbf{b}=b+b'i$.

Since $\qquad \mathbf{ab}=ab-a'b'+(ab'+a'b)\,i,$ $\qquad$ § 226

we have $\qquad |\mathbf{ab}|=\sqrt{(ab-a'b')^2+(ab'+a'b)^2}.$ $\qquad$ § 232

But on carrying out the indicated operations it will be found that

$$(ab-a'b')^2+(ab'+a'b)^2=(a^2+a'^2)(b^2+b'^2).$$

Hence $\sqrt{(ab-a'b')^2+(ab'+a'b)^2}=\sqrt{a^2+a'^2}\cdot\sqrt{b^2+b'^2}.$ $\qquad$ § 186

That is, $\qquad |\mathbf{ab}|=|\mathbf{a}|\cdot|\mathbf{b}|.$

Theorem 2. *The absolute value of a sum of two complex numbers cannot exceed the sum of their absolute values.* **235**

Employ the same notation as in § 234.

Then $\qquad \sqrt{a^2+a'^2}+\sqrt{b^2+b'^2}\geqq\sqrt{(a+b)^2+(a'+b')^2}$ $\qquad$ (1)

if $\quad a^2+a'^2+b^2+b'^2+2\sqrt{(a^2+a'^2)(b^2+b'^2)}$

$$\geqq a^2+b^2+a'^2+b'^2+2\,(ab+a'b') \qquad \text{§ 184}$$

$\therefore$ if $\sqrt{(a^2 + a'^2)(b^2 + b'^2)} \geqq ab + a'b'$ § 178

$\therefore$ if $a^2b^2 + a'^2b'^2 + a^2b'^2 + a'^2b^2 \geqq a^2b^2 + a'^2b'^2 + 2\,aba'b'$ § 184

$\therefore$ if $a^2b'^2 + a'^2b^2 \geqq 2\,aba'b'$ § 178

$\therefore$ if $(ab' - a'b)^2 \geqq 0.$ (2) § 178

But (2) is always true since the square of every real number is positive (or 0). Hence (1) is always true — which proves our theorem.

Thus, $|2 + i| = \sqrt{5}$ and $|1 + 3\,i| = \sqrt{10}.$

But $|(2 + i) + (1 + 3\,i)| = 5,$ and $5 < \sqrt{5} + \sqrt{10}.$

236 **Powers and roots.** 1. The nth *power* of $a + bi$, written $(a + bi)^n$, is to mean the product of n factors each of which is $a + bi$. It follows from § 226, 3, that this product is a complex number, as $c + di$.

It may be proved, as in § 185, that the laws of exponents hold good for powers of complex numbers as thus defined.

2. If $(a + bi)^n = c + di$, we call $a + bi$ an nth *root* of $c + di$, and we may indicate it thus, $\sqrt[n]{c + di}$.

We shall prove subsequently that every given complex number $c + di$ has n such nth roots: in other words, that in the system of complex numbers there are n different numbers whose nth powers equal $c + di$.

Thus, since $(1/\sqrt{2} + i/\sqrt{2})^2 = 1/2 + 2\,i/2 - 1/2 = i,$ the number $1/\sqrt{2} + i/\sqrt{2}$ is a *square root* of i, and therefore a *fourth root* of -1. The remaining three fourth roots of -1 are

$$-1/\sqrt{2} + i/\sqrt{2}, \quad 1/\sqrt{2} - i/\sqrt{2}, \quad -1/\sqrt{2} - i/\sqrt{2}.$$

237 **General conclusion.** No further enlargement of the number system is necessary. For, as has just been pointed out, §§ 226, 236, the complex system meets all the requirements of the four fundamental operations and evolution. And while certain other operations with numbers have a place in mathematics, — among them the operation of finding logarithms of numbers, § 140, — these operations admit of definition by infinite series, like $u_1 + u_2 + u_3 + \cdots$, whose terms are complex numbers; and if such a series have a sum, that sum is a complex number.

GRAPHICAL REPRESENTATION OF COMPLEX NUMBERS

238 Complex numbers may be pictured by points in a plane, the points being called the **graphs** of the corresponding numbers.

Take any two right lines $X'OX$, $Y'OY$ intersecting at right angles at the origin O; also some fixed unit segment s for measuring lengths.

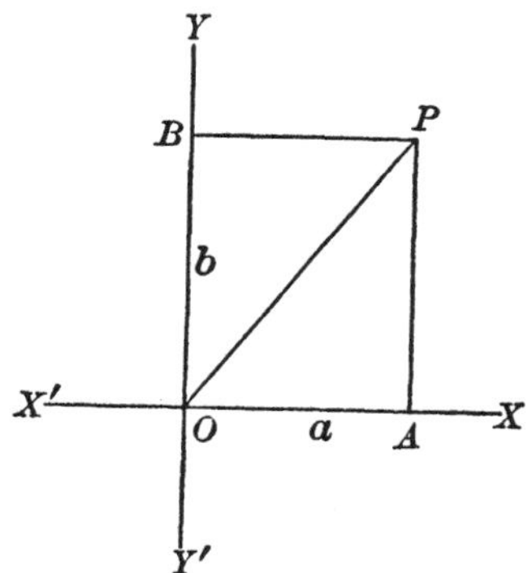

1. We represent each *real number*, a, by that point A on $X'OX$ whose distance from O, in terms of s, is $|a|$, § 209, taking A to the right or left of O, according as a is positive or negative.

2. We represent each *pure imaginary*, bi, by that point B on $Y'OY$ whose distance from O is $|b|$, taking B above or below O, according as b is positive or negative.

3. We represent each *complex number* $a + bi$ by the point P, which is obtained by the following construction:

Find A and B, the graphs of a and bi, as in 1 and 2. Then through A and B draw parallels to $Y'OY$ and $X'OX$ respectively. The point P in which these lines meet is the graph of $a + bi$.

We call $X'OX$ the *axis of real numbers* and $Y'OY$ the *axis of pure imaginaries*.

By this method we bring the system of complex numbers into a relation of *one-to-one correspondence*, § 2, with the assemblage of all points in a plane. Moreover we obtain a complete representation of the two-dimensional ordinal character of the complex system, § 223.

Observe that the graphs of all numbers which have the same imaginary part lie on the same parallel to $X'OX$, and that the graphs of all numbers which have the same real part lie on the same parallel to $Y'OY$.

239 *The absolute value of any complex number is the distance of its graph from the origin.*

For since the lengths of OA and AP in the figure of § 238 are a and b respectively, the length of OP is $\sqrt{a^2 + b^2}$ or $|a + bi|$, § 232.

240 The graphs of the *sum* and *product* of two complex numbers $\mathbf{a} = a + a'i$, $\mathbf{b} = b + b'i$, may be found as follows:

Given A *and* B, *the graphs of* **a** *and* **b** *respectively. Join* OA *and* OB *and complete the parallelogram* OACB. *Then* C *is the graph of* **a** + **b**.

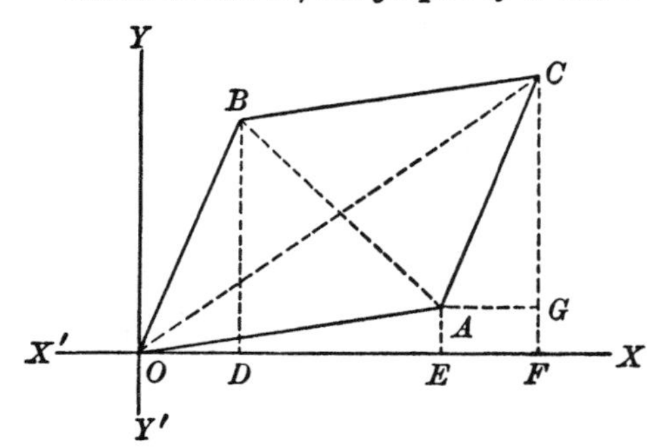

For, draw the perpendiculars BD, AE, CF, AG. Then a, a', b, b' are the lengths of OE, EA, OD, DB respectively, and the triangles ODB and AGC are congruent.

Hence $\quad OF = OE + EF = OE + OD = a + b$, in length,

and $\quad FC = FG + GC = EA + DB = a' + b'$, in length.

Therefore C is the graph of $a + b + (a' + b')i$, or **a** + **b**, § 226, 1.

When O, A, B are in a straight line (and always) C may be found by drawing AC equal in length and direction to OB.

Since $OC \leqq OA + AC$, *i.e.* $\leqq OA + OB$, we have $|\mathbf{a} + \mathbf{b}| \leqq |\mathbf{a}| + |\mathbf{b}|$.

The graph of the *difference* **a** − **b** is that of the sum **a** + (− **b**).

Given A *and* B, *the graphs of* **a** *and* **b**, *and let* I *denote the graph of* 1. *Join* OA, OB, IA, *and on* OB *construct the triangle* OBC *similar to* OIA *and such that were* OB *turned about* O *until it lay along* OX, OC *would lie along* OA. *Then* C *is the graph of* **ab**.

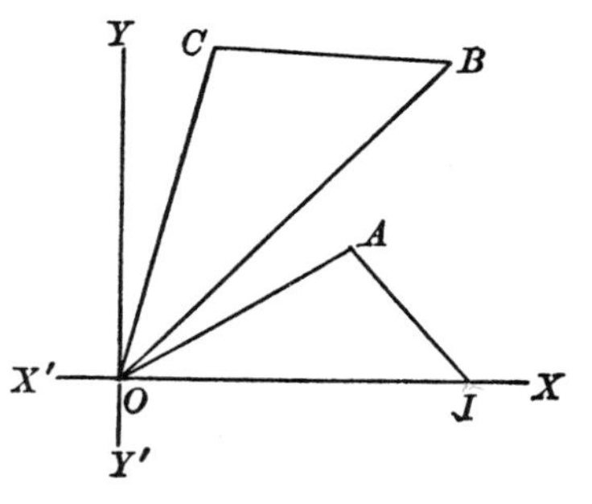

This rule will be proved later and rules derived from it for constructing graphs of *quotients* and *powers*.

When $\mathbf{b} = i$, OC is OA turned 90° "counter-clockwise" about O.

241 It follows that identical relations among complex numbers may be translated into geometrical theorems. Hence *imaginary numbers may express relations among real things.*

Thus, the identity $(\mathbf{a} + \mathbf{b})/2 = \mathbf{a} + (\mathbf{b} - \mathbf{a})/2$ shows that the diagonals of a parallelogram bisect each other; for the graphs of $(\mathbf{a} + \mathbf{b})/2$ and $\mathbf{a} + (\mathbf{b} - \mathbf{a})/2$ are the midpoints of OC and AB in the first figure, § 240.

PART SECOND — ALGEBRA

I. PRELIMINARY CONSIDERATIONS

ON THE USE OF LETTERS TO DENOTE NUMBERS

Constants and variables. In algebra a letter is often used to denote *any number whatsoever.* Thus, in the formula $ab = ba$ the letters a and b denote every two numbers, the meaning of the formula being: the product of *any* first number by *any* second number is the same as the product of the second by the first. **242**

In many algebraic discussions we find it convenient to make the following distinction between two letters which have this meaning, as between the letters b and x in the expression $x + b$.

First. We regard the one, b, as having had a *particular* value, but *any that we please,* assigned it at the outset, which it is then to retain throughout the discussion. Such a letter we shall call a *known* letter or number, or a *constant.*

Second. On the contrary, throughout the discussion we regard the other, x, as free to take every possible value and to change from any one value to any other. Such a letter we shall call a *variable.*

Unknown letters. But letters are also employed to denote *particular* numbers whose values *are to be found.* Such a letter we shall call an *unknown* letter or number. **243**

We are *not* at liberty to assign any value we please to an unknown letter, as we are to a constant or variable letter.

Thus, in the equation $2x - 5 = 0$, x is an unknown letter whose value we readily find to be $5/2$. In the expression $2x - 5$ we may assign x any value we please, but in the equation $2x - 5 = 0$ we can assign x no other value than $5/2$.

244 The choice of letters. The only *necessary* restriction on our choice of letters is that no single letter be made to represent two different numbers at the same time.

It is customary, however, to represent known numbers or constants by the earlier letters of the alphabet, as a, b, c; unknown numbers and variables by the later letters, as x, y, z.

Besides simple letters we sometimes use letters affected with accents or subscripts: thus, a', a'', a''' read "a prime," "a second," "a third"; and a_0, a_1, a_2 read "a sub-null," "a sub-one," "a sub-two."

245 On reckoning with letters. When we represent numbers by letters, as a, b, c, we can only *indicate* the results of combining them by the operations of arithmetic. Thus, to add b to a merely means to form the expression $a + b$, which we therefore call the *sum* of a and b. Similarly, the *product* of a by b is the expression ab.

Inasmuch as the literal expressions thus obtained denote numbers, we may reckon with them by the operations of arithmetic. But in such reckonings we cannot make any direct use of the *values* of the expressions, since these are not given. We can merely connect the expressions by the appropriate signs of operation and then simplify the *form* of the result by changes which we know will not *affect* its value, *no matter what this may be.*

Now, as we have seen, § 68, all the changes that can be made in the forms of sums and products without affecting their values are embodied in the following formulas:

1. $a + b = b + a.$ 2. $a + (b + c) = (a + b) + c.$

3. $ab = ba.$ 4. $a(bc) = (ab)c.$ 5. $a(b + c) = ab + ac.$

It may therefore be said that the formulas 1–5 are practically all the *definition* of addition and multiplication that we either need or can use when combining literal expressions; and the like is true of the remaining operations of arithmetic.

Thus, to add $2x + 3y$ and $4x + 5y$ merely means to find the *simplest form* to which the expression $2x + 3y + (4x + 5y)$ can be reduced by applying formulas 1–5, and adding given numbers. We thus obtain

$$2x + 3y + (4x + 5y) = 2x + 3y + 4x + 5y \qquad \text{by 2}$$
$$= 2x + (3y + 4x) + 5y = 2x + (4x + 3y) + 5y \qquad \text{by 2 and 1}$$
$$= 2x + 4x + 3y + 5y = (2x + 4x) + (3y + 5y) \qquad \text{by 2}$$
$$= (2 + 4)x + (3 + 5)y = 6x + 8y, \text{ the sum required.} \qquad \text{by 3 and 5}$$

THE FUNDAMENTAL RULES OF RECKONING

246 In accordance with what has just been said, we may regard addition, subtraction, multiplication, division, involution, and evolution as *defined* for algebra by the following rules and formulas, which we shall therefore call *the fundamental rules of reckoning.*

In these formulas — which have been established for numbers of all kinds in the first part of the book — the letters a, b, c denote any finite numbers whatsoever, and the sign of equality, $=$, means "represents the same number as."

247 **Addition.** The result of *adding* b to a is the expression $a + b$. We call this expression the *sum* of a and b. It has a value, and but one, for any given values of a and b. In particular, $a + 0 = 0 + a = a$.

248 Addition is a *commutative* and an *associative* operation; that is, it conforms to the two laws, §§ 34, 35:

$$a + b = b + a, \qquad a + (b + c) = (a + b) + c.$$

249 The following *rules of equality* are true of sums, § 39:

$$\text{If} \quad a = b, \quad \text{then } a + c = b + c.$$
$$\text{If } a + c = b + c, \text{ then} \quad a = b.^*$$

250 **Subtraction.** This is the *inverse of addition,* § 55. Given any two numbers, a and b, there is always a number, and but one,

* Later we shall find that this rule does not hold good when c is infinite.

from which a can be obtained by adding b. We call this number the *remainder* which results on subtracting b from a, and we represent it by the expression $a - b$. Hence, by definition,

$$(a - b) + b = a.$$

In particular, we represent $0 - b$ by $-b$.

251 **Multiplication.** The result of *multiplying* a by b is the expression ab. We call ab the *product* of a by b. It has a value, and but one, for any given values of a and b.

In particular, $a \cdot 0 = 0 \cdot a = 0$, whatever finite value a may have.

When b is a positive integer, $ab = a + a + \cdots$ to b terms.

252 Multiplication is a *commutative* and an *associative* operation, and it is *distributive* with respect to addition; that is, it conforms to the three laws, §§ 45–47:

$$ab = ba, \qquad a(bc) = (ab)c, \qquad a(b + c) = ab + ac.$$

253 The following *rules of equality* are true of products, §§ 75, 76:

If $a = b$, then $ac = bc$.

If $ac = bc$, then $a = b$, *unless* $c = 0$.*

If $ac = 0$, then $a = 0$, or $c = 0$.

254 **Division.** This is the *inverse of multiplication*, § 124. Given any two numbers, a and b; except when b is 0, there is always a number, and but one, from which a can be obtained by multiplying by b. We call this number the *quotient* which results on dividing a by b, and we represent it by the expression $\frac{a}{b}$ or a/b. Hence, by definition,

$$\left(\frac{a}{b}\right)b = a, \text{ except when } b = 0.$$

255 **Involution.** This is a case of *repeated multiplication.* We represent the continued product $a \cdot a \cdots$ to n factors by a^n and call it the nth *power* of a.

* Later we shall find that this rule does not hold good when c is infinite.

In the symbol a^n, n is called the *exponent*, and a the *base*.

Involution, or raising to a power, conforms to the following three laws, called the *laws of exponents*, § 185 : **256**

$$a^m \cdot a^n = a^{m+n}, \qquad (a^m)^n = a^{mn}, \qquad (ab)^n = a^n b^n.$$

The following *rules of equality* are true of powers, § 184 : **257**

$$\text{If} \quad a = b, \quad \text{then} \quad a^n = b^n.$$

$$\text{If} \quad a^2 = b^2, \quad \text{then} \quad a = b, \quad \text{or} \quad a = -b.$$

The second of these rules and the general rule of which it is a particular case will be demonstrated later.

Evolution. This is one of the *inverses of involution*, §§ 138, 140. Given any *positive* number a, there is a positive number, and but one, whose nth power is equal to a. We call this number the *principal* n*th root* of a, and we represent it by $\sqrt[n]{a}$, or, when $n = 2$, by $\sqrt{a}$. Hence, by definition, **258**

$$(\sqrt[n]{a})^n = a.$$

But this positive number, $\sqrt[n]{a}$, is not the only number whose nth power is equal to a. On the contrary, as will be shown subsequently, there are n different numbers whose nth powers are equal to a; and this is true not only when a is positive, but also when a is any other kind of number.

When a is positive and n is odd, the *principal* n*th root* of $-a$ is $-\sqrt[n]{a}$.

On the reversibility of the preceding rules. We have called certain of the rules just enumerated *rules of equality;* we may call the rest *rules of combination.* **259**

Observe that all the rules of combination and the rules of equality for sums are *reversible*, but that the rules of equality for products and powers are *not* completely reversible.

Thus, according to the distributive law, $a(b + c) = ab + ac$, which is one of the rules of combination, we can replace $a(b + c)$ by $ab + ac$, or *reversely*, $ab + ac$ by $a(b + c)$.

Again, if $a = b$, we may always conclude that $a + c = b + c$, and *reversely*, that, if $a + c = b + c$, then $a = b$.

But while, if $a = b$, we may always conclude that $ac = bc$; on the contrary, if $ac = bc$, we can conclude that $a = b$ only when we know that c is not 0.

And while from $a = b$ it always follows that $a^2 = b^2$, from $a^2 = b^2$ it only follows that *either* $a = b$ *or* $a = -b$.

260 **The rules of inequality.** The formula $a \neq b$ means "a is not equal to b."

Of two given unequal *real* numbers, a and b, the one is *algebraically* the *greater*, the other *algebraically* the *lesser*, § 62.

If a is the greater and b the lesser, we write

$$a > b \text{ or } b < a.$$

In particular, we have $a > 0$ or $a < 0$, according as a is positive or negative.

261 For any given real numbers a, b, c, we have the rules, §§ 178, 184:

1. If $a = b$ and $b = c$, then $a = c$.
   If $a = b$ and $b < c$, then $a < c$.
   If $a < b$ and $b < c$, then $a < c$.

2. According as $a <, =, \text{ or } > b$,
so is $a + c <, =, \text{ or } > b + c$,
and $ac <, =, \text{ or } > bc$, if $c > 0$;
but $ac >, =, \text{ or } < bc$, if $c < 0$.

3. When a and b are *positive*,
according as $a <, =, \text{ or } > b$,
so is $a^n <, =, \text{ or } > b^n$;
and $\sqrt[n]{a} <, =, \text{ or } > \sqrt[n]{b}$.

As has already been pointed out, the rules under 2 and 3 which involve only the sign = hold good of imaginary numbers also. This is also true of the rule: If $a = b$ and $b = c$, then $a = c$, which we may call the *general rule of equality*.

ADDITIONAL ALGEBRAIC SYMBOLS

Besides the symbols whose meanings have been explained in the preceding sections, the following are often employed in algebra: **262**

1. Various *signs of aggregation*, like the parentheses ( ) employed above, and [ ], { }, to indicate that the expression included by them is to be used as a single symbol.

2. The *double signs* $\pm$, read "plus or minus," and $\mp$, read "minus or plus."

Thus, in $a \pm b \mp c$, which means $a + b - c$ or $a - b + c$, the upper signs being read together and the lower signs similarly.

3. The symbol $\therefore$ for *hence* or *therefore.*

4. The symbol $\cdots$ for *and so on.*

5. Also, $\because$ for *since;* $\ngtr$ for *not greater than;* $\nless$ for *not less than;* $\gtrless$ for *greater or less than.*

ALGEBRAIC EXPRESSIONS

Any expression formed by combining letters, or letters and numbers, by the operations just described, is called an **algebraic expression.** **263**

Note. The number of times that an operation is involved in such an expression may be *limited,* as in $1 + x + x^2$, or *unlimited,* as in $1 + x + x^2 + \cdots$, supposed to be continued without end. In the one case we say that the expression is **finite**, in the other, **infinite.** For the present we shall have to do with finite expressions only. **264**

It is customary to classify algebraic expressions as follows, according to the manner in which the *variable* (or unknown) letters under consideration occur in them: **265**

An expression is called **integral** if it does not involve an indicated division by an expression in which a variable letter occurs; **fractional,** if it does. **266**

Thus, if x and y are the variable letters, but a, b, c constants,

then $\quad ax^2 + bx + c$ and $\dfrac{y}{b} + \sqrt{x}$ are integral,

but $\quad y + \dfrac{1}{x}$ and $\dfrac{2+x}{1-x}$ are fractional.

267 An expression is called **rational** if it does not involve an indicated root of an expression in which a variable letter occurs; **irrational**, if it does.

Thus, $a + \sqrt{b}x$ is rational, but $\sqrt{y} + \sqrt{y - x}$ is irrational.

268 **Notes.** 1. In applying these terms to an expression, we suppose it reduced to its *simplest form*. Thus, $\sqrt{x^2 + 2xy + y^2}$ is *rational*, since it can be reduced to the rational form $x + y$.

2. The terms *integral*, *rational*, and so on, have nothing to do with the *numerical values* of the expressions to which they are applied.

Thus, $x + 2$ is a rational integral expression, but it represents an integer only when x itself represents one. It represents a fraction for every fractional value of x, and an irrational number for every irrational value of x.

269 When an algebraic expression A is made up of certain parts connected by $+$ or $-$ signs, these parts with the signs immediately preceding them are called the **terms** of A.

Thus, the terms of the expression

$$a + a^2c - (b + c) + [d + e] - \{f + g\} + \overline{h + i} + \left.\begin{matrix} j \\ +k \end{matrix}\right| - \frac{l + m}{n + p}$$

are a, a^2c, $-(b + c)$, and so on, those of the terms which themselves consist of more terms than one being enclosed by *parentheses* or some equivalent *sign of aggregation*, § 262, 1.

270 Integral expressions are called **monomials, binomials, trinomials,** and in general **polynomials,** according to the number of their terms.

271 In any monomial, the product of the constant factors is called the **coefficient** of the product of the variable factors.

Thus, in $4\,ab^2x^3y^4$, $4\,ab^2$ is the coefficient of x^3y^4.

At the same time, it is proper to call *any* factor the coefficient of the rest of the product.

In every monomial the coefficient should be written first. When no coefficient is expressed, it is 1. Thus, 1 is the coefficient of x^2y.

Like terms are such as differ in their coefficients at most. **272**

Thus, $-2x^2y$ and bx^2y are like terms.

The **degree of a monomial** is the sum of the exponents of such of the *variables* under discussion as occur in it. **273**

Thus, if the variables are x and y, the degree of $4ab^2x^3y^4$ is *seven;* that of ax^3, *three;* that of b, *zero* (see § 595).

The **degree of a polynomial** is the degree of its term or terms of highest degree; and the *degree of any integral expression* is that of the simplest polynomial to which it can be reduced. **274**

Thus, the degree of $ax^3 + bx^2y + cy^3 + dx^2 + ey + f$ is *three;* and the degree of $(x-1)(x-2)$ is *two.*

It is convenient to *arrange the terms* of a polynomial in the order of their degrees, descending or ascending, and if there are several terms of the same degree, to arrange these in the order of their degrees in one of the variables. **275**

This order is observed in the polynomial given in § 274.

A polynomial is said to be **homogeneous** when all its terms are of the same degree. **276**

Thus, $5x^3 - 2x^2y + 4xy^2 + y^3$ is homogeneous.

Polynomials in a single variable. Rational integral expressions in a single variable, as x, are of especial importance. They play much the same rôle in algebra as integral numbers in arithmetic. In fact we shall find that they possess many properties analogous to those of integral numbers. They can always be reduced to the form of a *polynomial in* x, that is, one of the forms: **277**

$$a_0x + a_1, \quad a_0x^2 + a_1x + a_2, \quad a_0x^3 + a_1x^2 + a_2x + a_3, \cdots,$$

or, as we shall say, to the form:

$$a_0x^n + a_1x^{n-1} + a_2x^{n-2} + \cdots + a_{n-1}x + a_n,$$

where n denotes the degree of the expression, and the dots stand for as many terms as are needed to make the entire number of terms $n+1$.

The coefficients $a_0, a_1, \cdots$, denote constants, which may be of any kind. In particular, any of them except a_0 may be 0, the polynomial then being called *incomplete.*

Observe that in each term the sum of the subscript of a and the exponent of x is the degree of the polynomial.

Thus, in $5x^6 - x^3 + 2x^2 + x - 3$, we have $n = 6$, $a_0 = 5$, $a_1 = 0$, $a_2 = 0$, $a_3 = -1$, $a_4 = 2$, $a_5 = 1$, $a_6 = -3$.

278 **Functions.** Clearly an algebraic expression like $x + 2$ or $x^2 + y$, which involves one or more variables, is itself a variable. We call $x + 2$ a *function* of x because *its value depends on that of* x *in such a way that to each value of* x *there corresponds a definite value of* x + 2.

For a like reason we call $x^2 + y$ a function of x and y and, in general, we call every algebraic expression a *function* of all the variables which occur in it.

279 What we have just termed integral or fractional, rational or irrational expressions in x, x and y, and so on, we may also term integral or fractional, rational or irrational functions of x, x and y, and so on.

280 We shall often represent a given function of x by the symbol $f(x)$, read "function of x." We then represent the values of the function which correspond to $x = 0, 1, b$, by $f(0)$, $f(1)$, $f(b)$.

Thus, if $f(x) = x + 2$, we have $f(0) = 2$, $f(1) = 3$, $f(b) = b + 2$. And, in general, if $f(x)$ represent any given expression in x, $f(b)$ represents the result of substituting b for x in the expression.

When dealing with two or more functions of x, we may represent one of them by $f(x)$, the others by similar symbols, as $F(x)$, $\phi(x)$, $\psi(x)$.

In like manner, we may represent a function of two variables, x and y, by the symbol $f(x, y)$, and so on.

EXERCISE I

1. What is the degree of $x^2yz^3 + 2x^5y^4z^6 + 3x^7y^2z^8$ with respect to x, y, and z separately? with respect to y and z jointly? with respect to x, y, z jointly?

2. What is the degree of $(x+1)(2x^2+3)(x^4-7)$?

3. Given $3x^7 + x^6 - 4x^4 + x^3 - 12$; what are the values of n, a_0, a_1, $\cdots$ in the notation of § 277?

4. If $f(x) = 2x^3 - x^2 + 3$, find $f(0)$, $f(-1)$, $f(3)$, $f(8)$.

5. If $f(x) = (x^2 - 3x + 2)/(2x + 5)$, find $f(0)$, $f(-2)$, $f(6)$.

6. If $f(x) = x + \sqrt{x} + 3$, find $f(1)$, $f(4)$, $f(5)$.

7. If $f(x) = 2x + 3$, what is $f(x-2)$? $f(x^2+1)$?

8. If $f(x, y) = x^3 + x - y + 8$, find the following:

$$f(0,0),\quad f(1,0),\quad f(0,1),\quad f(1,1),\quad f(-2,-3).$$

IDENTICAL EQUATIONS OR IDENTITIES

281 If A denotes the very same expression as B, or one which can be transformed into B by the rules of reckoning, §§ 247–258, we say that A is **identically equal** to B.

The notation $A \equiv B$ means "A is identically equal to B."

Thus, $x(x+2)+4$ is identically equal to $x^2 + 2(x+2)$.

For
$$x(x+2)+4 \equiv (x^2+2x)+4$$
$$\equiv x^2 + (2x+4) \equiv x^2 + 2(x+2). \qquad \text{§§ 248, 252}$$

We call $A \equiv B$ an *identical equation*, or *identity*. Hence

282 *An* **identical equation** A $\equiv$ B *is a statement that a first expression,* A, *can be transformed into a second expression,* B, *by means of the rules of reckoning.*

283 In particular, an identical equation like

$$3 - 8 + 2 \equiv 4 + 7 - 14$$

in which no letters occur, is called a **numerical identity.**

The following very useful theorem is implied in § 282.

284 **Theorem.** *If two polynomials in* x *are identically equal, their corresponding coefficients are equal; that is,*

$$\textit{If}\quad a_0x^n + a_1x^{n-1} + \cdots + a_n \equiv b_0x^n + b_1x^{n-1} + \cdots + b_n,$$

$$\textit{then}\qquad a_0 = b_0,\quad a_1 = b_1, \cdots, a_n = b_n.$$

For were these coefficients different, the polynomials would be different as they stand and the first could not be transformed into the second by the rules of reckoning.

Thus, if $ax^2 + 3x - 3 \equiv 2x^2 + bx + c$, then $a = 2$, $b = 3$, $c = -3$.

If, instead of being constants, the coefficients $a_0, a_1, \cdots, b_0, b_1, \cdots$ denote *algebraic expressions which do not involve* x, it follows from the identity $a_0x^n + a_1x^{n-1} + \cdots \equiv b_0x^n + b_1x^{n-1} + \cdots$ that $a_0 \equiv b_0$, $a_1 \equiv b_1, \cdots$, in other words, that the expressions denoted by corresponding coefficients, a_0 and b_0, and so on, are identically equal.

285 A similar theorem holds good of two identically equal polynomials whose terms are products of powers of *two or more variables* with constant coefficients.

Thus, if $a + bx + cy + dx^2 + exy + fy^2 + \cdots$

$$\equiv a' + b'x + c'y + d'x^2 + e'xy + f'y^2 + \cdots,$$

then $\quad a = a', b = b', c = c', d = d', e = e', f = f', \cdots.$

286 **Properties of identical equations.** In algebraic reckoning we make constant use of the following theorems:

Theorem 1. *If* $A \equiv B$, *then* $B \equiv A$.

For the process by which A may be transformed into B is *reversible* since it involves only rules of combination, § 259. But the reverse process will transform B into A.

Thus, we may reverse the transformation in the example in § 281.

For $\quad x^2 + 2(x + 2) \equiv x^2 + (2x + 4)$

$$\equiv (x^2 + 2x) + 4 \equiv x(x + 2) + 4. \quad §§ 248, 252$$

Theorem 2. *If* $A \equiv C$ *and* $B \equiv C$, *then* $A \equiv B$.

For since $\quad B \equiv C$, we have $C \equiv B$. $\quad$ by Theorem 1

Hence $\quad A \equiv C$ and $C \equiv B$, and therefore $A \equiv B$.

Thus, since $x(x+2)+4 \equiv x^2+2x+4$, §§ 248, 252

and $x^2+2(x+2) \equiv x^2+2x+4$, §§ 248, 252

we have $x(x+2)+4 \equiv x^2+2(x+2)$.

Theorem 3. *An identity remains an identity when the same operation is performed on both its members.*

This follows from the rules of equality, §§ 249, 253, 257.

Thus, if $A \equiv B$, then $A + C \equiv B + C$, and so on.

287 **On proving identities.** To prove of two given expressions, A and B, that $A \equiv B$, it is not necessary actually to transform A into B. As § 286, 2, shows, *it is sufficient, if we can reduce* A *and* B *to the same form* C.

The following theorem supplies another very useful method.

288 *If from a supposed identity,* A ≡ B, *a known identity,* C ≡ D, *can be derived by a reversible process, the supposed identity* A ≡ B *is true.*

For since the process is reversible, $A \equiv B$ can be derived from $C \equiv D$. Therefore, since $C \equiv D$ is true, $A \equiv B$ is also true.

Example. Prove that $a + b - b$ is identically equal to a.

If we *suppose* $a + b - b \equiv a$ (1)

it will follow that $[(a+b)-b]+b \equiv a+b$. (2) § 249

But (2) is a known identity, § 250, and the step (1) to (2) is reversible. Therefore (1) is true.

That it is not safe to draw the conclusion $A \equiv B$ unless the process from $A \equiv B$ to $C \equiv D$ is *reversible* may be illustrated thus:

If we *suppose* $x \equiv -x$ (1)

it will follow that $x^2 \equiv (-x)^2$. (2)

Here (2) is true, but it does not follow from this that (1) is true, since the step (1) to (2) is *not* reversible, § 259. And in fact, (1) is false.

289 **Identity and equality.** It is important to remember that identity is primarily a relation of *form* rather than of *value*. At the same time,

If A *and* B *are finite expressions, and* A ≡ B, *then* A *and* B *have equal values for all values of the letters which may occur in them.*

For, by hypothesis, we can transform A into B by a limited number of applications of the rules $a+b=b+a$, and so on. But $a+b$ and $b+a$ have equal values whatever the values of a and b; and so on.

The reason for restricting the theorem to finite expressions will appear later.

Conversely, if A and B have equal values for all values of the letters in A and B, then $A \equiv B$. This will be proved subsequently.

Hence in the case of finite expressions we may always replace the sign of identity of form, $\equiv$, by the sign of equality of value, $=$, and when $A \equiv B$, write $A = B$. We shall usually follow this practice.

This use of the sign $=$ is to be carefully distinguished from that described in § 325.

ON CONVERSE PROPOSITIONS

290 Consider a **proposition** which has the form

$$\text{If A, then B,} \qquad (1)$$

or, more fully expressed: If a certain statement, A, is true, then a certain other statement, B, is also true.

Thus, If a figure is a square, then it is a rectangle.
If $x=1$, then $x-1=0$.

291 Interchanging the *hypothesis*, A, and the *conclusion*, B, of (1) we obtain the **converse** proposition

$$\text{If B, then A.*} \qquad (2)$$

Thus, the converses of the propositions just cited are:

If a figure is a rectangle, then it is a square.
If $x-1=0$, then $x=1$.

292 As the first of these examples illustrates, *the converse of a true proposition may be false.*

* A proposition like If A and B, then C, which has a *double* hypothesis, has *two* converses: namely, If C and B, then A, and If A and C, then B. Similarly, if there be a *triple* hypothesis there are *three* converses; and so on.

293 But the converse of a true proposition: If A, then B, is always *true* when the process of reasoning by which the conclusion, B, is derived from the hypothesis, A, is *reversible;* for by reversing the process we may derive A from B, in other words, prove If B, then A.

The method of proving a proposition by proving its converse by a reversible process is constantly employed in algebra. An illustration of this method has already been given in § 288.

294 When a proposition: If A, then B, is true, we call A a **sufficient condition** of B, and B a **necessary condition** of A.

Thus, the proposition If $x = 1$, then $(x - 1)(x - 2) = 0$ is true. Hence $x = 1$ is a *sufficient* condition that $(x - 1)(x - 2) = 0$, and $(x - 1)(x - 2) = 0$ is a *necessary* condition that $x = 1$.

295 When both the proposition If A, then B, and its converse If B, then A, are true, we say that A is the *sufficient and necessary condition* of B; and *vice versa.*

Thus, both (1) If $x = 1$, then $x - 1 = 0$, and (2) If $x - 1 = 0$, then $x = 1$, are true. Hence $x = 1$ is the sufficient and necessary condition that $x - 1 = 0$; and *vice versa.*

II. THE FUNDAMENTAL OPERATIONS

ADDITION AND SUBTRACTION

296 **Sum and remainder.** Let A and B denote any two algebraic expressions. By the *sum* of A and B, and by the *remainder* to be found by subtracting B from A, we shall mean the *simplest forms* to which the expressions $A + B$ and $A - B$ can be reduced by aid of the rules of reckoning, §§ 247–258.

297 **Some useful formulas.** In making these reductions the following formulas are very serviceable, namely:

1. $a + b - c = a - c + b$. 2. $a - (b + c) = a - b - c$.

3. $a + (b - c) = a + b - c$. 4. $a - (b - c) = a - b + c$.

5. $a(b - c) = ab - ac$.

These formulas may be described as extensions of the commutative, associative, and distributive laws to *subtraction.*

We may prove 1 and 2 by aid of the rule, § 249:

Two expressions are equal if the results obtained by adding the same expression to both are equal.

1. $a + b - c = a - c + b.$

For the result of adding c to each member is $a + b$.

Thus, $[(a + b) - c] + c = a + b,$ § 250

and $(a - c) + b + c = (a - c) + c + b = a + b.$ §§ 248, 250

2. $a - (b + c) = a - b - c.$

For the result of adding $b + c$ to each member is a.

Thus, $[a - (b + c)] + (b + c) = a,$ § 250

and $a - b - c + (b + c) = a - b - c + c + b$
$= a - b + b = a.$ §§ 248, 250

We may prove 3, 4, 5 as follows:

Since $b = (b - c) + c,$ § 250

we have, 3. $a + b - c = a + [(b - c) + c] - c$
$= a + (b - c) + c - c$ § 248
$= a + (b - c).$ by 1 and § 250

4. $a - b + c = a - [(b - c) + c] + c$
$= a - (b - c) - c + c$ by 2
$= a - (b - c).$ § 250

5. $ab - ac = a[(b - c) + c] - ac$
$= a(b - c) + ac - ac$ § 252
$= a(b - c).$ by 1 and § 250

Observe that it follows from § 248 and the formulas 1–4 that *a series of additions and subtractions may be performed in any order whatsoever.*

Thus, $a - b + c - d + e = a + c - b - d + e,$ by 1

$= a + c - (b + d) + e = a + c + e - (b + d),$ by 2 and 1

$= a + c + e - b - d.$ by 2

Rules of sign. The "rules of sign" which follow are particular cases of the formulas 3, 4, 5 just established. **298**

1. $a + (-c) = a - c.$ 2. $a - (-c) = a + c.$

3. $a(-c) = -ac.$ 4. $(-a)(-c) = ac.$

We obtain 1, 2, 3 at once by setting $b = 0$ in § 297, 3, 4, 5 respectively. We may prove 4 as follows:

$$(-a)(-c) = (-a)(0 - c) = (-a)0 - (-a)c \qquad \S\, 297,\ 5$$

$$= 0 - (-ac) = ac. \qquad \text{by 2 and 3}$$

Rule of parentheses. From the formulas § 248 and § 297, 2, 3, 4, we obtain the following important rule: **299**

Parentheses preceded by the + sign may be removed; parentheses preceded by the − sign may also be removed, if the sign of every term within the parentheses be changed.

Parentheses may be introduced in accordance with the same rule.

Thus, $a + b - c - d + e = a + b - (c + d - e).$

To simplify an expression which involves parentheses within parentheses, apply the rule to the several parentheses successively.

Thus, $a - \{b - [c - (d - e)]\} = a - b + [c - (d - e)]$

$= a - b + c - (d - e)$

$= a - b + c - d + e.$

Of course the parentheses may be removed in any order; but by beginning with the *outermost* one (as in the example) we avoid changing any sign more than once.

300 **Rules for adding and subtracting integral expressions.** From the formulas of §§ 248, 252, 297 we derive the rules:

To add (or subtract) two like terms, add (or subtract) their coefficients, and affix the common letters to the result.

To add two or more polynomials, write all their terms in succession with their signs unchanged, and then simplify by combining like terms.

To subtract one polynomial from another, change the sign of every term in the subtrahend and add.

Example 1. Add $4\,ab^2$ and $-5\,ab^2$; also subtract $-5\,ab^2$ from $4\,ab^2$.

We have $4\,ab^2 + (-5\,ab^2) = (4-5)\,ab^2 = -ab^2$; § 248

and $4\,ab^2 - (-5\,ab^2) = [4-(-5)]\,ab^2 = 9\,ab^2$. § 297, 5

Example 2. Add $x^3 + ax^2y + 2\,ab^3$ and $bx^2y - 5\,ab^3$.

We have
$$x^3 + ax^2y + 2\,ab^3 + (bx^2y - 5\,ab^3)$$
$$= x^3 + ax^2y + 2\,ab^3 + bx^2y - 5\,ab^3 \qquad \text{§ 299}$$
$$= x^3 + ax^2y + bx^2y + 2\,ab^3 - 5\,ab^3 \qquad \text{§ 248}$$
$$= x^3 + (a+b)\,x^2y - 3\,ab^3. \qquad \text{§§ 252, 297, 5}$$

Example 3. Subtract $2\,a^2b - ab^2 + b^3$ from $a^3 + a^2b + b^3$.

We have
$$a^3 + a^2b + b^3 - (2\,a^2b - ab^2 + b^3)$$
$$= a^3 + a^2b + b^3 - 2\,a^2b + ab^2 - b^3 \qquad \text{§ 299}$$
$$= a^3 - a^2b + ab^2. \qquad \text{§§ 252, 297}$$

When the polynomials to be added (or subtracted) have like terms, it is convenient to arrange these terms in columns and then to add (or subtract) by columns.

Example 4. Add $a^4 + a^3b - 2\,a^2b^2 - b^4$ and $ab^3 + 3\,a^2b^2 - a^3b$, and subtract $5\,a^2b^2 - ab^3$ from the result.

We have
$$\begin{array}{rrrrr} a^4 & + a^3b & - 2\,a^2b^2 & & - b^4 \\ & - a^3b & + 3\,a^2b^2 & + \ ab^3 & \\ & & - 5\,a^2b^2 & + \ ab^3 & \\ \hline a^4 & & - 4\,a^2b^2 & + 2ab^3 & - b^4 \end{array}$$

EXERCISE II

1. Add $4\,ax^2y,\quad -6\,ax^2y,\quad 5\,bx^2y$, and $-3\,bx^2y$.
2. Add $7\,a^2+2\,a-b^2,\quad 3\,a+b^2-2\,a^2$, and $b^2-4\,a-4\,a^2$.
3. Add $3\,x^2-5\,x+6,\quad x^2+2\,x-8$, and $-4\,x^2+3\,x-7$.
4. Add $4\,a^3+a^2b-5\,b^3,\quad \frac{5}{3}\,a^3-6\,ab^2-a^2b,\quad \frac{1}{3}\,a^3+10\,b^3$, and $6\,b^3-15\,ab^2-4\,a^2b-10\,a^3$.
5. Subtract $4\,a-2\,b+6\,c$ from $3\,a+b-c$.
6. Subtract $2\,x^2-5\,x+7$ from $x^3+6\,x^2+5$.
7. What must be added to $a^3+5\,a^2b$ to give a^3+b^3?
8. From $x^3+y^3-6\,x+5\,y$ take the sum of $-2\,x^2-6\,x+7\,y-8$ and $x^3+2\,x^2-5\,y+9$.
9. Simplify $-(a+b)+\{-a-(2\,a-b)\}-6\,(a-4\,b)$.
10. Simplify $6\,x-\{4\,x+[2\,x-(3\,x+\overline{5\,x+7}-1)+3]-8\}$.
11. Simplify $2\,a-[4\,a-c+\{3\,a-(4\,b-c)-(b+3\,c)\}-6\,c]$.
12. Subtract $x-(3\,y+2\,z)$ from $z-[3\,x+(y+5\,z)]$.
13. To what should $x^2+8\,x+5$ be added to give x^3-7?
14. To what should $x^4-9\,x^2+3\,y$ be added to give y^2+x-7?

MULTIPLICATION

Product. By the *product* of two algebraic expressions, A and B, we shall mean the *simplest form* to which the expression AB can be reduced by means of the rules of reckoning. **301**

Of especial importance in such reductions are:

1. The commutative, associative, and distributive laws.
2. The law of exponents $a^m \cdot a^n = a^{m+n}$.
3. The rules of sign:

$$a(-b)=(-a)b=-ab;\qquad (-a)(-b)=ab.$$

Rules for multiplying integral expressions. 1. *To find the product of two monomials, multiply the product of the numerical* **302**

factors by that of the literal factors, simplifying the latter by adding exponents of powers of the same letter.

Give the result the + or − sign, according as the monomials have like or unlike signs.

2. *To find the product of a polynomial by a monomial or polynomial, multiply each term of the multiplicand by each term of the multiplier and add the products thus obtained.*

The first rule follows from the commutative and associative laws and the law of exponents. The second rule follows from the distributive law; thus,

$$(a+b+c)(m+n) = (a+b+c)m + (a+b+c)n$$
$$= am + bm + cm + an + bn + cn.$$

The first rule applies also to products of *more than two monomials.* When an *odd number* of these monomials have − signs, the sign of the product is −; otherwise it is +.

A product of *more than two polynomials* may be found by repeated applications of the second rule.

Example 1. Find the product of $-4a^2b^2x^3$, $2bx^4$, and $-3a^3x$.

We have $-4a^2b^2x^3 \cdot 2bx^4 \cdot -3a^3x = 24a^2b^2x^3bx^4a^3x = 24a^5b^3x^8$.

Example 2. Find the product of $a-2b$ and $ab-b^2+a^2$.

For convenience we arrange both factors in descending powers of a, and choose the *simpler* factor as multiplier. We then have

$$(a^2+ab-b^2)(a-2b) = a^3+a^2b-ab^2-2a^2b-2ab^2+2b^3$$
$$= a^3-a^2b-3ab^2+2b^3.$$

303 The **degree of the product** with respect to any letter (or set of letters) is the *sum of the degrees of the factors* with respect to that letter (or set of letters).

This follows from § 302, 1, and the fact that the term of highest degree in any product is the product of the terms of highest degree in the factors.

Thus, the degrees of x^2+1 and x^3-1 are *two* and *three* respectively, and the degree of the product $(x^2+1)(x^3-1)$, or $x^5+x^3-x^2-1$, is *five*.

304 When both factors are *homogeneous*, § 276, the product is homogeneous.

For if all the terms of each factor are of the same degree, all the products obtained by multiplying a term of the one by a term of the other are of the same degree. Hence the sum of these products is a homogeneous polynomial.

305 **Arrangement of the reckoning.** When both factors are polynomials in x or any other single letter, or when both are homogeneous functions of two letters, it is convenient to arrange the reckoning as in the following examples.

Example 1. Multiply $2x^3 - x^2 + 5$ by $x - 3 + x^2$.

$$\begin{array}{rrrrrr}
2x^3 & - x^2 & & + 5 & & \\
x^2 & + x & - 3 & & & \\
\hline
2x^5 & - x^4 & & + 5x^2 & & \\
 & 2x^4 & - x^3 & & + 5x & \\
 & & - 6x^3 & + 3x^2 & & - 15 \\
\hline
2x^5 & + x^4 & - 7x^3 & + 8x^2 & + 5x & - 15
\end{array}$$

We arrange both factors in descending (or ascending) powers of x and place multiplier under multiplicand.

We then write in separate rows the "partial products" corresponding to the several terms of the multiplier, placing them so that like terms, that is, terms of the same degree, are in the same column.

Finally we add these like terms by columns.

Example 2. Multiply $x^2 - y^2 + 2xy$ by $2y + x$.

$$\begin{array}{rrrr}
x^2 & + 2xy & - y^2 & \\
x & + 2y & & \\
\hline
x^3 & + 2x^2y & - xy^2 & \\
 & 2x^2y & + 4xy^2 & - 2y^3 \\
\hline
x^3 & + 4x^2y & + 3xy^2 & - 2y^3
\end{array}$$

In this case both factors are homogeneous functions of x and y.

We arrange them both in descending powers of x, and therefore in ascending powers of y and then proceed as in Ex. 1.

306 **Detached coefficients.** In the reckoning illustrated in § 305, Ex. 1, the terms are so arranged that their *positions* suffice to indicate what powers of x occur in them. We may make use of this fact to abridge the reckoning by suppressing x and writing the coefficients only, and it is always worth while to do this when the given polynomials have numerical coefficients.

If either polynomial is incomplete, *care must be taken to indicate every missing term by a 0 coefficient.*

Example. Multiply $x^3 - 3x^2 + 2$ by $x^3 + 3x^2 - 2$.

$$\begin{array}{l} 1-3+0+2 \\ \underline{1+3+0-2} \\ 1-3+0+2 \\ \quad 3-9+0+\ 6 \\ \quad\quad\ \ \underline{-2+\ 6-0-4} \\ 1+0-9+0+12-0-4 \end{array}$$

We arrange the reckoning as in § 305, Ex. 1, but write the coefficients only, indicating the missing terms by 0 coefficients.

We omit the partial product corresponding to the 0 term of the multiplier. Inserting the appropriate powers of x in the final result — beginning with x^6 since the sum of the degrees of the factors is *six* — we obtain the product required, namely, $x^6 - 9x^4 + 12x^2 - 4$.

The degree of the product, *six*, is also indicated by the number of terms, *seven*, in the result $1 + 0 - 9 + 0 + 12 - 0 - 4$, § 277.

This is called the *method of detached coefficients.* It applies not only to polynomials in a single letter, — both arranged in descending or ascending powers of that letter, — but also to *homogeneous* polynomials in *two* letters. For in arranging two such polynomials in descending powers of one of the letters, we at the same time arrange them in ascending powers of the other letter, so that the position of any coefficient will indicate what powers of *both* letters go with it.

307 Formulas derived by the method of detached coefficients. Consider the following examples.

Example 1. Prove the truth of the identity

$$(a^4 + a^3b + a^2b^2 + ab^3 + b^4)(a - b) = a^5 - b^5.$$

$$\begin{array}{l} 1+1+1+1+1 \\ \underline{1-1} \\ 1+1+1+1+1 \\ \underline{\quad -1-1-1-1-1} \\ 1+0+0+0+0-1 \end{array}$$

We perform the multiplication indicated in the first member by detached coefficients, and so obtain the coefficients of the product arranged in descending powers of a and in ascending powers of b.

We know in advance that the degree of the product is *five*, which is also indicated by the number of terms, *six*, in the final result. Hence the product is

$$a^5 + 0 \cdot a^4b + 0 \cdot a^3b^2 + 0 \cdot a^2b^3 + 0 \cdot ab^4 - b^5, \text{ or } a^5 - b^5.$$

Example 2. Prove the truth of the identities

$$(a^2 - ab + b^2)(a + b) = a^3 + b^3. \quad (1)$$

$$(a^3 - a^2b + ab^2 - b^3)(a + b) = a^4 - b^4. \quad (2)$$

Proceeding precisely as in Ex. 1, we have

$$\begin{array}{l} 1 - 1 + 1 \quad (1) \\ \underline{1 + 1} \\ 1 - 1 + 1 \\ \underline{\quad 1 - 1 + 1} \\ 1 + 0 + 0 + 1, \ i.e.\ a^3 + b^3. \end{array} \qquad \begin{array}{l} 1 - 1 + 1 - 1 \quad (2) \\ \underline{1 + 1} \\ 1 - 1 + 1 - 1 \\ \underline{\quad 1 - 1 + 1 - 1} \\ 1 + 0 + 0 + 0 - 1, \ i.e.\ a^4 - b^4. \end{array}$$

By the method illustrated in these examples we may prove the truth of the following identities, of which the examples are special cases, namely:

For *every* positive integral value of n we have **308**

$$(a^{n-1} + a^{n-2}b + \cdots + ab^{n-2} + b^{n-1})(a - b) = a^n - b^n.$$

For every positive *odd* value of n, we have **309**

$$(a^{n-1} - a^{n-2}b + \cdots - ab^{n-2} + b^{n-1})(a + b) = a^n + b^n.$$

And for every positive *even* value of n, we have **310**

$$(a^{n-1} - a^{n-2}b + \cdots + ab^{n-2} - b^{n-1})(a + b) = a^n - b^n.$$

Powers of a binomial. We can compute successive powers **311** of $a + b$ by repeated multiplications. These multiplications are readily performed by detached coefficients.

As the coefficients of the multiplier are always $1 + 1$, it is only necessary to indicate for each multiplication the partial products and their sum. We thus obtain

$$\begin{array}{lll} (1)\ 1 + 1 & i.e.\ a + b. & \\ \quad\ \underline{\quad 1 + 1} & & \\ (2)\ 1 + 2 + 1 & i.e.\ a^2 + 2\,ab + b^2 & = (a + b)^2. \\ \quad\ \underline{\quad 1 + 2 + 1} & & \\ (3)\ 1 + 3 + 3 + 1 & i.e.\ a^3 + 3\,a^2b + 3\,ab^2 + b^3 & = (a + b)^3. \\ \quad\ \underline{\quad 1 + 3 + 3 + 1} & & \\ (4)\ 1 + 4 + 6 + 4 + 1 & i.e.\ a^4 + 4\,a^3b + 6\,a^2b^2 + 4\,ab^3 + b^4 & = (a + b)^4. \end{array}$$

Observe that in each multiplication the coefficients of the second partial product are those of the first shifted one place to the right. Hence when we add the coefficients of the two partial products and so obtain the coefficients of the next power, we are merely applying the rule:

312 *To any coefficient in a power already found add the coefficient which precedes it; the sum will be the corresponding coefficient in the next power.*

All the coefficients of this next power, except the first and last, can be found by this rule; these are 1 *and* 1.

Thus, the coefficients of (4) which correspond to 3, 3, 1 in (3) are $3 + 1$ or 4, $3 + 3$ or 6, $1 + 3$ or 4.

Applying the rule to (4), we obtain $4 + 1$ or 5, $6 + 4$ or 10, $4 + 6$ or 10, $1 + 4$ or 5. Hence

$$(a + b)^5 = a^5 + 5a^4b + 10a^3b^2 + 10a^2b^3 + 5ab^4 + b^5.$$

Evidently the coefficients of any *given* power of $a + b$ can be obtained by repeated applications of this rule.

Example. Find successively $(a + b)^6$, $(a + b)^7$, $(a + b)^8$.

313 **Products of two binomial factors of the first degree.** The student should accustom himself to obtaining products of this kind by *inspection*. We have

$$(x + a)(x + b) = x^2 + (a + b)x + ab. \quad (1)$$

$$(a_0x + a_1)(b_0x + b_1) = a_0b_0x^2 + (a_0b_1 + a_1b_0)x + a_1b_1. \quad (2)$$

In the product (1) the coefficient of x is the *sum* and the final term is the *product* of a and b.

In the product (2) the first and last coefficients are products of the first coefficients and of the last coefficients of the factors, and the middle coefficient is the sum of the "cross-products" a_0b_1 and a_1b_0.

Example 1. Find the product $(x + 5)(x - 8)$.

$$(x + 5)(x - 8) = x^2 + (5 - 8)x - 40 = x^2 - 3x - 40.$$

Example 2. Find the product $(x + 3y)(x + 10y)$.

$$(x + 3y)(x + 10y) = x^2 + (3 + 10)xy + 30y^2 = x^2 + 13xy + 30y^2.$$

Example 3. Find the product $(2x+3)(4x+7)$.

$(2x+3)(4x+7) = 2\cdot 4x^2 + (2\cdot 7 + 3\cdot 4)x + 3\cdot 7 = 8x^2 + 26x + 21.$

Example 4. By the methods just explained find the products

$(x-10)(x-15)$, $(3a+4b)(5a-6b)$, $(7x-y)(5x-3y)$.

Product of any two polynomials in x. Consider the product **314**

$$(a_0x^3 + a_1x^2 + a_2x + a_3)(b_0x^2 + b_1x + b_2)$$
$$= a_0b_0x^5 + (a_0b_1 + a_1b_0)x^4 + (a_0b_2 + a_1b_1 + a_2b_0)x^3$$
$$+ (a_1b_2 + a_2b_1 + a_3b_0)x^2 + (a_2b_2 + a_3b_1)x + a_3b_2.$$

The product is a polynomial in x whose degree is the sum of the degrees of the factors. And the coefficient of each term may be obtained by the following rule, in which a_h denotes one of the numbers a_0, a_1, a_2, a_3, and b_k one of the numbers b_0, b_1, b_2. *Find the difference between the degree of the product and the degree of the term, and then form and add all the products* a_hb_k *in which* $h+k$ *equals this difference.*

Thus, to obtain the coefficient of x^2, we find the difference $5-2$, or 3, and then form and add a_1b_2, a_2b_1, a_3b_0, these being all the products a_hb_k in which $h+k=3$.

This rule applies to the product of any two polynomials in x of the form $a_0x^m + \cdots + a_m$ and $b_0x^n + \cdots + b_n$. It also indicates how to obtain any particular coefficient of the product when the factors have numerical coefficients.

Example 1. Find the coefficient of x^{100} in the product

$(a_0x^{75} + a_1x^{74} + \cdots + a_{74}x + a_{75})(b_0x^{60} + b_1x^{59} + \cdots + b_{59}x + b_{60})$.

The degree of the product is $75+60$ or 135; and $135-100=35$. Hence the coefficient of x^{100} is $a_0b_{35} + a_1b_{34} + \cdots + a_{34}b_1 + a_{35}b_0$.

Similarly the coefficient of x^{35} is $a_{40}b_{60} + a_{41}b_{59} + \cdots + a_{74}b_{26} + a_{75}b_{25}$.

Example 2. Find the coefficient of x^3 in the product

$(3x^4 - 2x^3 + x^2 - 8x + 7)(2x^3 + 5x^2 + 6x - 3)$.

The required coefficient is $(-2)(-3) + 1\cdot 6 + (-8)5 + 7\cdot 2$, or -14.

Example 3. In the product of Ex. 1, find the coefficients of x^{110} and of x^{28}.

Example 4. In the product of Ex. 2, find separately the coefficients of x^6, x^5, x^4, x^2, and x.

315 **Products found by aid of known identities.** The following formulas or identities are very important and should be carefully memorized.

$$(a + b)^2 = a^2 + 2\,ab + b^2. \qquad (1)$$

$$(a - b)^2 = a^2 - 2\,ab + b^2. \qquad (2)$$

$$(a + b)(a - b) = a^2 - b^2. \qquad (3)$$

To this list may be added the formulas given in §§ 308, 309, 310, and the following, § 311:

$$(a + b)^3 = a^3 + 3\,a^2b + 3\,ab^2 + b^3. \qquad (4)$$

Inasmuch as the letters a and b may be replaced by any algebraic expressions whatsoever, these formulas supply the simplest means of obtaining a great variety of products. The following examples will make this clear.

Example 1. Find the product $(3\,x - 5\,y)^2$.

$$(3\,x - 5\,y)^2 = (3\,x)^2 - 2 \cdot 3\,x \cdot 5\,y + (5\,y)^2 = 9x^2 - 30\,xy + 25\,y^2. \qquad \text{by (2)}$$

Example 2. Find the product $(x^2 + xy + y^2)(x^2 - xy + y^2)$.

$$(x^2 + xy + y^2)(x^2 - xy + y^2) = [(x^2 + y^2) + xy]\,[(x^2 + y^2) - xy]$$
$$= (x^2 + y^2)^2 - x^2y^2 = x^4 + x^2y^2 + y^4. \quad \text{by (3), (1)}$$

Example 3. Explain the steps in the following process.

$$(x + y + z)(x - y + z)(x + y - z)(x - y - z)$$
$$= [x + (y + z)]\,[x - (y + z)] \cdot [x + (y - z)]\,[x - (y - z)]$$
$$= [x^2 - (y + z)^2] \cdot [x^2 - (y - z)^2]$$
$$= [(x^2 - y^2 - z^2) - 2\,yz] \cdot [(x^2 - y^2 - z^2) + 2\,yz]$$
$$= [x^2 - (y^2 + z^2)]^2 - 4\,y^2z^2$$
$$= x^4 - 2\,x^2(y^2 + z^2) + (y^2 + z^2)^2 - 4\,y^2z^2$$
$$= x^4 + y^4 + z^4 - 2\,x^2y^2 - 2\,y^2z^2 - 2\,z^2x^2.$$

Observe in particular that by this method we may derive from (1) and (4) the square and cube of any *polynomial.*

Thus, we have

$$(a+b+c)^2 = [(a+b)+c]^2 = (a+b)^2 + 2(a+b)c + c^2$$
$$= a^2 + b^2 + c^2 + 2ab + 2ac + 2bc.$$

$$(a+b+c)^3 = (a+b)^3 + 3(a+b)^2c + 3(a+b)c^2 + c^3$$
$$= a^3 + b^3 + c^3 + 3a^2b + 3b^2a + 3b^2c + 3c^2b + 3c^2a + 3a^2c + 6abc.$$

Generalizing the first of these results we have the theorem:

316 *The square of any polynomial is equal to the sum of the squares of all its terms together with twice the products of every two of its terms.*

Example 1. Find the product $(a - b + 2c - 3d)^2$.

Example 2. Find the product $(1 + 2x + 3x^2)^2$.

Example 3. Find the product $(x^3 - x^2y + xy^2 - y^3)^2$.

317 **Powers of monomial products.** By the nth power of any algebraic expression, A, we shall mean the simplest form to which the expression A^n can be reduced by the rules of reckoning.

From the laws of exponents $(a^m)^n = a^{mn}$ and $(ab)^n = a^nb^n$ we derive the following rule:

318 *To raise a monomial expression* A *to the* n*th power, raise its numerical coefficient to the* n*th power and multiply the exponent of each literal factor by* n.

If the sign of A *be* $-$, *give the result the sign* $+$ *or* $-$, *according as* n *is even or odd.*

Thus, $$(-2ax^2y^7)^4 = (-2)^4a^4x^8y^{28} = 16a^4x^8y^{28}.$$

For by repeated applications of the law $(ab)^n = a^nb^n$ we have

$$(-2ax^2y^7)^4 = (-2)^4a^4(x^2)^4(y^7)^4,$$

and by repeated applications of the law $(a^m)^n = a^{mn}$ we have

$$(-2)^4a^4(x^2)^4(y^7)^4 = 16a^4x^8y^{28}.$$

EXERCISE III

In the following examples perform each multiplication by the most expeditious method possible. In particular employ detached coefficients where this can be done with advantage; also the identities of § 315.

1. Multiply $3x^5 - 2x^4 - x^3 + 7x^2 - 6x + 5$ by $2x^2 - 3x + 1$.

2. Multiply $5x^3 - 3ax^2 + 2a^2x + a^3$ by $3x^2 - ax - 2a^2$.

3. Multiply $x^5 - x^4y + x^3y^2 - x^2y^3 + xy^4 - y^5$ by $x + y$.

4. Multiply $3x^3 - 2x^2 + 7$ by $2x^3 - 3x + 5$.

5. Multiply $7x - 2y$ by $4x - 5y$, by inspection.

6. Multiply $a^2 - ax + bx - x^2$ by $b + x$.

7. Multiply $x^4 - 2x + 5x^2 - x^3$ by $3 + x^2 - x$.

8. Multiply $2x^n - 3x^{n-2} + 5x^{n-3}$ by $x^{n-2} - x^{n-3}$.

9. Multiply $a^2 - ab + 3b^2$ by $a^2 + ab - 3b^2$.

10. Multiply $x + 3y - 2z$ by $x - 3y + 2z$.

11. Multiply $x^2 + xy + y^2 + x - y + 1$ by $x - y - 1$.

12. Multiply $a^2 + b^2 + c^2 + bc + ca - ab$ by $a + b - c$.

13. Multiply $3x - 2y + 5$ by $x - 4y + 6$.

14. Multiply $x + 7y - 3z$ by $2x + y - 8z$.

15. Find the product $(b + x)(b - x)(b^2 + x^2)$.

16. Also $(x^2 + x + 1)(x^2 - x + 1)(x^4 - x^2 + 1)$.

17. Also $(x + y + z)(-x + y + z)(x - y + z)(x + y - z)$.

18. Form a table of the coefficients of the first four powers of $x^2 + x + 1$.

19. Continue the table of coefficients of successive powers of $a + b$ as far as the *tenth* power.

20. Find $(4x - 3y)^2$ and $(4x - 3y)^3$.

21. Find $(x + 2y + 3z - 4u)^2$.

22. Find $(x + 2y + 3z)^3$; also $(x + 2y - 3z)^3$.

23. Multiply $(a + 2b)^2$ by $(a - 2b)^2$.

24. Find the coefficients of x^{29} and of x^{15} in the product

$$(a_0x^{27} + a_1x^{26} + \cdots + a_{26}x + a_{27})(b_0x^{19} + b_1x^{18} + \cdots + b_{18}x + b_{19}).$$

25. Find the coefficients of x^6, x^8, and x^4 in the product

$$(2x^6 - 3x^5 + 4x^4 - 7x^3 + 2x - 5)(3x^5 - x^3 + 2x^2 + 3x - 8).$$

26. Verify the following identities:

1. $(x + y + z)^3 - (x^3 + y^3 + z^3) = 3(y + z)(z + x)(x + y)$.
2. $(a^2 + b^2)(x^2 + y^2) = (ax + by)^2 + (bx - ay)^2$.
3. $(a^2 - b^2)(x^2 - y^2) = (ax + by)^2 - (bx + ay)^2$.
4. $(a + b + c)^3 = a^3 + b^3 + c^3 + 3a^2(b + c) + 3b^2(c + a) + 3c^2(a + b) + 6abc$.

27. Simplify the following powers:

$$(2a^2x^3y^7)^5, \quad (-x^5y^8z^9)^7, \quad (a^2b^mc^3)^{2n}, \quad (a^mb^nc^{2n})^n.$$

28. Simplify the following products:

$$(-ab^2c^3)(a^3b)^2(-ac^3)^5, \quad (-2x^2y^4)^3(ax^5y^{11})^2.$$

DIVISION

Quotient. Let A and B denote any two algebraic expressions of which B is *not equal to* 0. By the *quotient* of A divided by B, we shall mean the *simplest form* to which the fraction A/B can be reduced by the rules of reckoning. **319**

Formulas. In making such reductions the following formulas are especially useful, namely, **320**

1. $$\frac{ac}{bc} = \frac{a}{b}.$$

2. $$\frac{a^m}{a^n} = a^{m-n}, \text{ when } m > n; \quad \frac{a^m}{a^n} = \frac{1}{a^{n-m}}, \text{ when } n > m.$$

3. $$\frac{-a}{b} = -\frac{a}{b}, \quad \frac{a}{-b} = -\frac{a}{b}, \quad \frac{-a}{-b} = \frac{a}{b}.$$

4. $$\frac{a+b}{d} = \frac{a}{d} + \frac{b}{d}.$$

We may prove 1, 3, and 4 by aid of the rule, § 253:

Two expressions are equal if their products by any third expression (not 0) are equal.

For in 1 the product of each member by bc is ac.

Thus, $\frac{ac}{bc}bc = ac$; and $\frac{a}{b}bc = \frac{a}{b}b \cdot c = ac$. §§ 254, 252

Again, in 3 the product of each member of the first equation by b is $-a$, and the products of each member of the second and third equations by $-b$ are a and $-a$ respectively.

Thus, $\frac{-a}{b}b = -a$; and $\left(-\frac{a}{b}\right)b = -\frac{a}{b}b = -a$. §§ 298, 254

Finally, in 4 the product of each member by d is $a + b$.

Thus, $\frac{a+b}{d}d = a + b$; $\left(\frac{a}{d} + \frac{b}{d}\right)d = \frac{a}{d}d + \frac{b}{d}d = a + b$. §§ 254, 252

The formula 2 is a particular case of the formula 1.

Thus, if $m > n$, $a^m = a^{m-n} \cdot a^n$. § 256

Hence $\frac{a^m}{a^n} = \frac{a^{m-n} \cdot a^n}{a^n} = a^{m-n}$. by 1

321 Rules for simplifying A / B. The formulas 1, 2, and 3 give us the following rules for simplifying A / B.

1. *Cancel all factors common to numerator and denominator.*
2. *When numerator and denominator involve different powers of the same letter (or expression) as factors, cancel the lower power and subtract its exponent from that of the higher power.*
3. *Give the quotient the + or − sign, according as the numerator and denominator have the same or opposite signs.*

Thus, $\frac{bca^5}{ca^2} = ba^{5-2} = ba^3$, and $\frac{a^2}{-a^7} = -\frac{1}{a^{7-2}} = -\frac{1}{a^5}$.

322 Rules for dividing by a monomial. From the definition of division and § 320, 4, we derive the following rules.

1. *To divide one monomial by another, form a fraction by writing the dividend over the divisor, and simplify.*

2. *To divide a polynomial by a monomial, divide each term of the dividend by the divisor, and add the quotients so obtained.*

Thus, $-8a^3b^2c \div 6ab^6d = \dfrac{-8a^3b^2c}{6ab^6d} = -\dfrac{4a^2c}{3b^4d}$, by cancelling the common factor $2ab^2$ and applying the rule of signs.

Again, $(ax^3 - 4a^2x^2) \div ax = \dfrac{ax^3}{ax} - \dfrac{4a^2x^2}{ax} = x^2 - 4ax.$

But when d has no factor in common with a and b, we regard $(a+b)/d$ as a simpler form of the quotient than $a/d + b/d$.

Division of a polynomial by a polynomial. If A and B are polynomials which have common factors, the quotient is the expression to which A/B reduces when these factors are cancelled. **323**

Thus, if $A = x^2 - y^2$, $B = x^2 + 2xy + y^2$, the quotient is $(x-y)/(x+y)$.

For $\dfrac{A}{B} = \dfrac{x^2 - y^2}{x^2 + 2xy + y^2} = \dfrac{(x+y)(x-y)}{(x+y)^2} = \dfrac{x-y}{x+y}.$

In another chapter we shall give methods for finding the factors which are common to two polynomials. The process called *long division* is considered in Chapter V.

Complex expressions. Observe that $a \div b \times c$ means $\dfrac{a}{b}c$, while $a \div bc$, like $a \div (b \times c)$, means a/bc. **324**

In the chapter on fractions we shall consider complex expressions in which a number of indicated multiplications and divisions occur. In particular we shall find that

$$a \times (b \times c \div d) = a \times b \times c \div d. \qquad (1)$$

$$a \div (b \times c \div d) = a \div b \div c \times d. \qquad (2)$$

In (1) the signs $\times$ and $\div$ within the parentheses remain unchanged when the parentheses are removed; but in (2) each $\times$ is changed to $\div$, and each $\div$ to $\times$.

EXERCISE IV

1. Divide $15\,a^3bc^2$ by $10\,ab^2c^2$.

2. Divide $75\,x^2y^4z^{10}$ by $-100\,ax^7z^9$.

3. Divide $-35\,x^{2m}y^n$ by $28\,x^my^{m+n}$.

4. Divide $-54\,\{(ab^2)^2c\}^5$ by $-18\,\{a(b^2c)^2\}^3$.

5. Divide $x^2y - xy^2$ by $x^2 - y^2$.

6. Divide $(x^3 - y^3)(x^3 + y^3)$ by $(x - y)(x^2 - xy + y^2)$.

7. Simplify $\dfrac{(a-b)^2(b-c)^3(c-a)^4}{(b-a)(c-b)^2(a-c)^3}$.

8. Simplify $\dfrac{30\,a^2b^3c^4 - 25\,a^3b^2c^5 + 20\,a^4b^4c^7}{-5\,ab^2c^3}$.

9. Simplify $\dfrac{3(x-y)^4 - 2(x-y)^3 + 5(x-y)^2}{(y-x)^2}$.

10. Simplify $4\,a^7 \times (3\,ab^3c^2)^2 \div (abc)^2 \div 6\,bc$.

11. Simplify the following (1) by performing the operations in the order indicated, (2) by first removing the parentheses.

$$a^7 \div \{a^5 \div (a^4 \div a^2 \times a) \times (a^3 \times a \div a^2)\}.$$

12. By what must $2\,a(x^2y^3)^2$ be multiplied to give $-4\,a^2(x^3y^2)^2$?

III. SIMPLE EQUATIONS IN ONE UNKNOWN LETTER

CONDITIONAL EQUATIONS

325 The expressions $3\,x - 4$ and $x + 6$ are not identically equal, § 281, and therefore are not equal in value for all values of x. If asked, "For *what* value or values of x are the values of these expressions the same?", we begin by *supposing* them to be the same, and state the supposition thus:

$$3\,x - 4 = x + 6.$$

The expression thus formed is called a *conditional equation*, or an *equation of condition*, because it states a condition which the "unknown letter" x is to satisfy. It serves the purpose of restricting x to values which satisfy this condition, being true when the values of $3x - 4$ and $x + 6$ are the same and then only.

Similarly $x + y = 0$ is a conditional equation in the two unknown letters x and y, and, in general,

326 *When the expressions* A *and* B *are not identically equal,* A = B *is a* **conditional equation.** *This equation means: "*A *and* B *are supposed to have equal values." And it restricts the variable letters in* A *and* B *to values for which this supposition is true.*

The letters whose values the equation $A = B$ thus restricts are called the *unknown letters* of the equation.

In what follows, the word "equation" will mean "conditional equation."

327 If the only letters in an equation are the unknown letters, as x, y, z, we call it a **numerical equation**; but if there are also known letters, as a, b, c, we call it a **literal equation.**

Thus, $2x - 3y = 5$ is a numerical, but $ax + by = c$ is a literal equation. A literal equation does not restrict the values of the *known* letters.

328 If both A and B are rational and integral with respect to the unknown letters, the equation $A = B$ is said to be **rational and integral.** But if A or B is irrational or fractional, the equation is said to be **irrational** or **fractional.**

No account is taken of numbers or known letters in this classification. Thus, $\sqrt{2}x + y/b = c$ is both rational and integral.

329 In the case of a *rational integral* equation reduced to its simplest form, § 340, the degree of the term or terms of *highest* degree is called the **degree of the equation** itself.

Thus, the degree of $ax^2 + bx = c$ is *two*; that of $x^3z^2 + y^4 = b$ is *five*. The degree is measured with respect to *all* the unknown letters, but these letters only.

330 Equations of the first degree are often called **simple or linear** equations; those of the second, third, fourth degrees are called **quadratic, cubic, biquadratic** equations respectively.

331 An equation in *one* unknown letter, as x, restricts x to a *finite* number of values. We say that these values of x *satisfy* the equation, or that they are its *solutions* or *roots*. Hence

332 A **root** of an equation in x is any number or *known* expression which, if substituted for x, will make the equation an identity.

Thus, 1 and -2 are roots of the equation $x^2 + x = 2$; for $1^2 + 1 \equiv 2$ and $(-2)^2 + (-2) \equiv 2$.

Again, $a - b$ is a root of $x + b = a$; for $(a - b) + b \equiv a$.

333 **Notes.** 1. An equation may have *no* root; for it may state a condition which no number can satisfy.

Thus, no finite number can satisfy the equation $x + 2 = x + 3$.

2. In every equation in x which *has* roots, x is merely a symbol for one or other of these roots. In fact the equation itself is merely a *disguised identity*, a substitute for the several *actual* identities obtained by replacing x by each root in turn.

Thus, $x^2 + x = 2$ is merely a substitute for the two identities $1^2 + 1 \equiv 2$ and $(-2)^2 + (-2) \equiv 2$.

ON SOLVING EQUATIONS

334 To **solve** an equation in one unknown letter is to find all its roots, or to prove that it has no root.

The reasoning on which the process depends is illustrated in the following examples.

Example 1. Solve the equation $3x - 4 = x + 6$.

Starting with the *supposition* that x has a value for which this equation is true, we may reason as follows:

If	$3x - 4 = x + 6,$	(1)	
then	$3x - 4 + (-x + 4) = x + 6 + (-x + 4),$	(2)	§ 249
or	$2x = 10,$	(3)	§ 300
and therefore	$x = 5.$	(4)	§ 253
Hence, if	$3x - 4 = x + 6$, then $x = 5$.	(*a*)	

The proposition (*a*) thus proved states that if (1) is *ever* true, it is when $x = 5$, in other words, that the only number which *can be* a root of (1) is 5; but it does *not* state that 5 *is* a root of (1). That statement would be

If $\quad x = 5, \text{ then } 3x - 4 = x + 6. \quad (b)$

And (*b*) is not the same as (*a*) but its *converse*, § 291.

We may prove that 5 is a root of (1) by substituting 5 for x in (1); for we thus obtain the true identity $3 \cdot 5 - 4 \equiv 5 + 6$.

But this step is not necessary, except to verify the accuracy of our reckoning. For when a true proposition has been proved *by a reversible process*, we may always conclude that its converse is true, § 293. And this is the case with (*a*), since the process by which (4) was derived from (1) is made up of reversible steps and is therefore reversible as a whole. Thus,

If	$x = 5,$	(4)	
then	$2x = 10,$	(3)	§ 253
or	$3x - 4 + (-x + 4) = x + 6 + (-x + 4),$	(2)	§ 300
and therefore	$3x - 4 = x + 6.$	(1)	§ 249

Hence, in proving the proposition (*a*) by a reversible process, we have *at the same time* proved the converse proposition (*b*), that is, we have not only proved that no other number than 5 can be a root of (1), but also that 5 is itself a root of (1).

Example 2. Solve the equation $x^2 = 9$.

If $\quad x^2 = 9, \quad (1)$

then either $\quad x = 3, \quad (2) \quad$ or $\quad x = -3. \quad (3) \quad$ § 257

Hence (1) can have no other roots than 3 and -3.

But both 3 and -3 are roots of (1), since the step by which each of the equations (2) and (3) has been derived from (1) is reversible. Thus, (1) follows from (2) and also from (3), § 257.

These examples illustrate the following general principles:

335 *In seeking the roots of an equation in* x, *we treat the equation as if it were a known identity, and endeavor to find all the equations of the form* $x = c$ *which necessarily follow from it, when thus regarded, by the rules of reckoning.*

If the process by which one of these equations $x = c$ *has been derived is reversible when* x *has the value* c, *we may at once conclude that* c *is a root; and the process is reversible if it is made up of reversible steps.*

336 It is important to remember that the mere fact that a certain value of x has been derived from an equation by the rules of reckoning does not prove it to be a root. The process must be *reversible* to warrant this conclusion.

Thus, from $x - 2 = 0,$ (1)

it follows that $(x - 2)(x - 3) = 0,$ (2) § 253

and hence that either $x = 2,$ or $x = 3.$ (3) § 253

But we have no right to draw the absurd conclusion that 3 is a root of (1). For when $x = 3$ we *cannot reverse* the process, that is, divide both members of (2) by $x - 3$, since the divisor $x - 3$ is then 0.

On the other hand, when $x = 2$ we *can* reverse the process, since $x - 3$ is then not 0 but -1; and 2 *is* a root of (1).

TRANSFORMATION THEOREMS

337 In the light of what has just been said we may regard any correct application of the rules of reckoning to an equation as a legitimate *transformation* of the equation; and if such a transformation is reversible, we may conclude that it leaves the roots of the equation unchanged. Hence the following theorems.

338 **Theorem 1.** *The following transformations of an equation leave its roots unchanged, namely:*

1. *Applying the rules of combination,* § 259, *to each member separately.*

2. *Adding any expression which has a finite value to both members, or subtracting it from both.*

3. *Multiplying or dividing both members by the same constant (not* 0).

For all the rules of reckoning involved in these transformations are reversible, § 259.

We may also state the proofs of 2 and 3 as follows:

If A and B denote expressions in x, the roots of the equation $A = B$ are numbers which substituted for x in A and B make $A \equiv B$, § 332.

But any value of x which makes $A \equiv B$ and C finite will make $A + C \equiv B + C$, and conversely, § 249; hence the roots of $A = B$ are the same as those of $A + C = B + C$.

Again, if c denote any constant except 0, any value of x which makes $A \equiv B$ will make $cA \equiv cB$, and conversely, § 253; hence the roots of $A = B$ are the same as those of $cA = cB$.

Thus, in § 334, Ex. 1, the equations

$$3x - 4 = x + 6, \quad (1)$$

$$3x - 4 + (-x + 4) = x + 6 + (-x + 4), \quad (2)$$

$$2x = 10, \quad (3)$$

$$x = 5. \quad (4)$$

all have the same root, 5.

Here (2) is derived from (1) by the transformation 2, (3) from (2) by the transformation 1, and (4) from (3) by the transformation 3.

339 **Corollary.** *The following transformations of an equation leave its roots unchanged, namely:*

1. *Transposing a term, with its sign changed, from one member to the other.*
2. *Cancelling any terms that may occur in both members.*
3. *Changing the signs of all terms in both members.*

For 3 is equivalent to multiplying both members by -1. And 1 and 2 are equivalent to subtracting the term in question from both members of the equation.

Thus, if from both members of $x - a + b = c + b$ (1)

we subtract $-a + b = -a + b$

we obtain $x = c + a.$ (2)

The effect of the subtraction is to *cancel* b in both members of (1) and to *transpose* $-a$, with its *sign changed*, from the first member to the second.

340 By aid of these transformations, §§ 338, 339, every rational integral equation in x may, without changing its roots, be reduced to the **standard form**

$$a_0x^n + a_1x^{n-1} + \cdots + a_{n-1}x + a_n = 0.$$

We suppose such an equation reduced to this form when its degree is measured, § 329. The like is true of rational integral equations in more than one unknown letter.

Thus, $x^2 + 3x + 5 = x^2 - 4x + 7$ can be reduced to the form $7x - 2 = 0$. Its degree is therefore *one*, not two.

341 **Theorem 2.** *When* A, B, *and* C *are integral, the equation*

$$AC = BC$$

has the same roots as the two equations

$$A = B \text{ and } C = 0.$$

For any value of x which makes $AC \equiv BC$ must make either $A \equiv B$ or $C \equiv 0$; and, conversely, any value of x which makes either $A \equiv B$ or $C \equiv 0$ will make $AC \equiv BC$, §§ 251, 253.

In this proof it is assumed that A, B, C have finite values for the values of x in question. This is always true when, as is here supposed, A, B, C are integral; but it is not always true when A, B, C are fractional.

In particular, *when* A *and* C *are integral, the equation* AC = 0 *has the same roots as the equations* A = 0 *and* C = 0 *jointly.*

Thus, the roots of the equation $x^2 = 3x$ are the same as those of the two equations $x = 3$ and $x = 0$, that is, 3 and 0.

Similarly the roots of $(x - 1)(x - 2) = 0$ are the same as those of the two equations $x - 1 = 0$ and $x - 2 = 0$, that is, 1 and 2.

342 Hence the effect of multiplying both members of an integral equation $A = B$ by the same integral function C is *to introduce extraneous* roots, namely, the roots of the equation $C = 0$. Conversely, the effect of removing the same integral factor C from both members of an integral equation $AC = BC$, is *to lose* certain of its roots, namely, the roots of $C = 0$.

On the other hand, in a *fractional* equation, it is usually the case that *no* extraneous roots are introduced when both members are multiplied by the *lowest common denominator* of all the fractions.

Thus, if the equation be $1/x = 1/(2x - 1)$, and we multiply both members by $x(2x - 1)$, we obtain $2x - 1 = x$, whose root is 1. As 1 is not a root of $x(2x - 1) = 0$, we have introduced no extraneous root.

Corollary. *The integral equation* $A^2 = B^2$ *has the same roots as the equations* $A = B$ *and* $A = -B$ *jointly.* **343**

For $A^2 = B^2$ has the same roots as $A^2 - B^2 = 0$, § 339. And since $A^2 - B^2 \equiv (A - B)(A + B)$, the equation $A^2 - B^2 = 0$ has the same roots as the two equations $A - B = 0$ and $A + B = 0$, § 341, and therefore the same roots as the two equations $A = B$ and $A = -B$, § 339.

Thus, the roots of the equation $(2x - 1)^2 = (x - 2)^2$ are the same as those of the two equations $2x - 1 = x - 2$ and $2x - 1 = -(x - 2)$, that is, -1 and 1.

Hence the effect of squaring both members of the equation $A = B$ is to introduce extraneous roots, namely, the roots of the equation $A = -B$. Conversely, the effect of deriving from $A^2 = B^2$ the single equation $A = B$ is to lose certain of the roots, namely, the roots of the equation $A = -B$. **344**

Since $A^n - B^n \equiv (A - B)(A^{n-1} + A^{n-2}B + \cdots + B^{n-1})$, § 308, it follows by the reasoning of § 343 that the roots of $A^n = B^n$ are those of $A = B$ and $A^{n-1} + A^{n-2}B + \cdots + B^{n-1} = 0$ jointly. **345**

Thus, since $x^3 - 1 \equiv (x - 1)(x^2 + x + 1)$, the equation $x^3 = 1$ has the same roots as the equations $x = 1$ and $x^2 + x + 1 = 0$ jointly.

The theorems just demonstrated, §§ 338–345, hold good for equations in more than one unknown letter if the word *root* be replaced by the word *solution*, § 355. **346**

Thus, by § 339, the equation $x + 2y - 3 = 0$ (1) has the same solutions as the equation $x = -2y + 3$ (2), that is, every pair of values of x and y which satisfy (1) will also satisfy (2), and conversely.

Equivalent equations. When two or more equations have the same roots (or solutions), we say that they are *equivalent.* **347**

Thus, § 338, the equations $A = B$ and $A + C = B + C$ are equivalent. Again, § 341, the equation $AC = BC$ is equivalent to the two equations $A = B$ and $C = 0$.

But $x^2 = 9$ (1) and $x = 3$ (2) are *not* equivalent although both have the root 3. For (1) also has the root -3, which (2) does not have.

SOLUTION OF SIMPLE EQUATIONS

348 From the transformation theorems of §§ 338, 339 we may at once derive the following rule for solving a simple equation in one unknown letter, as x.

To solve a simple equation in x, *reduce it to the form* $ax = b$. *Then*

1. *If* $a \neq 0$, *the equation has the single root* b/a.
2. *If* $a = 0$, *and* $b \neq 0$, *the equation has no root.*
3. *If* $a = 0$, *and* $b = 0$, *the equation is an identity.*

If the equation has fractional coefficients, it is usually best to begin by multiplying both members by the lowest common denominator of these fractions. This process is called *clearing the equation of fractions.*

We then reduce the equation to the form $ax = b$ by *transposing* the unknown terms to the first member and the known terms to the second, and *collecting* the terms in each member. To *verify* the result, substitute it for x in the given equation.

Example 1. Solve $\frac{2x}{3} - \frac{x-2}{2} = \frac{x}{6} - (4 - x)$.

To clear of fractions, multiply both members by the l.c.d., 6.

Then $4x - 3(x - 2) = x - 6(4 - x)$,

or $4x - 3x + 6 = x - 24 + 6x$.

Transpose and collect terms, $-6x = -30$.

Therefore $x = 5$.

Verification. $\frac{2 \cdot 5}{3} - \frac{5-2}{2} \equiv \frac{5}{6} - (4 - 5)$.

Example 2. Solve $mx + n = px + q$.

Transpose and collect terms, $(m - p)x = q - n$.

Hence if $m \neq p$, the equation has the single root $(q - n)/(m - p)$.

If $m = p$ and $q \neq n$, it has no root.

If $m = p$ and $q = n$, it is an identity and every value of x satisfies it.

Example 3. Solve $(x+a)(x+b)=(x-a)^2$.

Expand, $x^2+(a+b)x+ab=x^2-2ax+a^2$.

Cancel x^2, and transpose and collect terms.

Then $(3a+b)x=a^2-ab$,

and therefore $x=\dfrac{a^2-ab}{3a+b}$.

349 Sometimes a root of an equation can be found by *inspection*. The equation is then completely solved if it be a simple equation, for it can have no other root than the one thus found.

Example. Solve $(x-a)^2-(x-b)^2=(a-b)^2$.

Evidently this is a simple equation, and when $x=b$ it reduces to the identity $(b-a)^2=(a-b)^2$. Hence its root is b.

350 The roots of an equation of the form $AB=0$, in which A and B denote integral expressions of *the first degree* in x, can be found by solving the two simple equations $A=0$ and $B=0$, § 341. In like manner, when A, B, C are of the first degree, the roots of $ABC=0$, $AC=BC$ and $A^2=B^2$ may be found by solving simple equations, §§ 341, 343.

Example 1. Solve $(x-2)(x+3)(2x-5)(3x+2)=0$.

This equation is equivalent, § 347, to the four equations

$$x-2=0,\ x+3=0,\ 2x-5=0,\ 3x+2=0.$$

Hence its roots are $2,\ -3,\ 5/2,\ -2/3$.

Example 2. Solve $4x^2-5x=3x^2+7x$.

This equation has the same roots as the two equations

$$x=0 \text{ and } 4x-5=3x+7.$$

Its roots are therefore 0 and 12.

EXERCISE V

Solve the following equations.

1. $15-(7-5x)=2x+(5-3x)$.
2. $x(x+3)-4x(x-5)=3x(5-x)-16$.
3. $(x+1)(x+2)-(x+3)(x+4)=0$.

4. $x = 1 + \frac{x}{2} + \frac{x}{4} + \frac{x}{8} + \frac{x}{16}$.

5. $x - 2[x - 3(x + 4) - 5] = 3\{2x - [x - 8(x - 4)]\} - 2$.

6. $2\{3[4(5x - 1) - 8] - 20\} - 7 = 1$.

7. $\frac{1}{2}\{\frac{1}{3}[\frac{1}{4}(\frac{1}{5}x - 1) - 6] + 4\} = 1$.

8. $3 - \frac{5 - 2x}{5} = 4 - \frac{4 - 7x}{10} + \frac{x + 2}{2}$.

9. $\frac{3x - 1}{3} + 3 = -\frac{x - 4}{6} + \frac{3x + 5}{4} - 2\frac{1}{2}$.

10. $\frac{5x - .4}{.3} + \frac{1.3x - .05}{2} = \frac{13.95 - 8x}{1.2}$.

11. $3cx - 5a + b - 2c = 6b - (a + 3bx + 2c)$.

12. $(b - c)(a - x) + (c - a)(b - x) + (a - b)(c - x) = 1 - x$.

13. $\frac{x + 1}{a + 1} + \frac{x}{a} = 2$, by inspection.

14. $\frac{x + 1}{a + b} + \frac{x - 1}{a - b} = \frac{2a}{a^2 - b^2}$.

15. $\left(\frac{m}{n} + \frac{n}{m}\right)x = \frac{m}{n} - \frac{n}{m} - 2x$.

16. $(2x - 1)(3x - 1)(4x + 1)(5x + 2) = 0$.

17. $(x^2 - x)(2x - 5) = (x^2 - x)(x + 9)$.

18. $(x + 2)^3 - (x - 2)^3 = 32x + 16$.

19. $[(a + b)x - c]^2 = [(a - b)x + c]^2$.

20. $(x^2 - 2x + 1)^2 - (x - 1)^2(x - 3)^2 = 0$.

PROBLEMS

351 **On solving problems.** In the following problems it is required to derive the values of certain *unknown numbers* from given relations, called the *conditions of the problem,* connecting these numbers with known numbers and one another.

In each case we represent one of the unknown numbers by a letter, as x. The given conditions then enable us to express the remaining unknown numbers in terms of x and to form a single equation connecting the expressions thus obtained. This equation is the statement of the problem in algebraic symbols. We solve it for x. If the problem have any solution, it will be the value thus found for x, together with the corresponding values of the other unknown numbers.

352 It may happen, however, that the value thus found for x is not an admissible solution of the problem. For the problem may be one which imposes a restriction on the *character* of the unknown numbers, as that they be integers, and the equation in x into which the statement of the problem has been translated does not express this restriction.

Having solved the equation in x, therefore, we must notice whether the result is a number of *the kind required* before we accept it as a solution of the problem. If it is not, we conclude that the problem is an impossible one.

Example 1. The sum of the digits of a certain number of two digits is 12. If we reverse the order of the digits we obtain a number which is 4/7 as great. What is the number?

Here there are four unknown numbers, namely, the tens digit, the units digit, the value of the number as it stands, and the value when the digits are reversed; but all four can be readily expressed in terms of either units or tens digit.

Thus, let $x =$ the tens digit.

Then $12 - x =$ the units digit,

$10x + (12 - x) =$ value of required number,

$10(12 - x) + x =$ value with digits reversed.

By the remaining condition of the problem, we have

$$10(12 - x) + x = \tfrac{4}{7}[10x + (12 - x)]. \qquad (1)$$

Solving this equation we obtain $x = 8$, which being an integer less than 10, is an admissible solution of the problem. The like is true of $12 - x$ or 4. Hence the required number is 84.

Notice that with a slight modification the problem becomes impossible. Thus, if we require that reversing the digits shall *double* the value of the number, we have, instead of (1), the equation

$$10(12 - x) + x = 2[10x + (12 - x)]. \qquad (2)$$

And solving (2) we obtain $x = 32/9$, which being fractional is not an *admissible* solution of the problem.

353 When dealing with a problem which has to do with certain *magnitudes*, as intervals of time, remember that the letters used in stating the problem algebraically are to represent not the magnitudes themselves, but the numbers which are their *measures* in terms of some given unit or units. Care must also be taken to express the measures of all magnitudes of the same kind, whether known or unknown, *in terms of the same unit.*

Example 1. A tank has a supply pipe A which will fill it in 3 hours, and a waste pipe B which will empty it in 3 hours and 40 minutes. If the tank be empty when both pipes are opened, how long will it be before the tank is full?

Let x denote the *number of hours* required.

Then $1/x$ is the part filled in one hour when both A and B are open.

But were A alone open, the part filled in one hour would be $1/3$.

And were B alone open (and water in the tank) the part emptied in one hour would be $1/3\frac{2}{3}$ or $3/11$.

Hence $$\frac{1}{x} = \frac{1}{3} - \frac{3}{11}, \text{ or } \frac{2}{33}.$$

Therefore $$x = 33/2 \text{ hours, or 16 hours 30 minutes.}$$

Example 2. A crew can row 2 miles against the current in a certain river in 15 minutes; with the current in 10 minutes. What is the rate of the current? And at what rate can the crew row in dead water?

Let x = rate of current in miles per minute.

As the rate of the crew against the current is $2/15$ in miles per minute, in dead water it would be $2/15 + x$.

And as the rate of the crew with the current is $2/10$, or $1/5$ in miles per minute, in dead water it would be $1/5 - x$.

Hence $$\tfrac{2}{15} + x = \tfrac{1}{5} - x,$$

whence $$x = \tfrac{1}{30} \text{ (miles per minute),}$$

and $$\tfrac{2}{15} + x = \tfrac{1}{6} \text{ (miles per minute).}$$

Example 3. At what time between two and three o'clock do the hour and minute hands of a clock point in opposite directions?

Let x = number of minutes past two o'clock at the time required.

Since the minute hand starts at XII it will then have traversed x minute spaces.

The hour hand starts at II, or 10 minute spaces in advance of the minute hand, but it moves only 1/12 as fast as the minute hand.

Therefore when the minute hand is at x minute spaces past XII, the hour hand is at $10 + x/12$ minute spaces past XII.

But by the conditions of the problem, at the time required the minute hand is 30 minute spaces in advance of the hour hand.

Hence $$x = \left(10 + \frac{x}{12}\right) + 30,$$

or solving, $$x = 43\tfrac{7}{11} \text{ minute spaces.}$$

Therefore the hands point in opposite directions at $43\frac{7}{11}$ minutes after two o'clock, or $16\frac{4}{11}$ minutes before three o'clock.

354 Sometimes in the statement of a problem the known numbers are denoted by letters, as a, b, c. The value found for x will then be an expression in a, b, c which may represent an admissible solution of the problem for certain values of these letters, but not for others. The discussion of the following problem, known as the *problem of couriers*, will illustrate this point.

Example. Two couriers A and B are traveling along the same road in the same direction at the rates of m and n miles an hour respectively. B is now d miles in advance of A. Will they ever be together, and if so, when?

Let $x =$ the number of hours hence when they will be together.

A will then have traveled mx miles, and B nx miles; and since B is now d miles in advance of A, we have

$$mx = nx + d, \quad (1)$$

whence $$(m - n)x = d, \quad (2)$$

and therefore $$x = \frac{d}{m - n} \text{ hours hence.} \quad (3)$$

1. If A is to overtake B, this value of x must be positive; and since by hypothesis d, m, n all denote positive numbers, this requires that $m > n$. Which corresponds to the obvious fact that if A is to overtake B, he must travel faster than B does.

2. At the same time we can interpret the negative value which x takes if we suppose $m < n$ as meaning that A and B *were* together $d/(n - m)$ hours *ago*.

3. If $m = n$, we cannot, properly speaking, derive (3) from (1), since the process involves dividing by $m - n$ which is 0. But we can derive (3) from (1) if m differs at all from n, it matters not how little. And if in (3) we regard m as a variable, which while greater than n is continually approaching equality with n, the fraction $d/(m - n)$ becomes a variable which continually increases, and that without limit, § 510. All of which corresponds to the obvious fact that the smaller the excess of A's rate over B's, the longer it will take A to overtake B, and that A will *never* overtake B if his rate be the same as B's rate.

4. Finally, if we suppose both $m = n$ and $d = 0$, the equation (1) is satisfied by every value of x. Which corresponds to the obvious fact that if A and B are traveling at the same rate and are now together, they will *always* be together.

EXERCISE VI

1. The sum of the digits of a certain number of two digits is 14. If the order of the digits be reversed, the number is increased by 18. What is the number?

2. By what number must 156 be divided to give the quotient 11 and the remainder 2?

3. There are two numbers whose difference is 298. And if the greater be divided by the less, the quotient and remainder are both 12. What are the numbers?

4. The tens digit of a certain number of two digits is twice the units digit. And if 1 be added to the tens digit and 5 to the units digit, the number obtained is three times as great as if the order of the digits be first reversed and then 1 be subtracted from the tens digit and 5 from the units digit. What is the number?

5. If 2 be subtracted from a certain number and the remainder be multiplied by 4, the same result is obtained as if twice the number and half a number one less be added together. What is the number?

6. A father is now *four* times as old as his son. If both he and his son live 20 years longer, he will then be *twice* as old as his son. What are the present ages of father and son, and how many years hence will the father be *three* times as old as the son?

7. A tank can be filled by one pipe in 3 hours, and emptied by a second in 2 hours, and by a third in 4 hours. How long will it take to empty the tank if it start full and all the pipes are opened?

8. A and B can do a certain piece of work in 10 days; but at the end of the seventh day A falls sick and B finishes the piece by working alone for 5 days. How long would it take each man to do the entire piece, working alone?

9. At what time between eight and nine o'clock do the hands of a watch point in the same direction? in opposite directions?

10. How soon after four o'clock are the hands of a watch at right angles?

11. In a clock which is not keeping true time it is observed that the interval between the successive coincidences of the hour and minute hands is 66 minutes. What is the error of the clock (in seconds per hour)?

12. Four persons, A, B, C, D, divide $1300 so that B receives $\frac{2}{3}$ as much as A, C $\frac{2}{3}$ as much as B, and D $\frac{2}{3}$ as much as C. How much does each receive?

13. A man leaves $\frac{1}{2}$ his property and $1000 besides to his oldest son; $\frac{1}{2}$ of the remainder and $1000 besides to his second son; $\frac{1}{2}$ of the sum still remaining and $1000 besides to his youngest son. If $3500 still remain, what is the amount of the entire property?

14. If 2 feet be added to both sides of a certain square, its area is increased by 100 square feet. What is the area of the square?

15. The height of a certain flagstaff is unknown; but it is observed that a flag rope fastened to the top of the staff is 2 feet longer than the staff, and that its end just reaches the ground when carried to a point 18 feet distant from the foot of the staff. What is the height of the staff?

16. A purse contains a certain number of dollar pieces, twice as many half-dollar pieces, and three times as many dimes. If the total value of the pieces is $11.50, how many pieces are there of each kind?

17. A man invests $5000, partly at 6% and partly at 4%, so that the average rate of interest on the entire investment is $5\frac{1}{2}$%. What sum does he invest at each rate?

18. In what proportions should two kinds of coffee worth 20 cts. and 30 cts. a pound respectively be combined to obtain a mixture worth 26 cts. a pound?

19. A pound of a certain alloy of silver and copper contains 2 parts of silver to 3 of copper. How much copper must be melted with this alloy to obtain one which contains 3 parts of silver to 7 of copper?

20. If a certain quantity of water be added to a gallon of a given liquid, it contains 30% of alcohol; if twice this quantity of water be added, it contains 20% of alcohol. How much water is added each time, and what percentage of alcohol did the original liquid contain?

21. A train whose rate of motion is 45 miles per hour starts on its trip from Philadelphia to Jersey City at 10 A.M., and at 10.30 A.M. another train whose rate is 50 miles an hour starts on its trip from Jersey City to Philadelphia. Assuming that the two cities are 90 miles apart, when will the trains pass each other, and at what distance from Jersey City?

22. If two trains start at the times mentioned in the preceding example and pass each other at a point half way between Jersey City and Philadelphia, and if the slower train moves $\frac{4}{5}$ as fast as the swifter one, what are their rates, and when do they pass each other?

23. A rabbit is now a distance equal to 50 of her leaps ahead of a fox which is pursuing her. How many leaps will the rabbit take before the fox overtakes her if she takes 5 leaps while the fox takes 4, but 2 of the fox's leaps are equivalent to 3 of her leaps?

24. If 19 ounces of gold weigh but 18 ounces when submerged in water, and 10 ounces of silver then weigh 9 ounces, how many ounces of silver and of gold are there in a mass of an alloy of the two metals which weighs 387 ounces in air and 351 ounces in water?

25. A traveler set out on a journey with a certain sum of money in his pocket and each day spent $\frac{1}{2}$ of what he began the day with and $2 besides. At the end of the third day his money was exhausted. How much had he at the outset?

26. The base of a certain pyramid is a square, and the altitude of each of the triangles which bound it laterally is equal to an edge of the base. Were this edge and altitude each increased by 3 inches, the area of the pyramid would be increased by 117 square inches. What is the area of the pyramid?

27. The sum of the digits of a certain number of two digits is a. If the order of the digits be reversed, the number is increased by b. What is the number? Show that the solution is admissible only when $9a \geqq b$ and when both $9a + b$ and $9a - b$ are exactly divisible by 18.

28. Two persons A and B are now a and b years old respectively. Is there a time when A was or when A will be c times as old as B, and if so, when?

Discuss the result for various values of a, b, c, as in § 354.

IV. SYSTEMS OF SIMULTANEOUS SIMPLE EQUATIONS

SIMULTANEOUS EQUATIONS

355 A conditional equation in two unknown letters, as x and y, will be satisfied by infinitely many pairs of values of these letters. We call every such pair a **solution** of the equation. The like is true of an equation in more than two unknown letters.

Thus, the equation $2x + y = 3$ (1) is satisfied if we give *any value whatsoever* to x and the corresponding value of $3 - 2x$ to y. For in (1) substitute any number, as b, for x and $3 - 2b$ for y, and we have the true identity $2b + (3 - 2b) \equiv 3$.

Thus, $x = 0$, $y = 3$; $x = 1$, $y = 1$; $x = 2$, $y = -1$; $\cdots$ are solutions of (1).

356 **Note.** When two unknown letters, x, y, are under consideration, the equation $x = 2$ means that x is to have the value 2, and y *any value whatsoever;* in other words, the equation $x = 2$ then has an infinite number of solutions. And the like is true of any equation which involves but one of the unknown letters.

357 It is therefore natural to inquire whether there may not be pairs of values of x and y which will satisfy *two* given equations in these letters. Such pairs usually exist.

Thus, both the equations $2x + y = 3$ and $4x + 3y = 5$ are satisfied when $x = 2$ and $y = -1$; for $2 \cdot 2 + (-1) \equiv 3$, and $4 \cdot 2 + 3(-1) \equiv 5$.

358 **Simultaneous equations.** Two or more equations involving certain unknown letters are said to be *simultaneous* when each unknown letter is supposed to stand for the same number in all the equations.

Thus, the equations $2x + y = 3$ (1) and $4x + 3y = 5$ (2) are simultaneous if we suppose x to denote the same number in (1) as in (2), and y similarly.

It is not necessary that all the unknown letters occur in every one of the equations. Thus, $x = 2$, $y = 3$ constitute a pair of simultaneous equations in x and y.

359 Generally speaking, the supposition that certain equations are simultaneous is allowable only when the number of equations is equal to, or less than, the number of unknown letters.

Thus, the two equations $x = 2$ and $x = 3$ cannot be simultaneous, since x must denote *different* numbers in the two.

360 A **solution** of a system of simultaneous equations is any set of values of the unknown letters which will satisfy *all* the equations of the system.

Thus, $x = 2$, $y = -1$ is a solution of the system

$$2x + y = 3, \; 4x + 3y = 5.$$

361 To **solve** a system of simultaneous equations is to find all its solutions or to prove that it has no solution.

362 The reasoning on which the process depends is similar to that described and illustrated in §§ 334, 335.

Thus, in the case of a pair of equations in x and y we begin by *supposing* that x and y actually have values which satisfy both equations. On this supposition the equations may be treated like identities and the rules of reckoning applied to them. By aid of these rules we endeavor to transform the equations into one or more pairs of equations of the form $x = a$, $y = b$. If the process by which such a pair $x = a$, $y = b$ has been derived is *reversible* when x, y have the values a, b, we may at once conclude that a, b is one of the solutions sought; and the process is reversible if it consists of reversible steps.

The only new principle involved in all this is the following:

363 **Principle of substitution.** *If from the supposition that all the given equations are actually satisfied it follows that the values of a certain pair of expressions,* A *and* B, *are the same, the one expression may be substituted for the other in any of the equations.*

Example. Solve the pair $2x + y = 12$, (1)

$$y = 8. \qquad (2)$$

From the supposition that x and y actually have values which satisfy both equations it follows that the value of y in (2) and therefore in (1) is 8.

Substituting this value, 8, for y in (1), we obtain

$$2x + 8 = 12, \qquad (3)$$

whence $$x = 2. \qquad (4)$$

Therefore, if (1), (2) have any solution, that solution is $x = 2$, $y = 8$.

But conversely, $x = 2$, $y = 8$ is a solution of (1), (2), inasmuch as the process from (1), (2) to (4), (2) is *reversible.*

Thus, (3) follows from (4), and then (1) from (3), (2).

Note 1. This principle of substitution is a consequence of the several rules of equality, §§ 249, 253, 257, and of the general rule of equality, If $a = b$, and $b = c$, then $a = c$, § 261. **364**

Thus, we may prove our right to make the preceding substitution as follows:

If $y = 8$, then $y + 2x = 8 + 2x$, or $2x + 8 = 2x + y$, § 249.

And if $2x + 8 = 2x + y$, and $2x + y = 12$, then $2x + 8 = 12$, § 261.

Note 2. Of course this principle can be applied only when we have a right to suppose the given equations to be simultaneous. **365**

Thus, from $x = 2$ and $x = 3$ we cannot draw the absurd conclusion $2 = 3$, because we have no right to suppose $x = 2$, $x = 3$ simultaneous.

TRANSFORMATION THEOREMS

In view of what has just been said we may regard any correct application of the rules of reckoning to a pair of equations as a legitimate transformation of the pair; and if such a transformation be reversible, we may conclude that it leaves the solutions of the pair unchanged. **366**

Hence the following theorems, which hold good for equations in any number of unknown letters.

Theorem 1. *The solutions of a pair of equations remain unchanged when the transformations of* §§ 338, 339 *are applied to the equations separately.* **367**

For the solutions of the *individual* equations remain unchanged by such transformations.

Thus, the pair of equations $3x - 2y = 1$ and $y - 2x = 5$
has the same solutions as $3x - 2y = 1$ and $y = 5 + 2x$.

368 **Theorem 2.** *The pair of equations*

$$y = X, \quad f(x, y) = 0$$

has the same solutions as the pair

$$y = X, \quad f(x, X) = 0.$$

Here X denotes any expression in x alone (or a constant), $f(x, y)$ any expression in x and y, and $f(x, X)$ the result of substituting X for y in $f(x, y)$, § 280.

The theorem is merely a special case of the principle of substitution.

Thus, the pair of equations $y = x + 2$ and $3x - 2y = 1$
has the same solutions as $y = x + 2$ and $3x - 2(x + 2) = 1$.

369 **Theorem 3.** *The pair of equations*

$$A = B, \qquad C = D$$

has the same solutions as the pair

$$A + C = B + D, \quad C = D.$$

For $A = B$, $C = D$ has the same solutions as $A + C = B + C$, $C = D$, § 338, and $A + C = B + C$, $C = D$ has the same solutions as $A + C = B + D$, $C = D$, § 363.

Thus, the pair $x + y = 5$ and $x - y = 1$
has the same solution as $x + y + (x - y) = 5 + 1$ and $x - y = 1$,
and therefore as $2x = 6$ and $x - y = 1$.

370 **Corollary.** Before applying the theorem of § 369 we may, without changing their solutions, multiply both the given equations — that is, both members of each equation — by any *constants* we please, except 0. Hence

If k *and* l *denote any constants except* 0, *the pair of equations*

$$A = B, \qquad C = D$$

has the same solutions as the pair

$$kA \pm lC = kB \pm lD, \quad C = D.$$

Theorem 4. *When* A, B, *and* C *are integral, the pair of equations* **371**

$$AB = 0, \qquad C = 0$$

has the same solutions as the two pairs

$$A = 0,\ C = 0 \text{ and } B = 0,\ C = 0.$$

For $AB = 0$ has the same solutions as the two equations $A = 0$ and $B = 0$ jointly, § 341.

Hence the solutions of the pair $AB = 0$, $C = 0$ are the same as those of the pairs $A = 0$, $C = 0$ and $B = 0$, $C = 0$ jointly.

Thus, the solutions of $xy = 0$ and $x + y = 2$

are that of $x = 0$ and $x + y = 2$,

together with that of $y = 0$ and $x + y = 2$.

Equivalent systems. Two systems of simultaneous equations are said to be *equivalent* when their solutions are the same. **372**

Thus, the pair of equations $x + 2y = 5$, $2x + y = 4$ is equivalent to the pair $3x + y = 5$, $4x + 3y = 10$, both pairs having the same solution 1, 2.

Again, the pair $xy = 0$, $x + y = 2$ is equivalent to the two pairs $x = 0$, $x + y = 2$ and $y = 0$, $x + y = 2$.

ELIMINATION. SOLUTION OF A PAIR OF SIMPLE EQUATIONS

Elimination. To *eliminate* an unknown letter, as x, from a pair of equations is to derive from this pair an equation in which x does not occur. **373**

We proceed to explain the more useful methods of eliminating x or y from a pair of simple equations in x and y, and of deriving the solution of the equations from the result.

Method of substitution. This method is based on the theorem of § 368. **374**

Example. Solve

$$x + 3y = 3, \quad (1)$$

$$3x + 5y = 1. \quad (2)$$

Solving (1) for x in terms of y, $x = 3 - 3y$. (3)

Substituting $3 - 3y$ for x in (2), $3(3 - 3y) + 5y = 1$. (4)

Solving (4), $y = 2$. (5)

Substituting 2 for y in (3), $x = -3$. (6)

Hence the solution, and the only one, of (1), (2) is $x = -3$, $y = 2$.

For, by §§ 367, 368, the following pairs of equations have the same solution, namely: (1), (2); (3), (2); (3), (4); (3), (5); (6), (5); and the solution of (6), (5) is $x = -3$, $y = 2$.

The same conclusion may be drawn directly from § 362. For the process from (1), (2) to (5), (6) is reversible.

Verification. $-3 + 3 \cdot 2 \equiv 3$, (1) $\qquad 3(-3) + 5 \cdot 2 \equiv 1.$ (2)

Here (4) was obtained by eliminating x by *substitution.*

To eliminate an unknown letter, as x, *from a pair of equations by substitution, obtain an expression for* x *in terms of the other letter (or letters) from one of the equations, and then in the other equation replace* x *by this expression.*

375 The following example illustrates a special form of this method, called **elimination by comparison.**

Example. Solve
$$x + 5y = 7, \quad (1)$$
$$x + 6y = 8. \quad (2)$$

Solving *both* (1) and (2) for x in terms of y,
$$x = 7 - 5y, \quad (3) \qquad x = 8 - 6y. \quad (4)$$

Equating these two expressions for x, $7 - 5y = 8 - 6y.$ (5)

Solving (5) $y = 1.$ (6)

Substituting 1 for y in (3), $x = 2.$ (7)

Hence the solution of (1), (2) is $x = 2$, $y = 1$.

376 **Method of addition or subtraction.** This method is based on the theorem of §§ 369, 370.

Example. Solve
$$2x - 6y = 7, \quad (1)$$
$$3x + 4y = 4. \quad (2)$$

Multiply (1) by 3, $6x - 18y = 21.$ (3)

Multiply (2) by 2, $6x + 8y = 8.$ (4)

Subtract (4) from (3), $-26y = 13.$ (5)

Whence, $y = -1/2.$ (6)

Substitute $-1/2$ for y in (1), $2x - 6(-1/2) = 7.$ (7)

Whence, $x = 2.$ (8)

Hence the solution of (1), (2) is $x = 2$, $y = -1/2$.

For by §§ 367, 368, 370, the following pairs of equations have the same solution, namely: (1), (2); (1), (5); (1), (6); (7), (6); (8), (6); and the solution of (8), (6) is $x = 2$, $y = -1/2$.

Verification. $2 \cdot 2 - 6(-1/2) \equiv 7$, (1) $3 \cdot 2 + 4(-1/2) \equiv 4$. (2)

Here x was eliminated by *subtraction.*

We can also find the value of x directly from (1), (2) by eliminating y by *addition.* Thus,

Multiply (1) by 2,	$4x - 12y = 14.$	(9)
Multiply (2) by 3,	$9x + 12y = 12.$	(10)
Add (9) and (10),	$13x \quad = 26.$	(11)
Whence, as before,	$x \quad = 2.$	(12)

To eliminate an unknown letter, as x, *from a pair of simple equations by addition or subtraction, multiply the equations by numbers which will make the coefficients of* x *in the resulting equations equal numerically. Then subtract or add according as these coefficients have like or unlike signs.*

377 **Exceptional cases.** Let $A = 0$, $B = 0$ denote a pair of simple equations in x and y. The preceding sections, §§ 374, 376, show that this pair $A = 0$, $B = 0$ has *one solution and but one,* unless the expressions A and B are such that in eliminating x *we shall at the same time eliminate* y. This can occur in the following cases only.

1. If the expressions A and B are such that $A \equiv kB$, where k denotes a constant, we say that the equations $A = 0$ and $B = 0$ are *not independent.*

Evidently if $A \equiv kB$, every solution of $B = 0$ is a solution of $A = 0$, and *vice versa,* so that the pair $A = 0$, $B = 0$ has *infinitely many* solutions.

Thus, let $A = 2x + 6y - 10 = 0$ (1), and $B = x + 3y - 5 = 0$ (2).

Here $A \equiv 2B$, so that $A = 0$ and $B = 0$ are not independent. Observe that if to eliminate x we multiply (2) by 2 and subtract the result from (1) we at the same time eliminate y.

2. If A and B are such that $A \equiv kB + l$, where k and l denote constants, l not 0, we say that the equations $A = 0$ and $B = 0$ are *not consistent.*

In this case the pair $A = 0$, $B = 0$ has *no* solution; for any values of x, y that make $B \equiv 0$ will make $A \equiv l$, not $A \equiv 0$.

Thus, let $A = 2x + 6y - 9 = 0$ (3), and $B = x + 3y - 5 = 0$ (4).
Here $A \equiv 2B + 1$, so that $A = 0$ and $B = 0$ are not consistent. If we eliminate x from (3), (4) we shall at the same time eliminate y.

378 Formulas for the solution. We may reduce any given pair of simple equations in x, y to the form

$$ax + by = c, \quad (1) \qquad a'x + b'y = c', \quad (2)$$

where a, b, c, a', b', c' denote known numbers or expressions.

By § 377, the pair (1), (2) has one solution, and but one, unless a constant k can be found such that $a' = ka$ and $b' = kb$, and therefore $ab' - a'b = k(ab - ab) = 0$.

To obtain this solution, eliminate y and x independently by the method of subtraction, § 376. The results are

$$(ab' - a'b)x = b'c - bc', \quad (3) \qquad (ab' - a'b)y = ac' - a'c. \quad (4)$$

Therefore, if $ab' - a'b \neq 0$, the solution of (1), (2) is

$$x = \frac{b'c - bc'}{ab' - a'b}, \quad y = \frac{ac' - a'c}{ab' - a'b}. \quad (5)$$

These formulas are more easily remembered if written

$$\frac{x}{bc' - b'c} = \frac{y}{ca' - c'a} = \frac{-1}{ab' - a'b}. \quad (6)$$

Did we not know in advance that the pair (1), (2) has a solution when $ab' - a'b \neq 0$, the argument here given would only prove that *if* the pair (1), (2) has *any* solution, it is (5).

EXERCISE VII

Solve the following pairs of equations for x and y.

1. $\begin{cases} x + y = 62, \\ x - y = 12. \end{cases}$ 2. $\begin{cases} 6x - 5y = 25, \\ 4x - 3y = 19. \end{cases}$ 3. $\begin{cases} 45x - 13y = 161, \\ 18x + 11y = 32. \end{cases}$

4. $\begin{cases} x - 3 = 7 - x, \\ 8x - 3y - 61 = 0. \end{cases}$ 5. $\begin{cases} 12x = 9 - 10y, \\ 8y = 7 - 9x. \end{cases}$ 6. $\begin{cases} 2y - 3x = 0, \\ 5x - 3y - 2 = 0. \end{cases}$

7. $\begin{cases} x/.3 + 5y = 3\frac{1}{2}, \\ 5x + 3y = 1.65. \end{cases}$

8. $\begin{cases} 2(2x+3y) = 3(2x-3y)+10, \\ 4x - 3y = 4(6y - 2x) + 3. \end{cases}$

9. $\begin{cases} (x+2)(y+1) = (x-5)(y-1), \\ x(4+y) = -y(8-x). \end{cases}$

10. $\begin{cases} ax + by = a^2 + 2a + b^2, \\ bx + ay = a^2 + 2b + b^2. \end{cases}$

11. $\begin{cases} ax + by = c, \\ px = qy. \end{cases}$

12. $\begin{cases} (a-b)x + (a+b)y = 2(a^2-b^2), \\ (a+b)x + (a-b)y = 2(a^2+b^2). \end{cases}$

13. $\begin{cases} \dfrac{x+y}{3} + \dfrac{y-x}{2} = 5, \\ \dfrac{x}{2} + \dfrac{x+y}{9} = 7 \end{cases}$

14. $\begin{cases} \dfrac{x-y}{4} - \dfrac{x+2y-5}{6} = \dfrac{y-3}{4} - \dfrac{y+2x-5}{6}, \\ 5x - 2y + 6 = 0. \end{cases}$

15. $\begin{cases} \dfrac{x}{a} + \dfrac{y}{b} = \dfrac{1}{c}, \\ \dfrac{x}{a'} - \dfrac{y}{b'} = \dfrac{1}{c'}. \end{cases}$

16. $\begin{cases} \dfrac{x}{a} + \dfrac{y}{b} = 1 + x, \\ \dfrac{x}{b} + \dfrac{y}{a} = 1 + y. \end{cases}$

17. Show that the following equations are inconsistent.

$$1\tfrac{1}{2}x - 2\tfrac{1}{2}y = 10, \quad 6x - 10y = 15.$$

18. In Ex. 15 assign values to a, b, c, a', b', c' for which the equations are (1) not consistent, (2) not independent.

PAIRS OF EQUATIONS NOT OF THE FIRST DEGREE WHOSE SOLUTIONS CAN BE FOUND BY SOLVING PAIRS OF SIMPLE EQUATIONS

379 A pair of equations which are not of the first degree with respect to x and y may yet be of the first degree with respect to a certain pair of functions of x and y. We can then solve the equations for this pair of functions, and from the result it is often possible to derive the values of x and y themselves.

Example 1. Solve $\dfrac{2}{x} + \dfrac{5}{3y} = 1, \quad \dfrac{9}{x} + \dfrac{10}{y} = 5.$

Both equations are of the first degree with respect to $1/x$ and $1/y$. Solving for $1/x$ and $1/y$, we find $1/x = 1/3$, $1/y = 1/5$. Hence $x = 3$, $y = 5$.

Example 2. Solve $3x + \frac{y}{x} = 6,\ 7x - \frac{2y}{x} = 1$.

Solving for x and y/x, we find $x = 1$, $y/x = 3$. Hence $x = 1$, $y = 3$.

380 Given a pair of equations reducible to the form $AB = 0$, $A'B' = 0$, where A, B, A', B' denote integral expressions of the first degree in x and y. It follows from the theorem of § 371 that all the solutions of this pair can be obtained by solving the four pairs of simple equations $A = 0$, $A' = 0$; $A = 0$, $B' = 0$; $B = 0$, $A' = 0$; $B = 0$, $B' = 0$.

Example. Solve

$$x^2 - 2xy = 0, \tag{1}$$

$$(x + y - 1)(2x + y - 3) = 0. \tag{2}$$

This pair is equivalent to the four pairs

$$x = 0, \quad x + y - 1 = 0, \tag{3}$$

$$x = 0, \quad 2x + y - 3 = 0, \tag{4}$$

$$x - 2y = 0, \quad x + y - 1 = 0, \tag{5}$$

$$x - 2y = 0, \quad 2x + y - 3 = 0. \tag{6}$$

Solving these four pairs (3), (4), (5), (6) we obtain the four solutions of (1), (2), namely: $x, y = 0, 1;\ 0, 3;\ 2/3, 1/3;\ 6/5, 3/5$.

381 And, in general, if $ABC \cdots$ and $A'B'C' \cdots$ denote products of m and n integral factors of the first degree in x and y, all the solutions of the pair of equations $ABC \cdots = 0$, $A'B'C' \cdots = 0$ can be found by solving the mn pairs of simple equations obtained by combining each factor of the first product equated to 0 with each factor of the second likewise equated to 0.

If all these pairs of simple equations are both independent and consistent, we thus obtain mn solutions, that is, *the number of solutions of the given equations is the product of their degrees.*

EXERCISE VIII

Solve the following pairs of equations.

1. $\frac{7}{2x} + \frac{1}{3y} = 0, \quad \frac{3}{x} + \frac{14}{y} + 3 = 0.$

2. $10x + \frac{6}{y} = 5, \quad 15x + \frac{10}{y} = 8.$

3. $\dfrac{y}{x} = \dfrac{2(3-y)}{x} + \dfrac{3}{2}, \quad \dfrac{y+3}{x} = \dfrac{3y-5}{x} + 1.$

4. $xy = 0, \quad (x + 2y - 1)(3x - y + 2) = 0.$

5. $xy - y = 0, \quad 3x - 8y + 5 = 0.$

6. $x(x - y)(x + y) = 0, \quad x + 2y - 5 = 0.$

7. $(x - 1)(y - 2) = 0, \quad (x - 2)(y - 3) = 0.$

8. $y^2 = (x - 1)^2, \quad 2x + 3y - 7 = 0.$

9. $(2x + y)^2 = (x - 3y + 5)^2, \quad (x + y)^2 = 1.$

10. $(x - 5y + 8)(x + 3y + 5) = 0, \quad (2x + y + 5)(5x + 2y - 14) = 0.$

GRAPHS OF SIMPLE EQUATIONS IN TWO VARIABLES

Graph of a pair of values of x and y. It is convenient to represent *pairs of values* of two variables, as x and y, by points in a plane. **382**

In the plane select as *axes of reference* two fixed straight lines, $X'OX$ and $Y'OY$, which meet at right angles at the point O, called the *origin*; and choose some convenient unit for measuring lengths.

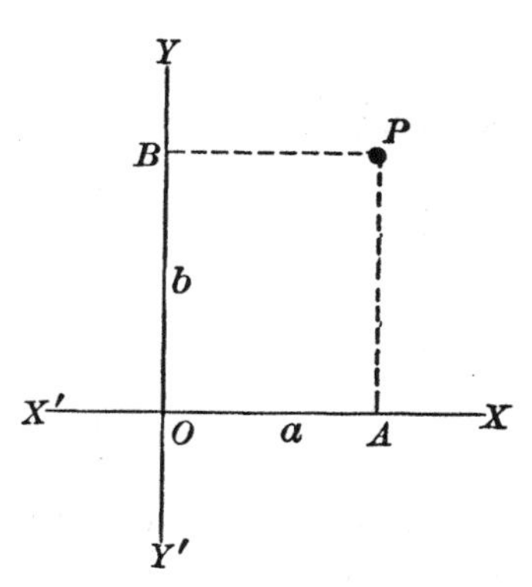

Then if the given pair of values be $x = a$, $y = b$, proceed as follows:

On $X'OX$ and to the right or left of O, according as a is positive or negative, measure off a segment, OA, whose length is $|a|$, the numerical value of a.

Similarly on $Y'OY$ and above or below O, according as b is positive or negative, measure off a segment, OB, whose length is $|b|$.

Then through A and B draw parallels to $Y'OY$ and $X'OX$ respectively. We take P, the point in which these parallels intersect, as the point-picture, or *graph*, of the pair of values $x = a$, $y = b$.

It is convenient to represent both the value-pair $x = a$, $y = b$ and its graph P by the symbol (a, b).

We call the number a, or one of the equal line segments OA or BP, the *abscissa* of P; and b, or one of the equal segments OB or AP, the *ordinate* of P. And we call the abscissa and ordinate together the *coördinates* of P.

We also call $X'OX$ the *x-axis* or the *axis of abscissas*, and $Y'OY$ the *y-axis* or the *axis of ordinates*.

Observe that this method brings the value-pairs of x, y into one-to-one correspondence, § 2, with the points of the plane; that is, for each value-pair (a, b) there is one point P, and reciprocally for each point P there is one value-pair (a, b) found by measuring the distances of P from $Y'OY$ and $X'OX$ respectively, and giving them their appropriate signs.

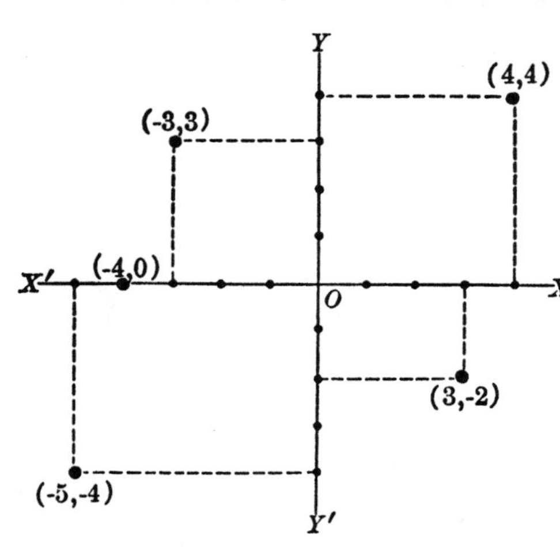

In particular, the graph of $(0, 0)$ is the origin, that of $(a, 0)$ is a point on the x-axis, and that of $(0, b)$ is a point on the y-axis.

Example. Plot the value-pairs $(4, 4)$, $(-3, 3)$, $(-4, 0)$, $(-5, -4)$, $(3, -2)$.

Carrying out the construction just described for each value-pair in turn, we obtain their graphs as indicated in the accompanying figure.

Notice particularly how the position of the graph depends on the *signs* of the coördinates.

383 The graph of an equation in x and y. If, as is commonly the case, a given equation in x and y has infinitely many real solutions, there will usually be a definite curve which contains the graphs of all these solutions and no other points. We call this curve the *graph of the equation*.

But the graph of an equation may consist of more than one curve. Observe that we here include straight lines among curves.

384 **Theorem.** *The graph of every simple equation in one or both of the letters* x *and* y *is a straight line.*

On this account simple equations are often called *linear* equations.

The student may readily convince himself of the truth of the theorem by selecting some particular equation and "plotting" a number of its solutions.

Thus, take the equation $y = -2x + 4$.

When $x = 0, \quad 1, \quad 2, \quad 3, \quad \cdots$

we have $y = 4, \quad 2, \quad 0, \quad -2, \quad \cdots$

And plotting these value-pairs (0, 4), (1, 2), (2, 0), (3, -2) $\cdots$ as in the accompanying figure, we find that their graphs all lie in the same straight line.

We may *prove* the theorem as follows:

1. When the equation has the form $x = a$, or $y = b$.

Example. Find the graph of $x = 2$.

This equation is satisfied by the value 2 of x and *every* value of y, § 356.

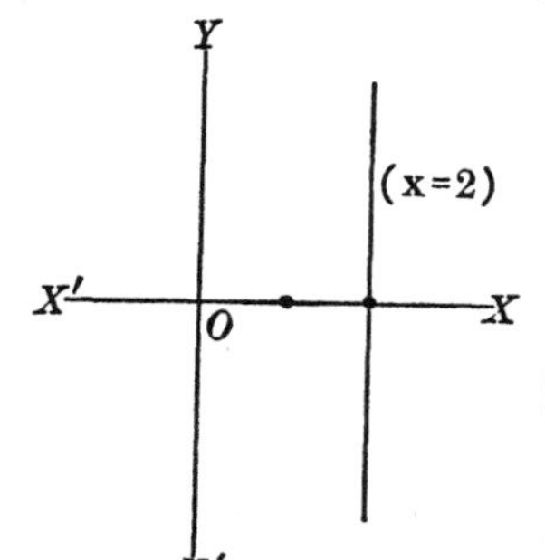

Hence the graph is a parallel to the y-axis at the distance 2 to its right. For this line contains all points whose abscissas are 2, and such points only.

And so, in general, the graph of $x = a$ is a parallel to the y-axis at the distance $|a|$ to the right or left according as a is positive or negative; and the graph of $y = b$ is a parallel to the x-axis at the distance $|b|$ above or below according as b is positive or negative.

In particular, the graph of $y = 0$ is the x-axis, and that of $x = 0$ is the y-axis.

2. When the equation has the form $y = mx$.

Example. Find the graph of $y = 2x$.

The graph is the right line which passes through the origin (0, 0) and the point (1, 2); for this line contains every point whose ordinate is twice its abscissa, and such points only.

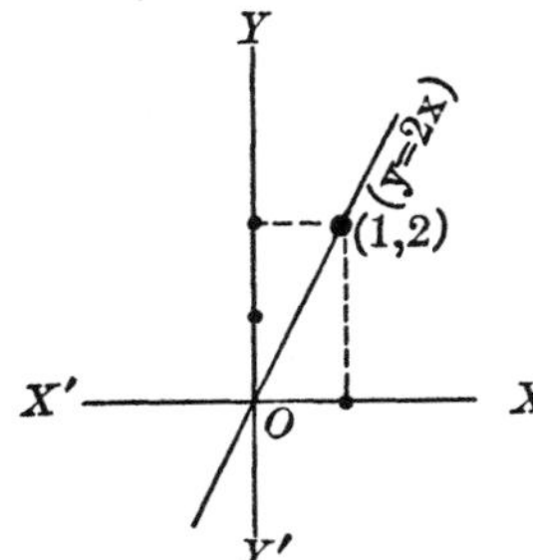

And so, in general, the graph of $y = mx$ is the right line which passes through the origin and the point $(1, m)$.

3. When the equation has the form $y = mx + c$.

Example. Find the graph of $y = 2x + 3$.

Evidently we shall obtain the graph of this equation if we increase the ordinate of every point of the graph of $y = 2x$ by 3. But that comes to the same thing as shifting the line $y = 2x$ upward parallel to itself until its point of intersection with the y-axis is 3 units above the origin.

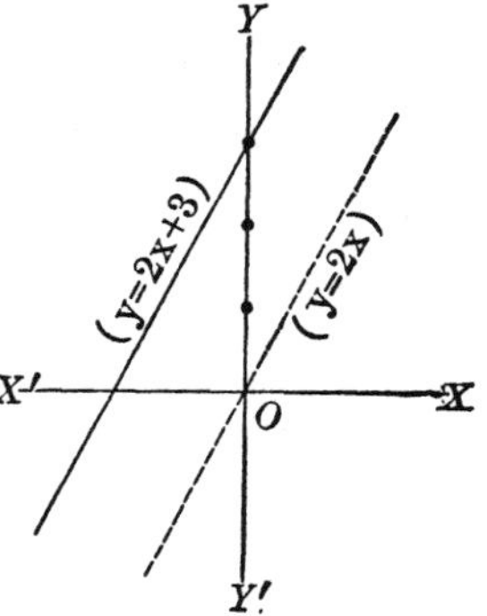

And so, in general, the graph of $y = mx + c$ is a right line parallel to the graph of $y = mx$ and meeting the y-axis at the distance $|c|$ from the origin, above or below, according as c is positive or negative.

385 **To find this line.** As any two of its points suffice to determine a right line, we may find the graph of any equation, $ax + by + c = 0$, as in the following example.

Example. Plot the graph of $3x + y - 6 = 0$.

First, when $y = 0$, then $x = 2$. Second, when $x = 0$, then $y = 6$.

Hence we have only to plot the points (2, 0) and (0, 6), that is, *the points where the line will meet the axes*, and draw the line which these points determine (see figure in § 386).

This method fails when the equation has one of the forms $x = a$, $y = b$, $y = mx$. We then find the line by the methods explained in § 384, 1 and 2.

Graph of the solution of a pair of simultaneous simple equations. **386**
This is the point of intersection of the two lines which are the graphs of the equations themselves; for this point, and this point only, is the graph of a solution of *both* equations.

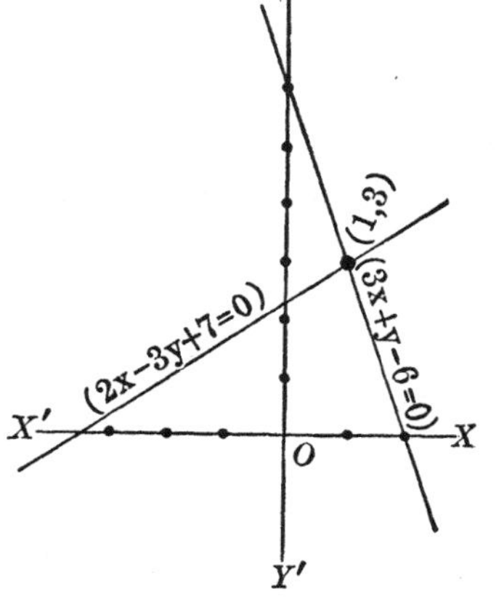

Thus, the solution of $2x - 3y + 7 = 0$ (1), and $3x + y - 6 = 0$ (2) is $x = 1$, $y = 3$. And, as the figure shows, the graphs of (1) and (2) intersect at the point (1, 3).

387 When the given equations are *not consistent*, § 377, 2, their graphs are lines which have *no* point in common, that is, *parallel* lines; when the equations are *not independent*, § 377, 1, their graphs are lines which have *all* their points in common, that is, *coincident lines.*

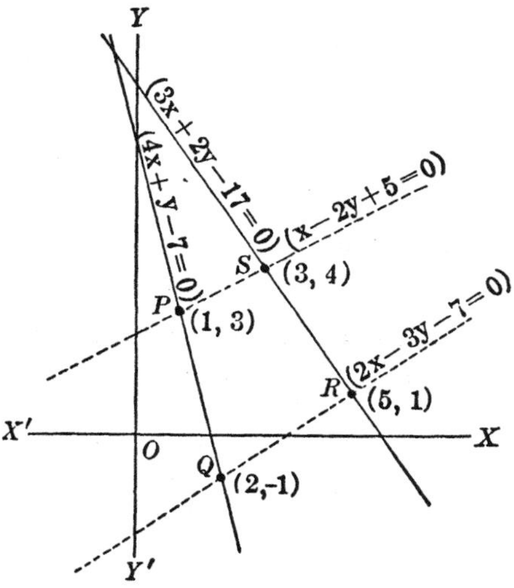

Thus, the equations $y = 2x$, $y = 2x + 3$ are not consistent, and the graphs of these equations, § 384, 3, are parallel lines.

Again, the equations $y = 2x$, $3y = 6x$, which are not independent, have the same graph.

388 The graph of an equation of the form $AB = 0$ consists of the graphs of $A = 0$ and $B = 0$ jointly; for the solutions of $AB = 0$ are those of $A = 0$ and $B = 0$ jointly, §§ 341, 346.

Example. Find the graphs of the equations

$$(4x + y - 7)(3x + 2y - 17) = 0, \quad (1)$$

$$(x - 2y + 5)(2x - 3y - 7) = 0, \quad (2)$$

and the graphs of their solutions.

The graph of (1) consists of the lines PQ and RS which are the graphs of $4x + y - 7 = 0$ and $3x + 2y - 17 = 0$ respectively.

The graph of (2) consists of the lines PS and QR which are the graphs of $x - 2y + 5 = 0$ and $2x - 3y - 7 = 0$ respectively.

The points P, Q, R, S in which the pair PQ, RS meets the pair PS, QR are the graphs of the solutions of (1), (2), namely, $(1, 3)$, $(2, -1)$, $(5, 1)$, $(3, 4)$.

389 Graph of an equation of higher degree in x and y. We find a number of the solutions of the equation, plot these solutions, and then with a free hand draw a curve which will pass through all the points thus found. By taking the solutions "near" enough together, we can in this way obtain a curve which differs from the true graph as little as we please.

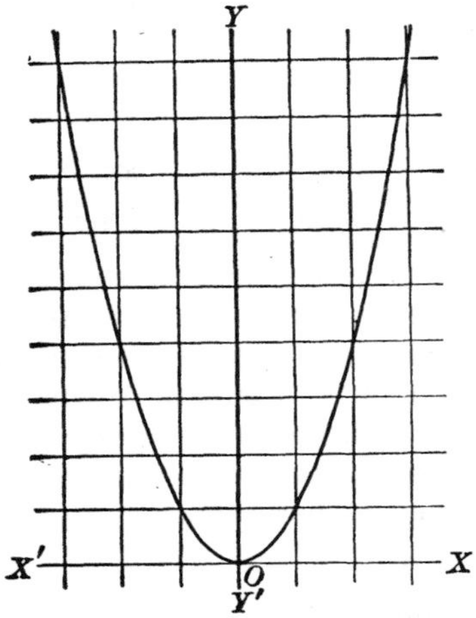

In work of this kind it is convenient to use paper ruled into small squares, as in the accompanying figure.

Example. Find the graph of the equation $y = x^2$.

When	$x = 0,$	$1,$	$2,$	$3,$	$4, \cdots$
we have	$y = 0,$	$1,$	$4,$	$9,$	$16, \cdots$
And when		$-1,$	$-2,$	$-3,$	$-4, \cdots$
we have		$1,$	$4,$	$9,$	$16, \cdots$

Taking the side of a square as the unit of length, plot the corresponding points $(0, 0)$, $(1, 1)$, $(2, 4) \cdots (-1, 1)$, $(-2, 4) \cdots$. A few of them suffice to indicate the general character of the graph, the curve in the figure, except between $x = -1$ and $x = +1$.

It lies wholly above the x-axis, extending upward indefinitely; and it lies symmetrically with respect to the y-axis, the same value of y corresponding to $x = a$ and $x = -a$.

When	$x = \pm 1/2,$	$\pm 1/4, \cdots$
we have	$y = \quad 1/4,$	$1/16, \cdots$

and plotting one or two of the corresponding points, we find that the graph touches the x-axis.

EXERCISE IX

1. Plot the following pairs of values of x and y.

$(0, 0)$, $(5, 0)$, $(0, -7)$, $(6, 2)$, $(-7, -1)$, $(-4, 3)$, $(5, -9)$.

2. Find the graphs of the following equations.

$x = 0$, $y = 0$, $2y + 7 = 0$, $3y + x = 0$, $x + y + 5 = 0$, $7x + 3y - 18 = 0$, $3x - 4y = 24$.

3. Find the graphs of the following.

$xy = 0$, $(x + y - 3)(x - 2y) = 0$, $x^2 - 1 = 0$, $x^2 = 4y^2$, $x^2 + y^2 = 0$.

4. Find the solutions of the following pairs of equations by the graphical method and verify the results algebraically.

$$(1) \begin{cases} x + y - 3 = 0, \\ x - 2y = 0. \end{cases} \qquad (2) \begin{cases} 3y + 2x + 19 = 0, \\ 2y - 3x + 4 = 0. \end{cases}$$

5. Do the same with each of the following pairs.

$$(1) \begin{cases} (x - 4y + 6)(x + 3y + 6) = 0, \\ (3x + 2y - 10)(2x - y + 5) = 0. \end{cases} \qquad (2) \begin{cases} (y - x - 2)x = 0, \\ (y - x + 2)y = 0. \end{cases}$$

6. Find the graphs of the following two equations.

$$y = -(x + 1)^2, \qquad y = x^3.$$

SYSTEMS OF SIMPLE EQUATIONS WHICH INVOLVE MORE THAN TWO UNKNOWN LETTERS

Method of solving a system of n simple equations in n unknown letters. A pair of equations in three unknown letters will ordinarily have infinitely many solutions. **390**

Thus, the pair $x = 2z$, $y = z + 1$ has infinitely many solutions; for both equations are satisfied if we assign any value whatsoever, as b, to z and the values $2b$ and $b + 1$ to x and y.

But a system of *three* simple equations in three unknown letters ordinarily has one, and but one, solution, which may be obtained as in the following example. **391**

Example. Solve the system of equations

$$3x - 2y + 4z = 13, \qquad (1)$$

$$2x + 5y - 3z = -9, \qquad (2)$$

$$6x + 3y + 2z = 7. \qquad (3)$$

Eliminate z between two pairs of these equations, thus:

Multiply (1) by 3,	$9x - 6y + 12z = 39$	(4)
Multiply (2) by 4,	$8x + 20y - 12z = -36$	(5)
Add	$17x + 14y = 3$	(6)
Again, (1) is	$3x - 2y + 4z = 13$	(7)
Multiply (3) by 2,	$12x + 6y + 4z = 14$	(8)
Subtract (7) from (8),	$9x + 8y = 1$	(9)

Eliminate y between the resulting equations (6), (9), thus:

Multiply (6) by 4,	$68x + 56y = 12$	(10)
Multiply (9) by 7,	$63x + 56y = 7$	(11)
Subtract (11) from (10),	$5x = 5$	(12)

Hence $x = 1.$

Substituting $x = 1$ in (9), we find $y = -1.$

Substituting $x = 1$, $y = -1$ in (1), we find $z = 2.$

Therefore, § 362, if (1), (2), (3) has any solution, it is $x = 1$, $y = -1$, $z = 2$. But the process by which we have derived $x = 1$, $y = -1$, $z = 2$ from (1), (2), (3) is *reversible*. In fact, it may readily be traced backward step by step. Hence $x = 1$, $y = -1$, $z = 2$ is the solution of (1), (2), (3).

We may also prove as follows that $x = 1$, $y = -1$, $z = 2$ is the solution of (1), (2), (3).

It is evident by § 368 that $x = 1$, $y = -1$, $z = 2$ is the solution of (12), (9), (1). We therefore have only to prove that the system (12), (9), (1) has the same solution as the given system (1), (2), (3).

Let us represent the equations (1), (2), (3), with the known terms transposed to the first members, thus:

$$A = 0, \quad (1) \qquad B = 0, \quad (2) \qquad C = 0. \quad (3)$$

It will then follow from the manner in which (9) and (12) were derived, that we may express the equations (1), (9), (12) thus:

$$A = 0, \quad (1) \qquad -A + 2C = 0, \quad (9) \qquad 19A + 16B - 14C = 0. \quad (12)$$

Evidently any set of values of x, y, z that makes $A \equiv 0$, $B \equiv 0$, $C \equiv 0$ will make $A \equiv 0$, $-A + 2C \equiv 0$, $19A + 16B - 14C \equiv 0$.

Conversely, when $A \equiv 0$ and $-A + 2C \equiv 0$, then $C \equiv 0$; and when also $19A + 16B - 14C \equiv 0$, then $B \equiv 0$.

Hence the system (1), (2), (3) has the same solution as the system (1), (9), (12), namely, $x = 1$, $y = -1$, $z = 2$.

392 In the case just considered, from the given system of *three* equations in the *three* unknown letters, x, y, z, we derived a system of *two* equations in *two* letters, x, y, and then from this system, a *single* equation in *one* letter, x.

And, in general, if we start with a system of n simple equations in n unknown letters and take $n-1$ of these steps, we shall arrive at a *single* equation in *one* of the letters, as x, of the form $ax - b = 0$.

Then, *unless* $a = 0$, the system has *one*, and *but one*, solution, in which the value of x is b/a and the values of the other unknown letters may be found by successive substitutions in the equations obtained in the process. This may always be proved as in the example.

On the other hand, if $a = 0$ the system ordinarily has *infinitely many* solutions when $b = 0$, and *no* solution when $b \neq 0$. This will be proved in § 394.

393 A much less laborious method of solving a system of simple equations is given in the chapter on *determinants*. In certain cases labor may be saved by special devices.

Example. Solve

$$x + y + z = 8, \qquad (1)$$

$$x + y + u = 12, \qquad (2)$$

$$x + z + u = 14, \qquad (3)$$

$$y + z + u = 14, \qquad (4)$$

Add (1), (2), (3), (4), $3x + 3y + 3z + 3u = 48$.

Hence

$$x + y + z + u = 16. \qquad (5)$$

And subtracting each of the equations (4), (3), (2), (1) in turn from (5), we obtain $x = 2$, $y = 2$, $z = 4$, $u = 8$.

394 **Exceptional cases.** Let $A = 0$, $B = 0$, $C = 0$ denote a system of simple equations in x, y, z, and, as in § 392, let $ax - b = 0$ denote the equation obtained by eliminating y and z.

1. If $a = 0$ and $b = 0$, it will be found that one of the functions A, B, C may be expressed in terms of the other two, thus: $A \equiv kB + lC$, where k and l denote constants. We then say that the equations $A = 0$, $B = 0$, $C = 0$ are *not independent*.

From the identity $A \equiv kB + lC$ it follows that every solution of $B = 0$ and $C = 0$ is a solution of $A = 0$. Hence if $B = 0$ and $C = 0$ are consistent, § 377, 2, the three equations $A = 0$, $B = 0$, $C = 0$ will have *infinitely many* solutions.

Thus, consider the system of equations

$$A = 3x - 2y + 4z - 13 = 0, \quad (1)$$

$$B = 2x + 5y - 3z + 9 = 0, \quad (2)$$

$$C = 7x + 8y - 2z + 5 = 0. \quad (3)$$

Eliminating z between (1) and (2),

$$3A + 4B \equiv 17x + 14y - 3 = 0. \quad (4)$$

Eliminating z between (1) and (3),

$$A + 2C \equiv 17x + 14y - 3 = 0. \quad (5)$$

Eliminating y between (4) and (5),

$$2A + 4B - 2C \equiv \quad 0 \cdot x \quad - 0 = 0. \quad (6)$$

Here the final equation $ax - b = 0$ has the form $0 \cdot x - 0 = 0$, and in deriving it, we find that the expressions A, B, C are connected by the identity $2A + 4B - 2C \equiv 0$, or $C \equiv A + 2B$.

And, in fact, we see on examining (1), (2), (3) that C may be obtained by multiplying B by 2 and adding the result to A.

Hence the system (1), (2), (3) has infinitely many solutions.

2. If $a = 0$ and $b \neq 0$, it will be found that one of the functions A, B, C may be expressed in terms of the other two, thus:

$$A \equiv kB + lC + m,$$

where k, l, m denote constants, m not 0. We then say that the equations $A = 0$, $B = 0$, $C = 0$ are *not consistent.*

From the identity $A \equiv kB + lC + m$ it follows that $A = 0$, $B = 0$, $C = 0$ have *no* solution. For any values of x, y, z that make $B \equiv 0$ and $C \equiv 0$ will make $A \equiv m$, not $A \equiv 0$.

Thus, consider the system of equations

$$A = 3x - 2y + 4z - 13 = 0, \quad (1)$$

$$B = 2x + 5y - 3z + 9 = 0, \quad (2)$$

$$C = 7x + 8y - 2z + 6 = 0. \quad (3)$$

Eliminating z and y as above, we obtain

$$2A + 4B - 2C \equiv 0 \cdot x - 2 = 0.$$

Hence the final equation $ax - b = 0$ has the form $0 \cdot x - 2 = 0$, and A, B, C are connected by the identity $C \equiv A + 2B + 1$. And, in fact, on examining (1), (2), (3) we find this to be the case.

Hence the system (1), (2), (3) has no solution.

395 **Systems of simple equations in general.** From the preceding discussion we may draw the conclusion:

Ordinarily a system of m *simple equations in* n *unknown letters has one solution when* m = n, *infinitely many solutions when* m < n, *no solution when* m > n.

Whenever exceptions to this rule occur, two or more of the equations are connected by identical relations of the kind described in §§ 377, 394.

In particular, a system of *three* simple equations, $A = 0$, $B = 0$, $C = 0$, in *two* unknown letters, x, y, has a solution when, and only when, A, B, C are connected by an identity of the form $A \equiv kB + lC$ and $B = 0$, $C = 0$ are consistent.

Thus, the system $x - y = 1$ (1), $x + y = 7$ (2), $3x - y = 10$ (3) has no solution; for the solution of (1), (2), namely, $x = 4$, $y = 3$, does not satisfy (3).

On the other hand, the system $x - y = 1$ (1), $x + y = 7$ (2), $3x - y = 9$ (4) has a solution; for (4) is satisfied by $x = 4$, $y = 3$. But observe that $3x - y - 9 \equiv 2(x - y - 1) + (x + y - 7)$.

Let the student draw the graphs of (1), (2), (4). He will find that they meet in a common point.

EXERCISE X

Solve the following systems of equations.

1. $\begin{cases} x + y = 11, \\ y + z = 13, \\ z + x = 12. \end{cases}$

2. $\begin{cases} x + y + z = 1, \\ x + 2y + 3z = 4, \\ x + 3y + 7z = 13. \end{cases}$

3. $\begin{cases} x + 2y - 3z = 3, \\ 3x - 5y + 7z = 19, \\ 5x - 8y - 11z = -13. \end{cases}$

4. $\begin{cases} 5x - 2y = -33, \\ x + y - 7z = 13, \\ x + 3y = -10. \end{cases}$

5. $\begin{cases} x + 2y - 4z = 11, \\ 2x - 3y = 0, \\ y - 4z = 0. \end{cases}$

6. $\begin{cases} 3x - 5 = 2(x - 2), \\ (x+1)(y-1) = (x+2)(y-2) + 5, \\ 2x + 3y + z = 6. \end{cases}$

7. $\begin{cases} \frac{1}{x} + \frac{1}{y} - \frac{6}{z} = 9, \\ \frac{1}{x} - \frac{1}{y} + \frac{4}{z} = 5, \\ \frac{3}{y} - \frac{2}{x} - \frac{1}{z} = 4. \end{cases}$

8. $\begin{cases} y + z + u = 4, \\ x + z + u = 3, \\ x + y + u = 1, \\ x + y + z = 10. \end{cases}$

9. $\begin{cases} 4x - 3z + u = 9, \\ 5y + z - 4u = 17, \\ 3y + u = 12, \\ x + 2y + 3u = 8. \end{cases}$

10. $\begin{cases} cx + by = l, \\ by + az = m, \\ az + cx = n. \end{cases}$

11. $\begin{cases} lx = my = nz, \\ ax + by + cz = d. \end{cases}$

12. $\begin{cases} 2x = 3y = 6z, \\ (x+2y+z-16)(3x-2y+20)=0. \end{cases}$

Show that the following systems are not independent and find the identity which connects the equations of each system.

13. $\begin{cases} x - y = 3, \\ y - z = -5, \\ z - x = 2. \end{cases}$

14. $\begin{cases} 3x - 8y + 7z = 10, \\ 2x + 5y - 3z = 12, \\ 16x + 9y - z = 80. \end{cases}$

PROBLEMS

396 The following problems can be solved by means of simple equations in two or more unknown letters, as $x, y, \cdots$. How many of these letters it is best to employ in any case will depend on the conditions of the problem. But when a choice has been made of the unknown numbers of the problem which $x, y, \cdots$ are to represent, and the remaining unknown numbers, if any, have been expressed in terms of these letters, it will be found that the conditions of the problem still unused will yield just as many independent and consistent equations connecting $x, y, \cdots$ as there are letters $x, y, \cdots$. In fact, if they gave more than this number of equations, the problem would have no solution; if less, an infinite number of solutions, § 395.

The remark of § 352 on the restrictions which the nature of the problem may impose on the character of the unknown numbers applies here also.

Example 1. In a certain number of three digits, the second digit is equal to the sum of the first and third, the sum of the second and third digits is 8, and if the first and third digits be interchanged, the number is increased by 99. Find the number.

Let $x =$ hundreds digit, $y =$ tens digit, $z =$ units digit.

Then the number is $100x + 10y + z$.

But, by the conditions of the problem, we have

$$x + z = y, \tag{1}$$

$$y + z = 8, \tag{2}$$

$$100z + 10y + x = 100x + 10y + z + 99. \tag{3}$$

Solving (1), (2), (3) we find $x = 2$, $y = 5$, $z = 3$.
Hence the number is 253.

Example 2. After walking a certain distance a pedestrian rests for 30 minutes. He then continues his journey, but at $\frac{7}{8}$ of his original rate, and on reaching his destination finds that he has accomplished the entire distance, 20 miles, in 6 hours. If he had walked 4 miles further at the original rate and then rested as before, the journey would have taken $5\frac{6}{7}$ hours. What was his original rate, and how far from the starting point did he rest?

Let $x =$ original rate in miles per hour, and let $y =$ number of miles from starting point to resting place.

Expressing in terms of x and y the number of hours taken by (1) the actual journey, (2) the supposed journey, we have

$$\frac{y}{x} + \frac{1}{2} + \frac{20 - y}{7x/8} = 6, \quad (1) \qquad \frac{y+4}{x} + \frac{1}{2} + \frac{16 - y}{7x/8} = 5\tfrac{6}{7}. \quad (2)$$

Solving (1), (2) for y/x and $1/x$, we find $y/x = 3/2$, $1/x = 1/4$.
Hence $x = 4$, $y = 6$.

Example 3. Two vessels, A and B, contain mixtures of alcohol and water. A mixture of 3 parts from A and 2 parts from B will contain 40% of alcohol; and a mixture of 1 part from A and 2 parts from B will contain 32% of alcohol. What are the percentages of alcohol in A and B respectively?

Let x and y denote the percentages of alcohol in A and B respectively. Then by the given conditions we shall have

$$\frac{3x}{5} + \frac{2y}{5} = \frac{40}{100}, \quad (1) \qquad \frac{x}{3} + \frac{2y}{3} = \frac{32}{100}. \quad (2)$$

Solving (1), (2) we find $x = 52/100$, or 52%, $y = 22/100$, or 22%.

EXERCISE XI

1. Find three numbers whose sum is 20 and such that (1) the first plus twice the second plus three times the third equals 44 and (2) twice the sum of the first and second minus four times the third equals -14.

2. The sum of three numbers is 51. If the first number be divided by the second, the quotient is 2 and the remainder 5; but if the second number be divided by the third, the quotient is 3 and the remainder 2. What are the numbers?

3. Find a number of two digits from the following data: (1) twice the first digit plus three times the second equals 37; (2) if the order of the digits be reversed, the number is diminished by 9.

4. A owes $5000 and B owes $3000. A could pay all his debts if besides his own money he had $\frac{2}{3}$ of B's; and B could pay all but $100 of his debts if besides his own money he had $\frac{1}{2}$ of A's. How much money has each?

5. Find the fortunes of three men, A, B, and C, from the following data: A and B together have p dollars; B and C, q dollars; C and A, r dollars. What conditions must p, q, and r satisfy in order that the solution found may be an admissible one?

6. A sum of money at simple interest amounts to $2556.05 in 2 years and to $2767.10 in 4 years. What is the sum of money, and what the rate of interest?

7. A man invested a certain sum of money partly in 4% bonds at par, partly in 5% bonds at 110, and his income from the investment was $650. If the 4% bonds had been at 80 and the 5% bonds at 110, his income from the investment would have been $100 greater. How much did he invest?

8. Find the area of a rectangle from the following data: if 6 inches be added to its length and 6 inches to its breadth, the one becomes $\frac{3}{2}$ of the other, and the area of the rectangle is increased by 84 square inches.

9. A gave B as much money as B had; then B gave A as much money as A had left; finally A gave B as much money as B then had left. A then had \$16 and B \$24. How much had each originally?

10. A and B together can do a certain piece of work in $5\frac{1}{7}$ days; A and C, in $4\frac{4}{5}$ days. All three of them work at it for 2 days when A drops out and B and C finish it in $1\frac{9}{17}$ days. How long would it take each man separately to do the piece of work?

11. Two points move at constant rates along the circumference of a circle whose length is 150 feet. When they move in opposite senses they meet every 5 seconds; when they move in the same sense they are together every 25 seconds. What are their rates?

12. It would take two freight trains whose lengths are 240 yards and 200 yards respectively 25 seconds to pass one another when moving in opposite directions; but were the trains moving in the same direction, it would take the faster one $3\frac{3}{4}$ minutes to pass the slower one. What are the rates of the trains in miles per hour?

13. Two steamers, A and B, ply between the cities C and D which are 200 miles apart. The steamer A can start from C 1 hour later than B, overtake B in 2 hours, and having reached D and made a 4 hours' wait there, on its return trip meet B 10 miles from D. What are the rates of A and B?

14. In a half-mile race A can beat B by 20 yards and C by 30 yards. By how many yards can B beat C?

15. A and B run two 440-yard races. In the first race A gives B a start of 20 yards and beats him by 2 seconds. In the second race A gives B a start of 4 seconds and beats him by 6 yards. What are the rates of A and B?

16. Two passengers together have 500 pounds of baggage. One pays \$1.25, the other \$1.75 for excess above the weight allowed. If the baggage had belonged to one person, he would have had to pay \$4. How much baggage is allowed free to a single passenger?

17. Given three alloys of the following composition: A, 5 parts (by weight) gold, 2 silver, 1 lead; B, 2 parts gold, 5 silver, 1 lead; C, 3 parts gold, 1 silver, 4 lead. To obtain 9 ounces of an alloy containing equal quantities (by weight) of gold, silver, and lead, how many ounces of A, B, and C must be taken and melted together?

18. A and B are alloys of silver and copper. An alloy which is 5 parts A and 3 parts B is 52% silver. One which is 5 parts A and 11 parts B is 42% silver. What are the percentages of silver in A and B respectively?

19. A marksman who is firing at a target 500 yards distant hears the bullet strike $2\frac{2}{5}$ seconds after he fires. An observer distant 600 yards from the target and 210 yards from the marksman hears the bullet strike $2\frac{1}{10}$ seconds after he hears the report of the rifle. Find the velocity of sound and the velocity of the bullet, assuming that both of these velocities are constant.

20. A tank is supplied by two pipes, A and B, and emptied by a third pipe, C. If the tank be full and all the pipes be opened, the tank will be emptied in 3 hours; if A and C alone be opened, in 1 hour; if B and C alone be opened, in 45 minutes. If A supplies 100 more gallons a minute than B does, what is the capacity of the tank, and how many gallons a minute pass through each of the pipes?

PROBLEMS ILLUSTRATING THE METHOD OF UNDETERMINED COEFFICIENTS

397 We proceed to consider one or two simple problems relating to the subject matter of algebra itself.

The inquiry may arise with regard to some particular function of the variables under consideration, Can this function be reduced to a certain specified form and, if so, what are its coefficients when reduced to this form?

The following example will illustrate the method of attacking a problem of this kind.

Example. Can the expression $x^2 + 4x + 6$ be reduced to the form of a polynomial of the second degree in $x + 1$, and, if so, what are its coefficients when reduced to this form?

The most general expression of the form in question may be written $a(x + 1)^2 + b(x + 1) + c$, where a, b, c denote constants.

Hence, if the reduction under consideration is possible, we must have

$$x^2 + 4x + 6 \equiv a(x + 1)^2 + b(x + 1) + c \tag{1}$$

or

$$x^2 + 4x + 6 \equiv ax^2 + (2a + b)x + (a + b + c). \tag{2}$$

By § 284, (2) and hence (1) will be an identity when, and only when, the coefficients of like powers of x in (2) are equal, that is, when $a = 1$, $2a + b = 4$, $a + b + c = 6$, or, solving for a, b, c, when $a = 1$, $b = 2$, $c = 3$

Hence $\quad x^2 + 4x + 6 \equiv (x + 1)^2 + 2(x + 1) + 3.$

Observe that we set the given expression equal to an expression of the required form but with *undetermined* coefficients. We then find that to make this supposed identity true, the coefficients must satisfy certain conditional equations. And by solving these equations we obtain the values of the coefficients.

398 The following is a more general kind of problem including that just considered.

Certain conditions are stated and the question is then asked, Does any function of a certain specified form exist which will satisfy these conditions, and, if so, what are its coefficients?

To solve such a problem, we construct an expression of the form in question with undetermined coefficients. These coefficients are the *unknown numbers* of the problem and the given conditions yield the system of equations which they must satisfy. If this system of equations has a single solution, we obtain a single function satisfying the given conditions; if the system has no solution, no such function exists; if the system has infinitely many solutions, the problem is indeterminate, there being infinitely many functions satisfying the given conditions. It is here supposed that the function under discussion is a *finite* expression, § 264.

Example. If possible, find a polynomial in x, of the second degree, which has the value 0 when $x = 1$ and when $x = 3$, and the value 6 when $x = 4$.

The polynomial in question must have the form $ax^2 + bx + c$. And by the conditions of the problem

$$a + b + c = 0, \qquad 9a + 3b + c = 0, \qquad 16a + 4b + c = 6.$$

Solving for a, b, c, we find $a = 2$, $b = -8$, $c = 6$.

Hence the required polynomial is $2x^2 - 8x + 6$.

Had the problem been to find a polynomial of the *first* degree satisfying the given conditions, there would have been no solution; had it been to find one of the *third* degree, there would have been infinitely many solutions.

399 The method illustrated in the preceding sections is called the method of *undetermined coefficients.* It is the principal method of investigation in algebra and we shall often have occasion to apply it as we proceed.

EXERCISE XII

1. Express $3x^3 - x^2 + 2x - 5$ as a polynomial in $x - 2$.

2. Express $4x^2 + 8x + 7$ as a polynomial in $2x + 3$.

3. Find $f(x) = ax^2 + bx + c$ such that

$$f(-1) = 11,\ f(1) = -5,\ f(5) = 6.$$

4. Find $f(x) = ax^3 + bx^2 + cx + d$ such that

$$f(0) = 5,\ f(-1) = 1,\ f(1) = 9,\ f(2) = 31.$$

5. Find $f(x, y) = ax + by + c$ such that

$$f(0, 0) = 4,\ f(4, 4) = 0,\ f(1, 0) = 6.$$

6. Find a simple equation $ax + by + 1 = 0$ two of whose solutions are $x = 3$, $y = 1$ and $x = 4$, $y = -1$.

7. Can a simple equation $ax + by + c = 0$ be found which has the three solutions $x = 3$, $y = 1$; $x = 4$, $y = -1$; $x = 1$, $y = 1$?

8. Find the simple equation whose graph is the straight line determined by the points $(2, 3)$, $(-4, 5)$.

9. Determine c so that the graph of $3x + y + c = 0$ will pass through the point $(-2, 3)$.

10. Find two simple equations, $ax + by + 1 = 0$ (1), $a'x + b'y + 1 = 0$ (2), such that both are satisfied by $x = 2$, $y = 3$ and also (1) by $x = 7$, $y = 5$ and (2) by $x = 3$, $y = 7$.

Plot the graphs of these equations.

11. Find the equation $x^3 + bx^2 + cx + d = 0$ whose roots are -2, 1, and 3.

12. Find an equation of the form $x^2 + bxy + cx + dy = 0$ which has the solutions $x = 1$, $y = 0$; $x = 2$, $y = 1$; $x = -2$, $y = 1$.

13. Express $3x + 2y - 3$ in the form

$$a(x + y - 1) + b(2x - y + 2) + c(x + 2y - 3),$$

where a, b, and c denote constants.

V. THE DIVISION TRANSFORMATION

THE GENERAL METHOD

Preliminary considerations. In § 319 we defined the quotient of A by B as the simplest form to which the fraction A / B can be reduced by the rules of reckoning. **400**

We are now to give a general method for finding the quotient as thus defined, when A and B are polynomials in the same letter, as x, and the degree of A is not less than that of B.

1. It is then *possible* that B is a factor of A, in other words, that A can be reduced to the form

$$A \equiv QB, \tag{1}$$

where Q is an *integral* function of x.

We then have
$$\frac{A}{B} \equiv Q,$$

that is, the quotient of A by B is the integral function Q; and we say that A is *exactly divisible* by B.

Thus, if $A = x^3 + 4x^2 - 2x - 5$ and $B = x^2 + 3x - 5$, it will be found that $x^3 + 4x^2 - 2x - 5 = (x + 1)(x^2 + 3x - 5)$, an identity of the form (1), Q being $x + 1$.

Hence
$$\frac{A}{B} = \frac{x^3 + 4x^2 - 2x - 5}{x^2 + 3x - 5} = x + 1.$$

2. But it will *usually* happen that B is *not* a factor of A. We cannot then reduce A to the form QB; but, as we shall show, § 401, we *can* reduce it to the form

$$A \equiv QB + R, \tag{2}$$

where both Q and R are *integral* functions of x, and the degree of R is *less* than that of B.

We then have
$$\frac{A}{B} \equiv Q + \frac{R}{B},$$

that is, the quotient of A by B is the sum of an integral function, Q, and a fraction, R/B, whose numerator is of *lower degree* than its denominator.

In this case we call Q the *integral part of the quotient*, and R the *remainder.*

Thus, if $A = x^3 + 2x^2 + 3x + 3$ and $B = x^2 + 2x + 2$, we can at once reduce A to the form (2) by writing it

$$x^3 + 2x^2 + 3x + 3 = x(x^2 + 2x + 2) + (x + 3),$$

where Q is x and R is $x + 3$, which is of lower degree than B.

Hence $$\frac{A}{B} = \frac{x^3 + 2x^2 + 3x + 3}{x^2 + 2x + 2} = x + \frac{x + 3}{x^2 + 2x + 2}.$$

401 **The division transformation.** It remains to show how to reduce A to the form $QB + R$, where R is of lower degree than B and may be 0. The process by which this is usually accomplished is called the *division transformation*, or "long division." It is illustrated in the following example.

Let $A = 2x^4 + 3x^3 + 4x^2 + x - 2$ and $B = x^2 - x + 1$.

Here the degree of B is *two*, and the problem is to find an integral function, Q, such that the remainder, R, obtained by subtracting QB from A, shall be of the *first* degree at most and may be 0; for if such a function, Q, be found, we shall have

$$A - QB \equiv R, \text{ and therefore } A \equiv QB + R.$$

Since the degree of A is *four* and that of R is to be not greater than *one*, Q must be such that the *first three terms of* A *are cancelled* when we subtract QB. This suggests the following method for finding Q.

$$\begin{array}{rl|l}
A = & 2x^4+3x^3+4x^2+x-2 & x^2-x+1=B \\
2x^2B = & 2x^4-2x^3+2x^2 & 2x^2+5x+7=Q \\
\hline
A-2x^2B = & 5x^3+2x^2+x-2 & (1) \\
5xB = & 5x^3-5x^2+5x & \\
\hline
A-(2x^2+5x)B = & 7x^2-4x-2 & (2) \\
7B = & 7x^2-7x+7 & \\
\hline
A-(2x^2+5x+7)B = & 3x-9=R & (3)
\end{array}$$

Evidently we shall cancel the *leading* term of A if we subtract any multiple of B which has the same leading term that A has. The simplest multiple of this kind is $2x^2B$, where the multiplier $2x^2$ is found by dividing the leading term of A, namely $2x^4$, by the leading term of B, namely x^2.

Subtracting $2x^2B$ from A, as above, we have

$$A - 2x^2B = 5x^3 + 2x^2 + x - 2. \tag{1}$$

We may cancel the leading term of the remainder (1) thus obtained, and with it the *second* term of A, by a similar process. The quotient of $5x^3$ by x^2 is $5x$; and multiplying B by $5x$ and subtracting we have

$$A - (2x^2 + 5x)B = 7x^2 - 4x - 2. \tag{2}$$

Finally, we shall cancel the leading term of the remainder (2), and with it the *third* term of A, by subtracting $7B$, where the multiplier, 7, is found by dividing $7x^2$ by x^2. The result is

$$A - (2x^2 + 5x + 7)B = 3x - 9. \tag{3}$$

The remainder (3) is of the *first* degree, and we obtain it by subtracting $(2x^2 + 5x + 7)B$ from A.

Hence the polynomials Q and R which we are seeking are

$$Q = 2x^2 + 5x + 7 \text{ and } R = 3x - 9.$$

And writing the identity (3) thus,

$$A = (2x^2 + 5x + 7)B + (3x - 9),$$

we have A in the form $QB + R$, where the degree of R is less than that of B.

We therefore have the following rule for finding Q and R when A and B are given.

Arrange both A *and* B *according to descending powers of* x.

Divide the leading term of A *by the leading term of* B*; the quotient will be the first term of* Q.

Multiply B *by this first term of* Q, *and subtract the product from* A.

Proceed in a similar manner with the remainder thus obtained, dividing its leading term by the leading term of B, *and so on.*

Continue the process until a remainder is reached which is of lower degree than B. *We shall then have found all the terms of* Q, *and the final remainder will be* A — QB *or* R.

It is customary to arrange the reckoning in the manner illustrated above. We can then use detached coefficients, as in multiplication.

Example 1. Given $A = 2x^5 - 6x^4 + 7x^3 + 8x^2 - 19x + 20$, and $B = x^2 - 3x + 4$; find Q and R.

$$\begin{array}{l|l} 2-6+7+8-19+20 & 1-3+4 \\ \underline{2-6+8} & 2+0-1+5 \\ \quad -1+8-19 & \\ \quad \underline{-1+3-4} & \text{Hence } Q = 2x^3 - x + 5 \\ \qquad\quad 5-15+20 & \text{and } R = 0. \\ \qquad\quad \underline{5-15+20} & \end{array}$$

Observe that instead of the first remainder, $-1+8-19+20$, we write only that part, $-1+8-19$, which is involved in the next subtraction; also that the second coefficient of Q is 0, because the *two* leading terms of A are cancelled by the first subtraction.

Example 2. Given $A = x^4 + 2x^3 + 3x^2 + 2x + 4$, $B = x^2 + 2x$, find Q and R.

402 **Remarks on this method.** 1. Observe that in the division transformation each intermediate remainder plays the rôle of a new dividend; also that if R_1 denote any such remainder, Q_1 the part of Q already obtained, and Q_2 the rest of Q, we have

$$A \equiv Q_1B + R_1 \text{ and } R_1 \equiv Q_2B + R.$$

2. The process by which Q and R are found is not itself the division of A by B, but a preliminary operation consisting of multiplications and subtractions, the aim of which is to reduce A to the form $A \equiv QB + R$. The division of A by B does not occur until we pass from the identity $A \equiv QB + R$ to the identity $A/B \equiv Q + R/B$.

At the same time it is customary to call the operation by which Q and R are found "division," and also to call Q the "quotient" instead of the "integral part of the quotient" even when R is not 0; and we shall usually follow this practice. But "dividing A by B" does not then mean, as in § 254, finding an expression which multiplied by B will produce A, but finding, first, what multiple of B we must subtract from A to obtain a remainder which is of lower degree than B, and, second, what this remainder is. Compare § 87.

3. The steps by which the integral expression A is reduced to the integral form $QB + R$ may be taken whatever the value of x. Hence A and $QB + R$ have equal values for all values of x, *even those values for which* B *is equal to* 0. On the other hand, neither A/B nor $Q + R/B$ has any meaning when $B = 0$.

Thus, if $\quad A = x^2 + x + 1$ and $B = x - 1$,

we find, by § 401, $\quad x^2 + x + 1 = (x + 2)(x - 1) + 3, \qquad (1)$

and therefore $\quad \dfrac{x^2 + x + 1}{x - 1} = x + 2 + \dfrac{3}{x - 1}. \qquad (2)$

Here $B = 0$ when $x = 1$. Substituting 1 for x in (1) and (2), we have $3 = 3$, which is true, but $3/0 = 3 + 3/0$, which is meaningless.

4. The transformation of A to the form $QB + R$ is *unique*, that is, there exists but *one* pair of integral functions Q and R (of which R is of a lower degree than B) such that

$$A \equiv QB + R.$$

For were there a second such pair, say Q', R', we should have

$$QB + R \equiv Q'B + R' \text{ and therefore } (Q - Q')B \equiv R' - R.$$

But this is impossible, since $R' - R$ would be of lower degree than B but $(Q - Q')B$ not of lower degree than B.

The effect of multiplying the dividend or divisor by a constant. 403 The following theorems will be of service further on.

1. *If we multiply the dividend by any constant, as* c, *we multiply both quotient and remainder by* c.

For if $\quad A \equiv QB + R$, then $cA \equiv cQ \cdot B + cR$.

2. *If we multiply the divisor by* c, *we divide the quotient by* c, *but leave the remainder unchanged.*

For if $$A \equiv QB + R, \text{ then } A \equiv \frac{Q}{c} \cdot cB + R.$$

3. *If we multiply both dividend and divisor by* c, *we multiply the remainder by* c, *but leave the quotient unchanged.*

For if $$A \equiv QB + R, \text{ then } cA \equiv Q \cdot cB + cR.$$

4. *If at any stage of a division transformation we multiply an intermediate remainder or the divisor by* c, *the final remainder, if changed at all, is merely multiplied by* c.

This follows from 1 and 2 and § 402, 1.

The student will do well to verify these theorems for some particular case.

Thus, we may verify the second theorem by dividing $A = 4x^2 + 6x + 1$ first by $B = 2x - 1$, and then by $2B = 4x - 2$.

$$\begin{array}{l|l} 4+6+1 & 2-1 \\ \hline \underline{4-2} & 2+4 \\ \quad 8+1 & \\ \quad \underline{8-4} & \therefore Q = 2x+4, \\ \qquad 5 & \quad R = 5. \end{array} \qquad \begin{array}{l|l} 4+6+1 & 4-2 \\ \hline \underline{4-2} & 1+2 \\ \quad 8+1 & \\ \quad \underline{8-4} & \therefore Q = x+2, \\ \qquad 5 & \quad R = 5. \end{array}$$

404 Division by the method of undetermined coefficients. We may also find Q and R when A and B are given, as follows:

Example 1. Divide $A = 2x^4 + 3x^3 + 4x^2 + x - 2$ by $B = x^2 - x + 1$.

Since the degree of A is *four* and that of B is *two*, we know in advance that the degree of Q is *two* and that the degree of R is *one* at most.

Hence let $$Q = c_0x^2 + c_1x + c_2 \text{ and } R = d_0x + d_1,$$

where such values are to be found for the coefficients c_0, c_1, c_2, d_0, d_1 that we shall have

$$\begin{aligned} 2x^4 + 3x^3 + 4x^2 + x - 2 &\equiv (c_0x^2 + c_1x + c_2)(x^2 - x + 1) + d_0x + d_1 \\ &\equiv c_0x^4 + \begin{array}{r|} -c_0 \\ +c_1 \\ \end{array}\, x^3 + \begin{array}{r|} c_0 \\ -c_1 \\ +c_2 \end{array}\, x^2 + \begin{array}{r|} c_1 \\ -c_2 \\ +d_0 \end{array}\, x + \begin{array}{r|} c_2 \\ +d_1 \\ \end{array} \end{aligned} \qquad (1)$$

But to make (1) an identity, we must have, § 284,

$$\begin{aligned}
c_0 &= 2. \\
-c_0 + c_1 &= 3, \quad \therefore c_1 = 3 + c_0 = 3 + 2 = 5. \\
c_0 - c_1 + c_2 &= 4, \quad \therefore c_2 = 4 - c_0 + c_1 = 4 - 2 + 5 = 7. \\
c_1 - c_2 + d_0 &= 1, \quad \therefore d_0 = 1 - c_1 + c_2 = 1 - 5 + 7 = 3. \\
c_2 + d_1 &= -2, \quad \therefore d_1 = -2 - c_2 = -2 - 7 = -9.
\end{aligned}$$

Hence $Q = 2x^2 + 5x + 7$ and $R = 3x - 9$, as in § 401.

Example 2. Divide $6x^5 + 13x^4 - 12x^3 - 11x^2 + 11x - 2$ by $2x^2 + x - 2$.

Exact division. Let A and B denote polynomials in x with literal coefficients, and suppose the degree of B to be m. For the division of A by B to be exact, the remainder R must equal 0 identically. This requires that all the coefficients of R be 0. Since the degree of R is $m - 1$, it has m coefficients, § 277, and evidently these coefficients are functions of the coefficients of A and B. Hence **405**

In order that a polynomial A *may be exactly divisible by a polynomial of the* m*th degree,* B, *the coefficients of* A *and* B *must satisfy* m *conditions.*

The following example will illustrate this fact.

Example 1. For what values of a and b is $x^3 + 3x^2 + bx + 2$ exactly divisible by $x^2 + ax + 1$?

We have

$$\begin{array}{l|l}
x^3 + 3x^2 + bx \quad + 2 & x^2 + ax + 1 \\
\underline{x^3 + ax^2 + x} & x + (3 - a) \\
(3 - a)x^2 + (b - 1)x \quad + 2 & \\
\underline{(3 - a)x^2 + (3a - a^2)x + (3 - a)} & \\
(b - 1 - 3a + a^2)x + (a - 1) &
\end{array}$$

Hence a and b must satisfy the *two* conditions

$$b - 1 - 3a + a^2 = 0, \quad a - 1 = 0; \text{ whence } a = 1 \text{ and } b = 3.$$

Example 2. Determine l and m so that $2x^3 + 3x^2 + lx + m$ may be exactly divisible by $x^2 + x - 6$.

Dividend and divisor arranged in ascending powers of x. Let A and B denote dividend and divisor arranged in *ascending* powers of x, and suppose that A does not begin with a term **406**

of lower degree than B begins with. We may then obtain an integral expression for A in terms by B by the process of cancelling leading terms described in § 401. If A is exactly divisible by B, this result is the same as when A and B are arranged in descending powers; but if A is not exactly divisible by B, the result is entirely different. The following examples will make this clear.

$$\begin{array}{l|l} 1+3x+3x^2+x^3 & 1+x \\ \underline{1+x} & \overline{1+2x+x^2} \\ \quad 2x+3x^2 & \\ \quad \underline{2x+2x^2} & \\ \qquad x^2+x^3 & \\ \qquad \underline{x^2+x^3} & \\ \qquad\quad 0 & \end{array} \qquad (1) \qquad\qquad \begin{array}{l|l} 1-2x+x^2 & 1+x \\ \underline{1+x} & \overline{1-3x+4x^2} \\ \quad -3x+x^2 & \\ \quad \underline{-3x-3x^2} & \\ \qquad 4x^2 & \\ \qquad \underline{4x^2+4x^3} & \\ \qquad\quad -4x^3 & \end{array} \qquad (2)$$

From this reckoning it follows by the reasoning of § 401 that

$$1+3x+3x^2+x^3 = (1+2x+x^2)(1+x) \qquad (1)$$

$$1-2x+x^2 = (1-3x+4x^2)(1+x)-4x^3. \qquad (2)$$

The result (1) is the same as that obtained when the given dividend and divisor are arranged in descending powers of x. This must always be the case when, as here, the division is exact, as follows from § 402, 4.

But the result (2) is entirely different from that obtained when we arrange $1-2x+x^2$ and $1+x$ in descending powers of x. We then get

$$x^2-2x+1 = (x-3)(x+1)+4. \qquad (3)$$

Both (2) and (3) are true identities, but they give us different expressions for x^2-2x+1 in terms of $x+1$ and lead to different expressions for the quotient of x^2-2x+1 by $x+1$, namely:

$$\frac{1-2x+x^2}{1+x} = 1-3x+4x^2-\frac{4x^3}{1+x},$$

$$\frac{x^2-2x+1}{x+1} = x-3+\frac{4}{x+1}.$$

407 Observe that in the arrangement according to ascending powers the degrees of the leading terms of the successive remainders *increase*, and, except when the division is exact, the process has no natural termination. By taking steps enough we can obtain, as the integral part of the quotient, a polynomial of as many terms and therefore of as high a degree as we please. Hence

If A *and* B *denote polynomials arranged in ascending powers of* x, A *not being exactly divisible by* B *nor beginning with a term of lower degree than* B *begins with, we can reduce the quotient of* A *by* B *to the form*

$$\frac{A}{B} \equiv Q' + \frac{R'}{B},$$

where Q′ *and* R′ *are integral functions arranged in ascending powers of* x, Q′ *ending with as high a power of* x *as we please and* R′ *beginning with a still higher power.*

If the number of terms in Q' is n, we call Q' the *quotient of* A *by* B *to* n *terms*, and R' the corresponding *remainder.*

When the value of x is small (how small will be shown subsequently) we can make the value of R'/B as small as we please by taking n great enough; that is, we can find a polynomial Q' whose value will differ as little as we please from that of A/B. On this account the polynomials Q' are sometimes called *approximate integral expressions* for the fraction A/B.

Thus "dividing" 1 by $1-x$ to n "steps," we obtain

$$\frac{1}{1-x} = 1 + x + x^2 + \cdots + x^{n-1} + \frac{x^n}{1-x}.$$

If we give x any value numerically less than 1, we can choose n so that $1 + x + \cdots + x^{n-1}$ will differ in value as little as we please from $1/(1-x)$. Thus, if $x = 1/3$, then $x^3/(1-x) = 1/18$, so that $1 + x + x^2$ differs from $1/(1-x)$ by only $1/18$. Similarly $1 + x + x^2 + x^3$ differs from $1/(1-x)$ by only $1/54$; and so on.

408 **Quotients to n terms found by the method of undetermined coefficients.** We proceed as in the following example.

Example 1. Find the quotient $(3-x)/(1-x+2x^2)$ to four terms.

Let
$$\frac{3-x}{1-x+2x^2} \equiv a_0 + a_1x + a_2x^2 + a_3x^3 + \cdots. \quad (1)$$

Multiplying both members by $1 - x + 2x^2$ and collecting terms, we have

$$\begin{array}{rl|l|l|l} 3 - x \equiv & a_0 + a_1 & x + a_2 & x^2 + a_3 & x^3 + \cdots \\ & \quad - a_0 & \; - a_1 & \; - a_2 & \\ & & + 2a_0 & + 2a_1 & \end{array} \qquad (2)$$

But to make (2) an identity, we must have, § 284,

$$\begin{array}{rll} a_0 = 3, & & \\ a_1 - a_0 = -1, & \therefore & a_1 = -1 + a_0 = 2. \\ a_2 - a_1 + 2a_0 = 0, & \therefore & a_2 = a_1 - 2a_0 = -4. \\ a_3 - a_2 + 2a_1 = 0, & \therefore & a_3 = a_2 - 2a_1 = -8. \end{array}$$

Hence $(3 - x)/(1 - x + 2x^2) = 3 + 2x - 4x^2 - 8x^3 + \cdots$.

Example 2. Find $(2 + x + 3x^2)/(1 + x - x^2)$ to five terms.

409 Polynomials involving more than one variable. Let two polynomials, A and B, be given which involve more than one variable. Unless A is of lower degree than B with respect to some one of the variables, it is *possible* that A is exactly divisible by B, in other words, that an integral function Q exists such that $A/B \equiv Q$. We may discover whether or not this is so, and if it is find Q, by first arranging both A and B as polynomials in some one of the variables and then applying the method of § 401.

Example 1. Divide $x^3 + y^3 + z^3 - 3xyz$ by $x + y + z$.

$$\begin{array}{l|l} x^3 - 3yz\cdot x + (y^3 + z^3) & x + (y + z) \\ \hline \underline{x^3 + (y + z)x^2} & x^2 - (y + z)x + (y^2 - yz + z^2) \\ \quad - (y + z)x^2 - 3yz\cdot x & \\ \quad \underline{- (y + z)x^2 - (y + z)^2 x} & \\ \qquad (y^2 - yz + z^2)x + (y^3 + z^3) & \\ \qquad \underline{(y^2 - yz + z^2)x + (y^3 + z^3)} & \end{array}$$

Hence $(x^3 + y^3 + z^3 - 3xyz)/(x + y + z) = x^2 + y^2 + z^2 - yz - zx - xy$.

Example 2. Divide $2x^2 + 5xy + 3y^2 + 7x + 11y - 4$ by $x + y + 4$.

If A is *not* exactly divisible by B, this method will yield an expression for A/B of the form $A/B \equiv Q + R/B$, where Q and R are integral with respect to the *letter of arrangement*, and R is of lower degree than B with respect to that letter. *But the form of this expression will depend on what choice is made of the letter of arrangement.*

Example. Divide $4x^2 + 6xy + y^2$ by $2x + y$.

(1) Choosing x as the letter of arrangement, we have

$$\begin{array}{l|l}
4x^2 + 6xy + y^2 & 2x + y \\
\underline{4x^2 + 2xy} & \overline{2x + 2y} \\
\phantom{4x^2+{}} 4xy + y^2 & \\
\phantom{4x^2+{}} \underline{4xy + 2y^2} & \\
\phantom{4x^2+4xy+{}} -y^2 &
\end{array}$$

Hence

$$\frac{4x^2 + 6xy + y^2}{2x + y} = 2x + 2y - \frac{y^2}{2x + y}.$$

(2) Choosing y as the letter of arrangement, we have

$$\begin{array}{l|l}
y^2 + 6yx + 4x^2 & y + 2x \\
\underline{y^2 + 2yx} & \overline{y + 4x} \\
\phantom{y^2+{}} 4yx + 4x^2 & \\
\phantom{y^2+{}} \underline{4yx + 8x^2} & \\
\phantom{y^2+4yx+{}} -4x^2 &
\end{array}$$

Hence

$$\frac{y^2 + 6yx + 4x^2}{y + 2x} = y + 4x - \frac{4x^2}{y + 2x}.$$

EXERCISE XIII

1. By the method of § 401 and using detached coefficients, divide
$$6x^4 - 7x^3 - 3x^2 + 24x - 20 \text{ by } 3x^2 + x - 6.$$

2. Also $3x^4 - 2x^3 - 32x^2 + 66x - 35$ by $x^2 + 2x - 7$.

3. Also $2x^5 - 5x^4 + 13x^3 - 15x^2 + 22x$ by $x^2 - 2x + 4$.

4. Also $4x^7 - 3x^5 + 19x^4 + 2x^3 + 4x^2 - 4x + 7$ by $x^3 - x + 5$.

5. By the method of undetermined coefficients, § 404, divide
$$2x^3 - 3x^2 + x - 5 \text{ by } x^2 - 3x + 2.$$

6. Also $2x^5 - 3x^4 + x^2 - 5$ by $x^3 - 3x + 2$.

7. Given $A = 3x^3 - 5x^2 - 7x + 12$ and $B = 3x^2 + x - 5$, reduce A to the form $A \equiv QB + R$, where R is of lower degree than B. Also write down the corresponding expression for A/B.

8. Determine a and b so that $x^4 + ax^3 + x^2 + bx + 1$ may be exactly divisible by $x^2 - 2x + 1$.

9. For what values of a and b is $(x^4 + 2x^3 + 3x^2 + ax + b)/(x^2 + 3x + 5)$ reducible to an integral expression?

10. Divide $x^6 + x^5 + x^3 + x + 1 + 2(x^4 + x^2)$ by $x^2 + x + 1$.

11. Divide $2x^2 + 5xy - 3y^2 - 5x + 13y - 12$ by $x + 3y - 4$.

12. Divide $2a^2 - b^2 - 6c^2 - ab + ac + 5bc$ by $2a + b - 3c$.

13. Divide $a^2(b + c) + b^2(c + a) - c^2(a + b) + abc$ by $ab + bc + ca$.

14. Divide $x^4 + (a - 3)x^3 + (4 - a)x^2 - 2ax + 8a$ by $x^2 - 3x + 4$.

15. Divide $8x^3 - 27y^3$ by $2x - 3y$, using detached coefficients.

16. Also $x^4 - 4xy^3 + 3y^4$ by $x - y$.

17. Also $6a^5 + a^4b - a^3b^2 + 11a^2b^3 - 5ab^4 + 4b^5$ by $2a^2 - ab + b^2$.

18. The dividend being $2x^3 + xy^2 + y^3$ and the divisor being $2x + y$, find Q and R, first, when x is taken as "letter of arrangement," second, when y is taken as that letter.

19. Arranging dividend and divisor in ascending powers of x, find quotient to *three* terms and remainder when dividend is $1 - 3x + 5x^2$ and divisor $1 + x + 3x^2$.

20. Also when dividend is $1 + x + 3x^2$ and divisor $1 - 3x + 5x^2$.

21. By the method of undetermined coefficients, § 408, find to *four* terms the quotient $1/(1 - 2x)$.

22. Also the quotient $(2 + 3x + 4x^2)/(1 - x + 2x^2)$ to *four* terms.

SYNTHETIC DIVISION AND THE REMAINDER THEOREM

410 **Synthetic Division.** We proceed to explain a very expeditious method of making the division transformation, § 401, when the divisor has the form $x - b$, that is, is a binomial of the first degree whose leading coefficient is 1.

Consider the result of dividing $a_0x^3 + a_1x^2 + a_2x + a_3$ by $x - b$.

$$\begin{array}{l|l}
a_0x^3 + a_1x^2 + a_2x + a_3 & x - b \\
\underline{a_0x^3 - a_0bx^2} & \overline{a_0x^2 + (a_0b + a_1)x + (a_0b^2 + a_1b + a_2)} \\
(a_0b + a_1)x^2 + a_2x & \\
\underline{(a_0b + a_1)x^2 - (a_0b^2 + a_1b)x} & \\
\qquad (a_0b^2 + a_1b + a_2)x + a_3 & \\
\qquad \underline{(a_0b^2 + a_1b + a_2)x - (a_0b^3 + a_1b^2 + a_2b)} & \\
\qquad\qquad a_0b^3 + a_1b^2 + a_2b + a_3 = R &
\end{array}$$

The coefficients of Q and R are

$$a_0, \quad a_0b + a_1, \quad a_0b^2 + a_1b + a_2, \quad a_0b^3 + a_1b^2 + a_2b + a_3.$$

Observe that the first of these coefficients is the leading coefficient of the dividend and that the rest may be obtained one after the other by the following rule:

Multiply the coefficient last obtained by b *and add the next unused coefficient of the dividend.*

Thus, $$a_0b^2 + a_1b + a_2 = (a_0b + a_1)\,b + a_2,$$

and $$a_0b^3 + a_1b^2 + a_2b + a_3 = (a_0b^2 + a_1b + a_2)\,b + a_3.$$

This rule applies whatever the degree of the dividend may be. For since the coefficient of the leading term of the divisor is 1, each new coefficient of Q will always be the same as the leading coefficient of the remainder last obtained. Like that coefficient, therefore, it is found by multiplying the preceding coefficient of Q by b and adding a new coefficient of the dividend. And for a like reason we shall obtain R, if we multiply the last coefficient of Q by b and add the last coefficient of the dividend.

411 Hence, when the divisor has the form $x - b$ and the dividend the form $a_0x^n + a_1x^{n-1} + \cdots + a_n$, we can find Q and R as follows, where $c_0, c_1, \cdots c_{n-1}$ denote the coefficients of Q.

$$\begin{array}{cccccc|l} a_0 & a_1 & a_2 & \cdots & a_{n-1} & a_n & b \\ \underline{} & \underline{c_0b} & \underline{c_1b} & \underline{\cdots} & \underline{c_{n-2}b} & \underline{c_{n-1}b}. & \\ c_0 & c_1 & c_2 & & c_{n-1}, & R & \end{array}$$

We first write down the coefficients of the dividend in their proper order and b at their right.

Under a_0 we write c_0, which we know to be the same as a_0.

We then multiply c_0 by b, set the product c_0b under a_1, add, and so obtain c_1.

In like manner we multiply c_1 by b, set the product c_1b under a_2, add, and so obtain c_2.

And we continue thus, multiplying and adding alternately, until all the coefficients $a_0, a_1, \cdots a_n$ are exhausted.

Example. Divide $3x^4 - 5x^3 - 4x^2 + 3x - 2$ by $x - 2$.

We have

$$\begin{array}{rrrrr|l} 3 & -5 & -4 & +3 & -2 & \underline{2} \\ & \underline{6} & \underline{2} & \underline{-4} & \underline{-2} & \\ \overline{3} & 1 & -2 & -1, & -4 & \end{array}$$

Hence $Q = 3x^3 + x^2 - 2x - 1$ and $R = -4$.

This very compact method is called *synthetic division.* The student should accustom himself to employ it whenever the divisor has the form $x - b$.

412 **Remarks on this method.** 1. In dividing synthetically when the dividend is an *incomplete* polynomial, care must be taken to indicate the missing powers of x by 0 coefficients.

2. Since $x + b = x - (-b)$, we may divide synthetically by a binomial of the form $x + b$. It is only necessary to replace b by $-b$ in the reckoning just explained.

Example 1. Divide $x^4 - 1$ by $x + 1$.

Here $x + 1 = x - (-1)$, and dividing by $x - (-1)$, we have

$$\begin{array}{rrrrr|l} 1 & +0 & +0 & +0 & -1 & \underline{-1} \\ & \underline{-1} & \underline{+1} & \underline{-1} & \underline{+1} & \\ \overline{1} & -1 & +1 & -1, & 0 & \end{array}$$

Hence $Q = x^3 - x^2 + x - 1$ and $R = 0$.

3. To divide by a binomial of the form $\alpha x - \beta$, write it thus: $\alpha(x - \beta/\alpha)$.

Then divide synthetically by $x - \beta/\alpha$, and let Q and R represent the quotient and remainder so obtained.

The quotient and remainder corresponding to the divisor $\alpha x - \beta$ will be Q/α and R, § 403, 2.

Example 2. Divide $3x^3 - 11x^2 + 18x - 3$ by $3x - 2$.

Here $3x - 2 = 3(x - 2/3)$, and dividing by $x - 2/3$, we have

$$\begin{array}{rrrr|l} 3 & -11 & +18 & -3 & \underline{2/3} \\ & \underline{2} & \underline{-6} & \underline{8} & \\ \overline{3} & -9 & 12, & 5 & \end{array}$$

Hence the required quotient is $(3x^2 - 9x + 12)/3$, or $x^2 - 3x + 4$ and the remainder is 5.

Example 3. Divide $5x^5 - x^3 + x + 2$ by $x - 3$.

Example 4. Divide $x^3 + 6x^2 + 11x + 6$ by $x + 3$.

Example 5. Divide $2x^3 - 3x^2 + 8x - 12$ by $2x - 3$.

413 **The Remainder Theorem.** *When a polynomial in* x *is divided by* x − b, *a remainder is obtained which is equal to the result of substituting* b *for* x *in the dividend; so that if* f(x) *denote the dividend,* f(b) *will denote the remainder.*

The demonstration of this theorem is given in § 410; for it is there shown that if we divide $a_0x^3 + a_1x^2 + a_2x + a_3$ by $x - b$, we obtain the remainder $a_0b^3 + a_1b^2 + a_2b + a_3$, and, in general, that if we divide $f(x) = a_0x^n + a_1x^{n-1} + \cdots + a_n$ by $x - b$, the remainder will be $a_0b^n + a_1b^{n-1} + \cdots + a_n$, or $f(b)$.

The theorem may also be proved as follows:

If $f(x)$ be the dividend, $x - b$ the divisor, $\phi(x)$ the quotient, and R the remainder, then, § 401,

$$f(x) \equiv \phi(x)(x - b) + R,$$

where R, being of lower degree than $x - b$, does not involve x at all and therefore has the same value for all values of x.

The two members of this identity have equal values whatever the value of x. In particular they have equal values when $x = b$. Hence

$$f(b) = \phi(b)(b - b) + R.$$

But $b - b = 0$; and since $\phi(x)$ is integral, $\phi(b)$ is finite.

Hence $\phi(b)(b - b) = 0$, and therefore

$$f(b) = R.$$

414 The following example will serve the double purpose of illustrating the truth of the remainder theorem and of showing that, when b and the coefficients of $f(x)$ are given numbers, usually the simplest method of computing the value of $f(b)$ is to divide $f(x)$ by $x - b$ *synthetically* — the remainder thus obtained being $f(b)$.

Example 1. What is the value of

$$f(x) = 5x^4 - 12x^3 - 20x^2 - 43x + 6, \text{ when } x = 4?$$

1. By the method of direct substitution we have

$$f(4) = 5 \cdot 4^4 - 12 \cdot 4^3 - 20 \cdot 4^2 - 43 \cdot 4 + 6 = 1280 - 768 - 320 - 172 + 6 = 26.$$

2. By the method of synthetic division we have

$$\begin{array}{rrrrrl} 5 & -12 & -20 & -43 & +6 & \underline{|4} \\ & 20 & 32 & 48 & 20 & \\ \hline 5 & 8 & 12 & 5, & 26 & = f(4). \end{array}$$

Example 2. Given $f(x) = 3x^4 - x^3 + 5x^2 - 8x + 4$. Find, by synthetic division, $f(2)$, $f(-2)$, $f(4)$, $f(-2/3)$.

415 **Corollary 1.** *If* $f(x)$ *vanishes when* $x = b$, *then* $f(x)$ *is exactly divisible by* $x - b$, *and conversely.*

For, by § 413, $f(b)$ is the remainder in the division of $f(x)$ by $x - b$, and the division is exact when the remainder is 0.

Thus, $f(x) = x^3 - 3x^2 + 2$ vanishes when $x = 1$, for $f(1) = 1 - 3 + 2 = 0$.

Hence $f(x)$ is exactly divisible by $x - 1$, as may be verified by actual division.

Again, $f(x) = x^n - b^n$ is exactly divisible by $x - b$, since $f(b) = b^n - b^n = 0$.

Example 1. If $x^3 + 3x^2 - m$ is exactly divisible by $x - 2$, what is the value of m?

We must have $\quad 2^3 + 3 \cdot 2^2 - m = 0$, or $m = 20$.

Example 2. Show that $x^n + b^n$ is exactly divisible by $x + b$ if n is odd, but not if n is even.

416 **Corollary 2.** *If an integral function of two or more variables vanishes when two of these variables, as* x *and* y, *are supposed equal, the function is exactly divisible by the difference of these variables, as* $x - y$.

For the function may be reduced to the form of a polynomial in x with coefficients involving the other variables. By hypothesis, this polynomial vanishes when $x = y$. It is therefore exactly divisible by $x - y$, § 415.

Thus, $x^2(y - z) + y^2(z - x) + z^2(x - y)$ vanishes when $x = y$; for substituting y for x, we have $y^2(y - z) + y^2(z - y) + z^2(y - y) \equiv 0$.

Hence $x^2(y-z)+y^2(z-x)+z^2(x-y)$ is exactly divisible by $x-y$. We may verify this conclusion by actual division, thus:

$$\begin{array}{l|l} (y-z)x^2-(y^2-z^2)x+(y^2z-z^2y) & x-y \\ \hline \underline{(y-z)x^2-(y^2-yz)x} & (y-z)x-(yz-z^2) \\ \quad -(yz-z^2)x+(y^2z-z^2y) & \\ \quad \underline{-(yz-z^2)x+(y^2z-z^2y)} & \end{array}$$

Example. Show that $(x-y)^3+(y-z)^3+(z-x)^3$ is exactly divisible by $x-y$, $y-z$, and $z-x$.

Theorem. *If a polynomial* f(x) *vanishes when* x = a *and also when* x = b, *then* f(x) *is exactly divisible by* (x − a)(x − b). **417**

For since $f(a)=0$ by hypothesis, $f(x)$ is exactly divisible by $x-a$, § 415, and if we call the quotient $\phi_1(x)$, we have

$$f(x)\equiv(x-a)\,\phi_1(x), \text{ where } \phi_1(x) \text{ is integral.} \qquad (1)$$

If in (1) we set $x=b$, we have

$$f(b)=(b-a)\,\phi_1(b). \qquad (2)$$

But by hypothesis $f(b)=0$, and $b-a\neq 0$.

Therefore, since when a product vanishes one of its factors must vanish, § 253, it follows from (2) that $\phi_1(b)=0$.

But if $\phi_1(b)=0$, then $\phi_1(x)$ is exactly divisible by $x-b$, § 415, and if we call the quotient $\phi_2(x)$ we have

$$\phi_1(x)\equiv(x-b)\,\phi_2(x), \text{ where } \phi_2(x) \text{ is integral.} \qquad (3)$$

Substituting this expression for $\phi_1(x)$ in (1), we have

$$f(x)\equiv(x-a)(x-b)\cdot\phi_2(x), \qquad (4)$$

which proves that $f(x)$ is exactly divisible by $(x-a)(x-b)$.

Continuing thus, we may prove the more general theorem

If f(x) *vanishes for* x = a, b, c, ⋯, *then* f(x) *is exactly divisible by* (x − a)(x − b)(x − c) ⋯. **418**

Thus, $2x^3+3x^2-2x-3$ vanishes when $x=1$, for $2+3-2-3=0$, and when $x=-1$, for $-2+3+2-3=0$.

Hence $2x^3+3x^2-2x-3$ is exactly divisible by $(x-1)(x+1)$, or x^2-1, as may be verified by actual division.

Example 1. Find a polynomial $f(x)$, of the second degree, which will take the value 0 when $x = 2$ and when $x = 3$, and the value 6 when $x = 4$.

Since $f(x)$ is of the second degree and is exactly divisible by $(x-2)(x-3)$, § 417, it may be expressed in the form $f(x) = a_0(x-2)(x-3)$, where a_0 denotes a constant.

And since $f(4) = 6$, we have $6 = a_0(4-2)(4-3)$, whence $a_0 = 3$.

Hence $f(x) = 3(x-2)(x-3) = 3x^2 - 15x + 18$.

Example 2. Find a polynomial $f(x)$ of the third degree which will vanish when $x = 2$ and when $x = 3$, and will take the value 6 when $x = 1$ and the value 18 when $x = 4$.

Reasoning as before we have $f(x) = (a_0x + a_1)(x-2)(x-3)$ where a_0, a_1 are constants.

Again, since $f(1) = 6$, and $f(4) = 18$, we have

$$6 = (a_0 + a_1)(1-2)(1-3), \quad \text{or} \quad a_0 + a_1 = 3, \tag{1}$$

$$18 = (4a_0 + a_1)(4-2)(4-3), \quad \text{or} \quad 4a_0 + a_1 = 9. \tag{2}$$

Solving (1) and (2), we obtain $a_0 = 2$, $a_1 = 1$.

Hence $f(x) = (2x+1)(x-2)(x-3) = 2x^3 - 9x^2 + 7x + 6$.

419 **Theorem.** *A polynomial* f(x), *whose degree is* n, *cannot vanish for more than* n *values of* x.

For if $f(x)$ could vanish for more than n values of x, it would be exactly divisible by the product of more than n factors of the form $x - a$, § 418, which is evidently impossible since the degree of such a product exceeds n.

420 **Theorem.** *If we know of a certain polynomial* f(x), *whose degree cannot exceed* n, *that it will vanish for more than* n *values of* x, *we may conclude that all its coefficients are* 0.

For if the coefficients were not all 0, the polynomial could not vanish for more than n values of x, § 419.

We say of such a polynomial that it *vanishes identically.*

421 **Theorem.** *If two polynomials of the* n*th degree,* f(x) *and* ϕ(x), *have equal values for more than* n *values of* x, *their corresponding coefficients are equal.*

For let $f(x) = a_0x^n + a_1x^{n-1} + \cdots + a_n$

and $\phi(x) = b_0x^n + b_1x^{n-1} + \cdots + b_n;$

also let $\psi(x) = f(x) - \phi(x)$

$$= (a_0 - b_0)x^n + (a_1 - b_1)x^{n-1} + \cdots + (a_n - b_n).$$

Then $\psi(x)$ is 0 whenever the values of $f(x)$ and $\phi(x)$ are the same, and by hypothesis these values are the same for more than n values of x.

Hence the polynomial $\psi(x) = (a_0 - b_0)x^n + \cdots + (a_n - b_n)$, whose degree does not exceed n, vanishes for more than n values of x, and therefore, § 420, all its coefficients are 0.

Therefore $a_0 - b_0 = 0,\ a_1 - b_1 = 0,\ \cdots,\ a_n - b_n = 0,$

whence $a_0 = b_0,\quad a_1 = b_1,\ \cdots,\quad a_n = b_n,$

that is, the corresponding coefficients of $f(x)$ and $\phi(x)$ are equal.

Thus, if $f(x) = 2x^2 + bx + 5$ and $\phi(x) = ax^2 + 3x + c$ have equal values when $x = 2, 4, 6$, we must have $a = 2$, $b = 3$, and $c = 5$.

EXERCISE XIV

1. Divide $x^4 - 3x^3 - x^2 - 11x - 4$ by $x - 4$ synthetically.

2. Also $5x^5 - 6x^4 - 8x^3 + 7x^2 + 6x + 3$ by $x - 3$.

3. Also $3x^4 + x^2 - 1$ by $x + 2$.

4. Also $3x^3 + 16x^2 - 13x - 6$ by $3x + 1$.

5. Also $3x^3 - 6x^2 + x + 2$ by $3x - 1$.

6. Also $x^3 - (a + b + c)x^2 + (ab + ac + bc)x - abc$ by $x - a$.

7. Also $2x^4 - x^3y - 7x^2y^2 + 7xy^3 - 10y^4$ by $x - 2y$.

8. Given $f(x) = 2x^3 - 5x + 3$. By the method of § 414, find

$$f(1),\ f(2),\ f(5),\ f(-1),\ f(-3),\ f(-6).$$

9. By aid of the remainder theorem, determine m so that

$$x^3 + mx^2 - 20x + 6$$

may be exactly divisible by $x - 3$.

10. In a similar manner, determine l and m so that $2x^3 - x^2 + lx + m$ may be exactly divisible by $(x + 2)(x - 4)$.

11. By § 416, show that $3bm + am - 2an - 6bn$ is exactly divisible by $m - 2n$, also by $a + 3b$.

12. By §§ 416, 417, show that $a(b-c)^3 + b(c-a)^3 + c(a-b)^3$ is exactly divisible by $(a-b)(b-c)(c-a)$.

13. Find the integral function of x of the third degree which vanishes when $x = 1, 4, -2$, and takes the value -16 when $x = 2$.

14. Find the integral function of x of the third degree which vanishes when $x = 2, 3$, and takes the value 6 when $x = 0$ and the value 12 when $x = 1$.

15. Show that $2x^3 - ax + 1$ and $x^3 + 5x + 2$ cannot have equal values for *four* values of x.

EXPRESSION OF ONE POLYNOMIAL IN TERMS OF ANOTHER

422 Let A and B denote two polynomials in x, A of higher degree than B.

Divide A by B and call the quotient Q, the remainder R; then

$$A \equiv QB + R. \qquad (1)$$

If Q is not of lower degree than B, divide Q by B and call the quotient Q_1, the remainder R_1; then

$$Q \equiv Q_1B + R_1. \qquad (2)$$

Similarly, if Q_1 is not of lower degree than B, divide Q_1 by B and call the quotient Q_2, the remainder R_2; then

$$Q_1 \equiv Q_2B + R_2. \qquad (3)$$

Suppose that Q_2 is of lower degree than B. We then have

$$\begin{aligned} A &\equiv QB + R && \text{by (1)} \\ &\equiv \{Q_1B + R_1\}B + R && \text{by (2)} \\ &\equiv \{(Q_2B + R_2)B + R_1\}B + R && \text{by (3)} \\ &\equiv Q_2B^3 + R_2B^2 + R_1B + R, \end{aligned}$$

where all the coefficients Q_2, R_2, R_1, R are of lower degree than B.

And, in general, if any polynomial A be given which is of higher degree than B, and we continue the process just described until a quotient is reached which is of lower degree than B, we shall have

$$A \equiv Q_{r-1}B^r + R_{r-1}B^{r-1} + \cdots + R_1B + R$$

where R, R_1, $\cdots$, R_{r-1}, Q_{r-1} denote the successive remainders and the final quotient, all being *of lower degree than* B.

Example. Reduce $x^5 - 4x^4 + 3x^3 - x^2 + x + 4$ to the form of a polynomial in $x^2 + x + 1$ with coefficients whose degrees are less than *two*.

Using detached coefficients, we may arrange the reckoning thus:

$$\begin{array}{l|l|l}
1-4+3-1+1+4 & 1+1+\ 1 & \\
\underline{1+1+1} & 1-5+\ 7-3 & 1+1+1 \\
\quad -5+2-1 & \underline{1+1+\ 1} & 1-6 \qquad \therefore Q_1 = x-6 \\
\quad \underline{-5-5-5} & \quad -6+\ 6-3 & \\
\qquad\quad 7+4+1 & \quad \underline{-6-\ 6-6} & \\
\qquad\quad \underline{7+7+7} & \qquad\quad 12+3 & \therefore R_1 = 12x+3 \\
\qquad\qquad -3-6+4 & & \\
\qquad\qquad \underline{-3-3-3} & & \\
\qquad\qquad\qquad -3+7 & \therefore R = -3x+7. &
\end{array}$$

Hence $x^5 - 4x^4 + 3x^3 - x^2 + x + 4$

$$\equiv (x-6)(x^2+x+1)^2 + (12x+3)(x^2+x+1) - (3x-7).$$

423 In particular, this method enables us to transform any polynomial in x into a polynomial of the same degree in $x - b$ with *constant* coefficients.

Example. Transform $2x^3 - x^2 + 4x - 5$ into a polynomial in $x - 2$.

We may perform the successive divisions synthetically and arrange the reckoning as follows:

$$\begin{array}{rrrrl}
2 & -\ 1 & +\ 4 & -\ 5 & \lfloor 2 \\
 & \underline{4} & \underline{6} & \underline{20} & \\
\bar{2} & +\ 3 & +10, & 15 & \therefore R = 15 \\
 & \underline{4} & \underline{14} & & \\
\bar{2} & +\ 7, & 24 & & \therefore R_1 = 24 \\
 & \underline{4} & & & \\
\bar{2}, & 11 & & & \therefore R_2 = 11 \text{ and } Q_2 = 2.
\end{array}$$

Hence $2x^3 - x^2 + 4x - 5 \equiv 2(x-2)^3 + 11(x-2)^2 + 24(x-2) + 15$.

EXERCISE XV

1. By the method of § 422 express $x^4 + x^3 - 1$ in terms of $x^2 + 1$.

2. Also $4x^4 + 2x^3 + 4x^2 + x + 6$ in terms of $2x^2 + 1$.

3. Also $2x^7 - 3x^6 + 2x^5 + 5x^4 - x^2 + 6$ in terms of $x^3 - x^2 + x + 3$.

4. Also $x^5 + x^3y^2 + x^2y^3 + y^5$ in terms of $x^2 + xy + y^2$.

5. By the method of § 423 express $2x^3 - 8x^2 + x + 6$ in terms of $x - 3$

6. Also $x^5 + 3x^4 - 6x^3 + 2x^2 - 3x + 7$ in terms of $x + 2$.

7. Also $x^3 + 9x^2 + 27x$ in terms of $x + 3$.

8. Also $x^3 + 3x^2 + x - 1$ in terms of $x + 1$.

VI. FACTORS OF RATIONAL INTEGRAL EXPRESSIONS

PRELIMINARY CONSIDERATIONS

424 **Factor.** Let A denote a rational integral function of one or more variables. Any *rational integral* function of these variables which exactly divides A is called a *factor* of A.

Hence in order that a given function, F, may be a factor of A, it is sufficient and necessary

1. That F be *rational* and *integral* with respect to the variables of which A is a function.

2. That A be reducible to the form $A \equiv GF$, where G also is *integral*.

Example 1. Since $2x^2 - 2xy = 2x(x - y)$, both x and $x - y$ are factors of $2x^2 - 2xy$.

Example 2. Since $3x^2 - 2y^2 = (\sqrt{3}x + \sqrt{2}y)(\sqrt{3}x - \sqrt{2}y)$, both $\sqrt{3}x + \sqrt{2}y$ and $\sqrt{3}x - \sqrt{2}y$ are factors of $3x^2 - 2y^2$.

Example 3. Although $x - y = (\sqrt{x} + \sqrt{y})(\sqrt{x} - \sqrt{y})$, we do not call $\sqrt{x} + \sqrt{y}$ and $\sqrt{x} - \sqrt{y}$ factors of $x - y$, because they are not *rational* with respect to x and y.

Note 1. The *coefficients* of a factor need not be either integral or rational. On the contrary, they may be numbers or expressions of any kind. In Ex. 2 they are irrational. **425**

Therefore, since $x^2 - y = (x + \sqrt{y})(x - \sqrt{y})$, the expression $x^2 - y$, regarded as a function of *both* x and y, cannot be factored; but regarded as a function of x *alone*, it has the factors $x + \sqrt{y}$ and $x - \sqrt{y}$. And the like may be said of other expressions which involve more than one letter.

Note 2. Except when dealing exclusively with functions having integral coefficients, it is not customary to include a mere "numerical factor," like 2 in Ex. 1, in a list of the factors of a given integral function, A; for if, as here, we do not require the coefficients of an integral function to be integers, *any* mere number (or constant) whatsoever may be said to divide A exactly. **426**

For a like reason, if F is a factor of A, and c is any constant (not 0), cF is also a factor of A; but we regard F and cF as essentially the *same* factor and include but one of them in a list of the factors of A.

Thus, in Ex. 1, it would be equally correct to say that $2x$ and $x - y$, or that $-2x$ and $y - x$, are the factors.

Theorem. *If* F *is a factor of* B, *and* B *is a factor of* A, *then* F *is a factor of* A. **427**

For, by § 424, A and B are reducible to the forms

$$A \equiv GB \quad \text{and} \quad B \equiv HF,$$

where G and H are integral.

Hence $$A \equiv G \cdot HF \equiv GH \cdot F,$$

that is, F is a factor of A, § 424.

Prime, composite. An integral function may have no other factor than itself (or a constant). In that case we call it *prime*. But if it have other factors, we call it *composite*. **428**

Thus, $x + y^2$ and $x - 2y$ are prime, but $x^2 - y^2$ is composite.

A composite function, A, of the nth degree, is the product of not less than *two* nor more than n prime functions, $B, C, \cdots$. These prime functions are called the **prime factors** of A. **429**

430 In what follows we shall assume that

1. *Any given function* A *has but one set of prime factors.*
2. *All other factors of* A *are products of these prime factors.*
3. *Two or more of these prime factors may be equal, but* A *can be expressed in only one way as a product of powers of its different prime factors.*

These theorems, of which 2 and 3 are corollaries to 1, will be proved in §§ 484, 485 for the case in which A is a function of a single variable, and they can be proved generally.

Thus, since $x^3y^3 - 2x^2y^4 = xxyyy(x - 2y)$, the prime factors of $x^3y^3 - 2x^2y^4$ are x, x, y, y, y, $x - 2y$. Its other factors, as x^2, xy, and so on, are products of two or more of these prime factors. Its *different* prime factors are x, y, $x - 2y$, and it can be expressed in but one way as a product of powers of these factors, namely thus: $x^2y^3(x - 2y)$.

431 **Factoring.** To *factor* a given function, A, *completely* is to "resolve it into its prime factors," that is, to reduce it to the form $A \equiv B \cdot C \cdot D \cdots$, where B, C, D, $\cdots$ denote prime functions.

But ordinarily we do not attempt to discover these prime factors at the outset. We endeavor first to resolve A into a product of some *two* of its factors, as F and G, next to resolve F and G, and so on, until the prime factors are reached. And even the first step in this process may be called "factoring" A.

Factoring is the reverse of multiplication. A multiplication usually involves two main steps: (1) a number of applications of the distributive law, in order to replace $(a + b)c$ by $ac + bc$, and so on; (2) the combination of like terms in the result thus obtained. To reverse the process, we must (1) separate the terms thus combined — it is in doing this that the difficulty of factoring consists — and then (2) apply the distributive law in order to replace $ac + bc$ by $(a + b)c$, and so on.

It must not be supposed that every composite function can be actually factored. Thus, while it can be proved that $x^5 + ax^3 + bx^2 + cx + d$ is composite, it can also be proved that the factors of this expression cannot be found by algebraic methods, that is, by applying, a finite number of times, the various algebraic operations.

EXPRESSIONS WHOSE TERMS HAVE A COMMON FACTOR

432 Expressions whose terms all have a common factor, monomial or polynomial, can be factored by a single application of the distributive law, namely:

$$ab + ac + ad + \cdots = a(b + c + d + \cdots).$$

Example 1. Factor $2a^2c + 2abc + 4ac^2 - 6acd$.

All the terms have the factor $2ac$. "Separating" it, we have

$$2a^2c + 2abc + 4ac^2 - 6acd = 2ac(a + b + 2c - 3d).$$

Example 2. Factor $a(c - d) + b(d - c)$.

Both terms have the factor $c - d$. Separating it, we have

$$a(c - d) + b(d - c) = a(c - d) - b(c - d) = (a - b)(c - d).$$

Factors such as these should be separated at the outset.

433 Some expressions which are not in this form as they stand can be reduced to it by combining such of their terms as have a common factor.

Example 1. Factor $ac + bd + ad + bc$.

Combining ac and ad, also bc and bd, we obtain $a(c + d) + b(c + d)$, a binomial whose terms have the common factor $c + d$.

Hence $$ac + bd + ad + bc = (a + b)(c + d).$$

Observe that the parts into which we separate the expression in applying this method must all have *the same number of terms.*

Example 2. Factor $a^2 + ab - bd - ad + ac - cd$.

We must have either *two* groups of *three* terms, or *three* groups of *two* terms. Four of the terms involve a, namely, a^2, ab, $-ad$, ac, and the remaining two involve d, namely, $-bd$ and $-cd$. To obtain groups which have the same number of terms we combine the term $-ad$ with the d-terms, and have

$$\begin{aligned} a^2 + ab + ac - ad - bd - cd &= a(a + b + c) - d(a + b + c) \\ &= (a - d)(a + b + c). \end{aligned}$$

EXERCISE XVI

Factor the following expressions.

1. $6x^4y^3z^2 - 12x^2y^4z + 8x^2y^3$.
2. $2n^2 + (n-3)n$.
3. $ab - a + b - 1$.
4. $mx - nx - mn + n^2$.
5. $3xy - 2x - 12y + 8$.
6. $10xy + 5y^2 + 6x + 3y$.
7. $x^3y^2 - x^2y^3 + 2x^2y - 2xy^2$.
8. $x^4 + x^3 + x^2 + x$.
9. $ac + bd - (bc + ad)$.
10. $a^2c - abd - abc + a^2d$.
11. $ad + ce + bd + ae + cd + be$.
12. $a^2 + cd - ab - bd + ac + ad$.

FACTORS FOUND BY AID OF KNOWN IDENTITIES

434 In the second chapter we derived a number of special products, as $(a+b)(a-b) = a^2 - b^2$. If a given function, A, can be reduced to the form of one of these products, we can write down its factors at once. The following sections will illustrate this method of factoring.

435 **Perfect trinomial squares.** This name is given to expressions which have one of the forms $a^2 \pm 2ab + b^2$. Such expressions can be factored by means of the formulas:

$$a^2 + 2ab + b^2 = (a+b)(a+b) = (a+b)^2.$$
$$a^2 - 2ab + b^2 = (a-b)(a-b) = (a-b)^2.$$

Observe that in a perfect trinomial square (properly arranged) *the middle term is twice the product of square roots of the extreme terms,* and that the factors, which are equal, are obtained by connecting the principal square roots of the extreme terms by the sign of the middle term.

To *extract the square root* of the perfect square is to find one of these equal factors.

Example 1. Factor $9x^2 - 12xy + 4y^2$.

This is a perfect square, since $12xy = 2\sqrt{9x^2}\cdot\sqrt{4y^2}$.

And since $\sqrt{9x^2} = 3x$, $\sqrt{4y^2} = 2y$, and the sign of the middle term is $-$, we have $9x^2 - 12xy + 4y^2 = (3x-2y)(3x-2y) = (3x-2y)^2$.

Example 2. Factor $a^2 + b^2 + c^2 + 2ab + 2ac + 2bc$.

We can reduce this to the form of a trinomial square by grouping the terms thus:

$$a^2 + 2ab + b^2 + 2ac + 2bc + c^2 = (a + b)^2 + 2(a + b)c + c^2$$
$$= (a + b + c)^2.$$

Example 3. Factor the following expressions.

1. $x^2 + 14x + 49$.
2. $9 - 6a + a^2$.
3. $9x^2y^2 + 30xy + 25$.
4. $x^2 - 4xy + 4y^2 + 6x - 12y + 9$.
5. $64a^8 - 48a^4 + 9$.
6. $a^2 + b^2 + c^2 - 2ab + 2ac - 2bc$.

436 **A difference of two squares.** Expressions of this form, or reducible to it, can be factored by aid of the formula:

$$a^2 - b^2 = (a + b)(a - b).$$

Thus, $x^2 - y^2 - z^2 + 2yz = x^2 - (y^2 - 2yz + z^2)$

$$= x^2 - (y - z)^2$$
$$= (x + y - z)(x - y + z).$$

A very useful device for reducing a *trinomial* expression to this form is that of making it a perfect square by adding a suitable quantity to one of its terms and then subtracting this quantity from the resulting expression.

Thus, $x^4 + x^2y^2 + y^4 = x^4 + 2x^2y^2 + y^4 - x^2y^2$

$$= (x^2 + y^2)^2 - x^2y^2$$
$$= (x^2 + y^2 + xy)(x^2 + y^2 - xy).$$

Example. Factor the following expressions.

1. $x^4 - y^6$.
2. $6a^3 - 6ab^2$.
3. $12a^3x^3 - 75axy^2$.
4. $25x^{2n} - 49x^{2m}$.
5. $36x^4 - 1$.
6. $x^4 - 3x^2y^2 + y^4$.

437 **A sum of two squares.** By making use of the imaginary unit $i = \sqrt{-1}$, §§ 218, 220, a *sum* of squares $a^2 + b^2$ can be reduced to the form of a difference of squares and then factored by § 436, the factors being imaginary.

For since $i^2 = -1$, we have $b^2 = -(-b^2) = -(ib)^2$.

Hence $a^2 + b^2 = a^2 - (ib)^2 = (a + ib)(a - ib)$.

As we have seen, §§ 219, 220, i conforms to all the ordinary rules of reckoning. One has only to remember when employing it that $i^2 = -1$.

438 Sums and differences of any two like powers. In §§ 308, 309, 310 we proved that

First. Whether n is *odd* or *even*,

$$a^n - b^n = (a - b)(a^{n-1} + a^{n-2}b + \cdots + ab^{n-2} + b^{n-1}). \quad (1)$$

Second. When n is *even*,

$$a^n - b^n = (a + b)(a^{n-1} - a^{n-2}b + \cdots + ab^{n-2} - b^{n-1}). \quad (2)$$

Third. When n is *odd*,

$$a^n + b^n = (a + b)(a^{n-1} - a^{n-2}b + \cdots - ab^{n-2} + b^{n-1}). \quad (3)$$

Hence the following theorems:

1. $a^n - b^n$ *is always divisible by* $a - b$.
2. $a^n - b^n$ *is divisible by* $a + b$ *when* n *is even.*
3. $a^n + b^n$ *is divisible by* $a + b$ *when* n *is odd.*
4. *In every case the quotient consists of the terms*

$$a^{n-1} \quad a^{n-2}b \cdots ab^{n-2} \quad b^{n-1}$$

connected by signs which are all + *when* $a - b$ *is the divisor, but are alternately* − *and* + *when* $a + b$ *is the divisor.*

Thus, 1. $x^6 - 1 = (x - 1)(x^5 + x^4 + x^3 + x^2 + x + 1)$.

2. $x^6 - 1 = (x + 1)(x^5 - x^4 + x^3 - x^2 + x - 1)$.

3.
$$\begin{aligned} 8a^3 + 27b^3c^3 &= (2a)^3 + (3bc)^3 \\ &= (2a + 3bc)[(2a)^2 - (2a)(3bc) + (3bc)^2] \\ &= (2a + 3bc)(4a^2 - 6abc + 9b^2c^2). \end{aligned}$$

Example. **Factor the following expressions.**

1. $64x^3 - 125y^3$. 2. $27x^3 + 1$. 3. $16x^4 - 81y^4$.

439 When n is composite. The following theorems are an immediate consequence of § 438, (1), (2), (3) and § 436.

1. *If* n *is a multiple of any integer,* p, *then* $a^n - b^n$ *is exactly divisible by* $a^p - b^p$.

Thus, $x^6 - y^6 = (x^2)^3 - (y^2)^3$
$= (x^2 - y^2)(x^4 + x^2y^2 + y^4).$

2. *If* n *is an even multiple of any integer,* p, *then* $a^n - b^n$ *is exactly divisible by* $a^p + b^p$.

Thus, $x^6 - y^6 = (x^3)^2 - (y^3)^2$
$= (x^3 + y^3)(x^3 - y^3).$

3. *If* n *is an odd multiple of any integer,* p, *then* $a^n + b^n$ is *exactly divisible by* $a^p + b^p$, *whether* n *itself is odd or even.*

Thus, $x^6 + y^6 = (x^2)^3 + (y^2)^3$
$= (x^2 + y^2)(x^4 - x^2y^2 + y^4).$

4. *If* n *is a power of* 2, *then* $a^n + b^n$ *can be resolved into factors of the second degree by repeated use of the device explained in* § 436.

Thus, $x^8 + y^8 = x^8 + 2x^4y^4 + y^8 - 2x^4y^4$
$= (x^4 + y^4)^2 - 2x^4y^4$
$= (x^4 + y^4 + \sqrt{2}x^2y^2)(x^4 + y^4 - \sqrt{2}x^2y^2).$

Again,

$$x^4 + y^4 + \sqrt{2}x^2y^2 = x^4 + 2x^2y^2 + y^4 - (2 - \sqrt{2})x^2y^2$$
$$= (x^2 + y^2)^2 - (2 - \sqrt{2})x^2y^2$$
$$= (x^2 + y^2 + \sqrt{2 - \sqrt{2}}xy)(x^2 + y^2 - \sqrt{2 - \sqrt{2}}xy),$$

and so on.

As each of these "quadratic" factors can be resolved into two (imaginary) factors of the first degree by § 444, the *complete* factorization of $a^n + b^n$ is always possible when n is a power of 2.

When n is composite, it is best to begin by resolving $a^n + b^n$, or $a^n - b^n$, into two factors whose *degrees are as nearly equal as possible.* It will always be possible to factor at least one of the factors thus obtained.

Thus, the factorization of $x^6 - y^6$ given under 2 is the best. Continuing, we have

$$x^6 - y^6 = (x^3 + y^3)(x^3 - y^3)$$
$$= (x + y)(x^2 - xy + y^2)(x - y)(x^2 + xy + y^2).$$

Example. Factor the following expressions.

1. $x^4 + y^4$. 2. $x^8 - y^8$. 3. $x^9 + y^9$.

440 The theorems of §§ 438, 439 also apply to expressions of the form $a^m \pm b^n$ when m and n are multiples of the same integer p.

Thus, $$x^6 - y^{15} = (x^2)^3 - (y^5)^3$$
$$= (x^2 - y^5)(x^4 + x^2y^5 + y^{10}).$$

EXERCISE XVII

In the following examples carry the factorization as far as is possible without introducing irrational or imaginary coefficients.

1. $4x^3y - 20x^2y^2 + 25xy^3$. 2. $28tx^2 - 63ty^2$.

3. $x^2 + 4y^2 + 9z^2 - 4xy - 12yz + 6zx$.

4. $(7a^2 + 2b^2)^2 - (2a^2 + 7b^2)^2$.

5. $(7x^2 + 4x - 3)^2 - (x^2 + 4x + 3)^2$. 6. $4(1 - b^2 - ab) - a^2$.

7. $x^4 + x^2 + 1$. 8. $a^4 - 6a^2b^2 + b^4$.

9. $a^4 + 4a^2 + 16$. 10. $9x^4 + 15x^2y^2 + 16y^4$.

11. $4(ab + cd)^2 - (a^2 + b^2 - c^2 - d^2)^2$.

12. $576x^6y^3 - 9y^{15}$. 13. $x^9 - y^9$. 14. $x^{12} - y^{12}$.

15. $x^{10} + y^{10}$. 16. $x^5 - 32$. 17. $x^7 + y^{14}$.

FACTORS FOUND BY GROUPING TERMS

441 Sometimes the terms of a polynomial in x can be combined in groups, all of which have some common factor, as F. This common factor F is then a factor of the entire expression. Compare § 433.

Example 1. Factor $x^3 + 3x^2 - 2x - 6$.

Noticing that the last two coefficients are equimultiples of the first two, we have

$$x^3 + 3x^2 - 2x - 6 = x^2(x + 3) - 2(x + 3)$$
$$= (x^2 - 2)(x + 3) = (x + \sqrt{2})(x - \sqrt{2})(x + 3).$$

Example 2. Factor $x^3 + 2x^2 + 2x + 1$.

Combining terms which have like coefficients, we have

$$\begin{aligned} x^3 + 2x^2 + 2x + 1 &= x^3 + 1 + 2x(x + 1) \\ &= (x^2 - x + 1)(x + 1) + 2x(x + 1) \\ &= (x^2 - x + 1 + 2x)(x + 1) = (x^2 + x + 1)(x + 1). \end{aligned}$$

Sometimes this can be accomplished by first separating one of the given terms into two terms.

Example 3. Factor $x^3 + 4x^2 + 5x + 6$.

We have

$$\begin{aligned} x^3 + 4x^2 + 5x + 6 &= x^3 + 3x^2 + x^2 + 3x + 2x + 6 \\ &= x^2(x + 3) + x(x + 3) + 2(x + 3) \\ &= (x^2 + x + 2)(x + 3). \end{aligned}$$

Consider also the following example.

Example 4. Factor $x^4 + 2x^3 + 3x^2 + 2x + 1$.

We have

$$\begin{aligned} x^4 + 2x^3 + 3x^2 + 2x + 1 &= x^4 + 2x^3 + x^2 + 2x^2 + 2x + 1 \\ &= (x^2 + x)^2 + 2(x^2 + x) + 1 \\ &= (x^2 + x + 1)^2. \end{aligned}$$

EXERCISE XVIII

Factor the following expressions.

1. $x^4 - x^3 + x - 1$.
2. $x^5 - x^3 - 8x^2 + 8$.
3. $x^4 - 2x^3 + 2x - 1$.
4. $x^3 - 7x^2 - 4x + 28$.
5. $x^6 - x^4y^2 - x^2y^4 + y^6$.
6. $x^3 + 2x^2 + 3x + 2$.
7. $x^5 + 2x^4 + 3x^3 + 3x^2 + 2x + 1$.
8. $x^4 + 4x^3 + 10x^2 + 12x + 9$.

FACTORIZATION OF QUADRATIC EXPRESSIONS

442 **The quadratic $x^2 + px + q$, factored by inspection.** This is sometimes possible, when p and q are integers.

Since $$(x + a)(x + b) = x^2 + (a + b)x + ab,$$

we shall know the factors of $x^2 + px + q$ if we can find two numbers, a and b, such that $a + b = p$ and $ab = q$.

Two such numbers always exist, § 444, though they are seldom rational. But when rational they are integers, § 454, and may be found by inspection, as in the following examples.

Example 1. Factor $x^2 + 13x + 42$.

We seek two integers, a and b, whose product is 42 and sum 13. As both ab and $a + b$ are positive, both a and b must be positive. Hence among the positive integers whose product is 42 — namely, 42 and 1, 21 and 2, 14 and 3, 7 and 6 — we seek a pair whose sum is 13, and find 7 and 6.

Hence $$x^2 + 13x + 42 = (x + 7)(x + 6).$$

Example 2. Factor $x^2 - 13x + 22$.

Here both a and b must be negative; for their product is positive and their sum negative. Hence, testing as before the pairs of *negative* integers whose product is 22, we find -11 and -2; for $-11 - 2 = -13$.

Hence $$x^2 - 13x + 22 = (x - 11)(x - 2).$$

Example 3. Factor $x^2 - 9x - 22$.

Here, since ab is negative, a and b must have opposite signs; and since $a + b$ is negative, the one which is numerically greater must be negative. Hence we set $-22 = -22 \times 1 = -11 \times 2$, and, testing as before, find $a = -11$ and $b = 2$; for $-11 + 2 = -9$.

Hence $$x^2 - 9x - 22 = (x - 11)(x + 2).$$

Example 4. Factor the following expressions.

1. $x^2 + 3x + 2$.
2. $x^2 - 16x + 15$.
3. $x^2 - 4x - 12$.
4. $x^2 + x - 30$.
5. $x^2 + 20x + 96$.
6. $x^2 - 21x + 80$.

443 The quadratic $ax^2 + bx + c$ factored by inspection. This is sometimes possible, when a, b, and c are integers.

By multiplying and dividing by a, we may reduce $ax^2 + bx + c$ to the form $[(ax)^2 + b(ax) + ac]/a$, and then factor the bracketed expression with respect to ax by the method just explained, namely, by finding two integers whose product is ac, and sum b.

Example 1. Factor $2x^2 + 7x + 3$.

We have $$2x^2 + 7x + 3 = \frac{(2x)^2 + 7(2x) + 6}{2}$$
$$= \frac{(2x + 6)(2x + 1)}{2} = (x + 3)(2x + 1).$$

Example 2. Factor $abx^2 + (a^2 + b^2)x + ab$.

We have

$$abx^2 + (a^2 + b^2)x + ab = \frac{(abx)^2 + (a^2 + b^2) \cdot abx + a^2b^2}{ab}$$
$$= \frac{(abx + a^2)(abx + b^2)}{ab}$$
$$= (bx + a)(ax + b).$$

Example 3. Factor $16x^2 + 72x - 63$.

In this case it is not necessary to multiply and divide by 16, for we have

$$16x^2 + 72x - 63 = (4x)^2 + 18(4x) - 63$$
$$= (4x + 21)(4x - 3).$$

Example 4. Factor the following expressions.

1. $6x^2 - 13x + 6$.
2. $5x^2 + 14x - 3$.
3. $14x^2 + x - 3$.
4. $18x^2 + 21x + 5$.
5. $49x^2 + 105x + 44$.
6. $abx^2 - (ac - b^2)x - bc$.

444 **The quadratic $x^2 + px + q$ or $ax^2 + bx + c$ factored by completing the square.** While the preceding methods apply in particular cases only, the following is perfectly general.

Since $$\left(x + \frac{p}{2}\right)^2 = x^2 + px + \frac{p^2}{4},$$

we can make $x^2 + px$ a perfect square by adding $\frac{p^2}{4}$, that is, *the square of half the coefficient of* x.

This process is called *completing the square* of $x^2 + px$.

1. We shall not affect the value of $x^2 + px + q$, if we both add and subtract $p^2/4$. But by this means we can transform the expression into the difference between two squares and then factor it by § 436. Thus,

$$x^2 + px + q = x^2 + px + \frac{p^2}{4} - \frac{p^2}{4} + q$$
$$= \left(x + \frac{p}{2}\right)^2 - \frac{p^2 - 4q}{4}$$
$$= \left(x + \frac{p}{2} + \frac{\sqrt{p^2 - 4q}}{2}\right)\left(x + \frac{p}{2} - \frac{\sqrt{p^2 - 4q}}{2}\right). \quad (1)$$

2. Since $$ax^2 + bx + c = a\left(x^2 + \frac{b}{a}x + \frac{c}{a}\right),$$

we may obtain the factors of this expression by substituting b/a for p and c/a for q in (1). Simplifying the result, we have

$$ax^2+bx+c=a\left(x+\frac{b}{2a}+\frac{\sqrt{b^2-4ac}}{2a}\right)\left(x+\frac{b}{2a}-\frac{\sqrt{b^2-4ac}}{2a}\right). \quad (2)$$

Example 1. Factor $x^2 - 6x + 2$.

We have
$$\begin{aligned} x^2 - 6x + 2 &= x^2 - 6x + 3^2 - 3^2 + 2 \\ &= (x-3)^2 - 7 \\ &= (x - 3 + \sqrt{7})(x - 3 - \sqrt{7}). \end{aligned}$$

Example 2. Factor $x^2 + 8x + 20$.

We have
$$\begin{aligned} x^2 + 8x + 20 &= x^2 + 8x + 4^2 - 4^2 + 20 \\ &= (x+4)^2 + 4 \\ &= (x+4)^2 - 4i^2 \\ &= (x + 4 + 2i)(x + 4 - 2i). \end{aligned}$$

Here we first obtain a *sum* of squares, $(x+4)^2 + 4$, and then transform this sum into a difference by replacing 4 by $-4i^2$, § 437. The factors are imaginary.

Example 3. Factor $3x^2 - 5x + 1$.

We have
$$\begin{aligned} 3x^2 - 5x + 1 &= 3\left[x^2 - \frac{5}{3}x + \frac{1}{3}\right] \\ &= 3\left[x^2 - \frac{5}{3}x + \left(\frac{5}{6}\right)^2 - \left(\frac{5}{6}\right)^2 + \frac{1}{3}\right] \\ &= 3\left[\left(x - \frac{5}{6}\right)^2 - \frac{13}{36}\right] \\ &= 3\left(x - \frac{5}{6} + \frac{\sqrt{13}}{6}\right)\left(x - \frac{5}{6} - \frac{\sqrt{13}}{6}\right). \end{aligned}$$

Example 4. Factor the following expressions.

1. $x^2 + 10x + 23$.
2. $x^2 - 10x + 24$.
3. $x^2 - 12x + 45$.
4. $x^2 + x + 1$.
5. $2x^2 + 3x + 2$.
6. $x^2 - 4ax - 4b^2 + 8ab$.

Homogeneous quadratic functions of two variables. The methods of §§ 442–444 are applicable to quadratics of the form **445**

$$ax^2 + bxy + cy^2.$$

Example 1. Factor $x^2 - 8xy + 14y^2$.

We have $x^2 - 8xy + 14y^2 = x^2 - 8xy + 16y^2 - 2y^2$

$$= (x - 4y)^2 - 2y^2$$

$$= [x - (4 + \sqrt{2})y][x - (4 - \sqrt{2})y].$$

Example 2. Factor the following expressions.

1. $x^2 + 5xy + 4y^2$. 2. $x^2 - xy + y^2$.

Non-homogeneous quadratic functions of two variables. Such functions are ordinarily *prime*. But when composite, they may be factored as in the following example. **446**

Example 1. Factor $A = x^2 + 2xy - 8y^2 + 2x + 14y - 3$.

If A is composite, it is the product of two polynomials of the first degree. Moreover its terms of the second degree, $x^2 + 2xy - 8y^2$, must be the product of the terms of the first degree in these polynomials.

We find by inspection that $x^2 + 2xy - 8y^2 = (x + 4y)(x - 2y)$.

Hence, if A is composite, there must be two numbers, l and m, such that we shall have

$$x^2 + 2xy - 8y^2 + 2x + 14y - 3$$

$$\equiv (x + 4y + l)(x - 2y + m)$$

$$\equiv x^2 + 2xy - 8y^2 + (l + m)x + (4m - 2l)y + lm. \quad (1)$$

But to make (1) an identity, we must have, § 285,

$$l + m = 2 \ (2), \qquad -2l + 4m = 14 \ (3), \qquad lm = -3 \ (4).$$

From (2) and (3) we find $l = -1$, $m = 3$. And these values satisfy (4); for $-1 \cdot 3 = -3$.

Therefore $x^2 + 2xy - 8y^2 + 2x + 14y - 3 = (x + 4y - 1)(x - 2y + 3)$.

Note. The example shows how exceptional these composite functions are. If, leaving A otherwise unchanged, we replace the last term, -3, by any other number, A becomes prime; for (4) will then not be satisfied by $l = -1$, $m = 3$.

This method is also applicable to *homogeneous* quadratic functions of *three* variables.

Thus, to factor $x^2 + 2xy - 8y^2 + 2xz + 14yz - 3z^2$, we set

$$x^2 + 2xy - 8y^2 + 2xz + 14yz - 3z^2 \equiv (x + 4y + lz)(x - 2y + mz)$$

and then proceed as above, again finding $l = -1$, $m = 3$.

Example 2. Factor $2x^2 - 7xy + 3y^2 + 5xz - 5yz + 2z^2$.

Example 3. Show that $x^2 - y^2 + 2x + y - 1$ is prime.

447 **Polynomials of the nth degree.** We have shown that every polynomial of the second degree, $a_0x^2 + a_1x + a_2$, is the product of factors of the first degree. The like is true of polynomials in x of every degree, though no general method exists for finding these factors; in other words,

Theorem. *Every polynomial in* x, *of the* n*th degree,*

$$f(x) = a_0x^n + a_1x^{n-1} + \cdots + a_{n-1}x + a_n$$

is the product of n *factors of the first degree; that is, there are* n *binomials,* $x - \beta_1$, $x - \beta_2$, $\cdots$, $x - \beta_n$, *such that*

$$f(x) \equiv a_0(x - \beta_1)(x - \beta_2)\cdots(x - \beta_n).$$

The proof of this theorem will be given later.

448 **Corollary.** *A homogeneous polynomial in two variables,* x *and* y, *of the* n*th degree, is the product of* n *homogeneous factors in* x *and* y, *of the first degree.*

Thus, the homogeneous polynomial $a_0x^3 + a_1x^2y + a_2xy^2 + a_3y^3$ may be derived from $a_0x^3 + a_1x^2 + a_2x + a_3$ by substituting x/y for x and multiplying the result by y^3.

But by § 447, $a_0x^3 + a_1x^2 + a_2x + a_3 \equiv a_0(x - \beta_1)(x - \beta_2)(x - \beta_3)$, and if we substitute x/y for x in this identity and then multiply both members by y^3, we obtain

$$a_0x^3 + a_1x^2y + a_2xy^2 + a_3y^3 \equiv a_0(x - \beta_1y)(x - \beta_2y)(x - \beta_3y).$$

EXERCISE XIX

Factor the following expressions.

1. $x^2 - 14x + 48$.
2. $x^2 - 21x - 120$.
3. $5x^2 - 53x - 22$.
4. $16x^2 + 64x + 63$.
5. $54x^2 - 21x + 2$.
6. $12x^2 + 20xy - 8y^2$.

7. $x^4 - 13x^2 + 36$.

8. $x^3y - 3x^2y^2 - 18xy^3$.

9. $x^2 - 3x + 3$.

10. $3x^2 + 2x - 3$.

11. $x^2 - 4xy - 2y^2$.

12. $x^2 - 6ax - 9b^2 - 18ab$.

13. $abx^2 - (a^2 + b^2)x - (a^2 - b^2)$.

14. $x^2 + bd + dx + bx + cx^2 + cdx$.

15. $x^2 - 8xy + 15y^2 + 2x - 4y - 3$.

16. $x^2 + 3xy + 2y^2 + 3zx + 5yz + 2z^2$.

APPLICATIONS OF THE REMAINDER THEOREM AND SYNTHETIC DIVISION

On finding factors by aid of the remainder theorem. Let $f(x)$ denote a polynomial in x. By the remainder theorem, § 415, if b denote a number such that $f(b) = 0$, then $x - b$ is a factor of $f(x)$. We can sometimes find such a number b by inspection. **449**

Example. Factor $f(x) = x^3 - 5x + 4$.

$$\begin{array}{rrrr|l} 1 & +0 & -5 & +4 & \underline{1} \\ & 1 & 1 & -4 & \\ \hline 1 & 1 & -4, & 0 & \end{array}$$

Since $f(1) = 1 - 5 + 4 = 0$, $x - 1$ is a factor of $f(x)$.

Dividing $f(x)$ by $x - 1$, we obtain the quotient $x^2 + x - 4$.

Hence $f(x) = (x - 1)(x^2 + x - 4)$.

Note. Observe that whenever, as in this example, the algebraic sum of the coefficients of $f(x)$ is 0, $x - 1$ is a factor of $f(x)$.

Polynomials with integral coefficients. If asked to factor a polynomial, $f(x)$, with integral coefficients, it is usually best to look first for any factors of the *first degree with integral coefficients* that it may have. These may always be found by aid of the following principles, §§ 451, 452. **450**

A polynomial $f(x) = a_0x^n + a_1x^{n-1} + \cdots + a_n$, with integral coefficients, may have a factor of the form $x - b$, where b is an integer. But if so, b *must be a factor of* a_n, *the constant term of* f (x). **451**

Thus, let $f(x) = a_0x^3 + a_1x^2 + a_2x + a_3$. If $x - b$ is to be a factor of $f(x)$, we must have, § 415,

$$f(b) = a_0b^3 + a_1b^2 + a_2b + a_3 \equiv 0,$$

and therefore

$$(a_0b^2 + a_1b + a_2)b \equiv -a_3.$$

Therefore, since $a_0b^2 + a_1b + a_2$ denotes an integer, b is a factor of a_3.

Hence all such factors $x - b$ may be found as in the following example.

Example. Factor $f(x) = 3x^5 - 3x^4 - 13x^3 - 11x^2 - 10x - 6$.

The factors of the constant term, -6, are ± 1, ± 2, ± 3, ± 6, and b must have one of these values if $x - b$ is to be a factor of $f(x)$. We test these values of b as follows by synthetic division.

$$\begin{array}{rrrrrrl}
3 & -3 & -13 & -11 & -10 & -6 & \underline{\lfloor -1} \\
 & -3 & +6 & +7 & +4 & +6 & \\
\hline
3 & -6 & -7 & -4 & -6, & 0 & \underline{\lfloor -1} \\
 & -3 & +9 & -2 & +6 & & \\
\hline
3 & -9 & +2 & -6, & 0 & & \underline{\lfloor 3} \\
 & 9 & 0 & +6 & & & \\
\hline
3 & 0 & 2, & 0 & & &
\end{array}$$

Since $f(1) \neq 0$, $x - 1$ is not a factor. We therefore begin by testing $x - (-1)$ or $x + 1$. The division proves to be exact, the quotient being $Q_1 = 3x^4 - 6x^3 - 7x^2 - 4x - 6$, and the remainder 0. Hence $x + 1$ is one factor of $f(x)$, and Q_1 is the product of the remaining factors.

We have next to factor Q_1. It also proves to be exactly divisible by $x + 1$, the quotient being $Q_2 = 3x^3 - 9x^2 + 2x - 6$.

To factor Q_2, whose constant term is also -6, we test successively $x + 1$, $x - 2$, $x + 2$, but in each case obtain a remainder which is not 0. Hence none of these are factors. But testing $x - 3$, we find that it divides Q_2 exactly, the quotient being $Q_3 = 3x^2 + 2$. Therefore

$$f(x) = (x + 1)^2 (x - 3)(3x^2 + 2).$$

452 A polynomial $f(x) = a_0x^n + a_1x^{n-1} + \cdots + a_n$, with integral coefficients, may have a factor of the form $\alpha x - \beta$, where α and β denote integers which have no common factor. But if so, *α must be a factor of* a_0, *and β a factor of* a_n. This theorem includes that of § 451.

Thus, let $f(x) = a_0x^3 + a_1x^2 + a_2x + a_3$. If $\alpha x - \beta$, or $\alpha(x - \beta/\alpha)$, is to be a factor of $f(x)$, we must have, § 415,

$$f\left(\frac{\beta}{\alpha}\right) = a_0\frac{\beta^3}{\alpha^3} + a_1\frac{\beta^2}{\alpha^2} + a_2\frac{\beta}{\alpha} + a_3 \equiv 0,$$

and therefore $$a_0\beta^3 + a_1\beta^2\alpha + a_2\beta\alpha^2 + a_3\alpha^3 \equiv 0. \tag{1}$$

From (1) we obtain $$a_0\beta^3 \equiv -(a_1\beta^2 + a_2\beta\alpha + a_3\alpha^2)\alpha. \tag{2}$$

Therefore, since $a_1\beta^2 + a_2\beta\alpha + a_3\alpha^2$ is an integer, α is a factor of $a_0\beta^3$. But α has no factor in common with β^3, § 492, 2. Hence α is a factor of a_0, § 492, 1.

Again from (1), $$(a_0\beta^2 + a_1\beta\alpha + a_2\alpha^2)\beta \equiv -a_3\alpha^3, \tag{3}$$

whence, reasoning as before, we conclude that β is a factor of a_3.

Hence all such factors $\alpha x - \beta$ may be found as in the following example.

Example. Factor $f(x) = 6x^4 + 5x^3 + 3x^2 - 3x - 2$.

If $\alpha x - \beta$ is to be a factor of $f(x)$, α must have one of the values ± 1, ± 2, ± 3, ± 6, and β one of the values ± 1, ± 2; therefore β/α must have one of the values ± 1, ± 2, $\pm 1/2$, $\pm 1/3$, $\pm 2/3$, $\pm 1/6$.

We may test $\alpha x - \beta$ for these various values of β/α by dividing $f(x)$ by $x - \beta/\alpha$ synthetically. If the division is exact and Q denotes the quotient, then $\alpha x - \beta$ is a factor of $f(x)$ and Q/α is the product of the remaining factors, § 412, 3.

$$\begin{array}{l}
6 + 5 + 3 - 3 - 2 \,\lfloor -1/2 \\
\underline{\quad - 3 - 1 - 1 \quad 2} \\
6 + 2 + 2 - 4, \quad 0 \\
3 + 1 + 1 - 2 \,\lfloor 2/3 \\
\underline{\quad\; 2 + 2 + 2} \\
3 + 3 + 3, \quad 0 \\
1 + 1 + 1
\end{array}$$

Testing $x - 1$, $x + 1$, $x - 2$, $x + 2$, successively, we find that none of them divides $f(x)$ exactly. But $x + 1/2$ does, the quotient being $Q_1 = 6x^3 + 2x^2 + 2x - 4$. Hence $2x + 1$ is one factor of $f(x)$ and the product of the remaining factors is $Q_1/2 = 3x^3 + x^2 + x - 2$.

We next proceed to factor $Q_1/2$. If $\alpha x - \beta$ is to be a factor, β/α must have one of the values ± 1, ± 2, $\pm 1/3$, $\pm 2/3$. But we already know that $x - 1$, $x + 1$, $x - 2$, $x + 2$ are not factors, since they are not factors of $f(x)$. Testing $x - 1/3$, $x + 1/3$, we find that neither of them divides $Q_1/2$ exactly; but $x - 2/3$ does, the quotient being $Q_2 = 3x^2 + 3x + 3$. Hence $3x - 2$ is a factor of $Q_1/2$, and the product of the remaining factors is $Q_2/3 = x^2 + x + 1$. Therefore

$$f(x) = (2x + 1)(3x - 2)(x^2 + x + 1).$$

453 **Note.** It often becomes evident before a division by $x - b$ or $x - \beta/\alpha$ is completed that the division cannot be exact.

$$\begin{array}{l}
5 - \;\, 4 + 1 + 8 \,\lfloor 2 \\
\underline{\quad\;\; 10} \\
5 \quad\;\; 6
\end{array}$$

Thus, the reckoning here given suffices to prove that $x - 2$ will not divide $5x^3 - 4x^2 + x + 8$ exactly; for since the "divisor" 2, the last coefficient of Q already found, namely 6, and the unused coefficients of the dividend, namely 1 and 8, are all *positive*, the remaining coefficients of Q and R must be positive, that is, R cannot be 0.

$$\begin{array}{l}
3 + 1 + 1 - 2 \,\lfloor 1/3 \\
\underline{\quad\; 1} \\
3 \quad 2
\end{array}$$

Similarly we may conclude from the reckoning here given that $x - 1/3$ will not divide $3x^3 + x^2 + x - 2$ exactly. For the number which occurs next in the reckoning, namely, $2 \cdot 1/3$, or $2/3$, is a *fraction*, and this will cause the remaining coefficients of Q and R to be fractional, so that R cannot be 0.

454 It follows from § 452 that *a polynomial* $f(x) = x^n + \cdots + a_n$, *whose leading coefficient is* 1, *the rest being integers, cannot vanish for a rational fractional value of* x.

For if $f(\beta/\alpha) = 0$, then $f(x)$ must be exactly divisible by $\alpha x - \beta$ and therefore α must be a factor of 1, which is only possible when $\alpha = \pm 1$.

455 **Factoring polynomials and solving equations.** It follows from § 350 that the problem of resolving a polynomial $f(x)$ into its factors of the first degree is essentially the same as that of solving the equation $f(x) = 0$.

Example 1. Solve $f(x) = 2x^4 + x^3 - 17x^2 - 16x + 12 = 0$.

By § 452 we find that $f(x) = (2x - 1)(x + 2)^2(x - 3)$.

Hence the equation $f(x) = 0$ is equivalent to the four equations

$$2x - 1 = 0, \qquad x + 2 = 0, \qquad x + 2 = 0, \qquad x - 3 = 0.$$

Therefore the roots of $f(x) = 0$ are $1/2$, -2, -2, and 3.

Example 2. Solve $x^3 + 3x^2 = 10x + 24$.

Transposing, $\quad x^3 + 3x^2 - 10x - 24 = 0$.

Factoring, $\quad (x + 2)(x - 3)(x + 4) = 0$.

Hence the required roots are -2, 3, and -4.

EXERCISE XX

Factor the following expressions.

1. $x^3 - 7x + 6$.
2. $x^3 + 6x^2 + 11x + 6$.
3. $x^4 - 10x^3 + 35x^2 - 50x + 24$.
4. $x^4 - 2x^2 + 3x - 2$.
5. $6x^3 - 13x^2 - 14x - 3$.
6. $2x^3 - 5x^2y - 2xy^2 + 2y^3$.
7. $2x^4 - x^3 - 9x^2 + 13x - 5$.
8. $4x^6 - 41x^4 + 46x^2 - 9$.
9. $6x^5 + 19x^4 + 22x^3 + 23x^2 + 16x + 4$.
10. $5x^6 - 7x^5 - 8x^4 - x^3 + 7x^2 + 8x - 4$.

Solve the following equations.

11. $x^2 - 4x - 12 = 0$.
12. $6x^2 - 7x + 2 = 0$.
13. $x^2 - 5x = 14$.
14. $x^2 + 6x = 2$.
15. $x^3 - 9x^2 + 26x = 24$.
16. $x^4 + 2x^3 - 4x^2 - 2x + 3 = 0$
17. $x^3 - 1 = 0$.
18. $10x^3 - 9x^2 - 3x + 2 = 0$.

EXERCISE XXI

The following expressions can be factored by methods explained in the present chapter. Carry the factorization as far as is possible without introducing irrational or imaginary coefficients.

1. $6xy + 15x - 4y - 10$.
2. $a^2bc - ac^2d - ab^2d + bcd^2$.
3. $a^3(a - b) + b^3(b - a)$.
4. $a^5 - 81ab^4$.
5. $a^4b - a^2b^3 + a^3b^2 - ab^4$.
6. $3abx^2 - 6axy + bxy - 2y^2$.
7. $3x^6 - 192y^6$.
8. $(x^2 + x)^3 - 8$.
9. $64x^6y^3 - y^{15}$.
10. $x^2 - (a - b)x - ab$.
11. $x^{2n} - 3x^n - 18$.
12. $x - x^2 + 42$.
13. $3x^4 + 3x^3 - 24x - 24$.
14. $x^5 - 9x^3 + 8x^2 - 72$.
15. $2xc - a^2 + x^2 - 2ab + c^2 - b^2$.
16. $x^2(x^2 - 20) + 64$.
17. $a^2 - 2ab + b^2 - 5a + 5b + 6$.
18. $x^4 - 10x^2y^2 + 9y^4$.
19. $6x^2 - 7xy - 5y^2 - 4x - 2y$.
20. $x^4 - (a^2 + b^2)x^2 + a^2b^2$.
21. $4(xz + uy)^2 - (x^2 - y^2 + z^2 - u^2)^2$.
22. $14x^2 + 19x - 3$.
23. $1 + 19y - 66y^2$.
24. $xy^3 + 55x^2y^2 + 204x^3y$.
25. $a^4 - 18a^2b^2c^2 + 81b^4c^4$.
26. $(x^2 - 7x)^2 + 6x^2 - 42x$.
27. $8(x + y)^3 - 27(x - y)^3$.
28. $(x - 2y)x^3 - (y - 2x)y^3$.
29. $x^2 + a^2 - bx - ab + 2ax$.
30. $x^5 - y^5 - (x - y)^5$.
31. $x^5 - x^4 - 2x^3 + 2x^2 + x - 1$.
32. $b^4 + b^2 + 1$.
33. $2x^2 + 7xy + 3y^2 + 9x + 2y - 5$.
34. $a^4 + 4$.
35. $x^2 - xy - 2y^2 + 4xz - 5yz + 3z^2$.
36. $4a^4 + 3a^2b^2 + 9b^4$.
37. $x^2 - 8ax - 40ab - 25b^2$.
38. $x^8 + x^4 + 1$.
39. $(x^2 + 2x - 1)^2 - (x^2 - 2x + 1)^2$.
40. $(ax + by)^2 - (bx + ay)^2$.
41. $x^3 - ax^2 - b^2x + ab^2$.
42. $x^4 + bx^3 - a^3x - a^3b$.
43. $a^2 - 9b^2 + 12bc - 4c^2$.
44. $8a^3 + 12a^2 + 6a + 1$.
45. $x^4 - 2x^3 + 3x^2 - 2x + 1$.
46. $(ax + by)^2 + (bx - ay)^2$.
47. $4x^5 + 4x^4 - 37x^3 - 37x^2 + 9x + 9$.
48. $x^4 - 4x + 3$.
49. $x^2 + 5ax + 6a^2 - ab - b^2$.
50. $15x^3 + 29x^2 - 8x - 12$.
51. $abcx^2 + (a^2b^2 + c^2)x + abc$.
52. $2x^3 - ax^2 - 5a^2x - 2a^3$.

53. $(a - b)x^2 + 2ax + (a + b)$.   54. $x^{15} - y^{15}$.

55. $x^4 - 6x^3 + 7x^2 + 6x - 8$.   56. $4x^3 - 3x - 1$.

57. $3x^5 - 10x^4 - 8x^3 - 3x^2 + 10x + 8$.

58. $5x^4 + 24x^3 - 15x^2 - 118x + 24$.

59. $a^2bc + ac^2 + acd - abd - cd - d^2$.

60. $x^4 + y^4 + z^4 - 2x^2y^2 - 2y^2z^2 - 2z^2x^2$.

VII. HIGHEST COMMON FACTOR AND LOWEST COMMON MULTIPLE

HIGHEST COMMON FACTOR

456 **Highest common factor.** Let $A, B, \cdots$ denote rational, integral functions of one or more variables, as x, or x and y.

If $A, B, \cdots$ have no factor in common, we say that they are *prime to one another*. But if they have any common factor, they have one whose degree is highest; we call it their *highest common factor* (H.C.F.).

Thus, $x^2 + y^2$ and $x + y$ are prime to one another.

But $4xyz^5$, $8xz^4$, and $4x^2yz^3$ have the common factors $x, z, z^2, z^3, xz, xz^2, xz^3$, and their *highest* common factor is xz^3.

457 **Notes.** 1. We here ignore common numerical factors.

2. It is sometimes said of two or more functions which are prime to one another that their H.C.F. is 1.

3. The numerical value of the H.C.F. of A and B is not necessarily the greatest common divisor of integral numerical values of A and B. Thus, the H.C.F. of $(2x + 1)x$ and $(x - 1)x$ is x. But when $x = 4$, the values of $(2x + 1)x$ and $(x - 1)x$ are 36 and 12, and the greatest common divisor of 36 and 12 is not 4, but 12.

458 **Theorem 1.** *The* H.C.F. *of* A, B, $\cdots$ *is the product of all the different common prime factors of* A, B, $\cdots$, *each raised to the lowest power in which it occurs in any of these functions.*

The truth of this theorem is obvious if we suppose each of the functions $A, B, \cdots$ expressed in the form of a product of powers of its different prime factors and, as in § 430, assume that there is but one such expression for each function.

Thus, the *different* common prime factors of xyz^5, xz^4, and x^2yz^3 are x and z, and the lowest powers of x and z in any of these functions are x and z^3. Hence the H.C.F. is xz^3.

Observe that if it were possible to express a given function in more than one way in terms of its prime factors, the process described in the theorem might lead to various results corresponding to the various ways of expressing $A, B, \cdots$, and there might be more than one common factor of highest degree.

459 **Applications of this theorem.** When the given functions can be completely factored, their H.C.F. may be written down at once by aid of the theorem of § 458.

Example 1. Find the H.C.F. of $x^5y^2 - 6x^4y^3 + 9x^3y^4$ and $x^4y - 9x^2y^3$.

We have $\quad x^5y^2 - 6x^4y^3 + 9x^3y^4 = x^3y^2(x - 3y)^2$

and $\quad x^4y - 9x^2y^3 = x^2y(x - 3y)(x + 3y)$.

Hence the H.C.F. is $\quad x^2y(x - 3y)$.

Example 2. Find the H.C.F. of the following.

1. $2x^4y^2z^5$, $3x^5y^3z$, and $4x^3y^4$.
2. $x^2 - y^2$, $x^2 + 2xy + y^2$, and $x^3 + y^3$.
3. $x^2 - x - 6$, $x^2 + 6x + 8$, and $x^2 + 5x + 6$.
4. $x^3 - 6x^2 + 11x - 6$ and $2x^3 - 9x^2 + 7x + 6$.

460 If the prime factors of one of the functions $A, B, \cdots$ are known, we can find by division or the remainder theorem which of them, if any, are factors of all the other functions. The H.C.F. may then be obtained by aid of § 458.

Example 1. Find the H.C.F. of

$$f(x) = x^2 - 3x + 2 \text{ and } \phi(x) = x^4 - 3x^3 + 5x^2 - 8x + 5.$$

By inspection we have $f(x) = (x - 1)(x - 2)$. Testing $x = 1$ and $x = 2$ in $\phi(x)$, we find $\phi(1) = 0$, but $\phi(2) \neq 0$. Hence the H.C.F. is $x - 1$.

Example 2. Find the H.C.F. of

$$f(x) = x^2 + 4x + 4 \quad \text{and} \quad \phi(x) = x^4 + 5x^3 + 9x^2 + 8x + 4.$$

Since $f(x) = (x + 2)^2$, we must find not only whether $x + 2$ is a factor of $\phi(x)$, but whether it is a factor once or twice. Dividing $\phi(x)$ by $x + 2$

(synthetically) we obtain $Q_1 = x^3 + 3x^2 + 3x + 2$ and $R_1 = 0$; dividing Q_1 by $x + 2$ we obtain $Q_2 = x^2 + x + 1$ and $R_2 = 0$. Hence the H.C.F. of $f(x)$ and $\phi(x)$ is $(x + 2)^2$.

Example 3. Find the H.C.F. of the following.

1. $x^2 + x - 6$ and $2x^3 + 7x^2 + 4x + 3$.
2. $x^2 + 5x + 6$ and $x^4 + 6x^3 + 13x^2 + 16x + 12$.
3. $(x - 1)^2(x - 3)^3(3x + 1)^2$ and $x^4 - 5x^3 + x^2 + 21x - 18$.

461 **Theorem 2.** *Let* A *and* B *denote two given integral functions, and* M *and* N *any two integral functions or constants. Then every common factor of* A *and* B *is a factor of* MA + NB.

For let F denote a common factor of A and B.

Then $A \equiv GF$ and $B \equiv HF$,

where G and H are integral.

Hence $MA + NB \equiv M \cdot GF + N \cdot HF \equiv (MG + NH)F$,

where $MG + NH$ is integral.

Therefore F is a factor of $MA + NB$, § 424.

462 **Applications of this theorem.** By aid of this theorem the problem of finding the H.C.F. of two polynomials in x, whose degrees are the same, may be reduced to that of factoring a single polynomial of a lower degree.

Example 1. Find the H.C.F. of $A = x^2 + 2x - 4$ and $B = x^2 + x - 3$.

Subtracting B from A, we obtain $A - B = x - 1$.

Hence, § 461, $x - 1$ is the only possible common factor of A and B.

But, since A does not vanish when $x = 1$, $x - 1$ is not a factor of A.

Therefore A and B are prime to one another.

Example 2. Find the H.C.F. of

$$A = 2x^3 - 3x^2 - 3x + 2 \text{ and } B = 3x^3 - 2x^2 - 7x - 2.$$

1. We begin by multiplying A and B by numbers which will give results having the *same leading term*, namely, A by 3 and B by 2.

Then subtracting $2B$ from $3A$, we obtain

$$3A - 2B = -5x^2 + 5x + 10 = -5(x^2 - x - 2) = -5(x + 1)(x - 2).$$

Hence the only possible common factors of A and B are $x + 1$ and $x - 2$.

By the remainder theorem we find that *both* are factors of A and B. Hence the H.C.F. of A and B is $(x+1)(x-2)$.

2. Or we may add A and B, thus obtaining

$$A + B = 5x^3 - 5x^2 - 10x = 5x(x^2 - x - 2) = 5x(x+1)(x-2).$$

It is at once evident that x is not a factor of A or B, so that as before we have only to test $x+1$ and $x-2$.

Example 3. Find the H.C.F. of the following.

1. $x^4 - x^3 + 3x^2 - 4x - 12$ and $x^4 - x^3 + 2x^2 + 3x - 22$.
2. $6x^3 + 25x^2 + 5x + 4$ and $4x^3 + 15x^2 - 2x + 8$.

Theorem 3. *If the four integral functions* A, B, Q, R *are so related that* $\mathrm{A} \equiv \mathrm{QB} + \mathrm{R}$, *the common factors of* A *and* B *are the same as the common factors of* B *and* R. **463**

We have $$A \equiv QB + R, \tag{1}$$
and therefore $$A - QB \equiv R. \tag{2}$$

It follows from (2), by § 461, that every common factor of A and B is a factor of R and therefore a common factor of B and R.

And, conversely, it follows from (1), by § 461, that every common factor of B and R is a factor of A, and therefore a common factor of A and B.

Hence the common factors of A and B are the same as those of B and R.

The general method for finding the H.C.F. of two polynomials in x. **464** When one polynomial in x is divided by another, the dividend, divisor, quotient, and remainder are connected by the identity $A \equiv QB + R$. Hence it follows from § 463 that

The common factors of dividend and divisor are always the same as those of divisor and remainder.

By making use of this fact, the H.C.F. of any two polynomials in x may always be found. The method is analogous to that employed in arithmetic to find the greatest common divisor of two integers. It is described in the following rule, where A and B represent the given polynomials, A the one of higher degree if their degrees are not the same.

465 **Rule.** *Divide* A *by* B *and call the quotient* q *and the remainder* R_1.

Next divide B *by* R_1 *and call the quotient* q_1 *and the remainder* R_2.

Next divide R_1 *by* R_2, *and so on, continually dividing each new remainder by the one last obtained, until a remainder is reached which does not involve* x.

If this final remainder is not 0, A *and* B *have no common factor. If it is* 0, *the divisor which yielded it is the* H.C.F. *of* A *and* B.

For suppose, for the sake of definiteness, that the final remainder is R_3. Then according as (1) $R_3 = c$, where c denotes a constant *not* 0, or (2) $R_3 = 0$, we shall have

$$(1)\quad \begin{aligned} A &\equiv qB + R_1 \\ B &\equiv q_1R_1 + R_2 \\ R_1 &\equiv q_2R_2 + c \end{aligned} \qquad \text{or} \qquad (2)\quad \begin{aligned} A &\equiv qB + R_1 \\ B &\equiv q_1R_1 + R_2 \\ R_1 &\equiv q_2R_2 \end{aligned}$$

(1) *In this case* A *and* B *have no common factor.*

For it follows from the identities (1), by § 463, that A and B have the same common factors as B and R_1; B and R_1, as R_1 and R_2; R_1 and R_2, as R_2 and c.

Hence the pairs of functions A and B, B and R_1, R_1 and R_2, R_2 and c, all have the same common factors.

But as c is a constant (not 0), R_2 and c have *no* common factor. Hence A and B have none.

(2) *In this case* R_2 *is the* H.C.F. *of* A *and* B.

For since $R_1 \equiv q_2R_2$, every factor of R_2 is a common factor of R_1 and R_2, and R_2 itself is the common factor *of highest degree.*

But as the common factors of R_1 and R_2 are the same as those of A and B, the factor of highest degree common to R_1 and R_2 is also the factor of highest degree common to A and B. Hence R_2 is the H.C.F. of A and B.

Example 1. Find the H.C.F. of $x^2 + x + 1$ and $x^3 + x^2 + 2x + 3$.

Writing divisors at the left of dividends, we have

$$
\begin{array}{rrrl}
B = x^2 + x + 1 \,| & x^3 + x^2 + 2x + 3 \,\underline{|\, x} = q & & \\
 & \underline{x^3 + x^2 + x } & & \\
R_1 = & x + 3 \,| & x^2 + x + 1 \,\underline{|\, x - 2} = q_1 & \\
 & & \underline{x^2 + 3x } & \\
 & & -2x + 1 & \\
 & & \underline{-2x - 6} & \\
 & R_2 = & 7 &
\end{array}
$$

As the final remainder, R_2, is not 0, $x^2 + x + 1$ and $x^3 + x^2 + 2x + 3$ have no common factor.

Example 2. Find the H.C.F. of

$$x^3 + x^2 + 2x + 2 \text{ and } x^3 + 2x^2 + 3x + 2.$$

Arranging the work as in Ex. 1, we have

$$
\begin{array}{rrrrl}
B = x^3 + x^2 + 2x + 2 \,| & x^3 + 2x^2 + 3x + 2 \,\underline{|\, 1} & & & \\
 & \underline{x^3 + x^2 + 2x + 2} & & & \\
R_1 = & x^2 + x \,| & x^3 + x^2 + 2x + 2 \,\underline{|\, x} & & \\
 & & \underline{x^3 + x^2 } & & \\
 & R_2 = & 2x + 2 \,| & x^2 + x \,\underline{|\, x/2} & \\
 & & & \underline{x^2 + x} & \\
 & & R_3 = & 0 &
\end{array}
$$

Here the division by R_2 is exact, R_3 being 0. Hence, discarding the numerical factor 2 in R_2, we have H.C.F. $= x + 1$.

466 Here for the first time we have an actual proof — for functions of a single variable — that if two integral functions have any common factor, they have a *highest* common factor; for in §§ 463, 465 it is not assumed as in § 458 that an integral function can be expressed in but one way in terms of its prime factors.

467 Observe that in the proof in § 465 it is shown that

1. *Every two consecutive functions in the list* A, B, R_1, R_2, $\cdots$ *have the same* H.C.F. *as* A *and* B.

2. *Every common factor of* A *and* B *is a divisor of the* H.C.F. *of* A *and* B.

468 **Abridgments of this method.** 1. If any of the prime factors of A or B are obvious by inspection, begin by removing these factors and then find the H.C.F. of the resulting expressions. The result thus obtained, multiplied by such of the factors removed at the outset as are common to A and B, will be the H.C.F. of A and B, § 458.

The same course may be followed with any two consecutive functions in the list $A, B, R_1, R_2, \cdots$, since every two such functions have the same H.C.F. as A and B, § 467.

Thus, $A = x^4 + x^3 + 2x^2 + 2x$ and $B = x^4 + 2x^3 + 3x^2 + 2x$ obviously have the common factor x. Removing it we have $x^3 + x^2 + 2x + 2$ and $x^3 + 2x^2 + 3x + 2$, whose H.C.F. we have just found to be $x + 1$ (see § 465, Ex. 2). Hence the H.C.F. of A and B is $x(x + 1)$.

Again in Ex. 2, since x is a factor of R_1, but not of B, it cannot be a factor of the H.C.F. of B and R_1, and is therefore not a factor of the H.C.F. of A and B. Hence we may discard this factor x of R_1 and divide B by the remaining factor $x + 1$, so lessening the number of divisions.

2. In any of the divisions we may multiply or divide the divisor or dividend or any intermediate remainder by a numerical factor; for this will affect the subsequent remainders by a numerical factor at most, § 403, and therefore the H.C.F. not at all. This device enables us to avoid fractional coefficients when the given coefficients are rational.

3. It is advantageous to employ detached coefficients.

Example 1. Find the H.C.F. of

$$A = x^4 + 3x^3 + 2x^2 + 3x + 1 \text{ and } B = 2x^3 + 5x^2 - x - 1.$$

Multiplying A by 2 and using detached coefficients, we have

```
2 + 5 - 1 - 1 | 2 + 6 +  4 +  6 + 2 |_1
                2 + 5 -  1 -  1
                ---------------
                    1 +  5 +  7 + 2
                    2
                    ---------------
                    2 + 10 + 14 + 4 |_1
                    2 +  5 -  1 - 1
                    ---------------
                        5)5 + 15 + 5
                          1 +  3 + 1 | 2 + 5 - 1 - 1 |_2 - 1
                                       2 + 6 + 2
                                       ---------
                                       - 1 - 3 - 1
                                       - 1 - 3 - 1
                                       -----------
```

Hence the H.C.F. of A and B is $x^2 + 3x + 1$.

This reckoning may be more compactly arranged, as follows:

$2+5-1-1$	$2+6+4+6+2$	$1+1$
$2+6+2$	$2+5-1-1$	
$-1-3-1$	$1+5+7+2$	
$-1-3-1$	$2+10+14+4$	
	$2+5-1-1$	
	$5+15+5$	
	$1+3+1$	$2-1$

Example 2. Find the H.C.F. of the following.

$$2x^4+3x^3+4x^2+2x+1 \text{ and } 2x^4-x^3+2x^2+1.$$

Theorem. *If the coefficients of* A *and* B *are rational, so are those of their* H.C.F.; *and if the coefficients of* A *and* B *are real, so are those of their* H.C.F. **469**

For the H.C.F. of A and B can be found by the method of § 465. Hence its coefficients are rational combinations of the coefficients of A and B, and are therefore rational when these are rational, real when these are real.

This theorem has important consequences, some of which will be noticed later. By applying it, we can often shorten the work of finding a H.C.F.

Example. Find the H.C.F. of x^2-2 and x^3+x^2-5x+6.

Either these polynomials are prime to one another, or their H.C.F. is x^2-2 itself; for since the factors of x^2-2, namely $x+\sqrt{2}$ and $x-\sqrt{2}$, have irrational coefficients, neither of them can be the H.C.F.

But by trial we find that x^3+x^2-5x+6 is not exactly divisible by x^2-2. Hence these polynomials are prime to one another.

The H.C.F. of more than two polynomials in x. This may be obtained by the following rule. **470**

First find the H.C.F. *of two of the polynomials, next the* H.C.F. *of the result and the third polynomial, and so on. The final result will be the* H.C.F. *required.*

Thus, if D_1 is the H.C.F. of A and B, and D_2 is the H.C.F. of D_1 and C, then D_2 is the H.C.F. of A, B, and C.

For (1), D_2 is a factor of D_1 and C; and D_1 is a factor of A and B. Hence, § 427, D_2 is a factor of A, B, and C.

And (2), every common factor of A, B, and C is a common factor of D_1 and C, and therefore a factor of D_2, § 467. Hence D_2 is the *highest* common factor of A, B, and C.

The same conclusion follows from § 458.

Example. Find the H.C.F. of $A = x^4 + x^3 - x^2 + x - 2$,

$$B = 2x^4 + 5x^3 - 2x^2 - 7x + 2, \text{ and } C = 3x^4 - x^3 - x^2 - 2.$$

By § 465, we find that the H.C.F. of A and B is $D_1 = x^2 + x - 2$; and that the H.C.F. of D_1 and C is $x - 1$.

Hence the H.C.F. of A, B, and C is $x - 1$.

471 **The H.C.F. of polynomials in more than one variable.** The general problem of finding the H.C.F. of two such polynomials is too complicated to be considered here. But the H.C.F. of two polynomials which are *homogeneous* functions of *two* variables, as x and y, may readily be found by aid of the rule given in § 465.

EXERCISE XXII

Find the H.C.F. of the following.

1. $10x^3y^2z^5$, $4x^5yz^3$, $6x^4y^3z^5$, and $8x^4y^4z^4u$.
2. $(a+b)^2(a-b)$, $(a+b)(a-b)^2$, and $a^3b - ab^3$.
3. $y^4 + y^2 + 1$ and $y^2 - y + 1$.
4. $a^2 - 1$, $a^2 + 2a + 1$, and $a^3 + 1$.
5. $x^3 - 1$ and $x^3 + ax^2 - ax - 1$.
6. $x^4 - y^4$, $x^6 + y^6$, and $x^3 + x^2y + xy^2 + y^3$.
7. $x^2 + 5x + 6$, $x^2 + x - 2$, and $x^2 - 14x - 32$.
8. $(x-1)(x-2)$ and $5x^4 - 15x^3 + 8x^2 + 6x - 4$.
9. $x^3 - 1$ and $x^3 - 4x^2 - 4x - 5$.
10. $(x^2-1)^2(x+1)^2$ and $(x^3 + 5x^2 + 7x + 3)(x^2 - 6x - 7)$.
11. $(x-1)^2(x-2)^2$ and $(x^2 - 3x + 2)(2x^3 - 5x^2 + 5x - 6)$.
12. $2x^3 - 3x^2 - 11x + 6$ and $4x^3 + 3x^2 - 9x + 2$.
13. $x^3 - 2x^2 - 2x - 3$ and $2x^3 + x^2 + x - 1$.

14. $3x^3 + 2x^2 - 19x + 6$ and $2x^3 + x^2 - 13x + 6$.

15. $x^4 - x^3 - 3x^2 + x + 2$ and $2x^4 + 3x^3 - x^2 - 3x - 1$.

16. $3x^3 - 13x^2 + 23x - 21$ and $6x^3 + x^2 - 44x + 21$.

17. $3x^3 + 8x^2 - 4x - 15$ and $6x^4 + 10x^3 - 3x^2 - 2x + 5$.

18. $6x^5 + 7x^4 - 9x^3 - 7x^2 + 3x$ and $6x^5 + 7x^4 + 3x^3 + 7x^2 - 3x$.

19. $6x^4 - 3x^3 + 7x^2 + x - 3$ and $2x^4 + 3x^3 + 7x^2 + 3x + 9$.

20. $6x^5 - 4x^4 - 11x^3 - 3x^2 - 3x - 1$ and $4x^4 + 2x^3 - 18x^2 + 3x - 5$.

21. $x^5 - x^3 - 4x^2 - 3x - 2$ and $5x^4 - 3x^2 - 8x - 3$.

22. $3x^3 - x^2 - 12x + 4$, $x^3 - 2x^2 - 5x + 6$, and $7x^3 + 19x^2 + 8x - 4$.

23. $x^3 + ax^2 - 3x - 3a$, $x^3 - x^2 - 3x + 3$, and $x^3 + x^2 - 3x - 3$.

24. $7x^4y - 6x^3y^2 - 18x^2y^3 + 4xy^4$ and $14x^3y - 19x^2y^2 - 32xy^3 + 28y^4$.

25. $x(x-1)(x^3 + 4x^2 + 4x + 3)$ and $(x-1)(x+3)(12x^3 + x^2 + x - 1)$.

26. $4x^3 - 8x^2 - 3x + 9$ and $(2x^2 - x - 3)(2x^2 - 7x + 6)$.

LOWEST COMMON MULTIPLE

Lowest common multiple. A *common multiple* of two or more integral functions, A, B, $\cdots$, is an integral function which is exactly divisible by each of the functions A, B, $\cdots$. **472**

Among such common multiples there is one whose degree is lowest. We call this the *lowest common multiple* (L.C.M.) of A, B, $\cdots$.

Theorem 1. *The* L.C.M. *of two or more integral functions,* A, B, $\cdots$, *is the product of all the different prime factors of* A, B, $\cdots$, *each raised to the highest power in which it occurs in any of these functions.* **473**

This follows from the fact that a common multiple of A, B, $\cdots$ must contain every prime factor of each function A, B, $\cdots$ at least as often as it occurs in that function, hence all the factors mentioned in the theorem. And the common multiple *of lowest degree,* that is, the L.C.M., is the one which contains no factors besides these.

Here, as previously, we ignore numerical factors and assume that an integral function can be expressed in only one way as a product of powers of its different prime factors.

474 **Finding the L.C.M. by inspection.** If we can resolve $A, B, \cdots$ into their prime factors, we may obtain their L.C.M. at once by applying the theorem just demonstrated.

Example 1. Find the L.C.M. of $3x^5y^2z$, xy^4z^3, and $2x^2yz^6$.

Here the different prime factors, each raised to the highest power in which it occurs in any of the functions, are x^5, y^4, z^6.

Hence the L.C.M. is $x^5y^4z^6$.

Example 2. Find the L.C.M. of $x^2y^2 - 4xy^2 + 4y^2$ and $x^2y - 4y$.

We have $x^2y^2 - 4xy^2 + 4y^2 = y^2(x-2)^2$ and $x^2y - 4y = y(x-2)(x+2)$.

Hence the L.C.M. is $y^2(x-2)^2(x+2)$.

475 **Theorem 2.** *The* L.C.M. *of two integral functions,* A *and* B, *is the product of the two, divided by their* H.C.F.

For let D denote the H.C.F. of A and B, and let A_1 and B_1 denote the quotients obtained by dividing A and B by D, so that

$$A \equiv A_1D \text{ and } B \equiv B_1D.$$

Then if M denote the L.C.M. of A and B, we have

$$M \equiv A_1B_1D \equiv AB/D.$$

For evidently a common multiple of A and B must contain (1) the product of all prime factors common to A and B, namely D, (2) the product of all prime factors of A not belonging to B, namely A_1, (3) the product of all prime factors of B not belonging to A, namely B_1; and the *lowest* common multiple will contain no factors besides these.

476 **Corollary.** *The product of two integral functions,* A *and* B, *is equal to the product of their* L.C.M. *and their* H.C.F.

477 **General method for finding the L.C.M. of two polynomials in x.** It follows from §§ 465, 475 that the L.C.M. of two such polynomials, A and B, may always be obtained by the rule:

To find the L.C.M. *of* A and B, *divide* A *by the* H.C.F. *of* A *and* B, *and multiply the result by* B.

Observe that this is equivalent to multiplying B by all those prime factors of A which are not already present in B.

Example. Find the L.C.M. of

$$x^4 + 3x^3 + 2x^2 + 3x + 1 \text{ and } 2x^3 + 5x^2 - x - 1.$$

By § 465, we find that the H.C.F. $= x^2 + 3x + 1$.

Again $(2x^3 + 5x^2 - x - 1)/(x^2 + 3x + 1) = 2x - 1$.

Hence the L.C.M. is $(x^4 + 3x^3 + 2x^2 + 3x + 1)(2x - 1)$.

478 **The L.C.M. of more than two polynomials in x.** This may be obtained by the following rule.

First find the L.C.M. *of two of the polynomials, next the* L.C.M. *of the result and the third polynomial, and so on. The final result will be the* L.C.M. *required.*

This follows from the fact that each step in the process is equivalent to multiplying the L.C.M. last obtained by those prime factors of the next function which are not already present in that L.C.M.

Example. Find the L.C.M. of $A = x^4 + 3x^3 + 2x^2 + 3x + 1$, $B = 2x^3 + 5x^2 - x - 1$, and $C = 2x^3 - 3x^2 + 2x - 3$.

As we have just shown, § 477, Ex., the L.C.M. of A and B is

$$M_1 = (x^4 + 3x^3 + 2x^2 + 3x + 1)(2x - 1).$$

We have next to find the L.C.M. of M_1 and C.

By division we find that $2x - 1$ is prime to C, and by § 465 we find that the H.C.F. of $x^4 + 3x^3 + 2x^2 + 3x + 1$ and C is $x^2 + 1$.

Furthermore, $C/(x^2 + 1) = 2x - 3$.

Hence the L.C.M. of M_1 and C, and therefore of A, B, C, is

$$M = (x^4 + 3x^3 + 2x^2 + 3x + 1)(2x - 1)(2x - 3).$$

Observe that we do not multiply the factors of M_1 together before proceeding to find the H.C.F. of M_1 and C.

EXERCISE XXIII

Find the L.C.M. of the following.

1. $3x - 1$, $9x^2 - 1$, and $9x^2 + 1$.
2. $(a + b)(a^5 - b^5)$ and $(a - b)(a^5 + b^5)$.

3. $a^3 + a^2 + a$, $a^5 - a^3$, and $a^6 - a^3$.

4. $(x^3 - y^3)(x - y)^3$, $(x^4 - y^4)(x - y)^2$, and $(x^2 - y^2)^2$.

5. $x^2 - 3x + 2$, $x^2 - 5x + 6$, and $x^2 - 4x + 3$.

6. $x^2 - (y + z)^2$, $y^2 - (z + x)^2$, and $z^2 - (x + y)^2$.

7. $2x^2 + 3xy - 9y^2$, $3x^2 + 8xy - 3y^2$, and $6x^2 - 11xy + 3y^2$.

8. $x^3 + x^2 + x + 1$ and $x^3 - x^2 + x - 1$.

9. $2a^2x + 2x^2y + 3y^2x + 3a^2y$ and $(2x^2 - 3a^2)y + (2a^2 - 3y^2)x$.

10. $8x^3 - 18xy^2$, $8x^3 + 8x^2y - 6xy^2$, and $8x^2 - 2xy - 15y^2$.

11. $x^3 + y^3$, $x^3 - y^3$, and $x^4 + x^2y^2 + y^4$.

12. $x^6 - 1$, $3x^3 - 5x^2 - 3x + 5$, and $x^4 - 1$.

13. $8x^3 + 27$, $16x^4 + 36x^2 + 81$, and $6x^2 + 5x - 6$.

14. $x^2 - 4a^2$, $x^3 + 2ax^2 + 4a^2x + 8a^3$, and $x^3 - 2ax^2 + 4a^2x - 8a^3$.

15. $x^2 + 2x$, $x^2 + bx + 2x + 2b$, and $x^3 + ax^2 - b^2x - ab^2$.

16. $(x^2 + 3x + 2)(x^2 + 7x + 12)$ and $(x^2 + 5x + 6)(2x^2 - 3x - 5)$.

17. $(x^3 - 8)(27x^3 + 1)$ and $(2x^3 + 5x^2 + 10x + 4)(x^3 - x^2 - x - 2)$.

18. $x^3 - 6x^2 + 11x - 6$, $2x^3 - 7x^2 + 7x - 2$, and $2x^4 + x^2 - 13x + 6$.

19. $x^4 + 5x^2 + 4x + 5$, $2x^4 - x^3 + 10x^2 + 4x + 5$, and
$2x^4 + x^3 + 7x^2 + 3x + 3$.

20. $2x^4 - x^3 + 2x^2 + 3x - 2$, $2x^4 + 3x^3 - 4x^2 + 13x - 6$, and
$x^4 + 3x^3 + x^2 + 5x + 6$.

ON THE PRIME AND IRREDUCIBLE FACTORS OF FUNCTIONS OF A SINGLE VARIABLE

In the following theorems A and B denote polynomials in x.

479 **Fundamental Theorem.** *If* A *is prime to* B, *two integral functions,* M *and* N, *can be found such that*

$$\text{MA} + \text{NB} \equiv 1.$$

For if we apply the method of § 465 to A and B, we shall obtain as final remainder a constant, c, different from 0.

If we suppose, as in § 465, that c is the *third* remainder, and use the notation there explained, we have

1. $A \equiv qB + R_1$, and therefore 4. $R_1 \equiv A - qB$,
2. $B \equiv q_1R_1 + R_2$, 5. $R_2 \equiv B - q_1R_1$,
3. $R_1 \equiv q_2R_2 + c$, 6. $c \equiv R_1 - q_2R_2$.

Substitute in 6 the value of R_2 given by 5, collecting the R_1 and the B terms, and in the result substitute the value of R_1 given by 4, collecting the A and B terms. We thus obtain

$$c \equiv R_1 - q_2R_2$$
$$\equiv (1 + q_1q_2) R_1 - q_2B$$
$$\equiv (1 + q_1q_2) A - (q + q_2 + qq_1q_2) B.$$

Divide both sides of this last identity by c, and for $(1 + q_1q_2)/c$ and $-(q + q_2 + qq_1q_2)/c$, which are integral functions since c is a constant, write M and N. We obtain

$$1 \equiv MA + NB,$$

where, as just said, M and N are integral functions.

And we may demonstrate the theorem in the same way when the constant remainder, c, is obtained earlier or later than the third division.

Conversely, *If* $MA + NB \equiv 1$, *where* M *and* N *are integral, then* A *is prime to* B. **480**

For a common factor of A and B would be a factor of $MA + NB$, § 461, and therefore of 1, which is impossible.

The following theorems are some of the more important consequences of the fundamental theorem just demonstrated.

Theorem 1. *If* A *is prime to* B, *and the product* AC *is divisible by* B, *then* C *is divisible by* B. **481**

For since A is prime to B, we can find M and N, § 479, such that

$$MA + NB \equiv 1,$$

and therefore $$M \cdot AC + NC \cdot B \equiv C.$$

But B is a factor of both AC and B. Hence it is a factor of C, § 461.

Theorem 2. *If* A *is prime to each of the functions* B *and* C, *it is prime to their product,* BC. **482**

For since A is prime to B, we can find M and N, § 479, such that

$$MA + NB \equiv 1,$$

and therefore $$MC \cdot A + N \cdot BC \equiv C.$$

Hence, if A and BC had a common factor, it would be contained in C, § 461. But this is impossible, since A is prime to C.

483 **Corollary.** *If* A *is prime to each of the functions,* B, C, D, *and so on, it is prime to their product,* $B \cdot C \cdot D \cdots$.

For, as just demonstrated, A is prime to BC.

And since A is prime to BC and also to D, it is prime to the product BCD; and so on.

484 **Theorem 3.** *A composite function has one and but one set of prime factors.*

For let P denote the given function and n its degree.

If P is composite it has some factor A. If A, in turn, is composite, it has some factor B. Continuing thus, we must ultimately come upon a prime function; for the degrees of the successive functions $P, A, B, \cdots$ begin with the finite number n, decrease, and cannot fall below 1.

Let F denote this prime function. It is one of the prime factors of P, § 427, and we have $P \equiv FM$, where M is integral.

Similarly if M is composite, a prime function F' exists such that $M \equiv F'M'$, and therefore $P = FF'M'$, where M' is integral.

Continuing thus, we reach the conclusion that a series of prime functions $F, F', F'', \cdots$ exists, whose number cannot exceed n, such that

$$P \equiv F \cdot F' \cdot F'' \cdots.$$

Hence P has at least one set of prime factors.

Moreover P can have but one such set of factors. For, suppose that

$$P \equiv FF'F'' \cdots \equiv G \cdot G' \cdot G'' \cdots$$

where $G, G', G'', \cdots$ also denote prime functions.

Then G cannot be prime to *all* the functions $F, F', F'', \cdots$, for, if so, it would be prime to their product P, § 483, whereas it is a factor of P.

Suppose, therefore, that G is not prime to F, for example. Then G and F have a common factor. But G and F are prime functions, and two prime functions can have no factor in x but themselves in common. Hence G differs from F by a numerical factor, as c, at most, and we have $G \equiv cF$.

But substituting this value of G in the identity $FF'F'' \cdots \equiv GG'G'' \cdots$, and dividing both members by F, we have

$$F'F'' \cdots \equiv cG'G'' \cdots,$$

from which it follows by a mere repetition of our reasoning that G' differs from one of the functions F', F'', $\cdots$ by a numerical factor at most.

Continuing thus, we reach the conclusion that the set of functions G, G', G'', $\cdots$ differs from the set F, F', F'', $\cdots$ at most by numerical factors or in the order in which they are arranged.

Corollary. *A composite function can be expressed in only one way as a product of powers of its different prime factors.* **485**

This follows at once from the identity $P \equiv F \cdot F' \cdot F'' \cdots$, if we replace each set of equal factors in the product $F \cdot F' \cdot F'' \cdots$ by the corresponding power of one of these factors.

Irreducible factors. By the *irreducible* factors of an integral function with *rational* coefficients, we usually mean the factors of lowest degree with *rational coefficients.* **486**

Thus, while the *prime* factors of $(x-1)(x^2-2)$ are $x-1$, $x-\sqrt{2}$, $x+\sqrt{2}$, the *irreducible* factors are $x-1$ and x^2-2.

From the theorems just demonstrated and the theorem of § 469, it follows that **487**

A reducible integral function with rational coefficients can be expressed in only one way as a product of powers of its different irreducible factors.

DIGRESSION IN THE THEORY OF NUMBERS

Theorems analogous to those just demonstrated hold good for integral numbers. **488**

We shall employ the letters a, b, and so on, to represent integers, positive or negative (not 0), and shall mean by a **factor** of a any integer which exactly divides a.

A **prime number** is an integer which has no other factors than itself and 1. **489**

490 If two integers, a and b, have no common factor except 1, a is said to be **prime** to b.

491 **Theorem.** *If* a *is prime to* b, *two integers,* m *and* n, *can always be found such that*

$$ma + nb = 1.$$

For since a is prime to b, if we apply the usual method for finding the greatest common divisor, we shall obtain 1 as the final remainder. We may deduce the theorem from this fact by the reasoning of § 479.

Thus, let $a = 325$, $b = 116$. Applying the method for finding G.C.D., we have

$$\begin{array}{rrrl}
116 | 325 \lfloor 2 & & & \\
232 & & & \\
r_1 = 93 | 116 \lfloor 1 & i.e. & 325 = 2 \cdot 116 + 93, \text{ or } 93 = 325 - 2 \cdot 116 & (1) \\
93 & & & \\
r_2 = 23 | 93 \lfloor 4 & & 116 = 1 \cdot 93 + 23, \text{ or } 23 = 116 - 1 \cdot 93 & (2) \\
92 & & & \\
r_3 = 1 & & 93 = 4 \cdot 23 + 1, \text{ or } 1 = 93 - 4 \cdot 23 & (3)
\end{array}$$

Hence, starting with (3), and substituting first the value of 23 given by (2), and then the value of 93 given by (1), we have

$$\begin{aligned} 1 &= 93 - 4 \cdot 23 \\ &= 5 \cdot 93 - 4 \cdot 116 \\ &= 5 \cdot 325 - 14 \cdot 116. \end{aligned}$$

Therefore $5 \cdot 325 + (-14) \cdot 116 = 1.$

Hence we have found two integers, $m = 5$ and $n = -14$, such that

$$m \cdot 325 + n \cdot 116 = 1.$$

And similarly in every case.

Example. Find integers m and n such that $223m + 125n = 1$.

492 **Corollaries.** From this fundamental theorem we may derive for integral numbers theorems analogous to those derived for integral functions in §§ 481–485, and by the same reasoning. In particular we may prove that

1. *If* a *is prime to* b, *and the product* ac *is divisible by* b, *then* c *is divisible by* b.

2. *If* a *is prime to* b *and* c, *then* a *is prime to* bc.

3. *A composite number can be expressed in one way, and but one, as a product of powers of its different prime factors.*

VIII. RATIONAL FRACTIONS

REDUCTION OF FRACTIONS

Fractions. Let A and B denote any two algebraic expressions, of which B is not 0. The quotient of A by B, expressed in the form A/B, is called a *fraction;* and A is called the *numerator,* B the *denominator,* and A and B together the *terms* of this fraction. **493**

When both A and B are rational, A/B is called a **rational fraction.** **494**

When both A and B are integral, A/B is called a **simple fraction**; but if A or B is fractional, A/B is called a **complex fraction.** **495**

A simple fraction is called a **proper** or an **improper fraction,** according as the degree of its numerator is or is not less than that of its denominator. **496**

Thus, $\dfrac{x-y}{x^2+y^2}$ and $\dfrac{2x^2-3}{x^3+1}$ are proper, $\dfrac{2x^2+1}{x^2+1}$ and $\dfrac{x^3-3}{x^2+1}$ improper.

An improper fraction whose terms are functions of a single variable can be reduced to the sum of an integral expression and a proper fraction, § 400. This sum is called a **mixed expression.** **497**

Thus, $$\frac{2x^2+1}{x^2+1}=2-\frac{1}{x^2+1},\quad \frac{x^3-3}{x^2+1}=x-\frac{x+3}{x^2+1}.$$

Allowable changes in the form of a fraction. These depend on the following theorem, § 320, 1. **498**

The value of a fraction remains unchanged when its numerator and denominator are multiplied or divided by the same expression (*not* 0).

In particular, *we may change the signs of both numerator and denominator,* this being equivalent to multiplying both numerator and denominator by -1. Changing the sign of the numerator or of the denominator alone will change the sign of the fraction itself, § 320, 3.

If the numerator or denominator be a polynomial, changing its sign is equivalent to changing the signs of *all* its terms.

Thus, $$\frac{a+b-c}{a-b+c}=\frac{c-a-b}{b-c-a}=-\frac{c-a-b}{a-b+c}=-\frac{a+b-c}{b-c-a}.$$

If the numerator, denominator, or both, are products of certain factors, we may change the signs of an *even* number of these factors; but changing the signs of an *odd* number of them will change the sign of the fraction.

Thus, $$\frac{(a-b)(c-d)}{(e-f)(g-h)}=\frac{(b-a)(c-d)}{(e-f)(h-g)}=-\frac{(b-a)(d-c)}{(f-e)(g-h)}.$$

499 **Reduction of fractions.** To *simplify* a fraction is to cancel all factors which are common to its numerator and denominator, this being a change in the form of the fraction which will not affect its value, § 498.

When this has been done the fraction is said to be in its *lowest terms*, or to be *irreducible.*

We discover what these common factors are, or show that there are none, by the methods of Chapter VII. We look first for common monomial factors and other common factors which are obvious by inspection or which can be found by aid of the remainder theorem, and when these simpler methods fail we apply the general method of § 465.

The following examples will illustrate some of these methods.

Example 1. Simplify $(aec-ade)/(bde-ebc)$.

We have $$\frac{aec-ade}{bde-ebc}=\frac{ae(c-d)}{be(d-c)}=-\frac{a(c-d)}{b(c-d)}=-\frac{a}{b}.$$

Example 2. Simplify $(x^3+x^2+x+6)/(x^2+3x+2)$.

By inspection, the factors of the denominator are $x+1$ and $x+2$. Hence if numerator and denominator have any common factor, it must be one of these. Testing by synthetic division, we find that the numerator is not divisible by $x+1$, but is divisible by $x+2$, the quotient being x^2-x+3.

Hence $$\frac{x^3+x^2+x+6}{x^2+3x+2}=\frac{x^2-x+3}{x+1}.$$

Example 3. Simplify $(x^3 + 7x + 10)/(x^3 + 5x + 6)$.

Subtracting denominator from numerator, we have

$$x^3 + 7x + 10 - (x^3 + 5x + 6) = 2(x + 2).$$

Hence, if the numerator and denominator have any common factor, it must be $x + 2$, § 461. But the numerator does not vanish when $x = -2$. Hence, § 415, the fraction is already in its lowest terms.

Example 4. Simplify $\dfrac{a^2(b-c) + b^2(c-a) + c^2(a-b)}{(a-b)(b-c)(c-a)}$.

Here the only possible common factors are $a - b$, $b - c$, and $c - a$. Setting $a = b$ in the numerator, we have $b^2(b-c) + b^2(c-b)$, or 0. Hence, § 417, the numerator is divisible by $a - b$. And we may show in the same way that it is divisible by $b - c$ and $c - a$.

Therefore the numerator is exactly divisible by the denominator. But the two are of the same degree, namely *three*, in a, b, c. Their quotient must therefore be a mere number; and since the a^2 terms in the two, when arranged as polynomials in a, are $a^2(b-c)$ and $-a^2(b-c)$ respectively, this number is -1.

Hence the given fraction is equal to -1.

Example 5. Simplify $(2x^3+13x^2-6x+7)/(2x^4+5x^3+8x^2-2x+5)$.

By § 465, we find that the H.C.F. of numerator and denominator is $2x^2 - x + 1$. And dividing both numerator and denominator by $2x^2 - x + 1$, we obtain

$$\frac{2x^3 + 13x^2 - 6x + 7}{2x^4 + 5x^3 + 8x^2 - 2x + 5} = \frac{x+7}{x^2 + 3x + 5}.$$

EXERCISE XXIV

Reduce the following fractions to their lowest terms.

1. $\dfrac{x^5y^3 - 4x^3y^5}{x^3y^2 - 2x^2y^3}$.

2. $\dfrac{(x^6 - y^6)(x + y)}{(x^3 + y^3)(x^4 - y^4)}$.

3. $\dfrac{x^2 - 4x - 21}{x^2 + 2x - 63}$.

4. $\dfrac{3x^2 - 8x - 3}{3x^2 + 7x + 2}$.

5. $\dfrac{3x^2 - 18bx + 27b^2}{2x^2 - 18b^2}$.

6. $\dfrac{5x^2 + 6ax + a^2}{5x^2 + 2ax - 3a^2}$.

7. $\dfrac{(x^2 - 25)(x^2 - 8x + 15)}{(x^2 - 9)(x^2 - 7x + 10)}$.

8. $\dfrac{15x^2 - 46x + 35}{10x^2 - 29x + 21}$.

9. $\dfrac{x^4 + x^2y^2 + y^4}{(x^3 + y^3)(x^3 - y^3)}$.

10. $\dfrac{x^2 - y^2 + z^2 + 2xz}{x^2 + y^2 - z^2 + 2xy}$.

11. $\dfrac{(1 + xy)^2 - (x + y)^2}{1 - x^2}$.

12. $\dfrac{2mx - my - 12nx + 6ny}{6mx - 3my - 2nx + ny}$.

13. $\dfrac{2x^3 + 7x^2 - 7x - 12}{2x^3 + 3x^2 - 14x - 15}$.

14. $\dfrac{x^3 - 8x^2 + 19x - 12}{2x^3 - 13x^2 + 17x + 12}$.

15. $\dfrac{x^4 + x^3 + 5x^2 + 4x + 4}{2x^4 + 2x^3 + 14x^2 + 12x + 12}$.

16. $\dfrac{x^3 - 2x^2 - x - 6}{x^4 + 3x^3 + 8x^2 + 8x + 8}$.

17. $\dfrac{(x^2 + c^2)^2 - 4b^2x^2}{x^4 + 4bx^3 + 4b^2x^2 - c^4}$.

18. $\dfrac{(a - b)^3 + (b - c)^3 + (c - a)^3}{(a - b)(b - c)(c - a)}$.

OPERATIONS WITH FRACTIONS

500 **Lowest common denominator.** To add or subtract fractions, we first reduce them to equivalent fractions having a common denominator.

Evidently the lowest common multiple of the given denominators will be the common denominator of lowest degree. It is therefore called the *lowest common denominator* (L.C.D.) of the given fractions.

Example. Reduce $\dfrac{a}{bc}$, $\dfrac{b}{ca}$, and $\dfrac{c}{ab}$ to a lowest common denominator.

The L.C.M. of the given denominators is abc.

To reduce a/bc to an equivalent fraction having the denominator abc, we must multiply both its terms by a.

Similarly we must multiply both terms of b/ca by b, and both terms of c/ab by c.

Thus, $$\frac{a}{bc} = \frac{a^2}{abc},\quad \frac{b}{ca} = \frac{b^2}{abc},\quad \frac{c}{ab} = \frac{c^2}{abc}.$$

501 *To reduce two or more fractions to equivalent fractions having a lowest common denominator, find the lowest common multiple of the given denominators.*

Then in each fraction replace the denominator by this lowest common multiple, and multiply the numerator by the new factor thus introduced in the denominator

502 **Addition and subtraction.** For fractions which have a common denominator the rule of addition and subtraction is contained in the formula, § 320, 4,

$$\frac{a}{d}+\frac{b}{d}-\frac{c}{d}=\frac{a+b-c}{d}.$$

Hence to find the algebraic sum of two or more fractions,

If necessary, reduce them to a lowest common denominator.

Connect the numerators of the resulting fractions by the signs which connect the given fractions, and write the result over the common denominator.

Finally, simplify the result thus obtained.

This rule applies when *integral* expressions take the place of one or more of the fractions; for such an expression may be regarded as a *fraction whose denominator is* 1.

It is best to reduce each of the given fractions to its lowest terms, unless a factor which would thus be cancelled occurs in one of the other denominators.

Care should be taken that the expression selected as the lowest common denominator actually is this denominator. A frequent mistake is to treat factors like $a-b$ and $b-a$, which differ only in sign, as distinct, and to introduce both of them in the lowest common denominator.

It is often better to combine the given fractions by pairs.

Example 1. Simplify $\frac{1}{a+b}+\frac{1}{a-b}-\frac{2b}{a^2-b^2}$.

Here the lowest common denominator is a^2-b^2, and we have

$$\frac{1}{a+b}+\frac{1}{a-b}-\frac{2b}{a^2-b^2}=\frac{a-b}{a^2-b^2}+\frac{a+b}{a^2-b^2}-\frac{2b}{a^2-b^2}$$

$$=\frac{a-b+a+b-2b}{a^2-b^2}=\frac{2a-2b}{a^2-b^2}=\frac{2}{a+b}.$$

Observe that the denominator of the sum, when reduced to its lowest terms, may be but a factor of the lowest common denominator.

Example 2. Simplify $x - \dfrac{1}{1-x} - \dfrac{x^3 - 3x + 1}{x^2 - 1}$.

Since the first denominator is 1, and the second is $-(x-1)$, the lowest common denominator is $x^2 - 1$. We therefore have

$$x - \frac{1}{1-x} - \frac{x^3 - 3x + 1}{x^2 - 1} = \frac{x^3 - x}{x^2 - 1} + \frac{x+1}{x^2-1} - \frac{x^3 - 3x + 1}{x^2 - 1}$$

$$= \frac{x^3 - x + x + 1 - x^3 + 3x - 1}{x^2 - 1} = \frac{3x}{x^2 - 1}.$$

Example 3. Simplify $\dfrac{1}{x-2} + \dfrac{2}{x+1} - \dfrac{2}{x-1} - \dfrac{1}{x+2}$.

Here it is simpler to combine the fractions by pairs. Thus,

$$\frac{1}{x-2} - \frac{1}{x+2} = \frac{x+2-(x-2)}{x^2-4} = \frac{4}{x^2-4}.$$

$$\frac{2}{x+1} - \frac{2}{x-1} = 2\frac{x-1-(x+1)}{x^2-1} = -\frac{4}{x^2-1}.$$

$$\frac{4}{x^2-4} - \frac{4}{x^2-1} = 4\frac{x^2-1-(x^2-4)}{(x^2-1)(x^2-4)} = \frac{12}{x^4 - 5x^2 + 4}.$$

Example 4. Simplify $\dfrac{x^2-1}{x^4 + x^2 - 2x} + \dfrac{2x^2 + 3x - 2}{2x^3 + x^2 + 3x - 2}$.

By aid of the remainder theorem, § 415, we find that $x - 1$ is a common factor of the two terms of the first fraction, but is not a factor of the second denominator. We therefore simplify the first fraction by cancelling $x - 1$ in both terms, thus obtaining $(x+1)/(x^3 + x^2 + 2x)$.

By § 465, the H.C.F. of $x^3 + x^2 + 2x$ and $2x^3 + x^2 + 3x - 2$ is $x^2 + x + 2$; and $x^3 + x^2 + 2x = (x^2 + x + 2)x$, $2x^3 + x^2 + 3x - 2 = (2x-1)(x^2 + x + 2)$.

Before reducing to a common denominator, we inquire whether $2x - 1$ is also a factor of the numerator $2x^2 + 3x - 2$. We find that it is, and cancelling it, reduce the second fraction to $(x+2)/(x^2 + x + 2)$.

Hence
$$\frac{x^2-1}{x^4 + x^2 - 2x} + \frac{2x^2 + 3x - 2}{2x^3 + x^2 + 3x - 2}$$

$$= \frac{x+1}{x(x^2+x+2)} + \frac{x+2}{x^2+x+2}$$

$$= \frac{x + 1 + x^2 + 2x}{x^3 + x^2 + 2x} = \frac{x^2 + 3x + 1}{x^3 + x^2 + 2x}.$$

503 Multiplication. The product of two or more fractions may be found by applying the following theorem.

The product of two or more fractions is the fraction whose numerator is the product of the numerators of the given fractions, and its denominator the product of their denominators.

Thus, $$\frac{a}{b}\cdot\frac{c}{d}=\frac{ac}{bd}.$$

For the product of each member by bd is ac (see § 253).

Thus, $$\frac{ac}{bd}bd = ac\,;\ \text{and}\ \frac{a}{b}\cdot\frac{c}{d}\cdot bd = \frac{a}{b}b\cdot\frac{c}{d}d = ac. \qquad \text{§§ 252, 254}$$

In particular, to multiply a fraction by an *integral* expression, multiply its numerator by that expression.

The fraction ac/bd should always be reduced to its lowest terms before the multiplications indicated in its numerator and denominator are actually performed.

Example 1. Multiply $(x^3-1)/(x^3+1)$ by $(x+1)/(x-1)$.

We have $$\frac{x^3-1}{x^3+1}\cdot\frac{x+1}{x-1}=\frac{(x^3-1)(x+1)}{(x^3+1)(x-1)}=\frac{x^2+x+1}{x^2-x+1}.$$

Example 2. Multiply $1-(x-2)/(x^2+x-2)$ by $(x+2)/x$.

We have $$\left(1-\frac{x-2}{x^2+x-2}\right)\cdot\frac{x+2}{x}=\frac{x^2}{x^2+x-2}\cdot\frac{x+2}{x}=\frac{x}{x-1}.$$

Involution. From § 503 we derive the rule: **504**

To raise a fraction to any given power, raise both numerator and denominator to that power.

Thus, $$\left(\frac{a}{b}\right)^n=\frac{a^n}{b^n}.$$

For $$\left(\frac{a}{b}\right)^n=\frac{a}{b}\cdot\frac{a}{b}\cdots\text{to } n \text{ factors}=\frac{a\cdot a\cdots\text{to } n\text{ factors}}{b\cdot b\cdots\text{to } n\text{ factors}}=\frac{a^n}{b^n}.$$

Example. Find the cube of $-ab^2c^3/efg^4$.

We have $$\left(-\frac{ab^2c^3}{efg^4}\right)^3=-\frac{a^3b^6c^9}{e^3f^3g^{12}}.$$

Division. To *invert* a fraction, as a/b, is to interchange its numerator and denominator. The fraction b/a thus obtained is called the *reciprocal* of a/b. **505**

To divide one fraction by another, multiply the dividend by the reciprocal of the divisor.

Thus, $$\frac{a}{b} \div \frac{c}{d} = \frac{a}{b} \cdot \frac{d}{c} = \frac{ad}{bc}.$$

For the product of each member by c/d is a/b (see § 253).

Thus, $$\frac{a}{b} \div \frac{c}{d} \times \frac{c}{d} = \frac{a}{b}; \text{ and } \frac{ad}{bc} \cdot \frac{c}{d} = \frac{adc}{bcd} = \frac{a}{b}. \qquad \S\S\ 252,\ 254,\ 503$$

In particular, to divide a fraction by an *integral* expression, multiply its denominator by that expression.

Example 1. Divide

$$(x^2 - xy + y^2)/(x^2 - y^2) \text{ by } (x^4 + x^2y^2 + y^4)/(x^4 - y^4).$$

We have $$\frac{x^2 - xy + y^2}{x^2 - y^2} \div \frac{x^4 + x^2y^2 + y^4}{x^4 - y^4} = \frac{x^2 - xy + y^2}{x^2 - y^2} \cdot \frac{x^4 - y^4}{x^4 + x^2y^2 + y^4}$$

$$= \frac{x^2 + y^2}{x^2 + xy + y^2}.$$

Example 2. Divide $(x^2 + 5x + 6)/(x + 1)$ by $x^2 + 6x + 8$.

We have

$$\frac{x^2 + 5x + 6}{x + 1} \div (x^2 + 6x + 8) = \frac{x^2 + 5x + 6}{(x + 1)(x^2 + 6x + 8)}$$

$$= \frac{(x + 2)(x + 3)}{(x + 1)(x + 2)(x + 4)} = \frac{x + 3}{x^2 + 5x + 4}.$$

506 Rational expressions in general. Complex and continued fractions. We have shown that the sum, difference, product, and quotient of two fractions are themselves fractions. It therefore follows that *every rational expression,* § 267, *can be reduced to the form of a simple rational fraction.*

No general rule can be given as to the best method for reducing a given rational expression to its simplest form. Seek to avoid all unnecessary steps. In particular be on the watch for opportunities to reduce fractions to their lowest terms, and perform no multiplications of polynomials until the reduction can be carried no further without so doing.

507 As has already been said, § 495, the fraction A/B is called *complex* when A or B, either or both, is fractional.

In simplifying a complex fraction, A/B, it is sometimes better to divide A by B at the outset by the rule of § 505, sometimes better first to multiply both A and B by the L.C.M. of all the denominators in A and B. Before taking either of these steps it is often best to simplify A and B separately.

Example 1. Simplify $\left(\dfrac{a+b}{a-b}+1\right)\Big/\left(\dfrac{a-b}{a+b}+1\right)$.

We have
$$\frac{\dfrac{a+b}{a-b}+1}{\dfrac{a-b}{a+b}+1}=\frac{\dfrac{a+b+a-b}{a-b}}{\dfrac{a-b+a+b}{a+b}}=\frac{\dfrac{2a}{a-b}}{\dfrac{2a}{a+b}}$$
$$=\frac{2a}{a-b}\cdot\frac{a+b}{2a}=\frac{a+b}{a-b}.$$

Observe that when the terms of a complex fraction are simple fractions we may cancel any factors which are common to the numerators or to the denominators of these fractions. Thus, in the third expression above, we may cancel $2a$.

Example 2. Simplify $\left(\dfrac{a}{a-b}-\dfrac{b}{a+b}\right)\Big/\left(\dfrac{a}{a+b}+\dfrac{b}{a-b}\right)$.

We may proceed as in Ex. 1; but a simpler method is to begin by multiplying both terms by $(a+b)(a-b)$. We thus obtain

$$\frac{\dfrac{a}{a-b}-\dfrac{b}{a+b}}{\dfrac{a}{a+b}+\dfrac{b}{a-b}}=\frac{a(a+b)-b(a-b)}{a(a-b)+b(a+b)}=\frac{a^2+ab-ba+b^2}{a^2-ab+ba+b^2}=1.$$

Example 3. Simplify $\dfrac{a}{b+\dfrac{c}{d+\dfrac{e}{f}}}$

Working from the bottom upwards, we have

$$\frac{a}{b+\dfrac{c}{d+\dfrac{e}{f}}}=\frac{a}{b+\dfrac{cf}{df+e}}=\frac{a(df+e)}{b(df+e)+cf}=\frac{adf+ae}{bdf+be+cf}.$$

Complex fractions like that in Ex. 3 are called *continued fractions.*

EXERCISE XXV

Simplify the following expressions.

1. $\dfrac{1}{2a-3b}+\dfrac{1}{2a+3b}-\dfrac{6b}{4a^2-9b^2}$.

2. $\dfrac{1}{x+1}+\dfrac{1}{x^2-1}+\dfrac{1}{x^3+1}$.

3. $\dfrac{1}{x^2-3x+2}+\dfrac{1}{x^2-5x+6}+\dfrac{1}{x^2-4x+3}$.

4. $\dfrac{x+1}{(x-1)(x-2)}-\dfrac{x+2}{(2-x)(x-3)}+\dfrac{x+3}{(3-x)(1-x)}$.

5. $\dfrac{1}{x+b}-\dfrac{1}{x+c}+\dfrac{1}{x-b}-\dfrac{1}{x-c}$.

6. $\dfrac{a}{(a-b)(a-c)}+\dfrac{b}{(b-c)(b-a)}+\dfrac{c}{(c-a)(c-b)}$.

7. $\dfrac{yz(x+a)}{(x-y)(x-z)}+\dfrac{zx(y+a)}{(y-z)(y-x)}+\dfrac{xy(z+a)}{(z-x)(z-y)}$.

8. $x+\dfrac{1}{3-2x}-\dfrac{8x^4-33x}{8x^3-27}-\dfrac{2x+6}{4x^2+6x+9}$.

9. $\left(x+\dfrac{1}{x}\right)^2+\left(y+\dfrac{1}{y}\right)^2+\left(xy+\dfrac{1}{xy}\right)^2-\left(x+\dfrac{1}{x}\right)\left(y+\dfrac{1}{y}\right)\left(xy+\dfrac{1}{xy}\right)$

10. $\dfrac{(a+b)^3-c^3}{a+b-c}+\dfrac{(b+c)^3-a^3}{b+c-a}+\dfrac{(c+a)^3-b^3}{c+a-b}$.

11. $\dfrac{x^2-4}{x^3-3x^2-x+6}-\dfrac{3x^2-14x-5}{3x^3-2x^2-10x-3}$.

12. $\dfrac{1}{x^4-4x^2-x+2}+\dfrac{1}{2x^4-3x^3-5x^2+7x-2}+\dfrac{1}{2x^4+3x^3-2x^2-2x+1}$.

13. $\left(a^4-\dfrac{1}{a^4}\right)\div\left(a-\dfrac{1}{a}\right)$. 14. $\left(\dfrac{1}{a^3}-\dfrac{1}{a^2}+\dfrac{1}{a}\right)(a^4+a^3)$.

15. $\dfrac{x^2-5x+6}{x^2+3x-4}\cdot\dfrac{x^2+7x+12}{x^2-8x+15}\div\dfrac{x^2+x-6}{x^2-4x-5}$.

16. $\dfrac{1}{x}-\left\{1-\left[\dfrac{x-1}{x}+\dfrac{1}{2}\left(\dfrac{x-1}{x+1}-\dfrac{(x-2)(x-3)}{x(x+1)}\right)\right]\right\}$.

17. $\dfrac{ax+x^2}{2b-cx}\cdot\dfrac{2bx^2-cx^3}{(a+x)^2}$. 18. $(x^2-y^2-z^2+2yz)\div\dfrac{x-y+z}{x-y-z}$.

19. $\left(\dfrac{a+b}{a-b}-\dfrac{a^3+b^3}{a^3-b^3}\right)\left(\dfrac{a+b}{a-b}+\dfrac{a^2+b^2}{a^2-b^2}\right)$.

20. $\dfrac{\dfrac{1}{x}-\dfrac{1}{y+z}}{\dfrac{1}{x}+\dfrac{1}{y+z}}\div\dfrac{\dfrac{1}{y}-\dfrac{1}{x+z}}{\dfrac{1}{y}+\dfrac{1}{x+z}}$. 21. $\dfrac{\dfrac{a}{b}+\dfrac{b}{a}}{\dfrac{a}{b}-\dfrac{b}{a}}\div\dfrac{\dfrac{1}{a^4}-\dfrac{1}{b^4}}{\left(\dfrac{1}{a}+\dfrac{1}{b}\right)^2}$.

22. $\dfrac{x-2}{x-2-\dfrac{x}{x-\dfrac{x-1}{x-2}}}$. 23. $x+\dfrac{1}{x+\dfrac{1}{x+\dfrac{1}{x}}}$.

INDETERMINATE FORMS

Limits. Suppose the variable x to be taking successively **508** the values $1/2$, $3/4$, $7/8$, $15/16$, and so on without end; then evidently x is approaching the value 1, and in such a manner that the difference $1-x$ will ultimately become and remain less than every positive number that we can assign, no matter how small that number may be. We indicate all this by saying that as x runs through the never-ending sequence of values $1/2, 3/4, 7/8, 15/16, \cdots$, it *approaches* 1 *as limit.*

And in general, if x *denote a variable which is supposed to be running through some given but never-ending sequence of values, and if there be a number* a *such that the difference* a − x *will ultimately become and remain numerically less than every positive number that we can assign, we say that* x *approaches this number* a *as limit.*

To indicate that x is approaching the limit a, we write $x \doteq a$, or $a = \lim x$.

It will be noticed that the word *variable* has here a more restricted meaning than in § 242.

Whether or not a variable of the kind here under consideration approaches a limit depends on the sequence of values through which it is supposed to be running. Thus, if the sequence be 1, 2, 1, 2, ⋯, the variable will *not* approach a limit.

A full discussion of variables and limits will be found in §§ 187–205, which the student is advised to read, at least in part, in this connection.

509 **Theorems respecting limits.** In § 203 it is proved that if the variables x and y approach limits, then their sum, difference, product, and quotient also approach limits, and

$$\lim (x + y) = \lim x + \lim y.$$
$$\lim (x - y) = \lim x - \lim y.$$
$$\lim xy = \lim x \cdot \lim y.$$
$$\lim \frac{x}{y} = \frac{\lim x}{\lim y}, \text{ unless } \lim y = 0.$$

From these theorems it follows that if $F(x)$ denote any given rational function of x, and $F(a)$ its value when $x = a$, then F(x) *will approach* F(a) *as limit whenever* x *approaches* a *as limit,* that is,

$$\lim_{x \doteq a} F(x) = F(a),$$

where $\lim\limits_{x \doteq a} F(x)$ is read "limit of $F(x)$, as x approaches a."

Thus, $\lim\limits_{x \doteq a} (2x^2 - 3x + 1) = 2a^2 - 3a + 1.$

510 **Infinity.** Evidently if x be made to run through the never ending sequence 1, 2, 3, 4, ⋯, it will ultimately become and remain greater than every number that we can assign.

A variable x *which will thus ultimately become and remain numerically greater than every number that we can assign is said to approach infinity.*

For the word *infinity* we employ the symbol ∞, and, to indicate that x is approaching infinity, write $x \doteq \infty$, or $\lim x = \infty$.

511 **Note.** It is important to notice that ∞, as thus defined, does not denote a definite number, and that the rules for reckoning with numbers do not apply to it. Illustrations of this will be found below.

The phrase "x is approaching infinity" is merely an abbreviation for "x is a variable which will ultimately become and remain greater than every number that we can assign."

When, as is sometimes convenient, we write $\lim x = \infty$, of course the word *limit* does not have the definite meaning given it in § 508.

Theorem. *Given any fraction whose numerator is a constant and its denominator a variable.* **512**

If the denominator approach 0 *as limit, the fraction will approach* ∞*; and if the denominator approach* ∞, *the fraction will approach* 0 *as limit.*

Thus, consider the fraction $1/x$.

If x approach 0 by running through the sequence of values 1, .1, .01, .001, $\cdots$, then $1/x$ will run through the sequence 1, 10, 100, 1000, $\cdots$, and will therefore approach ∞.

And if x approach ∞ by running through the sequence 1, 10, 100, 1000, $\cdots$, then $1/x$ will run through the sequence 1, .1, .01, .001, $\cdots$, and will therefore approach 0 as limit.

And so in general.

Indeterminate forms. A rational fraction of the form $f(x)/\phi(x)$ has a definite value for any given value of x except one for which $\phi(x) = 0$. But when $\phi(x) = 0$, the fraction takes one of the forms $0/0$ or $a/0$, which are arithmetically meaningless, § 103. It is convenient, nevertheless, to assign a meaning to the fraction in both of these cases. **513**

The form $0/0$. The fraction $(x^2 - 1)/(x - 1)$ takes the form $0/0$ when $x = 1$. **514**

Now, *except* when $x = 1$, we can divide $x^2 - 1$ by $x - 1$, and have

$$(x^2 - 1)/(x - 1) = x + 1.$$

This is true however little x may differ from 1. Hence if, without actually giving x the value 1, we make it approach 1 as *limit*, we shall have

$$\lim (x^2 - 1)/(x - 1) = \lim (x + 1) = 2.$$

Thus, while the rules of reckoning give us no meaning for $(x^2 - 1)/(x - 1)$ when $x = 1$, they enable us to prove that this fraction always approaches the definite limit 2 when x approaches 1 as limit.

Now we have just shown, § 509, that $F(a) = \lim_{x \doteq a} F(x)$ whenever $F(x)$ represents a rational function and $F(a)$ *has* a meaning. It is therefore convenient when, as here, the rules of reckoning give us no meaning for $F(a)$, to *assign* to it the value $\lim_{x=a} F(x)$; in other words, to make the formula $F(a) = \lim_{x=a} F(x)$ our *definition* of $F(a)$.

We therefore give $(x^2 - 1)/(x - 1)$ the value 2 when $x = 1$. We then have $(x^2 - 1)/(x - 1) = x + 1$ for all values of x, *the value* 1 *included.*

And for a like reason, to every fraction which can be written in the form $(x - a)f(x)/(x - a)\phi(x)$, where $f(x)$ and $\phi(x)$ are integral and $\phi(x)$ is not divisible by $x - a$, we assign the value $f(a)/\phi(a)$ when $x = a$, and so have

$$\frac{(x-a)f(x)}{(x-a)\phi(x)} = \frac{f(x)}{\phi(x)}$$

for all values of x, the value a included.

515 **The form a/0.** The fraction $1/x$ takes the form $1/0$ when $x = 0$.

While we cannot divide 1 by 0, we can divide 1 by a value of x which differs as little as we please from 0. Moreover we have shown, § 512, that if x be made to approach 0 as limit, then $1/x$ will approach ∞.

We therefore assign to $1/0$, and in general to $a/0$, when $a \neq 0$, the "value" ∞, writing

$$\frac{a}{0} = \infty.$$

And for a like reason, to every fraction of the form

$$f(x)/(x-a)\phi(x),$$

where $f(x)$ and $\phi(x)$ are integral and $f(x)$ is not divisible by $x - a$, we assign the "value" ∞ when $x = a$; our meaning being that the fraction will always approach ∞ when x is made to approach a as limit.

Conclusion as to the values of a fraction. From §§ 514, 515 **516** we draw the following conclusions regarding a simple fraction of the form $f(x)/\phi(x)$.

1. If $f(x)/\phi(x)$ is in its lowest terms, it will *vanish* for values of x which make its *numerator* $f(x)$ vanish, and become *infinite* for values of x which make its *denominator* $\phi(x)$ vanish. For all other finite values of x it has a value different from both 0 and ∞.

2. But if $f(x)$ and $\phi(x)$ have the common factor $x - a$, and $f(x)$ contains this factor m times and $\phi(x)$ contains it n times, then $f(x)/\phi(x)$ will vanish for $x = a$ when $m > n$, become infinite when $m < n$, and have a value different from both 0 and ∞ when $m = n$.

Thus when $x = 2$, we have

$$\frac{x-2}{x+1} = 0,\ \frac{x+1}{x-2} = \infty,\ \frac{(x-2)^3}{x(x-2)} = 0,\ \frac{(x-2)}{x(x-2)^2} = \infty,\ \frac{(x-2)^2}{x(x-2)^2} = \frac{1}{2}.$$

The form ∞/∞. It is often important to know what limit **517** the value of a fraction $f(x)/\phi(x)$ approaches when x is indefinitely increased, that is, when $x \doteq \infty$.

Consider the following examples.

We have shown, § 512, that $1/x,\ 1/x^2,\ \cdots \doteq 0$, when $x \doteq \infty$.

Hence, when $x \doteq \infty$,
$$\frac{x^2 - x + 3}{2x^2 + x - 4} = \frac{1 - 1/x + 3/x^2}{2 + 1/x - 4/x^2} \doteq 1/2, \qquad (1)$$

$$\frac{x+2}{x^2 + x + 5} = \frac{1 + 2/x}{x + 1 + 5/x} \doteq 0, \qquad (2)$$

$$\frac{x^2 + x - 7}{2x + 3} = \frac{x + 1 - 7/x}{2 + 3/x} \doteq \infty. \qquad (3)$$

And in general, when $x \doteq \infty$, the fraction

$$(a_0x^m + a_1x^{m-1} + \cdots + a_m)/(b_0x^n + b_1x^{n-1} + \cdots + b_n)$$

approaches the limit a_0/b_0, if, as in (1), the degrees of numerator and denominator are the same; the limit 0, if, as in (2), the degree of the denominator is the greater; the limit ∞, if, as in (3), the degree of the numerator is the greater.

And in each case the limit is called the "value which the fraction takes when $x = \infty$," that is, when the fraction itself assumes the indeterminate form ∞ / ∞.

518 **The forms $0 \cdot \infty$ and $\infty - \infty$.** A rational function of x may take one of the indeterminate forms $0 \cdot \infty$ or $\infty - \infty$ for some particular value of x. But the expression can then be reduced to one which will take one of the forms $0/0$, $a/0$, or ∞/∞ already considered.

1. Thus, $(x^2 - 1) \cdot \dfrac{1}{x-1}$ takes the form $0 \cdot \infty$ when $x = 1$. But, except when $x = 1$, we have $(x^2 - 1) \cdot \dfrac{1}{x-1} \equiv \dfrac{x^2-1}{x-1}$, and therefore

$$\lim_{x \doteq 1} \left[(x^2 - 1) \cdot \frac{1}{x-1} \right] = \lim_{x \doteq 1} \frac{x^2-1}{x-1} = \lim_{x \doteq 1} (x+1) = 2.$$

Hence we assign to the given expression the value 2, when $x = 1$.

2. Again, $\dfrac{1}{x} - \dfrac{2}{x(x+2)}$ takes the form $\infty - \infty$ when $x = 0$. But, except when $x = 0$, we have $\dfrac{1}{x} - \dfrac{2}{x(x+2)} \equiv \dfrac{x}{x(x+2)}$, and therefore

$$\lim_{x \doteq 0} \left[\frac{1}{x} - \frac{2}{x(x+2)} \right] = \lim_{x \doteq 0} \frac{x}{x(x+2)} = \lim_{x \doteq 0} \frac{1}{x+2} = \frac{1}{2}.$$

Hence we assign to the given expression the value $1/2$ when $x = 0$.

519 **General conclusion.** Therefore, if a given function of a single variable, as $F(x)$, assumes an indeterminate form when $x = a$, proceed as follows:

Reduce the expression to its simplest form, and then find what limit its value approaches when x *is made to approach* a *as limit. Call this limit the value which the function has when* x = a.

520 **Note.** This method is restricted to functions of a *single* variable, as $F(x)$. For the reason that the method yields definite results is this: *the value of* $\lim_{x \doteq a} F(x)$ *depends solely on the value of* a *and not on the values which* x *may take in approaching* a; and the like is *not* true of functions of more than one variable.

Thus, suppose that x and y are unrelated variables, and consider the fraction x/y when $x=0$ and $y=0$.

The limit, if any, which x/y will approach when $x \doteq 0$ and $y \doteq 0$, depends on the sequences of values through which x and y may run.

For example, a variable will approach 0 as limit if it runs through either of the following sequences:

$$1/2,\ 1/3,\ 1/4,\ \cdots \quad (1); \qquad 1/2^2,\ 1/3^2,\ 1/4^2,\ \cdots \quad (2)$$

If x runs through (1), and y through (2), then x/y will run through the sequence $2, 3, 4, \cdots$, and approach ∞. But if x runs through (2), and y through (1), then x/y will run through the sequence $1/2, 1/3, 1/4, \cdots$, and approach 0.

Therefore, if x and y are *unrelated* variables, we regard x/y as *absolutely indeterminate* when $x=0$ and $y=0$. And so in general.

Infinity in relation to the rules of reckoning. If we take infinite values of the letters into account, we must state the rules of §§ 249, 251, 253 as follows: **521**

1. $a \cdot 0 = 0$, unless $a = \infty$.
2. If $ac = bc$, then $a = b$, unless $c = 0$ or ∞.
3. If $a + c = b + c$, then $a = b$, unless $c = \infty$.

It is important to keep these exceptional cases in mind when applying the rules to the solution of equations.

Thus, consider the product $\dfrac{1}{x^2-1} \cdot x - 1$. When $x = 1$, the second factor, $x - 1$, is 0; but as the first factor, $1/(x^2-1)$, is then ∞, it does not follow that the product is 0. The product is $1/2$ in fact, § 518.

Infinite roots of equations. Instead of saying, as we have been doing, that the equation $x + 2 = x + 3$ and other simple equations which will reduce to the form $0 \cdot x = b$, have *no* root, we sometimes say that *they have the root* ∞. **522**

For however small a may be, if not actually 0, $ax = b$ has the root b/a. And if, keeping b constant and different from 0, we make a approach 0 as limit, b/a will approach ∞, § 512. In other words, as $ax = b$ *approaches* the form $0x = b$, its root b/a *approaches* the value ∞. It is therefore quite in agreement with the practice explained in § 515 to say that when $ax = b$ *has* the form $0x = b$, it *has* the root ∞.

Observe that if we regard $x + 2 = x + 3$ as a true equation whose root is ∞, we are not driven to the absurd conclusion that $2 = 3$. For since $x = \infty$ we have no right to infer that the result of subtracting x from both members is a true equation, § 521, 3.

523 **Infinite solutions of simultaneous equations.** In like manner, instead of saying of a system of inconsistent simple equations, § 377, 2, § 394, 2, that it has *no* solution, we sometimes say that it has an *infinite solution;* for from such a system we can obtain by elimination a single equation of the form $0x = b$, and, by § 522, this equation has the root ∞.

Thus, we may say that the pair of inconsistent equations $y - x = 0$ (1), $y - x = 1$ (2) has an infinite solution.

Observe that this pair (1), (2) is the limiting case, as $m \doteq 1$, of the pair

$$y - mx = 0 \ (3), \quad y - x = 1 \ (2).$$

The solution of the pair (3), (2) is

$$x = 1/(m-1), \quad y = m/(m-1),$$

and when $m \doteq 1$, both $1/(m-1)$ and $m/(m-1)$ approach infinity.

The same thing may be shown by the graphical method, §§ 386, 387. For, when $m \doteq 1$, the graph of (3) approaches parallelism with that of (1), and the point of intersection of the two graphs recedes to an infinite distance in the plane.

EXERCISE XXVI

Assign the appropriate values to the following expressions.

1. $\dfrac{x^2 - 5x + 6}{x^2 - 6x + 8}$, when $x = 2$.

2. $\dfrac{x^3 - 3x^2 + 2}{x^3 - 2x + 1}$, when $x = 1$.

3. $\dfrac{x^2 - 1}{x^2 - 2x + 1}$, when $x = 1$.

4. $\dfrac{x^2 - 2ax + a^2}{x^2 - (a+b)x + ab}$, when $x = a$.

5. $\dfrac{(3x+1)(x+2)^2}{(x^2-4)(x^2+3x+2)}$, when $x = -2$.

6. $\dfrac{x^3 - x^2 - x + 1}{x^3 - 3x^2 + 3x - 1}$, when $x = 1$.

7. $\dfrac{3x^2 - x + 5}{2x^2 + 6x - 7}$, $\dfrac{x^2+1}{x}$, $\dfrac{3x}{x^2+1}$, $\dfrac{(2x^2+1)(x^3-5)}{(x^4+1)(x-6)}$, when $x = \infty$.

8. $\dfrac{x-1}{x^2-9} - \dfrac{x-2}{x(x-3)}$, when $x = 3$.

9. $\dfrac{1}{x-1} + \dfrac{2}{x(x-1)}$, when $x = 1$.

10. $\dfrac{x^2 + \dfrac{x+1}{x-2}}{x^2 + \dfrac{x-1}{x-2}}$, when $x = 2$.

11. $\dfrac{\dfrac{x}{x-1} - \dfrac{x}{x+1}}{\dfrac{3x+1}{x^2+1}}$, when $x = \infty$.

FRACTIONAL EQUATIONS

524 **On solving a fractional equation.** Any given fractional equation may be transformed into one which is integral by multiplying both its members by D, the lowest common denominator of all its fractions. We call this process *clearing the equation of fractions.*

It follows from §§ 341, 342 that the integral equation which is thus derived will have all the roots of the given equation, and, if it has any roots besides these, that they must be roots of the equation $D = 0$ and so may readily be detected and rejected.

Example 1. Solve $\dfrac{3}{x} + \dfrac{6}{x-1} - \dfrac{x+13}{x(x-1)} = 0.$ (1)

Clear of fractions by multiplying by $D = x(x-1)$.

We obtain $3(x-1) + 6x - (x+13) = 0.$ (2)

Solving (2), $x = 2.$ (3)

Therefore, since 2 is not a root of $D = x(x-1) = 0$, it is a root of (1), and the only root.

Example 2. Solve $\dfrac{3}{x} + \dfrac{6}{x-1} - \dfrac{x+5}{x(x-1)} = 0.$ (1)

Clearing of fractions, $3(x-1) + 6x - (x+5) = 0.$ (2)

Solving (2), $x = 1.$ (3)

As 1 is a root of $D = x(x-1) = 0$, it is *not* a root of (1). In fact, when $x = 1$, the first member of (1) has the form $3 + 6/0 - 6/0$; and by § 518 we find that its value is 8, not 0.

Hence (1) has no root.

We may sum up the method thus illustrated in the rule:

525 *To solve a fractional equation, clear of fractions by multiplying both members by* D, *the lowest common denominator of all the fractions.*

Solve the resulting integral equation.

The roots of this equation — except those, if any, for which D *vanishes — are the roots of the given equation.*

526 **Note.** We may also establish this rule as follows:

Let $N/D = 0$ represent the result of collecting all the terms of the given equation in one member and adding them; then $N = 0$ will be the integral equation obtained by clearing of fractions.

1. If N/D is *in its lowest terms*, the roots of $N/D = 0$ and $N = 0$ are the same; for a fraction in its lowest terms vanishes when its numerator vanishes, and then only, § 516.

Thus, in § 524, Ex. 1, $\dfrac{3}{x} + \dfrac{6}{x-1} - \dfrac{x+13}{x(x-1)} = \dfrac{8(x-2)}{x(x-1)} = \dfrac{N}{D}$.

Here N/D is in its lowest terms and the root of $N/D = 0$ is the same as the root of $N = 0$, namely, 2.

2. If N/D is *not in its lowest terms*, $N = 0$ will have roots which $N/D = 0$ does not have, namely, the roots for which the factor common to N and D vanishes.

Thus, in § 524, Ex. 2, $\dfrac{3}{x} + \dfrac{6}{x-1} - \dfrac{x+5}{x(x-1)} = \dfrac{8(x-1)}{x(x-1)} = \dfrac{N}{D}$.

Here N/D is not in its lowest terms, and the root of $N = 0$, namely 1, is not a root of $N/D = 0$; for when $x = 1$, $N/D = 8$, § 514.

Evidently 1. is the *general* case and 2. is exceptional.

Thus, consider the equation $\dfrac{3}{x} + \dfrac{6}{x-1} - \dfrac{x+a}{x(x-1)} = 0$.

Here $N/D = [8x - (a+3)]/x(x-1)$, and this is in its lowest terms except when $a = 5$ or -3.

3. It must not be inferred from what has just been said that the given equation will *never* be satisfied by a root of $N = 0$ which is also a root of $D = 0$.

Thus, consider the equation $\dfrac{x}{x-1} - \dfrac{2}{x} - \dfrac{1}{x(x-1)} = 0$.

Here $N/D = (x-1)^2/x(x-1)$ and, by § 516, this expression vanishes when $x = 1$. But observe that $N = (x-1)^2 = 0$ has the root 1 *a greater number of times* than $D = x(x-1) = 0$ has this root.

527 In applying the rule of § 525, care must be taken not to introduce extraneous factors in the expression selected as the lowest common denominator.

If any fraction in the equation is not in its lowest terms, begin by simplifying this fraction, unless the factors thus cancelled occur in other denominators.

Before clearing of fractions it is sometimes best to combine certain of the fractions, or to reduce certain of them to mixed expressions.

Example 1. Solve $\dfrac{x^2 - 6x + 5}{x^2 - 8x + 15} - \dfrac{x^2}{6x - 2x^2} = \dfrac{11}{5}$.

Here the terms of the first fraction have the common factor $x - 5$, and those of the second the common factor x. Cancelling these factors, we have

$$\frac{x-1}{x-3} - \frac{x}{6-2x} = \frac{11}{5}, \text{ or } \frac{x-1}{x-3} + \frac{x}{2(x-3)} = \frac{11}{5}.$$

Clearing of fractions, $\qquad 10x - 10 + 5x = 22x - 66$.

Solving, $\qquad x = 8$.

sec. 497

Example 2. Solve $\dfrac{x+1}{x+2} + \dfrac{x+6}{x+7} = \dfrac{x+2}{x+3} + \dfrac{x+5}{x+6}$.

Reducing each fraction to a mixed expression and simplifying,

$$\frac{1}{x+2} + \frac{1}{x+7} = \frac{1}{x+3} + \frac{1}{x+6}.$$

Transposing so that the terms in each member may be connected by minus signs,

$$\frac{1}{x+2} - \frac{1}{x+3} = \frac{1}{x+6} - \frac{1}{x+7}.$$

Combining the terms of each member separately,

$$\frac{1}{x^2 + 5x + 6} = \frac{1}{x^2 + 13x + 42}.$$

Clearing of fractions, $x^2 + 13x + 42 = x^2 + 5x + 6.$

Solving, $x = -9/2.$

The given equation may be solved by clearing it of fractions as it stands, but that method is much more laborious.

528 **Simultaneous fractional equations.** The general method of solving such a system is to clear the several equations of fractions and then to find, if possible, the solutions of the resulting system of integral equations. The solutions thus found — except those, if any, which make denominators in the given equations vanish — are the solutions of these equations, § 371.

But if the equations are of the form described in § 379, or if they can be reduced to this form, they should be solved by the method explained in that section.

Example 1. Solve the following pair of equations for x, y.

$$\frac{x-1}{y-2} - \frac{x-3}{y-4} = 0, \quad \frac{1}{xy - 2x} + \frac{1}{4y - 2y^2} - \frac{2}{xy} = 0.$$

Clearing both equations of fractions and simplifying, we obtain

$$x - y + 1 = 0, \quad x + 2y - 8 = 0.$$

Solving $x = 2, \ y = 3.$

Therefore, since none of the denominators in the given equations vanish when $x = 2$ and $y = 3$, these equations have the solution $x = 2$, $y = 3$.

Example 2. Solve the following system for x, y, z.

$$\frac{x+y}{xy} = \frac{5}{6}, \quad \frac{yz}{y+z} = -\frac{3}{2}, \quad 2(z+x) + xz = 0.$$

These equations can be reduced to the form, § 379,

$$\frac{1}{x} + \frac{1}{y} = \frac{5}{6}, \quad \frac{1}{y} + \frac{1}{z} = -\frac{2}{3}, \quad \frac{1}{z} + \frac{1}{x} = -\frac{1}{2}.$$

Solving for $1/x$, $1/y$, $1/z$, we find $1/x = 1/2$, $1/y = 1/3$, $1/z = -1$. Hence the required solution is $x = 2$, $y = 3$, $z = -1$.

EXERCISE XXVII

Solve the following equations for x.

1. $\dfrac{6x-1}{3x+2}-\dfrac{4x-7}{2x-5}=0.$

2. $\dfrac{6x}{5x-1}+\dfrac{8}{3-15x}=\dfrac{1}{6}.$

3. $\dfrac{4}{x-2}-\dfrac{1}{x-4}=\dfrac{4}{x^2-6x+8}.$

4. $\dfrac{3}{2x+3}+\dfrac{1}{x-5}-\dfrac{8}{2x^2-7x-15}=0.$

5. $\dfrac{1}{(x+1)(x-3)}+\dfrac{2}{(x-3)(x+2)}+\dfrac{3}{(x+2)(x+1)}=0.$

6. $\dfrac{2}{x^2-1}-\dfrac{2}{x^2+4x-5}+\dfrac{3}{x^2+6x+5}=0.$

7. $\dfrac{x+1}{3x+1}+\dfrac{2x}{5-6x}=\dfrac{5}{5+9x-18x^2}.$

8. $\dfrac{x+a}{b(x+b)}+\dfrac{x+b}{a(x+a)}=\dfrac{a+b}{ab}.$

9. $\dfrac{x^3+1}{x+1}-\dfrac{x^3-1}{x-1}=20.$

10. $\dfrac{x^2+2x+1}{x^2+5x+4}+\dfrac{x-1}{x^2+3x-4}=0.$

11. $\dfrac{x-8}{x-3}-\dfrac{x-9}{x-4}=\dfrac{x+7}{x+8}-\dfrac{x+2}{x+3}.$

12. $\dfrac{x+7}{x+6}+\dfrac{x+9}{x+8}=\dfrac{x+10}{x+9}+\dfrac{x+6}{x+5}.$

13. $\dfrac{x^3+2}{x-2}-\dfrac{x^3-2}{x+2}-\dfrac{15}{x^2-4}=4x.$

14. $\dfrac{1}{x-1}-\dfrac{x-2}{x^2-1}+\dfrac{3x^2+x}{1-x^4}=0.$

15. $\dfrac{3}{x^3-8}+\dfrac{2x+5}{2x^2+4x+8}-\dfrac{1}{x-2}=0.$

16. $\dfrac{ax+c}{x-p}+\dfrac{bx+d}{x-q}=a+b.$

17. $\dfrac{x^2+7x-8}{x-1}+\dfrac{x^2+x+3}{x+2}+\dfrac{2x^2-x+7}{x+3}=4x.$

18. $\dfrac{x^2-ax+2bx-2ab}{x-a}+\dfrac{b^2-x^2}{x-2b}+\dfrac{3c^2}{x-2c}=0.$

19. $\dfrac{(x-a)^2}{(x-b)(x-c)}+\dfrac{(x-b)^2}{(x-c)(x-a)}+\dfrac{(x-c)^2}{(x-a)(x-b)}=3.$

20. $\dfrac{3x+2}{x^2+x}-\dfrac{x-5}{x^2-1}-\dfrac{x-2}{x^2-x}=0.$

21. $\dfrac{a}{x+2}+\dfrac{2}{x-2}-\dfrac{x+6}{x^2-4}=0.$

Solve the following for x and y, or for x, y, and z.

22. $\begin{cases}\dfrac{3x+y-1}{x-y+2}=\dfrac{6}{7},\\[2mm] \dfrac{x+9}{y+4}=\dfrac{x+3}{y+3}.\end{cases}$

23. $\begin{cases}\dfrac{y-2}{x-3}+\dfrac{x-y}{x^2-9}=\dfrac{y-4}{x+3},\\[2mm] \dfrac{2}{x^2-2x}+\dfrac{3}{xy-2y}+\dfrac{9}{xy}=0.\end{cases}$

24. $\begin{cases}\dfrac{xy}{x+y}=a,\\[2mm] \dfrac{yz}{y+z}=b,\\[2mm] \dfrac{zx}{z+x}=c.\end{cases}$

25. $\begin{cases}\dfrac{2}{x+2y}+2y+2z=3,\\[2mm] \dfrac{y+z}{2}-\dfrac{5}{z-3x}=\dfrac{7}{2},\\[2mm] \dfrac{4}{z-3x}-\dfrac{2}{x+2y}=-1.\end{cases}$

PARTIAL FRACTIONS

529 It follows from § 506 that every rational function of a single variable, as x, can be reduced to the form of an integral function, or a proper fraction, or the sum of an integral function and a proper fraction.

For certain purposes it is useful to carry this reduction further and, *when a proper fraction* A/B *is given, find the simplest set of fractions of which* A/B *is the sum.* The method depends on the following theorems in which the letters A, B, P, Q, and so on, denote rational integral functions of x.

(496)

Theorem 1. *The sum and the difference of two proper fractions* A / B *and* C / D *are themselves proper fractions.* **530**

For $$\frac{A}{B} \pm \frac{C}{D} = \frac{AD \pm BC}{BD}.$$

Since A is of lower degree than B, AD is of lower degree than BD.

And since C is of lower degree than D, BC is of lower degree than BD.

Hence $AD \pm BC$ is of lower degree than BD.

Theorem 2. *Let* I *and* I′ *denote integral functions, and* A / B *and* A′ / B′ *proper fractions.* **531**

If $I + A/B \equiv I' + A'/B'$, *then* $I \equiv I'$ *and* $A/B \equiv A'/B'$.

For, by hypothesis, $I - I' \equiv A'/B' - A/B$.

But $I - I'$ denotes an integral function (or 0), and, § 530, $A'/B' - A/B$ denotes a proper fraction (or 0).

Therefore, since an integral function cannot be identically equal to a proper fraction, we have

$$I - I' \equiv 0 \text{ and } A'/B' - A/B \equiv 0,$$

or $$I \equiv I' \text{ and } A/B \equiv A'/B'.$$

Theorem 3. *Let* A / PQ *denote a proper fraction whose denominator has been separated into two factors,* P *and* Q, *which are prime to one another.* **532**

This fraction can be reduced to a sum of two proper fractions of the forms, B / P *and* C / Q.

For, since Q is prime to P, we can find, § 479, two integral functions M and N, such that

$$1 \equiv MQ + NP, \text{ and therefore } A \equiv AMQ + ANP.$$

Hence $$\frac{A}{PQ} \equiv \frac{AMQ + ANP}{PQ} \equiv \frac{AM}{P} + \frac{AN}{Q}. \qquad (1)$$

If AM/P and AN/Q are proper fractions, our theorem is already demonstrated.

If AM/P and AN/Q are not proper fractions, reduce them to sums of integral functions and proper fractions, and let the results be

$$\frac{AM}{P} \equiv I + \frac{B}{P} \text{ and } \frac{AN}{Q} \equiv K + \frac{C}{Q}. \quad (2)$$

Substituting these expressions for AM/P and AN/Q in (1), we have

$$\frac{A}{PQ} \equiv \frac{B}{P} + \frac{C}{Q} + I + K. \quad (3)$$

But since A/PQ, B/P, and C/Q are proper fractions, it follows from (3), by §§ 530, 531, that $I + K \equiv 0$ and

$$\frac{A}{PQ} \equiv \frac{B}{P} + \frac{C}{Q},$$

as was to be demonstrated.

533 **Note.** The fraction A/PQ can be reduced to but *one* such sum $B/P + C/Q$.

For suppose
$$\frac{A}{PQ} \equiv \frac{B}{P} + \frac{C}{Q} \equiv \frac{B'}{P} + \frac{C'}{Q}, \quad (1)$$

where B'/P and C'/Q also denote proper fractions.

Then $\dfrac{B - B'}{P} \equiv \dfrac{C' - C}{Q}$, and therefore $\dfrac{(B - B')\,Q}{P} \equiv C' - C. \quad (2)$

But (2) is impossible unless $B - B' \equiv 0$ and $C' - C \equiv 0$. For otherwise (2) would mean that $(B - B')\,Q$ is exactly divisible by P, and this cannot be the case since Q is prime to P and $B - B'$ is of lower degree than P, § 481.

534 **Partial fractions.** We call the fractions B/P and C/Q, whose existence we have just proved, *partial fractions* of A/PQ.

To resolve a given fraction of the form A/PQ into its partial fractions B/P and C/Q, it is not necessary to carry out the process indicated in § 532; we may apply the method of undetermined coefficients, § 397, as in the following example.

Example 1. Resolve $(2x^2 + 1)/(x^3 - 1)$ into a sum of two partial fractions.

This is a proper fraction, and its denominator is a product of two factors, $x - 1$ and $x^2 + x + 1$, which are prime to each other.

Hence, § 532, $(2x^2+1)/(x^3-1)$ is equal to a sum of two proper fractions whose denominators are $x-1$ and x^2+x+1 respectively. The numerator of the first of these fractions must be a constant, that of the second an expression whose degree is *one* at most. Hence we must have

$$\frac{2x^2+1}{x^3-1} \equiv \frac{a}{x-1} + \frac{bx+c}{x^2+x+1} \tag{1}$$

where a, b, and c denote constants.

To find a, b, c, clear (1) of fractions.

We obtain $2x^2+1 \equiv a(x^2+x+1)+(bx+c)(x-1)$,

or $$2x^2+1 \equiv (a+b)x^2+(a-b+c)x+(a-c). \tag{2}$$

As (2) is an identity, the coefficients of like powers of x are equal, § 284.

Hence $$a+b=2,\ a-b+c=0,\ a-c=1, \tag{3}$$

or, solving (3), $a=1,\ b=1,\ c=0.$

Therefore $$\frac{2x^2+1}{x^3-1} = \frac{1}{x-1} + \frac{x}{x^2+x+1}.$$

Example 2. Resolve $(5x+4)/(x^4+x^3+x^2-x)$ into a sum of two partial fractions.

General theorem regarding partial fractions. From the theorem of § 532 we may draw the following conclusions. **535**

1. Let A/PQR denote a proper fraction in which the three factors of the denominator P, Q, R are prime to one another. This fraction can be reduced to a sum of the form

$$\frac{A}{PQR} \equiv \frac{B}{P} + \frac{D}{Q} + \frac{E}{R}$$

where B/P, D/Q, and E/R denote proper fractions.

For since P is prime to QR, § 482, A/PQR is the sum of *two* proper fractions of the form $B/P + C/QR$, § 532; and since Q is prime to R, C/QR is itself the sum of two proper fractions of the form $D/Q + E/R$, § 532.

The like is true when the denominator is the product of any number of factors *all prime to one another*.

2. Consider the proper fraction A/PQ^3 in which P is prime to Q. By § 532 it can be resolved into the sum

$$\frac{A}{PQ^3} \equiv \frac{B}{P} + \frac{C}{Q^3}.$$

We cannot apply the theorem of § 532 to the fraction C/Q^3, since the factors Q, Q, Q are not prime to one another.

But since C is of lower degree than Q^3, it can be reduced, § 422, to a polynomial in Q of the form

$$C \equiv C_1Q^2 + C_2Q + C_3,$$

where C_1, C_2, and C_3 are of lower degree than Q.

Dividing each member of this identity by Q^3, we have

$$\frac{C}{Q^3} \equiv \frac{C_1}{Q} + \frac{C_2}{Q^2} + \frac{C_3}{Q^3}.$$

Hence the given fraction can be reduced to the sum

$$\frac{A}{PQ^3} \equiv \frac{B}{P} + \frac{C_1}{Q} + \frac{C_2}{Q^2} + \frac{C_3}{Q^3},$$

where B is of lower degree than P, and C_1, C_2, C_3 are of lower degree than Q.

And so in general when a factor, as Q, occurs more than once in the denominator.

We therefore have the following theorem.

Suppose that the denominator of a given proper fraction has been separated into factors — some occurring once, some, it may be, more than once — which are all prime to one another.

The fraction itself can then be resolved into one and but one sum of proper fractions in which (1) *for each factor,* P, *which occurs but once, there is a single fraction of the form* B/P, *and* (2) *for each factor,* Q, *which occurs* r *times, there is a group of* r *fractions of the form* $C_1/Q + C_2/Q^2 + \cdots + C_r/Q^r$, *where* $C_1, C_2, \cdots, C_r$ *are all of lower degree than* Q.

536 **Simplest partial fractions.** It can be proved that every polynomial in x with real coefficients is the product of factors of one or both the types $x - a$ and $x^2 + px + q$, where a, p, and q are real, but where the factors of $x^2 + px + q$ have imaginary coefficients.

Moreover it follows from §§ 469, 532 that, if the numerator of a given proper fraction and the factors into which its denominator has been separated have real coefficients, so will the numerators of the corresponding partial fractions. Hence, by § 534,

Every proper fraction whose numerator and denominator have real coefficients is equal to a definite sum of partial fractions related as follows to the factors $x - a$ *and* $x^2 + px + q$ *of its denominator.*

1. *For every factor* $x - a$ *occurring once there is a single fraction of the form* $A/(x - a)$, *where* A *is a real constant.*

2. *For every factor* $x - a$ *occurring* r *times there is a group of* r *fractions of the form*

$$A_1/(x - a) + A_2/(x - a)^2 + \cdots + A_r/(x - a)^r,$$

where $A_1, A_2, \cdots A_r$ *are real constants.*

3. *For every factor* $x^2 + px + q$ *occurring once there is a single fraction of the form* $(Dx + E)/(x^2 + px + q)$, *where* D *and* E *are real constants.*

4. *For every factor* $x^2 + px + q$ *occurring* r *times there is a group of* r *fractions of the form*

$$(D_1x + E_1)/(x^2 + px + q) + \cdots + (D_rx + E_r)/(x^2 + px + q)^r,$$

where $D_1, E_1, D_2, E_2, \cdots D_r, E_r$ *denote real constants.*

537 The fractions here described are usually called the *simplest partial fractions* of the given fraction. They are best found by the method of undetermined coefficients.

Example 1. Resolve $\dfrac{x^2 + x - 3}{(x-1)(x-2)(x-3)}$ into its simplest partial fractions.

By § 536, we have $\dfrac{x^2 + x - 3}{(x-1)(x-2)(x-3)} \equiv \dfrac{A}{x-1} + \dfrac{B}{x-2} + \dfrac{C}{x-3}$ (1)

where A, B, C are constants.

Clearing (1) of fractions, we obtain

$$x^2 + x - 3 \equiv A(x-2)(x-3) + B(x-3)(x-1) + C(x-1)(x-2). \quad (2)$$

We may find A, B, C by arranging the second member of (2) according to powers of x and equating coefficients of like powers; but, since A, B, C are constants, the same results will be obtained by the following method, which is simpler.

In (2) let $x = 1$, and we have $-1 = A(-1)(-2)$, $\therefore A = -1/2$;
next let $x = 2$, and we have $3 = B(-1)\cdot 1$, $\therefore B = -3$;
finally let $x = 3$, and we have $9 = C\cdot 2\cdot 1$, $\therefore C = 9/2$.

Hence $$\frac{x^2+x-3}{(x-1)(x-2)(x-3)} = -\frac{1}{2(x-1)} - \frac{3}{x-2} + \frac{9}{2(x-3)}.$$

Example 2. Resolve $\dfrac{x+1}{x(x-1)^3}$ into its simplest partial fractions

By § 536, we have $$\frac{x+1}{x(x-1)^3} \equiv \frac{A}{x} + \frac{B}{x-1} + \frac{C}{(x-1)^2} + \frac{D}{(x-1)^3},$$

and therefore $$x+1 \equiv A(x-1)^3 + Bx(x-1)^2 + Cx(x-1) + Dx. \quad (1)$$

In (1) let $x = 0$, and we have $1 = A(-1)^3$, $\therefore A = -1$;
next let $x = 1$, and we have $2 = D$, $\therefore D = 2$.

Substitute these values of A and D in (1), transpose to the first member the terms thus found, namely $-(x-1)^3$ and $2x$, and simplify the result. We obtain

$$x^3 - 3x^2 + 2x \equiv Bx(x-1)^2 + Cx(x-1). \quad (2)$$

Dividing both members of (2) by $x(x-1)$, we have

$$x - 2 \equiv B(x-1) + C. \quad (3)$$

Equating coefficients of like powers of x in (3), we have

$$1 = B \text{ and } -2 = -B + C, \quad \therefore B = 1 \text{ and } C = -1.$$

Hence $$\frac{x+1}{x(x-1)^3} = -\frac{1}{x} + \frac{1}{x-1} - \frac{1}{(x-1)^2} + \frac{2}{(x-1)^3}.$$

Or we may arrange (1) according to powers of x, obtaining

$$x+1 \equiv (A+B)x^3 - (3A+2B-C)x^2 + (3A+B-C+D)x - A.$$

Equating coefficients of like powers of x, we have

$$A+B=0,\ 3A+2B-C=0,\ 3A+B-C+D=1,\ -A=1.$$

And from these equations we find, as before,

$$A=-1,\ B=1,\ C=-1,\ D=2.$$

Example 3. Resolve $\dfrac{5x^2-4x+16}{(x^2-x+1)^2(x-3)}$ into its simplest partial fractions.

The factors of $x^2 - x + 1$ being imaginary, we have, § 536,

$$\frac{5x^2-4x+16}{(x^2-x+1)^2(x-3)} \equiv \frac{Ax+B}{(x^2-x+1)^2} + \frac{Cx+D}{x^2-x+1} + \frac{E}{x-3},$$

where A, B, C, D, E are constants.

Clearing of fractions,

$$5x^2 - 4x + 16 \equiv (Ax + B)(x - 3) + (Cx + D)(x^2 - x + 1)(x - 3) + E(x^2 - x + 1)^2. \quad (1)$$

We may find A, B, C, D, E by arranging (1) according to powers of x and then equating coefficients of like powers; but the following method is simpler.

In (1) let $x = 3$, and we have $49 = 49E$, $\therefore E = 1$.

Substitute this value of E in (1), transpose the term $(x^2 - x + 1)^2$ thus found to the first member, simplify, and divide both members by $x - 3$. We obtain

$$-(x^3 + x^2 + x + 5) \equiv Ax + B + (Cx + D)(x^2 - x + 1). \quad (2)$$

Next divide both members of (2) by $x^2 - x + 1$. We obtain

$$-x - 2 - \frac{2x + 3}{x^2 - x + 1} \equiv \frac{Ax + B}{x^2 - x + 1} + Cx + D. \quad (3)$$

By § 531, the fractional parts and the integral parts in (3) are equal.

Hence $-x - 2 \equiv Cx + D$, and therefore $C = -1$, $D = -2$,

and $-2x - 3 \equiv Ax + B$, and therefore $A = -2$, $B = -3$.

Therefore $\dfrac{5x^2 - 4x + 16}{(x^2 - x + 1)^2(x - 3)} = -\dfrac{2x + 3}{(x^2 - x + 1)^2} - \dfrac{x + 2}{x^2 - x + 1} + \dfrac{1}{x - 3}$.

538 When the denominator of the given fraction has the form $(x - a)^r$, it is best to begin by expressing the numerator in powers of $x - a$, § 423. Similarly when the denominator has the form $(x^2 + px + q)^r$, the factors of $x^2 + px + q$ being imaginary, we express the numerator in powers of $x^2 + px + q$.

Example. Resolve $\dfrac{x^4 + x^3 - 8x^2 + 6x - 32}{(x - 2)^5}$ into its simplest partial fractions.

By § 423, we find

$$x^4 + x^3 - 8x^2 + 6x - 32 = (x - 2)^4 + 9(x - 2)^3 + 22(x - 2)^2 + 18(x - 2) - 28.$$

Dividing both members by $(x - 2)^5$, we have

$$\frac{x^4 + x^3 - 8x^2 + 6x - 32}{(x - 2)^5} = \frac{1}{x - 2} + \frac{9}{(x - 2)^2} + \frac{22}{(x - 2)^3} + \frac{18}{(x - 2)^4} - \frac{28}{(x - 2)^5}.$$

539 If given an improper fraction, we may first reduce it to the sum of an integral expression and a proper fraction and then resolve the latter into its partial fractions.

Example. Apply this method to the fraction $\dfrac{x^3-2x^2-6x-21}{x^2-4x-5}$.

We have
$$\frac{x^3-2x^2-6x-21}{x^2-4x-5}=x+2+\frac{7x-11}{x^2-4x-5}$$
$$=x+2+\frac{7x-11}{(x+1)(x-5)};$$

and proceeding as above we find
$$\frac{7x-11}{(x+1)(x-5)}=\frac{3}{x+1}+\frac{4}{x-5}.$$

EXERCISE XXVIII

Resolve the following into the simplest partial fractions whose denominators have real coefficients.

1. $\dfrac{2x+11}{(x-2)(x+3)}$.

2. $\dfrac{6x-1}{(2x+1)(3x-1)}$.

3. $\dfrac{4x}{(x+1)(x+2)(x+3)}$.

4. $\dfrac{x^2+2x+3}{(x-1)(x-2)(x-3)(x-4)}$.

5. $\dfrac{x^2+2}{1+x^3}$.

6. $\dfrac{8x+2}{x-x^3}$.

7. $\dfrac{x^3-x^2-5x+4}{x^2-3x+2}$.

8. $\dfrac{2x^3-x^2+1}{(x-2)^4}$.

9. $\dfrac{x-1}{2x^3-5x^2-12x}$.

10. $\dfrac{6}{2x^4-x^2-1}$.

11. $\dfrac{2x^3-3x^2+4x-5}{(x+3)^5}$.

12. $\dfrac{x^2+x+1}{(x^2+1)(x^2+2)}$.

13. $\dfrac{x^2+6x-1}{(x-3)^2(x-1)}$.

14. $\dfrac{3x-1}{(x-2)(x^2+1)}$.

15. $\dfrac{2x^5-x+1}{(x^2+x+1)^3}$.

16. $\dfrac{2x^2-x+1}{(x^2-x)^2}$.

17. $\dfrac{3x^2-x+2}{(x^2+2)(x^2-x-2)}$.

18. $\dfrac{x^2+px+q}{(x-a)(x-b)(x-c)}$.

19. $\dfrac{2x^2-3x-2}{x(x-1)^2(x+3)^3}$.

20. $\dfrac{x^3+x+3}{x^4+x^2+1}$.

IX. SYMMETRIC FUNCTIONS

ABSOLUTE SYMMETRY AND CYCLO-SYMMETRY

Absolute symmetry. In the expression $x^2 + y^2 + z^2$ the letters x, y, z are involved in such a manner that if any two of them be interchanged $x^2 + y^2 + z^2$ is transformed into an identically equal expression, namely, $y^2 + x^2 + z^2$, or $z^2 + y^2 + x^2$, or $x^2 + z^2 + y^2$. To indicate this, we say that $x^2 + y^2 + z^2$ is *symmetric* with respect to x, y, z. **540**

And, in general, a function of a certain set of letters is said to be *symmetric* with respect to these letters when every interchange of two of the letters will transform the function into an identically equal function.

Other examples of symmetric functions are

$(xy + xz + yz)/(x + y)(x + z)(y + z)$ with respect to x, y, z,

$a + b + c$ and $(x + a)(x + b)(x + c)$ with respect to a, b, c.

On the other hand, $x + y - z$ is *not* symmetric; for if we interchange y and z we obtain $x + z - y$, which is not equal to $x + y - z$.

We call $2x^2y$ and $3y^2z$ *terms of the same type* with respect to the variables x, y, z, because the variable parts of these terms, namely, x^2y and y^2z, can be transformed into one another by interchanges of pairs of the letters x, y, z. And so in general. **541**

The sufficient and necessary condition that an integral function of certain letters, as x, y, z, *be symmetric with respect to these letters is that all its terms of the same type shall have the same coefficients.* **542**

This implies that if a symmetric function contains one term of a certain type, it must contain *all* terms of that type; that is, all terms that can be derived from the term in question by making *every possible* interchange of the letters.

Thus, if $ax^2 + by^2 + cz^2$ is to be symmetric, we must have $a = b = c$.

Again, if a symmetric function of x, y, z contains the term x^2y, it must contain all the terms $x^2y + y^2x + x^2z + z^2x + y^2z + z^2y$.

543 This theorem will indicate the *general form* of a symmetric function of given degree with respect to a given set of letters.

Thus, the general form of a symmetric function of the first degree with respect to x, y, z, u is $a(x + y + z + u) + b$, where a and b denote constants.

Again, the most general symmetric and *homogeneous* functions of the first, second, and third degrees with respect to x, y, z are

1. $a(x + y + z)$. 2. $a(x^2 + y^2 + z^2) + b(xy + xz + yz)$.

3. $a(x^3 + y^3 + z^3) + b(x^2y + y^2x + x^2z + z^2x + y^2z + z^2y) + cxyz$.

544 **On expressing a symmetric function.** The notation Σx^2 means the sum of all terms of the same type as x^2; that is, if x, y, z are the letters under consideration, $\Sigma x^2 = x^2 + y^2 + z^2$. Similarly $\Sigma x^2y = x^2y + y^2x + x^2z + z^2x + y^2z + z^2y$; and so on.

Any given symmetric function may be represented by selecting from its terms one of each type, and writing the symbol Σ before their sum.

Thus,

$$\Sigma(2x - x^3y^2) = 2x + 2y + 2z - x^3y^2 - y^3x^2 - x^3z^2 - z^3x^2 - y^3z^2 - z^3y^2.$$

545 When writing out symmetric functions at length, it is best to arrange the terms in accordance with some fixed rule. The following examples will indicate a convenient rule for the arrangement of *integral* symmetric functions.

Suppose that the letters under consideration are a, b, c, d, and by the *normal order* of these letters understand the order a, b, c, d.

We shall then write Σab and Σabc as follows:

$$\Sigma ab = ab + ac + ad + bc + bd + cd, \quad \Sigma abc = abc + abd + acd + bcd.$$

Observe that in each term we write the letters in their normal order.

In forming Σab we take each letter a, b, c in turn and after it write each subsequent letter. The terms of Σabc are derived in a similar manner from those of Σab.

We shall arrange the terms of Σa^mb^n, when $m \neq n$, as follows:

$$\Sigma a^2b^3 = a^2b^3 + b^2a^3 + a^2c^3 + c^2a^3 + \cdots + c^2d^3 + d^2c^3.$$

Observe that we keep the order of the *exponents* fixed; then under the exponents we write the letters of the first term of Σab in both the orders ab and ba, and so on.

In like manner we may write

$$\Sigma a^2b^3c^4 = a^2b^3c^4 + a^2c^3b^4 + b^2c^3a^4 + b^2a^3c^4 + c^2a^3b^4 + c^2b^3a^4$$

+ (terms similarly derived from the remaining terms of Σabc).

546 **A general theorem regarding symmetry.** It follows from the definition of symmetry, § 540, that a symmetric function will remain symmetric when its form is changed by the rules of reckoning. In particular,

The sum, difference, product, and quotient of two symmetric functions are themselves symmetric.

By aid of this theorem we may obtain the result of combining given symmetric functions by algebraic operations without actually carrying out these operations. It is only necessary to compute the various typical terms of the result.

Example 1. Find $(\Sigma a)^2 = (a + b + c + \cdots)^2$.

Evidently the required result is a homogeneous symmetric function of the second degree consisting of terms of the two types a^2 and $2\,ab$.

Hence $$(\Sigma a)^2 = \Sigma a^2 + 2\,\Sigma ab.$$

Example 2. Find $\Sigma x^2 \cdot \Sigma x = (x^2 + y^2 + z^2)(x + y + z)$.

Evidently this product is a sum of terms of the two types x^3 and x^2y.

Hence $$\Sigma x^2 \cdot \Sigma x = \Sigma x^3 + \Sigma x^2y = x^3 + y^3 + z^3$$
$$+ x^2y + y^2x + x^2z + z^2x + y^2z + z^2y.$$

Example 3. Find $(\Sigma x)^3 = (x + y + z)^3$.

The required result is homogeneous, symmetric, and of the third degree with respect to x, y, z. We must therefore have, § 543,

$$(x + y + z)^3 = a(x^3 + y^3 + z^3) + b(x^2y + y^2x + x^2z + z^2x + y^2z + z^2y) + cxyz.$$

To find the values of the constants a, b, c, assign any three sets of values to x, y, z which will yield equations in a, b, c, and solve these equations.

Thus, putting $x = 1,\ y = 0,\ z = 0$, we have $1 = a$. (1)

Again, putting $x = 1,\ y = 1,\ z = 0$, we have $8 = 2\,a + 2\,b$. (2)

Finally, putting $x = 1,\ y = 1,\ z = 1$, we have $27 = 3\,a + 6\,b + c$. (3)

Solving (1), (2), (3), we obtain $a = 1$, $b = 3$, $c = 6$.

Hence $$(\Sigma x)^3 = \Sigma x^3 + 3\,\Sigma x^2y + 6\,\Sigma xyz.$$

547 **Cyclo-symmetry.** In the expression $x^2y + y^2z + z^2x$ the letters x, y, z are involved in such a manner that if we replace x by y, y by z, and z by x, we obtain an identically equal expression, namely, $y^2z + z^2x + x^2y$. To indicate this, we say that $x^2y + y^2z + z^2x$ is *cyclo-symmetric*, or *cyclic*, with respect to the letters x, y, z, *taken in the order* x, y, z.

And, in general, an expression is said to be *cyclo-symmetric*, or *cyclic*, with respect to certain letters *arranged in a given order*, if it is transformed into an identically equal expression when we replace the first of these letters by the second, the second by the third, and so on, and the last by the first.

Such an interchange of the letters is called a *cyclic interchange.*

548 Observe that the terms of $x^2y + y^2z + z^2x$ are themselves arranged *cyclicly;* that is, so that the first changes into the second, the second into the third, and the third into the first, when we replace x by y, y by z, and z by x. Cyclic expressions are of frequent occurrence and reckoning with them is greatly facilitated by arranging them cyclicly.

549 Evidently every symmetric function is cyclic, but not every cyclic expression is symmetric.

Thus, $x^2y + y^2z + z^2x$, though cyclic, is not symmetric. Its value changes if x and y are interchanged. To make it symmetric we must add the group of terms $y^2x + z^2y + x^2z$.

550 As the example shows, a cyclic function will ordinarily not contain *all* the terms of a given type, but *such of these terms as it does contain will have the same coefficients.*

551 **Theorem.** *The sum, difference, product, and quotient of two cyclic functions are themselves cyclic.*

This follows at once from the definition of cyclo-symmetry.

Example. Find the product $(x^2y + y^2z + z^2x)(x + y + z)$.

Evidently the product is cyclic but not symmetric. Moreover it contains the terms x^3y, x^2y^2, x^2yz, each once, and terms of these types only.

Hence the product is

$$x^3y + y^3z + z^3x + x^2y^2 + y^2z^2 + z^2x^2 + x^2yz + y^2zx + z^2xy.$$

EXERCISE XXIX

1. State the letters with respect to which the expression

$$x^4 - 2y^4 + z^4 + 4(x^3 - y^3)(y^3 - z^3)(x^2 + z^2)$$

is symmetric.

2. Write out in full the following symmetric functions of a, b, c.

$$\Sigma a^2b^2, \quad \Sigma a^3b^4, \quad \Sigma a^2/b, \quad \Sigma a^2b^3c^5, \quad \Sigma a^2b^2c^4,$$
$$\Sigma(a+b)c, \quad \Sigma(a+b^2)c^3, \quad \Sigma(a+2b+3c).$$

3. Show that $(a-b)(b-c)(c-a)$ is cyclic but not symmetric with respect to a, b, c; also that $(a-b)^2(b-c)^2(c-a)^2$ is symmetric.

4. Is $(a-b)^2(b-c)^2(c-d)^2(d-a)^2$ symmetric with respect to a, b, c, d?

5. Arrange the following sets of expressions cyclicly.

$$y^2 - x^2, \; z^2 - y^2, \; x^2 - z^2; \quad a^2bc, \; abd^2, \; ac^2d, \; b^2cd;$$
$$(a-c)(b-a), \; (a-c)(c-b), \; (a-b)(b-c).$$

6. Write out in full the cyclic functions of a, b, c, d whose first terms are

$$ab^3c^2, \quad a(b-c), \quad (b+2c)(a+d), \quad a^2/(a-b)(a-c).$$

7. Prove the truth of the following identities.

$$\Sigma a^3 \cdot \Sigma a = \Sigma a^4 + \Sigma a^3b; \quad \Sigma ab \cdot \Sigma a = \Sigma a^2b + 3\,\Sigma abc.$$

FACTORIZATION OF SYMMETRIC AND CYCLIC FUNCTIONS

552 By aid of the remainder theorem and the principles just explained it is often possible to factor a complicated symmetric or cyclic function with comparatively little reckoning.

Example 1. Factor $x^3(y-z) + y^3(z-x) + z^3(x-y)$.

This function vanishes when $y = z$, for

$$x^3(z-z) + z^3(z-x) + z^3(x-z) \equiv 0.$$

Hence the function is exactly divisible by $y - z$, § 416; and for a like reason it is exactly divisible by $z - x$ and by $x - y$, and therefore by the product $(y-z)(z-x)(x-y)$.

Both dividend and divisor are *cyclic* and *homogeneous*, and their degrees are *four* and *three* respectively. Hence the quotient must be a cyclic and homogeneous function of the *first* degree and therefore have the form $k(x+y+z)$, where k denotes some constant. Hence

$$x^3(y-z)+y^3(z-x)+z^3(x-y)\equiv k(y-z)(z-x)(x-y)(x+y+z) \quad (1)$$

To find k, assign to x, y, z any set of values for which the coefficient of k will not vanish.

Thus, putting $x=2$, $y=1$, $z=0$, we have $6=-6k$, or $k=-1$.

Or we may find k by equating the coefficients of like powers of x in the two members of (1) arranged as polynomials in x. Thus, the x^3 term in the first member is $x^3(y-z)$, and in the second member it is $-kx^3(y-z)$, whence as before $k=-1$. We therefore have

$$x^3(y-z)+y^3(z-x)+z^3(x-y)=-(y-z)(z-x)(x-y)(x+y+z).$$

Example 2. Factor $(x+y+z)^5-x^5-y^5-z^5$.

This function vanishes when $x=-y$, for

$$(-y+y+z)^5+y^5-y^5-z^5\equiv 0.$$

Hence the function is exactly divisible by $x+y$; and for a like reason it is divisible by $y+z$ and by $z+x$, and therefore by $(x+y)(y+z)(z+x)$.

As the dividend and divisor are symmetric and homogeneous and of the *fifth* and *third* degrees respectively, the quotient must be of the form $k(x^2+y^2+z^2)+l(xy+yz+zx)$, § 543.

Hence $\qquad (x+y+z)^5-x^5-y^5-z^5$

$$\equiv(x+y)(y+z)(z+x)[k(x^2+y^2+z^2)+l(xy+yz+zx)].$$

Putting $x=1$, $y=1$, $z=0$, we obtain $15=2k+l$.

Putting $x=2$, $y=1$, $z=0$, we obtain $35=5k+2l$.

Solving for k and l, we find $k=5$, $l=5$, and therefore have

$$(x+y+z)^5-x^5-y^5-z^5$$
$$=5(x+y)(y+z)(z+x)(x^2+y^2+z^2+xy+yz+zx).$$

Example 3. Factor

$$(x+y+z)^3-(y+z-x)^3-(z+x-y)^3-(x+y-z)^3.$$

This function vanishes when $x=0$, for

$$(y+z)^3-(y+z)^3-(z-y)^3-(y-z)^3\equiv 0.$$

Hence the function is exactly divisible by $x-0$ or x; and for a like reason it is divisible by y and by z and therefore by xyz.

Since both dividend and divisor are of the *third* degree, the quotient is some constant, k. Hence

$$(x+y+z)^3-(y+z-x)^3-(z+x-y)^3-(x+y-z)^3\equiv kxyz.$$

Putting $x=1$, $y=1$, $z=1$, we find $k=24$, and therefore have

$$(x+y+z)^3-(y+z-x)^3-(z+x-y)^3-(x+y-z)^3=24xyz.$$

553 The method just explained is sometimes useful in simplifying cyclic *fractional* expressions.

Example. Simplify $\dfrac{a^3}{(a-b)(a-c)} + \dfrac{b^3}{(b-c)(b-a)} + \dfrac{c^3}{(c-a)(c-b)}$.

This expression is cyclic with respect to a, b, c.

The lowest common denominator is $(b-c)(c-a)(a-b)$.

On reducing the fractions to this lowest common denominator we obtain as the first numerator $-a^3(b-c)$. Hence, by cyclo-symmetry, the second and third numerators are $-b^3(c-a)$ and $-c^3(a-b)$.

Adding these three numerators and factoring the result, § 552, **Ex. 1,** we have $(a+b+c)(b-c)(c-a)(a-b)$.

Hence the given expression reduces to $a+b+c$.

EXERCISE XXX

Factor the following expressions.

1. $x^2(y-z)+y^2(z-x)+z^2(x-y)$.
2. $yz(y-z)+zx(z-x)+xy(x-y)$.
3. $(y-z)^3+(z-x)^3+(x-y)^3$.
4. $x(y-z)^3+y(z-x)^3+z(x-y)^3$.
5. $x^2(y-z)^3+y^2(z-x)^3+z^2(x-y)^3$.
6. $x^4(y^2-z^2)+y^4(z^2-x^2)+z^4(x^2-y^2)$.
7. $(x+y+z)^3-x^3-y^3-z^3$.
8. $(y-z)^5+(z-x)^5+(x-y)^5$.
9. $(x+y+z)^5-(y+z-x)^5-(z+x-y)^5-(x+y-z)^5$.
10. $(y-z)(y+z)^3+(z-x)(z+x)^3+(x-y)(x+y)^3$.
11. $x(y+z)^2+y(z+x)^2+z(x+y)^2-4xyz$.
12. $x^5(y-z)+y^5(z-x)+z^5(x-y)$.

Simplify the following fractional expressions.

13. $\dfrac{a^4}{(a-b)(a-c)} + \dfrac{b^4}{(b-c)(b-a)} + \dfrac{c^4}{(c-a)(c-b)}$.

14. $\dfrac{x+a}{(a-b)(a-c)} + \dfrac{x+b}{(b-c)(b-a)} + \dfrac{x+c}{(c-a)(c-b)}$.

15. $\dfrac{a^2 - bc}{(a-b)(a-c)} + \dfrac{b^2 - ca}{(b-c)(b-a)} + \dfrac{c^2 - ab}{(c-a)(c-b)}$.

16. $\dfrac{(b+c)^2}{(a-b)(a-c)} + \dfrac{(c+a)^2}{(b-c)(b-a)} + \dfrac{(a+b)^2}{(c-a)(c-b)}$.

17. $\dfrac{a^2}{(a-b)(a-c)(x-a)} + \dfrac{b^2}{(b-c)(b-a)(x-b)} + \dfrac{c^2}{(c-a)(c-b)(x-c)}$.

X. THE BINOMIAL THEOREM

554 **Structure of continued products.** To obtain the product

$$(a+b+c+d)(e+f+g)(h+k)$$

we may multiply each term of $a+b+c+d$ by each term of $e+f+g$, then multiply every product thus obtained by each term of $h+k$, and finally add the results of these last multiplications.

Hence we shall obtain one term of the product if we select one term from each of the three given factors and multiply these terms together. And we shall obtain all the terms of the product if we make this selection of terms from the three given factors in all possible ways.

Thus, selecting b from the first factor, g from the second, and k from the third, we have the term bgk of the product; and so on.

Since we can select a term from $a+b+c+d$ in *four* ways, a term from $e+f+g$ in *three* ways, and a term from $h+k$ in *two* ways, the number of terms in the complete product is $4 \cdot 3 \cdot 2$, or 24. And so, in general,

The product of any number of polynomials is the sum of all the products that can be obtained by selecting one term from each factor and multiplying these terms together.

And if the first factor has m *terms, the second* n, *the third* p, *and so on, the number of terms in the complete product — before like terms, if any, have been collected — is* mnp $\cdots$.

555 This theorem supplies a useful check on the correctness of a multiplication. It may be applied to a product in which like terms have been collected, provided its terms represent sums of terms of like sign and without numerical coefficients, the coefficient of a term then indicating how many uncollected terms it represents.

Thus, by our theorem, the product $(a + b + c)(a + b + c)$ should have $3 \cdot 3$ or 9 terms, which are all of the same sign. But, as we have shown, $(a + b + c)(a + b + c) = a^2 + b^2 + c^2 + 2ab + 2ac + 2bc$, and this represents an uncollected product of $1 + 1 + 1 + 2 + 2 + 2$ or 9 terms, as it should.

Similarly, the product $(a + b)(a + b)(a + b)$ should have $2 \cdot 2 \cdot 2$ or 8 terms. But this product when simplified is $a^3 + 3a^2b + 3ab^2 + b^3$, which means, as it should, an uncollected product of $1 + 3 + 3 + 1$ or 8 terms.

556 One should bear this theorem in mind when reckoning with symmetric functions by the methods of the last chapter.

Thus, the student has proved, p. 249, Ex. 7, $\Sigma ab \cdot \Sigma a = \Sigma a^2b + 3\Sigma abc$.

To test this formula, suppose that only the letters a, b, c are involved.

Then Σab has 3 terms, Σa has 3 terms, Σa^2b has 6 terms, and Σabc has 1 term; and $3 \cdot 3 = 6 + 3 \cdot 1$, as it should.

557 **Products of binomial factors of the first degree.** The theorem of § 554 enables one to obtain the product of any number of factors of the form $x + b$ by inspection. Thus,

$$(x + b_1)(x + b_2)(x + b_3)$$
$$= x^3 + (b_1 + b_2 + b_3)x^2 + (b_1b_2 + b_1b_3 + b_2b_3)x + b_1b_2b_3.$$

For, selecting x from each factor, we have the term x^3.

Selecting in all possible ways x's from two of the factors and a b from the third, we have the terms b_1x^2, b_2x^2, b_3x^2.

Selecting in all possible ways x from one of the factors and b's from the other two, we have the terms b_1b_2x, b_1b_3x, b_2b_3x.

Selecting b's from all three factors, we have the term $b_1b_2b_3$.

Observe that when the terms of the product are collected, as in the formula, the coefficient of x^2 is the *sum* of the three letters b_1, b_2, b_3, the coefficient of x is the *sum of the products of every two* of these letters, and the final term is the *product of all three.*

Hence these coefficients are *symmetric* functions of b_1, b_2, b_3, as was to be expected since $(x + b_1)(x + b_2)(x + b_3)$ is itself symmetric with respect to b_1, b_2, b_3.

Observe also that since $(x + b_1)(x + b_2)(x + b_3)$ is symmetric with respect to b_1, b_2, b_3, we may obtain the product by finding one term of each type, as x^3, b_1x^2, b_1b_2x, $b_1b_2b_3$, and then writing out all the terms of these several types.

558 By the same reasoning we may prove the general formula

$$(x + b_1)(x + b_2)(x + b_3) \cdots (x + b_n)$$
$$= x^n + B_1x^{n-1} + B_2x^{n-2} + \cdots + B_n,$$

where
$$B_1 = \Sigma b_1 \quad = b_1 + b_2 + b_3 + \cdots + b_n,$$
$$B_2 = \Sigma b_1b_2 \quad = b_1b_2 + b_1b_3 + \cdots + b_2b_3 + \cdots + b_{n-1}b_n,$$
$$B_3 = \Sigma b_1b_2b_3 = b_1b_2b_3 + b_1b_2b_4 + \cdots + b_{n-2}b_{n-1}b_n,$$
$$\cdot \quad \cdot \quad \cdot \quad \cdot \quad \cdot \quad \cdot \quad \cdot$$
$$B_n = \quad b_1b_2b_3 \cdots b_n;$$

that is, B_1 is the *sum*, and B_n is the *product* of all the letters b_1, b_2, $\cdots b_n$, and the intermediate coefficients are: B_2, the sum of the products of every *two* of these letters; B_3, the sum of the products of every *three*; and so on.

Thus, we obtain one term of the product each time that we select b's from *three* of the factors and x's from the rest. Making the selection in all possible ways, we obtain the terms $b_1b_2b_3x^{n-3}$, $b_1b_2b_4x^{n-3}$, $\cdots$, and their sum is B_3x^{n-3}.

Observe that, as indicated above, the coefficients

$$B_1, B_2, \cdots, B_n$$

are *symmetric* functions of the letters b_1, b_2, $\cdots$, b_n.

559 In like manner, we have

$$(x - b_1)(x - b_2)(x - b_3) \cdots (x - b_n)$$
$$= x^n - B_1x^{n-1} + B_2x^{n-2} - \cdots + (-1)^nB_n,$$

where B_1, B_2, $\cdots B_n$ have the same meanings as in § 558. and the signs connecting the terms are alternately $-$ and $+$, the last sign, that of $(-1)^n B_n$, being $+$ when n is even and $-$ when n is odd.

We obtain this formula by merely changing the signs of all the letters $b_1, b_2, \cdots b_n$ in the formula of § 558. For this leaves unchanged every B whose terms are products of an even number of b's, and merely changes the sign of every B whose terms are products of an odd number of b's.

Example. By the method of §§ 557–559 find the following products.

1. $(x+1)(x+2)(x+3)$.
2. $(x+2)(x-3)(x+4)$.
3. $(x+a)(x+b)(x+c)(x+d)$.
4. $(x-y)(x+2y)(x-3y)(x+4y)$.

560 **The number of terms in the sums Σb_1, $\Sigma b_1 b_2$, $\cdots$.** Let $n_1, n_2, \cdots$ denote the number of terms in Σb_1, $\Sigma b_1 b_2$, $\cdots$ respectively.

1. Since $\Sigma b_1 = b_1 + b_2 + \cdots + b_n$, evidently $n_1 = n$.

2. If we multiply each of the n letters $b_1, b_2, \cdots, b_n$ by each of the other $n-1$ letters in turn, we obtain $n(n-1)$ products all told. But these $n(n-1)$ products are the terms of $\Sigma b_1 b_2$, each counted *twice*. Hence n_2, the number of terms in $\Sigma b_1 b_2$, is $n(n-1)/2$, or

$$n_2 = n_1 \frac{n-1}{2} = \frac{n(n-1)}{1 \cdot 2}.$$

Thus, we have

$$b_1 b_2,\ b_1 b_3,\ \cdots,\ b_1 b_n;\ b_2 b_1,\ b_2 b_3,\ \cdots,\ b_2 b_n;\ \cdots;\ b_n b_1,\ b_n b_2,\ \cdots,\ b_n b_{n-1}.$$

There are n groups of these products, and $n-1$ products in each group, hence $n(n-1)$ products all told.

But the product $b_1 b_2$ here occurs twice, namely once in the form $b_1 b_2$ and once in the form $b_2 b_1$; and so on.

3. Again, if we multiply each of the n_2 terms of $\Sigma b_1 b_2$ by each of the $n-2$ letters which do not occur in that term, we obtain $n_2(n-2)$ products all told. But these $n_2(n-2)$ products are the terms of $\Sigma b_1 b_2 b_3$, each counted *three* times. Hence n_3, the number of terms in $\Sigma b_1 b_2 b_3$, is $n_2(n-2)/3$, or

$$n_3 = n_2 \cdot \frac{n-2}{3} = \frac{n(n-1)(n-2)}{1 \cdot 2 \cdot 3}.$$

Thus, we have

$b_1b_2b_3$, $b_1b_2b_4$, $\cdots$, $b_1b_2b_n$; $b_1b_3b_2$, $b_1b_3b_4$, $\cdots$, $b_1b_3b_n$;

$\cdots$; $b_{n-1}b_nb_1$, $b_{n-1}b_nb_2$, $\cdots$, $b_{n-1}b_nb_{n-2}$.

There are n_2 groups of these products, and $n-2$ products in each group, hence $n_2(n-2)$ products all told.

But the product $b_1b_2b_3$ here occurs three times, namely in three forms, $b_1b_2 \cdot b_3$, $b_1b_3 \cdot b_2$, $b_2b_3 \cdot b_1$. And similarly every term of $\Sigma b_1b_2b_3$ here occurs three times, once for each of the three ways in which a product of three letters may be obtained by multiplying the product of two of the letters by the remaining letter.

4. By the same reasoning we can show that

$$n_4 = n_3\frac{n-3}{4} = \frac{n(n-1)(n-2)(n-3)}{1 \cdot 2 \cdot 3 \cdot 4},$$

and, in general, that

$$n_r = \frac{n(n-1)(n-2)\cdots \text{ to } r \text{ factors}}{1 \cdot 2 \cdot 3 \cdots r}.$$

Thus, the numbers of products of four letters b_1, b_2, b_3, b_4, taken *one*, *two*, *three*, *four* at a time, are

$$n_1 = 4, \quad n_2 = \frac{4 \cdot 3}{1 \cdot 2} = 6, \quad n_3 = \frac{4 \cdot 3 \cdot 2}{1 \cdot 2 \cdot 3} = 4, \quad n_4 = \frac{4 \cdot 3 \cdot 2 \cdot 1}{1 \cdot 2 \cdot 3 \cdot 4} = 1.$$

561 **Binomial theorem.** If in the formula of § 558, namely,

$$(x+b_1)(x+b_2)\cdots(x+b_n) = x^n + B_1x^{n-1} + B_2x^{n-2} + \cdots + B_n,$$

we replace all the n different letters $b_1, b_2, \cdots b_n$ by the same letter b, and x by a, the first member becomes $(a+b)^n$.

Again, since each of the n terms of B_1 becomes b, and each of the n_2 terms of B_2 becomes b^2, and so on, we have, § 560,

$$B_1 = nb,\ B_2 = \frac{n(n-1)}{1 \cdot 2}b^2,\ B_3 = \frac{n(n-1)(n-2)}{1 \cdot 2 \cdot 3}b^3, \cdots.$$

Our formula therefore reduces to the following:

$$(a+b)^n = a^n + \frac{n}{1}a^{n-1}b + \frac{n(n-1)}{1 \cdot 2}a^{n-2}b^2 + \frac{n(n-1)(n-2)}{1 \cdot 2 \cdot 3}a^{n-3}b^3 + \cdots,$$

where

1. *The number of terms on the right is* n + 1.

2. *The exponents of* a *decrease by one and those of* b *increase by one from term to term, their sum in each term being* n.

3. *The first coefficient is* 1, *the second is* n, *and the rest of them may be found by the following rule:*

Multiply the coefficient of any term by the exponent of a *in the term and divide by the exponent of* b *increased by* 1; *the result will be the coefficient of the next term.*

This formula is known as the *binomial theorem*, the expression on the right being called the *expansion* of $(a+b)^n$ by this theorem.

Thus, $$(a+b)^3 = a^3 + 3a^2b + \frac{3\cdot 2}{1\cdot 2}ab^2 + \frac{3\cdot 2\cdot 1}{1\cdot 2\cdot 3}b^3$$
$$= a^3 + 3a^2b + 3ab^2 + b^3.$$

562 Since $a+b$ is symmetric with respect to a and b, it follows from § 542 that the terms *of the same type* in the expansion of $(a+b)^n$ — namely those involving a^n and b^n, $a^{n-1}b$ and ab^{n-1}, and so on — must have the same coefficients. But these are the first and last terms, the second and next to last terms, and in general every two terms which are equally removed from the beginning and the end of the expansion.

Hence the last term is b^n, the next to last is nab^{n-1}, and so on. Since the number of terms is $n+1$, there will be one middle term when n is even, two when n is odd. When there are two middle terms, they are of the same type and have the same coefficients. And by what has just been said, *the coefficients of the terms which follow the middle term or terms are the same as those which precede them — but in reverse order.*

563 It may also readily be shown that the coefficients increase up to the middle term or terms and then decrease, so that *the middle coefficient or coefficients are the greatest.*

This follows from the rule of coefficients, § 561, 3, since in the terms which precede the middle term or terms the exponent of a is greater than the exponent of b increased by 1, while in the terms which follow it is less.

564 By changing the sign of b in the preceding formula and simplifying, we obtain

$$(a-b)^n = a^n - na^{n-1}b + \frac{n(n-1)}{1\cdot 2}a^{n-2}b^2 - \frac{n(n-1)(n-2)}{1\cdot 2\cdot 3}a^{n-3}b^3 + \cdots,$$

where the terms which involve odd powers of b have $-$ signs, and those which involve even powers have $+$ signs.

Example. Find the expansion of $(2x - y^3)^6$.

Substituting $2x$ for a, and y^3 for b, in the formula, and remembering that the last three coefficients are the same as the first three in reverse order, § 562, we have

$$(2x-y^3)^6 = (2x)^6 - 6(2x)^5y^3 + \frac{6\cdot 5}{1\cdot 2}(2x)^4(y^3)^2 - \frac{6\cdot 5\cdot 4}{1\cdot 2\cdot 3}(2x)^3(y^3)^3 + \cdots$$

$$= (2x)^6 - 6(2x)^5y^3 + 15(2x)^4(y^3)^2 - 20(2x)^3(y^3)^3 + 15(2x)^2(y^3)^4 - 6(2x)(y^3)^5 + (y^3)^6$$

$$= 64x^6 - 192x^5y^3 + 240x^4y^6 - 160x^3y^9 + 60x^2y^{12} - 12xy^{15} + y^{18}.$$

565 **The general term.** From § 561 it follows that the $(r+1)$th term in the expansion of $(a+b)^n$ is

$$\frac{n(n-1)(n-2)\cdots \text{to } r \text{ factors}}{1\cdot 2\cdot 3\cdots r}a^{n-r}b^r.$$

This, with a *minus* sign before it when r is odd, is also the $(r+1)$th term in the expansion of $(a-b)^n$.

Example 1. Find the eighth term in the expansion of $(x-y)^{16}$.

Here $n = 16$ and $r + 1 = 8$, or $r = 7$. Hence the required term is

$$-\frac{16\cdot 15\cdot 14\cdot 13\cdot 12\cdot 11\cdot 10}{1\cdot 2\cdot 3\cdot 4\cdot 5\cdot 6\cdot 7}x^9y^7 = -11440x^9y^7.$$

Example 2. Does any term in the expansion of $(x^3 + 1/x)^{12}$ contain x^{20}? If so, find this term.

Let $r+1$ denote the number of the term. Then, since $n = 12$, $a = x^3$, and $b = 1/x$, we must have

$$a^{n-r}b^r = (x^3)^{12-r}/x^r = x^{36-4r} = x^{20}.$$

This condition is satisfied if $36 - 4r = 20$, or $r = 4$.

Hence the *fifth* term contains x^{20}, and substituting in the formula we find this term to be $\frac{12 \cdot 11 \cdot 10 \cdot 9}{1 \cdot 2 \cdot 3 \cdot 4} x^{20} = 495 x^{20}$.

EXERCISE XXXI

Expand the following by aid of the binomial theorem.

1. $(3x + 2y)^8$. 2. $(a - b)^8$. 3. $(1 + 2x^2)^7$.

4. $(2 + 1/x)^4$. 5. $(x - 3/x)^6$. 6. $(x/y - y/x)^5$.

7. $(1 - x + 2x^2)^4$. 8. $(a^2 + ax - x^2)^3$.

9. Find the sixth term in $(1 + x/2)^{11}$.

10. Find the eighth term in $(3a - 4b)^{12}$.

11. Find the middle term in $(a^2 - 2bc)^{10}$.

12. Find the two middle terms in $(1 - x)^9$.

13. Find the coefficient of x^5 in $(1 + x)^8$.

14. Find the coefficient of x^4 in $(3 - 2x)^7$.

15. Find the coefficient of x^8 in $(1 - x^2)^6$.

16. Find the coefficient of x^3 in $(1 + 2x)^9 + (1 - 2x)^{11}$.

17. Find the constant term in $(x + 1/x)^{12}$.

18. Find the coefficient of x^7 in $(2x - 1/x)^{15}$.

19. Find $(x + 2y)(x - 3y)(x - 5y)$ by inspection.

20. Find $(x + 2)(x + 3)(x - 4)(x - 5)$ by inspection.

21. What is the number of terms in the product

$$(a + b + c + d)(f + g + h)(k + l)(m + n + p + q)?$$

22. Find the sum of the coefficients in the following products.

1. $(1 + x^2 + x^3 + x^4)^3$. 2. $(1 + 2x + x^2)^2(1 + x + 3x^3)^2$.

23. What is the sum of the coefficients in the following symmetric functions of *four* letters a, b, c, d when expanded?

1. $\Sigma a^2 \cdot \Sigma a$. 2. $\Sigma a^3 \cdot \Sigma abc$. 3. $\Sigma ab \cdot \Sigma abc$.

24. Show that the sum of the coefficients in the expansion of $(a + b)^n$ is 2^n.

25. Show that in the expansion of $(a - b)^n$ the sum of the positive coefficients is numerically equal to the sum of the negative coefficients.

XI. EVOLUTION

566 **Perfect powers.** Given a rational function P. It is possible that P is a *perfect* n*th power;* in other words, that a second rational function Q exists such that $P = Q^n$. If so, this rational function Q will be an n*th root* of P.

In the present chapter we consider the problem: A rational function P being given, it is required to determine whether or not P is a perfect nth power, and, if it is, to find its nth root Q. We suppose n to denote a given positive integer.

567 **Roots of monomials.** Let P denote a rational monomial reduced to its simplest form. If P is a perfect nth power, an nth root of P may be obtained by the following rule.

Divide the exponents of the several literal factors of P *by* n, *and multiply the result by the principal* n*th root of the numerical coefficient of* P.

This follows at once from the rule for involution, § 318.

Thus, $(a^kb^l/c^m)^n = a^{kn}b^{ln}/c^{mn}$, § 318. Hence a^kb^l/c^m is an nth root of $a^{kn}b^{ln}/c^{mn}$, § 566, and it is obtained by dividing the exponents of $a^{kn}b^{ln}/c^{mn}$ by n.

568 The root thus obtained is called the *principal* n*th root* of P (compare § 258). We shall mean this root when we speak of *the* nth root of P, or when we use the symbol $\sqrt[n]{P}$.

Example 1. Find the cube root of $-8a^3b^6/27x^3y^9$.

We have $$\sqrt[3]{\frac{-8a^3b^6}{27x^3y^9}} = -\frac{2ab^2}{3xy^3}.$$

Example 2. Find the following roots.

1. $\sqrt{\dfrac{64a^4b^6}{100c^8d^{12}}}$. 2. $\sqrt[4]{81x^4y^8z^{12}}$. 3. $\sqrt[5]{\dfrac{32x^{10}y^{30}}{a^5z^{25}}}$.

569 **Roots of polynomials.** Consider the following examples.

Example 1. Determine whether or not $4x^4 - 4x^3 + 13x^2 - 6x + 9$ is a perfect square, and, if it is, find its square root.

If this is a perfect square, evidently it must have a square root of the form $2x^2 + px + q$, where p and q are constants. We must therefore have

$$4x^4 - 4x^3 + 13x^2 - 6x + 9 \equiv (2x^2 + px + q)^2$$
$$\equiv 4x^4 + 4px^3 + (p^2 + 4q)x^2 + 2pqx + q^2,$$

which requires, § 284, that p and q satisfy the equations:

$$4p = -4 \quad (1), \qquad p^2 + 4q = 13 \quad (2), \qquad 2pq = -6 \quad (3), \qquad q^2 = 9 \quad (4).$$

From (1) and (2) we find $p = -1$, $q = 3$, and these values of p and q satisfy (3) and (4); for $2(-1)3 = -6$, and $3^2 = 9$.

Hence $4x^4 - 4x^3 + 13x^2 - 6x + 9$ is a perfect square and $2x^2 - x + 3$ is its square root.

Example 2. Find the cube root of

$$x^6 + 6x^5 + 21x^4 + 44x^3 + 63x^2 + 54x + 27.$$

If this be a perfect cube, it will have a cube root of the form $x^2 + px + q$. We must therefore have

$$x^6 + 6x^5 + 21x^4 + 44x^3 + 63x^2 + 54x + 27 \equiv (x^2 + px + q)^3$$
$$\equiv x^6 + 3px^5 + 3(p^2 + q)x^4 + (p^3 + 6pq)x^3$$
$$+ 3(p^2q + q^2)x^2 + 3pq^2x + q^3,$$

which requires, § 284, that p and q satisfy the six equations:

$$3p = 6 \quad (1), \qquad 3(p^2 + q) = 21 \quad (2), \cdots \qquad q^3 = 27 \quad (6).$$

From (1) and (2) we obtain $p = 2$, $q = 3$. And these values of p and q will be found to satisfy the remaining equations $(3) \cdots (6)$.

Hence $x^6 + 6x^5 + \cdots + 54x + 27$ is a perfect cube, and its cube root is $x^2 + 2x + 3$.

By the method illustrated in these examples it is always possible to determine whether or not a given polynomial in x is a perfect nth power, and, if it is, to find its nth root.

Let the polynomial be $a_0x^m + a_1x^{m-1} + \cdots + a_m$. If this be a perfect nth power, its degree m must be a multiple of n so that $m = kn$, where k is an integer; and it must have an nth root of the form $\alpha x^k + A_1x^{k-1} + \cdots + A_k$, where α denotes the principal nth root of a_0, and $A_1, \cdots, A_k$ are unknown constants. We call this root the *principal* nth root.

To determine whether $a_0x^m + \cdots + a_m$ has any such root, and to find this root if it exists, we set

$$a_0x^m + a_1x^{m-1} + \cdots + a_m \equiv (\alpha x^k + A_1x^{k-1} + \cdots + A_k)^n.$$

reduce the second member to the form of a polynomial in x, and then equate its coefficients to those of the like powers of x in the first member. We thus obtain a system of nk equations in $A_1, A_2, \cdots, A_k$. The first k of these equations will give a single set of values for $A_1, A_2, \cdots, A_k$; and this set of values must satisfy the rest of the equations if $a_0x^m + \cdots + a_m$ is to be a perfect nth power.

Example 3. Find the cube root of

$$8x^6 - 12x^5 + 18x^4 - 13x^3 + 9x^2 - 3x + 1.$$

570 Square roots of polynomials. If a polynomial P is a perfect square, its *square root* may also be obtained by the following method.

As in the preceding section, let P denote a polynomial in x of even degree and arranged in descending powers of x.

Let us *suppose* that P is a perfect square and that $a, b, c, \cdots$ denote the terms of its square root arranged in descending powers of x, so that $P \equiv (a + b + c + \cdots)^2$.

The problem is, knowing P, to find $a, b, c, \cdots$.

Now, whatever the values of $a, b, c, \cdots$ may be, we have

$$(a+b)^2 = a^2 + 2ab + b^2 = a^2 + (2a + b)b,$$

$$\begin{aligned}(a+b+c)^2 &= (a+b)^2 + 2(a+b)c + c^2 \\ &= a^2 + (2a+b)b + [2(a+b)+c]c,\end{aligned}$$

$$\begin{aligned}(a+b+c+d)^2 = a^2 + (2a+b)b &+ [2(a+b)+c]c \\ &+ [2(a+b+c)+d]d,\end{aligned}$$

and so on, a new group of terms being added on the right with each new letter on the left, namely, *a group formed by adding the new letter to twice the sum of the old letters and multiplying the result by the new letter.*

Therefore, since by hypothesis $P \equiv (a + b + c + \cdots)^2$, we have

$$\begin{aligned}P \equiv a^2 + (2a+b)b &+ [2(a+b)+c]c \\ &+ [2(a+b+c)+d]d + \cdots,\end{aligned}$$

where the leading terms of the several groups on the right, namely, a^2, $2ab$, $2ac$, $2ad$, $\cdots$, are all of higher degree in x than any of the terms which follow them.

From this identity we may find $a, b, c, \cdots$ as follows:

1. Evidently a is the square root of the leading term of P.

2. Subtract a^2 from P. As the leading term of the remainder, R_1, must equal $2ab$, we may find b by dividing this term by $2a$.

3. Having found b, form $(2a+b)b$ and subtract it from R_1. As the leading term of the remainder, R_2, must equal $2ac$, we may find c by dividing this term by $2a$.

4. Continue thus until a remainder of lower degree than a is reached.

If this final remainder is 0, P is, as was supposed, a perfect square and its square root is $a+b+c+\cdots$.

If this final remainder is not 0, P is not a perfect square; but we shall have reduced P to the form

$$P \equiv (a+b+c+\cdots)^2 + R,$$

that is, to the sum of a perfect square and an integral function which is of lower degree than a.

It is convenient to arrange the reckoning just described as in the following example.

Example. Find the square root of $4x^4 - 4x^3 + 13x^2 - 6x + 9$.

$$\begin{array}{rl|l}
 & P = 4x^4 - 4x^3 + 13x^2 - 6x + 9 & 2x^2 - x + 3 = a + b + c \\
 & a^2 = 4x^4 & \\
2a + b = 4x^2 - x & -4x^3 + 13x^2 - 6x + 9 = R_1 = P - a^2 & \\
 & -4x^3 + x^2 \qquad = (2a+b)b & \\
2(a+b) + c = 4x^2 - 2x + 3 & 12x^2 - 6x + 9 = R_2 = P - (a+b)^2 & \\
 & 12x^2 - 6x + 9 = [2(a+b)+c]c & \\
 & 0 \qquad = R = P - (a+b+c)^2 &
\end{array}$$

Since the final remainder is 0, P is a perfect square and its square root is $2x^2 - x + 3$. Compare § 569, Ex. 1.

Observe that as each new remainder $R_1, R_2, \cdots$ is found we divide its leading term by $2a$ and so get the next term of the root. Then at the left of the remainder we write twice the part of the root previously obtained plus the new term of the root. We multiply this sum by the new term of the root, subtract the result from the remainder under consideration, and thus obtain the next remainder.

Example. Find the square root of $25x^4 - 40x^3 + 46x^2 - 24x + 9$.

571 This method is applicable to a polynomial P which involves more than one letter, provided it be a perfect square. We first arrange P in descending powers of one of the letters, with coefficients involving the rest, and then proceed as in § 570, it being understood that x now denotes the letter of arrangement.

572 **Approximate square roots.** We may also apply this method to a polynomial in x arranged in *ascending* powers of this letter. But $a, b, c, \cdots$ will then be arranged in ascending powers of x, and the degrees of the successive remainders will *increase*. Hence, § 570, 4, if P is not a perfect square but has a constant term, we can reduce it to the form

$$P \equiv (a + b + c + \cdots)^2 + R',$$

that is, to the sum of a perfect square and a polynomial, R', *whose lowest term is of as high a degree as we please.*

For small values of x we can make the value of R' as small as we choose by carrying this reckoning far enough. Hence in this case we call $a + b$, $a + b + c, \cdots$ the *approximate square roots* of P to two terms, three terms, and so on.

It should be added that these approximate roots are found more readily by the method of § 569.

Example 1. Find the square root of $1 + x$ to *four* terms.

By § 569, we write $\sqrt{1+x} \equiv 1 + px + qx^2 + rx^3 + \cdots$.

Squaring, $1 + x \equiv 1 + 2px + (p^2 + 2q)x^2 + 2(pq + r)x^3 + \cdots$

Hence, § 284, $2p = 1, \quad p^2 + 2q = 0, \quad pq + r = 0,$

or solving, $p = 1/2, \quad q = -1/8, \quad r = 1/16.$

Therefore the required result is $1 + x/2 - x^2/8 + x^3/16$.

Let the student verify this by the method of §§ 570, 571.

Example 2. Find the square root of $4 - x + x^2$ to *three* terms.

573 **Square roots of numbers.** From the formulas in § 570 we also derive the ordinary method of finding the square root of a *number*.

Example. Find the square root of 53361.

Let a denote the greatest integer *with but one significant figure*, whose square is contained in 53361. Its significant figure will be the leading figure of the root and its remaining figures will be 0's. We find a as follows:

Remembering that for each 0 at the end of a there will be two 0's at the end of a^2, we mark off in 53361, from right to left, as many *periods of two figures* as we can, thus: 5'33'61.

Each of the periods 61 and 33 calls for one 0 at the end of a, and the remaining period, 5, calls for the initial figure 2, since 2 is the greatest integer whose square is less than 5. Hence $a = 200$.

Having found a, we proceed quite as when seeking the square root of a polynomial. This is indicated in the scheme below at the left, where b denotes the second figure of the root multiplied by 10, and c the units figure. The scheme at the right gives the reckoning as abridged in common practice.

$$
\begin{array}{r|l}
 & \quad a + b + c \\
5'\,33'\,61 & \underline{200 + 30 + 1} \\
4\;00\;00 & = a^2 \\
2a = 400 & 1\;33\;61 = R_1 \\
2a + b = 430 & \underline{1\;29\;00} = (2a + b)b \\
2(a + b) = 460 & 4\;61 = R_2 \\
2(a + b) + c = 461 & \underline{4\;61} = [2(a + b) + c]c \\
 & \quad 0 \;\; = R
\end{array}
\qquad\qquad
\begin{array}{r|l}
5'\,33'\,61 & \underline{231} \\
4 & \\
43 & 1\;33 \\
 & \underline{1\;29} \\
461 & 4\;61 \\
 & \underline{4\;61} \\
 & \quad 0
\end{array}
$$

We first subtract a^2, then find the significant figure of b by dividing the remainder R_1 by $2a$, next find R_2 by subtracting $(2a + b)b$ from R_1, and finally c by dividing R_2 by $2(a + b)$.

The simplest way of accomplishing all this, as indicated in the abridged scheme at the right, is to omit final 0's and to bring down one period at a time. Then, as each new remainder is obtained, we write at its left twice the part of the root already found as a "trial divisor," obtain the next figure of the root by dividing the remainder by this trial divisor, and complete the divisor by affixing this figure to it. We then multiply the complete divisor by the new figure of the root, subtract, and so obtain the next remainder. If too large a figure is obtained at any stage in the process, that is, a figure which makes the product just described greater than the remainder in question, we try the next smaller figure.

574 Approximate square roots of numbers. The method just explained also enables us to obtain *approximate values* of the square roots of numbers which are *not* perfect squares.

Example. Find an approximate value of the square root of 7.342 correct to the third place of decimals.

```
       7.34' 20' 00 |2.709
       4
  47 | 3 34
     | 3 29
5409 | 5 20 00
     | 4 86 81
          33 19
```

Hence
$$\sqrt{7.342} = 2.709\cdots$$

Evidently for each *decimal* figure in the root there are two such figures in the number. Hence we separate the decimal part of the number into periods of two figures, proceeding from the decimal point to the *right*. The integral part of the number we separate into periods, as in § 573, proceeding from the decimal point to the *left*.

Observe that a decimal number cannot be a perfect square if it has an *odd* number of decimal figures.

575 Cube roots of polynomials. There is also a special method for finding the *cube root* of a polynomial P, when P is a perfect cube, analogous to that just given for finding a square root.

Let P denote a polynomial in x whose degree is some multiple of 3 and which is arranged in descending powers of x.

Let us suppose that P is a perfect cube and that $a, b, c, \cdots$ denote the terms of its cube root, arranged in descending powers of x, so that $P \equiv (a + b + c + \cdots)^3$.

The problem is, knowing P, to find $a, b, c, \cdots$.

Now, whatever the values of $a, b, c, \cdots$ may be, we have

$$(a+b)^3 = a^3 + (3a^2 + 3ab + b^2)b,$$
$$(a+b+c)^3 = a^3 + (3a^2 + 3ab + b^2)b$$
$$+ [3(a+b)^2 + 3(a+b)c + c^2]c,$$

and so on, a new group of terms being added on the right with each new letter on the left, namely, *a group formed by adding together three times the square of the sum of the old letters, three times the product of the sum of the old letters by the new letter, and the square of the new letter, and then multiplying the result by the new letter.*

Therefore, since by hypothesis $P \equiv (a + b + c + \cdots)^3$, we have

$$P \equiv a^3 + (3a^2 + 3ab + b^2)b + [3(a+b)^2 + 3(a+b)c + c^2]c + \cdots,$$

where the leading terms of the several groups on the right, namely, a^3, $3a^2b$, $3a^2c$, $\cdots$, are all of higher degree in x than any of the terms which follow them.

From this identity we may find a, b, c, $\cdots$ as follows:

1. Evidently a is the cube root of the leading term of P.

2. Subtract a^3 from P. As the leading term of the remainder, R_1, must equal $3a^2b$, we may find b by dividing this term by $3a^2$.

3. Having found b, form $(3a^2 + 3ab + b^2)b$ and subtract it from R_1. As the leading term of the remainder, R_2, must equal $3a^2c$, we may find c by dividing this term by $3a^2$.

4. Continue thus until a remainder is reached which is of lower degree than a^2.

If this final remainder is 0, then P is, as was supposed, a perfect cube and its cube root is $a + b + c + \cdots$.

If this final remainder is not 0, P is not a perfect cube, but we shall have reduced it to the form

$$P \equiv (a + b + c + \cdots)^3 + R,$$

where R is of lower degree than a^2.

It is convenient to arrange this reckoning as follows:

Example. Find the cube root of

$$x^6 + 6x^5 + 21x^4 + 44x^3 + 63x^2 + 54x + 27.$$

$$\begin{array}{r|l}
 & \underline{x^2 + 2x + 3} \\
3a^2 = 3x^4 & x^6 + 6x^5 + 21x^4 + 44x^3 + 63x^2 + 54x + 27 \\
 & \underline{x^6} \\
3(x^2)^2 = 3x^4 & 6x^5 + 21x^4 + 44x^3 + 63x^2 + 54x + 27 = R_1 \\
3x^2 \cdot 2x + (2x)^2 = \underline{6x^3 + 4x^2} & \\
3x^4 + 6x^3 + 4x^2 & \underline{6x^5 + 12x^4 + 8x^3} \\
3(x^2 + 2x)^2 = 3x^4 + 12x^3 + 12x^2 & 9x^4 + 36x^3 + 63x^2 + 54x + 27 = R_2 \\
3(x^2 + 2x)3 + 3^2 = \underline{9x^2 + 18x + 9} & \\
3x^4 + 12x^3 + 21x^2 + 18x + 9 & \underline{9x^4 + 36x^3 + 63x^2 + 54x + 27} \\
 & 0 \qquad = R
\end{array}$$

Since the final remainder is 0, $x^6 + 6x^5 + \cdots + 54x + 27$ is a perfect cube and its cube root is $x^2 + 2x + 3$. Compare § 569, Ex. 2.

Observe that as each new remainder $R_1, R_2, \cdots$ is found, we divide its leading term by $3a^2$ and so get the next term of the root. Then at the left of the remainder we write the sum of three times the square of the part of the root previously obtained, three times the product of this part by the new term, and the square of the new term. We multiply this sum by the new term, subtract the result from the remainder under consideration, and thus obtain the next remainder.

576 This method is also applicable to a polynomial which involves more than one letter, if it be a perfect cube (compare § 571).

The method may also be applied to a polynomial in x arranged in *ascending* powers of this letter, — if it does not lack a constant term. If the polynomial is not a perfect cube, we thus obtain *approximate cube roots* (compare § 572).

577 **Cube roots of numbers.** We may also find the cube root of a *number* by aid of the formulas of § 575.

Example. Extract the cube root of 12487168.

$$\begin{array}{rr|l}
 & & a + b + c \\
 & N = 12'\,487'\,168 & \underline{200 + 30 + 2} = 232 \\
 & 8\ 000\ 000 & \\
3a^2 = 120000 & 4\ 487\ 168 & = R_1 = N - a^3 \\
3ab = 18000 & & \\
b^2 = \underline{900} & & \\
138900 & 4\ 167\ 000 & = (3a^2 + 3ab + b^2)b \\
\hline
3(a+b)^2 = 158700 & 320\ 168 & = R_2 = N - (a+b)^3 \\
3(a+b)c = 1380 & & \\
c^2 = \underline{4} & & \\
160084 & \underline{320\ 168} & = [3(a+b)^2 + 3(a+b)c + c^2]c \\
 & 0 & = R = N - (a+b+c)^3.
\end{array}$$

In order to find a, the greatest number with one significant figure whose cube is contained in N, we begin by marking off periods of *three* figures in N from right to left (also from the decimal point to the right when there are decimal figures in N), thus: 12' 487' 168. Each of the periods 168 and 487 calls for one 0 at the end of a, and the remaining period, 12, calls for the initial figure 2, 2 being the greatest integer whose cube is contained in 12. Hence $a = 200$.

The rest of the reckoning is fully indicated above.

Observe that each new figure of the root is found by *dividing the remainder last obtained by three times the square of the part of the root already found;* thus, we find the significant figure of b by dividing R_1 by $3a^2$, and c by dividing R_2 by $3(a+b)^2$. If too large a figure is thus obtained, we test the next smaller figure.

The process may be abbreviated in the same way as that for finding the square root of a number.

Approximate cube roots of numbers which are not perfect cubes may also be found by this process (compare § 574).

578 **Higher roots of polynomials.** The *fourth* root of a polynomial which is a perfect fourth power may be obtained by finding the square root of its square root; similarly the *sixth* root of a polynomial which is a perfect sixth power may be obtained by finding the cube root of its square root.

It is also possible to develop special methods, analogous to those of §§ 570, 575, for finding any root that may be required. But the general method of § 569 makes this unnecessary. In fact we have given the special methods for square and cube roots explained in §§ 570, 575 only because of their historic interest and their relation to the problem of finding square and cube roots of numbers.

EXERCISE XXXII

Simplify the following expressions.

1. $\sqrt[3]{-\dfrac{27x^6y^{15}}{125a^9z^{12}}}$. 2. $\sqrt{\dfrac{529a^4b^6}{625c^2d^8}}$. 3. $\sqrt[6]{(x^4y^2-2x^3y^3+x^2y^4)^3}$.

By § 569 or § 570 find the square roots of the following.

4. $x^4-2x^3+3x^2-2x+1$.

5. $x^2-2x^4+6x^3-6x+x^6+9$.

6. $4x^6+12x^5y+9x^4y^2-4x^3y^3-6x^2y^4+y^6$.

7. $4x^2-20x+13+30/x+9/x^2$.

8. $49 - 84x - 34x^2 + 60x^3 + 25x^4$.

9. $x^8 + 2x^7 - x^6 - x^4 - 6x^3 + 5x^2 - 4x + 4$.

10. $(x^2 + 1)^2 - 4x(x^2 - 1)$.

11. $4x^4 + 9x^2y^2 - 12x^3y + 16x^2 - 24xy + 16$.

12. $x^2/y^2 + y^2/x^2 + 2 + 2x^2 + 2y^2 + x^2y^2$.

Find approximate square roots to *four* terms of the following.

13. $1 - 2x$.

14. $4 - x + 3x^2$.

By § 569 or § 575 find the cube roots of the following.

15. $x^6 + 3x^5 + 6x^4 + 7x^3 + 6x^2 + 3x + 1$.

16. $27x^{12} + 27x^{10} - 18x^8 - 17x^6 + 6x^4 + 3x^2 - 1$.

17. $8x^6 - 36ax^5 + 90a^2x^4 - 135a^3x^3 + 135a^4x^2 - 81a^5x + 27a^6$.

18. $x^3/y^3 + y^3/x^3 + 3x^2/y^2 + 3y^2/x^2 + 6x/y + 6y/x + 7$.

19. Find the approximate cube root to *three* terms of the expression $1 - x + x^2$.

20. By § 569 or § 578 find the fourth root of

$$x^8 - 4x^7 + 10x^6 - 16x^5 + 19x^4 - 16x^3 + 10x^2 - 4x + 1.$$

21. By § 569 find the fifth root of

$$x^{10} + 5x^9 + 15x^8 + 30x^7 + 45x^6 + 51x^5 + 45x^4 + 30x^3 + 15x^2 + 5x + 1.$$

22. To make $x^4 + 6x^3 + 11x^2 + ax + b$ a perfect square, what values must be assigned to a and b?

Find the square roots of the following numbers.

23. 27889.

24. 2313.61.

25. 583.2225.

26. 4149369.

27. .00320356.

28. 9.024016.

Find approximate square roots of the following numbers correct to the third decimal figure.

29. 2.

30. 55.5.

31. 234.561.

Find the cube roots of the following numbers.

32. 1860867.

33. 167284.151.

34. 1036.433728.

XII. IRRATIONAL FUNCTIONS. RADICALS AND FRACTIONAL EXPONENTS

REDUCTION OF RADICALS

Roots. In what follows the letters $a, b, \cdots$ will denote *positive* numbers or literal expressions supposed to have positive values. **579**

Again, $\sqrt[n]{a}$ will denote the *principal* nth root of a, that is, the *positive* number whose nth power is a; in other words, the positive number which is defined by the formula $(\sqrt[n]{a})^n = a$.

Finally, when n is *odd*, $\sqrt[n]{-a}$ will denote the principal nth root of $-a$, namely $-\sqrt[n]{a}$.

And when we use the word *root* we shall mean *principal* root.

Note. This is a restricted use of the word *root;* for *any* number whose nth power equals a is itself an nth root of a, and there are always n such numbers, as will be proved subsequently. **580**

Thus, since $2^2 = 4$ and $(-2)^2 = 4$, both 2 and -2 are square roots of 4. We shall indicate the principal root 2 by $\sqrt{4}$, the other root -2 by $-\sqrt{4}$.

When n is *odd* and a is *real*, one of the nth roots of a is real and of the same sign as a, and the rest are imaginary.

When n is *even* and a is *positive*, two of the nth roots of a are real, equal numerically, but of contrary sign, and the rest are imaginary.

When n is *even* and a is *negative*, all the nth roots of a are imaginary.

In the higher mathematics $\sqrt[n]{a}$ usually denotes *any* nth root of a, not, as here, the *principal* root only.

Radicals. Any expression of the form $\sqrt[n]{a}$ or $b\sqrt[n]{a}$ is called a *radical;* and n is called the *index*, a the *radicand*, and b the *coefficient* of the radical. **581**

When both a and b are *rational* numbers or expressions, $b\sqrt[n]{a}$ is called a *simple radical.*

Thus $5\sqrt[3]{4}$ is a simple radical whose index is 3, its radicand 4, and its coefficient 5.

582 Formulas for reckoning with radicals. The rules for reckoning with radicals are based on the following formulas, in which m, n, p denote positive integers.

$$1.\quad \sqrt[n]{a^m} = \sqrt[np]{a^{mp}}.$$

$$2.\quad \sqrt[n]{ab} = \sqrt[n]{a} \cdot \sqrt[n]{b}. \qquad 3.\quad \sqrt[n]{\frac{a}{b}} = \frac{\sqrt[n]{a}}{\sqrt[n]{b}}.$$

$$4.\quad (\sqrt[n]{a})^m = \sqrt[n]{a^m}. \qquad 5.\quad \sqrt[m]{\sqrt[n]{a}} = \sqrt[mn]{a}.$$

Observe in particular that, by 1, the value of a radical is not changed if its index and the exponent of its radicand are multiplied by the same positive integer or if any factor common to both is cancelled; thus, $\sqrt[9]{a^6} = \sqrt[3]{a^2}$. The similarity of this rule to the rule for simplifying a fraction is obvious.

These formulas may be proved by aid of the definition $(\sqrt[n]{a})^n = a$, the laws of exponents $(a^m)^n = a^{mn}$, $(ab)^n = a^n b^n$, and the rule of equality, § 261, 3,

Two positive numbers are equal if any like powers of these numbers are equal.

Thus,

1. $\sqrt[n]{a^m} = \sqrt[np]{a^{mp}}$, since their npth powers are equal.

For $\quad (\sqrt[np]{a^{mp}})^{np} = a^{mp}$; and $(\sqrt[n]{a^m})^{n \cdot p} = (a^m)^p = a^{mp}$.

2. $\sqrt[n]{ab} = \sqrt[n]{a} \cdot \sqrt[n]{b}$, since their nth powers are equal.

For $\quad (\sqrt[n]{ab})^n = ab$; and $(\sqrt[n]{a} \cdot \sqrt[n]{b})^n = (\sqrt[n]{a})^n \cdot (\sqrt[n]{b})^n = ab$.

3. $\sqrt[n]{\dfrac{a}{b}} = \dfrac{\sqrt[n]{a}}{\sqrt[n]{b}}$, since their nth powers are equal.

For $\quad \left(\sqrt[n]{\dfrac{a}{b}}\right)^n = \dfrac{a}{b}$; and $\left(\dfrac{\sqrt[n]{a}}{\sqrt[n]{b}}\right)^n = \dfrac{(\sqrt[n]{a})^n}{(\sqrt[n]{b})^n} = \dfrac{a}{b}$.

4. $(\sqrt[n]{a})^m = \sqrt[n]{a^m}$, since their nth powers are equal.

For $\quad (\sqrt[n]{a^m})^n = a^m$; and $[(\sqrt[n]{a})^m]^n = [(\sqrt[n]{a})^n]^m = a^m$.

5. $\sqrt[m]{\sqrt[n]{a}} = \sqrt[mn]{a}$, since their mnth powers are equal.

For $(\sqrt[mn]{a})^{mn} = a$; and $\left(\sqrt[m]{\sqrt[n]{a}}\right)^{m \cdot n} = (\sqrt[n]{a})^n = a$.

The following examples will show the usefulness of these formulas.

1. $\sqrt[6]{8} = \sqrt[6]{2^3} = \sqrt{2}$.

2. $\sqrt{8ab^3} = \sqrt{4b^2} \cdot \sqrt{2ab} = 2b\sqrt{2ab}$.

3. $\sqrt[3]{\dfrac{3c}{d^3e^6}} = \dfrac{\sqrt[3]{3c}}{\sqrt[3]{d^3e^6}} = \dfrac{\sqrt[3]{3c}}{de^2}$.

4. $\sqrt[5]{\sqrt{32x^{15}y^5}} = \sqrt[10]{32x^{15}y^5} = \sqrt{2x^3y}$.

5. $(\sqrt[3]{2xy^2})^2 = \sqrt[3]{(2xy^2)^2} = \sqrt[3]{4x^2y^4} = y\sqrt[3]{4x^2y}$.

On simplifying radicals. That form of a radical is regarded as simplest in which the radicand is the simplest *integral* expression possible. Hence for simplifying radicals we have the following rules, which are immediate consequences of the formulas just demonstrated. **583**

1. *If the radicand be a power whose exponent has a factor in common with the index, cancel that factor in both exponent and index.*

Thus, $\sqrt[9]{27x^3y^6} = \sqrt[9]{(3xy^2)^3} = \sqrt[3]{3xy^2}$.

2. *If any factor of the radicand be a power whose exponent is divisible by the index, divide the exponent by the index and then remove the factor from under the radical sign.*

Thus, $\sqrt[4]{16x^7y^9} = \sqrt[4]{2^4x^4x^3y^8y} = 2xy^2\sqrt[4]{x^3y}$.

3. *If the radicand be a fraction; multiply its numerator and denominator by the simplest expression which will render it possible to remove the denominator from under the radical sign.*

Thus, $\sqrt[3]{\dfrac{xy}{2z^2}} = \sqrt[3]{\dfrac{4xyz}{8z^3}} = \dfrac{1}{2z}\sqrt[3]{4xyz}$.

Similar radicals. Radicals which, when reduced to their simplest forms, differ in their coefficients only are said to be *similar*. **584**

Thus, $\sqrt{4x^3y}$ and $\sqrt{81x^5y^3}$ are similar; for their simplest forms, namely $2x\sqrt{xy}$ and $9x^2y\sqrt{xy}$, differ in their coefficients only.

585 **On bringing the coefficient of a radical under the radical sign.** Since $b\sqrt[n]{a} = \sqrt[n]{b^n a}$, the coefficient of a radical may be brought under the radical sign if its exponent be multiplied by the index of the radical.

EXERCISE XXXIII

Reduce each of the following radicals to its simplest form.

1. $\sqrt{18}$. 2. $\sqrt{588}$. 3. $\sqrt[3]{-27^2}$. 4. $\sqrt[9]{-1000}$.

5. $\sqrt{3/2}$. 6. $\sqrt[3]{3/2}$. 7. $\sqrt[3]{3/4}$. 8. $\sqrt[5]{3/16}$.

9. $\sqrt[5]{25\,a^5b^{10}c^{15}d^6}$. 10. $\sqrt[6]{128\,a^2b^4c^8}$. 11. $\sqrt[12]{8\,x^6y^9z^{15}}$.

12. $\sqrt[2n]{25\,a^2b^4c^6}$. 13. $\sqrt[3n]{a^nb^{2n}c^{3n}}$. 14. $\sqrt[n]{a^{2n+1}b^{3n+2}c^{4n}}$.

15. $\sqrt{x^2y^2 - x^2z^2}$. 16. $\sqrt{(x^2 - y^2)(x + y)}$.

17. $\sqrt[3]{x^6 - x^3y^3}$. 18. $\sqrt[4]{a^4b^4 - 2\,a^3b^5 + a^2b^6}$.

19. $\sqrt[3]{\dfrac{a^3 + b^3}{32\,ab^2}}$. 20. $\sqrt{\dfrac{a + b}{a - b}}$. 21. $\sqrt[3]{\dfrac{x^2 - x + 1}{9\,(x + 1)^2}}$.

22. $\sqrt[3]{1 - \dfrac{a^3}{b^3}}$. 23. $\sqrt[3]{\dfrac{c^{n+3}}{a^{3n}b^{3n+2}}}$. 24. $\sqrt{\dfrac{a^2x^2}{b^3} - \dfrac{2\,ax}{b^2} + \dfrac{1}{b}}$.

Bring the coefficients of the following under the radical sign.

25. $3\,a\sqrt{3\,a}$. 26. $\dfrac{a + b}{a - b}\sqrt{\dfrac{a - b}{a + b}}$. 27. $3\,ax\sqrt[4]{1/27\,a^3x^3}$

Show that the following sets of radicals are similar.

28. $\sqrt{18}$, $\sqrt{50}$, and $\sqrt{1/8}$. 29. $\sqrt[3]{24}$, $\sqrt[3]{192}$, and $\sqrt[3]{8/9}$.

30. $\sqrt{(x^3 - y^3)(x - y)}$ and $\sqrt{x^4y^2 + x^3y^3 + x^2y^4}$.

OPERATIONS WITH RADICALS

586 **Addition and subtraction.** We have the rule :

To reduce the algebraic sum of two or more radicals to its simplest form, simplify each radical and then combine such of them as are similar by adding their coefficients.

Example. Add $\sqrt{16a^2b}$, $-\sqrt{9a^2b}$, $3\sqrt{2}$, and $-2\sqrt{1/2}$.

We have $\sqrt{16a^2b} - \sqrt{9a^2b} + 3\sqrt{2} - 2\sqrt{1/2}$.

$$= 4a\sqrt{b} - 3a\sqrt{b} + 3\sqrt{2} - \sqrt{2} = a\sqrt{b} + 2\sqrt{2}.$$

Observe that a sum of two *dissimilar* radicals cannot be reduced to a single radical.

Thus, we cannot have $\sqrt{x} + \sqrt{y} = \sqrt{x+y}$ except when x or y is 0; for squaring, we have $x + y + 2\sqrt{xy} = x + y$, $\therefore 2\sqrt{xy} = 0$, $\therefore xy = 0$, $\therefore$ either $x = 0$ or $y = 0$.

Reduction of radicals to a common index. It follows from the formula $\sqrt[n]{a^m} = \sqrt[np]{a^{mp}}$ that we can always reduce two or more radicals to equivalent radicals having a common index. The *least common index* is the least common multiple of the given indices. **587**

Example. Reduce $\sqrt[6]{a^5}$ and $\sqrt[8]{b^3}$ to their least common index.

The least common multiple of the given indices, 6 and 8, is 24. And $\sqrt[6]{a^5} = \sqrt[24]{a^{20}}$ and $\sqrt[8]{b^3} = \sqrt[24]{b^9}$.

Comparison of radicals. We make the reduction to a common index when we wish to compare given radicals. **588**

Example 1. Compare $\sqrt[15]{16}$, $\sqrt[10]{6}$, and $\sqrt[6]{3}$.

The least common multiple of the given indices, 15, 10, 6, is 30; and

$$\sqrt[15]{16} = \sqrt[30]{16^2} = \sqrt[30]{256}\,;\ \sqrt[10]{6} = \sqrt[30]{6^3} = \sqrt[30]{216}\,;\ \sqrt[6]{3} = \sqrt[30]{3^5} = \sqrt[30]{243}.$$

Therefore, since $256 > 243 > 216$, we have $\sqrt[15]{16} > \sqrt[6]{3} > \sqrt[10]{6}$.

Example 2. Compare $2\sqrt{3}$ and $\sqrt[3]{41}$.

Bringing the coefficient of the first radical under the radical sign, § 585, and then reducing both radicals to the common index 6, we have

$$2\sqrt{3} = \sqrt{12} = \sqrt[6]{12^3} = \sqrt[6]{1728}\,;\ \sqrt[3]{41} = \sqrt[6]{41^2} = \sqrt[6]{1681}.$$

Therefore, since $1728 > 1681$, we have $2\sqrt{3} > \sqrt[3]{41}$.

Multiplication and division. From the formulas **589**

$$\sqrt[n]{a} \cdot \sqrt[n]{b} = \sqrt[n]{ab} \text{ and } \sqrt[n]{a} / \sqrt[n]{b} = \sqrt[n]{a/b}$$

we derive the following rule:

To multiply or divide one radical by another, reduce them if necessary to radicals having the least common index. Then find the product or quotient of their coefficients and radicands separately.

Example 1. Multiply $4\sqrt{xy}$ by $2\sqrt[3]{x^2y^2}$.

We have $4\sqrt{xy}\cdot 2\sqrt[3]{x^2y^2} = 8\sqrt[6]{x^3y^3}\cdot\sqrt[6]{x^4y^4} = 8\sqrt[6]{x^7y^7} = 8\,xy\sqrt[6]{xy}$.

Example 2. Divide $6\sqrt{xy}$ by $2\sqrt[4]{xy}$.

We have $6\sqrt{xy}/2\sqrt[4]{xy} = 3\sqrt[4]{x^2y^2}/\sqrt[4]{xy} = 3\sqrt[4]{xy}$.

590 **Involution.** From the formulas

$$(\sqrt[n]{a})^m = \sqrt[n]{a^m} \text{ and } \sqrt[np]{a^{mp}} = \sqrt[n]{a^m}$$

we derive the following rule:

To raise a radical of the form $\sqrt[n]{a^q}$ *to the* m*th power, cancel any factor which may be common to* m *and the index of the radical, and then multiply the exponent of the radicand by the remaining factor of* m.

Example. Raise $2\sqrt[6]{xy^2}$ to the 9th power.

We have

$$(2\sqrt[6]{xy^2})^9 = 2^9(\sqrt[6]{xy^2})^9 = 128(\sqrt{xy^2})^3 = 128\sqrt{x^3y^6} = 128\,xy^3\sqrt{x}.$$

591 **Evolution.** From the formulas

$$\sqrt[m]{\sqrt[n]{a}} = \sqrt[mn]{a} \text{ and } \sqrt[np]{a^{mp}} = \sqrt[n]{a^m}$$

we derive the following rule:

To find the m*th root of a radical of the form* $\sqrt[n]{a^q}$, *cancel any factor which may be common to* m *and the exponent of the radicand and multiply the index of the radical by the remaining factor of* n.

Example 1. Find the sixth root of $\sqrt[5]{x^2y^4}$.

We have $$\sqrt[6]{\sqrt[5]{x^2y^4}} = \sqrt[3]{\sqrt[5]{xy^2}} = \sqrt[15]{xy^2}.$$

Example 2. Find the cube root of $54\,a\sqrt{b}$.

We have

$$\sqrt[3]{54\,a\sqrt{b}} = \sqrt[3]{3^3\cdot 2\,a\sqrt{b}} = 3\sqrt[3]{\sqrt{4\,a^2b}} = 3\sqrt[6]{4\,a^2b}.$$

Simple radical expressions. By a *simple radical expression* we shall mean any expression which involves simple radicals only. Thus, $\sqrt{a}+\sqrt{b}$ is a simple radical expression. We call such an expression *integral* when it involves no fraction with a radical in its denominator. **592**

By the rules just given, *sums, differences, products,* and *powers* of simple integral radical expressions can be reduced to algebraic sums of simple radicals. In § 607 we shall show that the like is true of quotients. But ordinarily a *root* of a simple radical expression, as $\sqrt[3]{a+\sqrt{b}}$, cannot be reduced to a simple radical expression.

Example 1. Multiply $3\sqrt{6}+2\sqrt{5}$ by $2\sqrt{3}-\sqrt{10}$.

We have

$$\begin{aligned}(3\sqrt{6}+2\sqrt{5})(2\sqrt{3}-\sqrt{10}) &= 6\sqrt{18}+4\sqrt{15}-3\sqrt{60}-2\sqrt{50}\\ &= 8\sqrt{2}-2\sqrt{15}.\end{aligned}$$

Example 2. Square $\sqrt{2}+\sqrt[3]{4}$.

We have

$$(\sqrt{2}+\sqrt[3]{4})^2 = 2+2\sqrt{2}\sqrt[3]{4}+\sqrt[3]{16} = 2+4\sqrt[6]{2}+2\sqrt[3]{2}.$$

EXERCISE XXXIV

Reduce the following to their least common index.

1. $\sqrt[6]{3}$, $\sqrt[10]{3}$, and $\sqrt[15]{3}$.
2. $\sqrt[3]{a^2}$, $\sqrt[4]{2a^3b^2}$, and $\sqrt[6]{7b^5}$.

Compare the following.

3. $3\sqrt{2}$ and $2\sqrt[3]{3}$.
4. $\sqrt{3}$, $\sqrt[3]{4}$, and $\sqrt[4]{5}$.

Reduce each of the following to a simple radical in its simplest form.

5. $\sqrt{35}\div\sqrt{7/5}$.
6. $10\div\sqrt{5}$.
7. $4\div\sqrt[3]{2}$.
8. $\sqrt{6}\cdot\sqrt{10}\cdot\sqrt{15}$.
9. $\sqrt[3]{60}\cdot\sqrt[3]{90}\cdot\sqrt[3]{15}$.
10. $2\sqrt{3}\div 3\sqrt{2}$.
11. $\sqrt{2}\cdot\sqrt[3]{2}\cdot\sqrt[4]{2}$.
12. $\sqrt[6]{3}\div\sqrt[4]{5}$.
13. $2\sqrt{35}\cdot\sqrt{65}\div\sqrt{91}$.
14. $\sqrt{a^3b^5c^7}\cdot\sqrt[3]{a^2b^4c^8}$.
15. $\sqrt[2n]{a}\cdot\sqrt[3n]{a}$.
16. $\sqrt{a^3b^3}\div\sqrt[6]{a^5b^5}$.
17. $\sqrt[3]{a^2bc^2}\cdot\sqrt[3]{ab^2c^4}$.

18. $\sqrt[3]{a} \cdot \sqrt[6n]{a}$. **19.** $\sqrt[6]{a/b} \div \sqrt[9]{a/b}$.

20. $\sqrt[3]{ab^2} \cdot \sqrt[6]{ab^5} \div (\sqrt[10]{a^7b^9} \cdot \sqrt[15]{a^{12}b^{14}})$.

21. $(\sqrt{12})^3$. **22.** $(\sqrt[3]{a^2})^6$. **23.** $(2\sqrt[4]{xy^2z^3})^6$.

24. $\sqrt[4]{\sqrt[3]{a^2}}$. **25.** $\sqrt[3]{\sqrt{8}}$. **26.** $\sqrt[6]{\sqrt[5]{a^3b^6/c^9}}$.

27. $\sqrt[4]{\sqrt[3]{256}}$. **28.** $\sqrt{2\sqrt{2}}$. **29.** $\sqrt{2\sqrt[3]{2}}$.

30. $\sqrt{\sqrt{2} \cdot \sqrt[3]{2}}$. **31.** $\sqrt[2m/n]{\sqrt[n]{a^m}}$. **32.** $\left(\sqrt[2m]{\sqrt[2n]{a}}\right)^{mnp}$.

Simplify each of the following as far as possible.

33. $\sqrt{12} + \sqrt{75} - \sqrt{48} + \sqrt{147}$. **34.** $\sqrt{125} + \sqrt{175} - \sqrt{28} + \sqrt{1/20}$.

35. $\sqrt[3]{500} - \sqrt[3]{108} + \sqrt[3]{1/2}$. **36.** $\sqrt{a/bc} + \sqrt{b/ca} + \sqrt{c/ab}$.

37. $\sqrt{50} - \sqrt{4\frac{1}{2}} + \sqrt[3]{-24} + \sqrt[3]{7\frac{1}{9}}$. **38.** $\sqrt{(a+b)^2c} - \sqrt{a^2c} - \sqrt{b^2c}$.

39. $\sqrt{ax^3 + 6\,ax^2 + 9\,ax} - \sqrt{ax^3 - 4\,a^2x^2 + 4\,a^3x}$.

40. $(x+y)\sqrt{\dfrac{x-y}{x+y}} - (x-y)\sqrt{\dfrac{x+y}{x-y}} + \sqrt{\dfrac{1}{x^2-y^2}}$.

41. $(\sqrt{2} + \sqrt{3} + \sqrt{6}) \cdot \sqrt{6}$. **42.** $(\sqrt{6} + \sqrt{10} + \sqrt{14}) \div \sqrt{2}$.

43. $(\sqrt{6} + \sqrt{5})(\sqrt{2} + \sqrt{15})$. **44.** $\sqrt{5 + 2\sqrt{2}} \cdot \sqrt{5 - 2\sqrt{2}}$.

45. $(1 + \sqrt{3})^3$. **46.** $(\sqrt{a} + \sqrt[4]{a} + 1)(\sqrt{a} - \sqrt[4]{a} + 1)$.

FRACTIONAL AND NEGATIVE EXPONENTS

593 In many cases reckoning with radicals is greatly facilitated by the use of *fractional exponents.*

Thus far we have attached a meaning to the expression a^n only when n denotes a positive integer. The rules for reckoning with such expressions, namely,

1. $a^m \cdot a^n = a^{m+n}$, 2. $(a^m)^n = a^{mn}$, 3. $(ab)^n = a^nb^n$,

are among the simplest in algebra. It is therefore natural to inquire: Can we find useful meanings for a^n, *in agreement with these rules,* when n is *not* a positive integer?

594 **The definition $a^{p/q} = \sqrt[q]{a^p}$.** Take $a^{\frac{1}{2}}$ for instance. We wish, if possible, to find a meaning for this symbol which will be in agreement with the rules 1, 2, 3.

But, to be in agreement with 1, we must have

$$(a^{\frac{1}{2}})^2 = a^{\frac{1}{2}} \cdot a^{\frac{1}{2}} = a^{\frac{1}{2}+\frac{1}{2}} = a^1 = a,$$

that is, $a^{\frac{1}{2}}$ must mean either $\sqrt{a}$ or $-\sqrt{a}$.

We choose the more convenient of these two meanings, and *define* $a^{\frac{1}{2}}$ as $\sqrt{a}$.

We thus find that *one* of the conditions which we wish $a^{\frac{1}{2}}$ to satisfy suffices to fix its meaning.

Similar reasoning leads us to define $a^{\frac{1}{3}}$ as $\sqrt[3]{a}$, $a^{\frac{2}{3}}$ as $\sqrt[3]{a^2}$, and in general $a^{\frac{p}{q}}$ as $\sqrt[q]{a^p}$, that is, as *the principal* q*th root* of a^p.

Observe that since $a^{\frac{p}{q}} = \sqrt[q]{a^p} = \sqrt[qm]{a^{pm}} = a^{\frac{pm}{qm}}$, the value of $a^{\frac{p}{q}}$ is not changed when p/q is replaced by an equivalent fraction.

Thus, $a^{\frac{2}{3}} = a^{\frac{4}{6}} = a^{\frac{6}{9}}$; also $a^2 = a^{\frac{4}{2}} = a^{\frac{6}{3}}$.

595 **The definition $a^0 = 1$.** Again, to be in agreement with 1, we must have

$$a^0 a^m = a^{0+m} = a^m,$$

and therefore
$$a^0 = a^m / a^m = 1.$$

We are therefore led to *define* a^0 as 1.

596 **The definition $a^{-s} = 1/a^s$.** Finally, to be in agreement with 1, we must have, § 595,

$$a^{-s} \cdot a^s = a^{-s+s} = a^0 = 1,$$

and therefore
$$a^{-s} = 1 / a^s.$$

We are therefore led to define a^{-s} as $1/a^s$.

Thus, by definition, $a^{-3} = 1/a^3$, $a^{-\frac{5}{6}} = 1/a^{\frac{5}{6}} = 1/\sqrt[6]{a^5}$.

It remains to prove that the meanings thus found for $a^{\frac{p}{q}}$, a^0, and a^{-s} are in complete agreement with the rules of exponents.

597 **Theorem 1.** *The law* $a^m \; a^n = a^{m+n}$ *holds good for all rational values of* m *and* n.

Let p, q, r, s denote any positive integers. Then

1. When $m = p/q$ and $n = r/s$, we have, § 582,

$$a^{\frac{p}{q}} \cdot a^{\frac{r}{s}} = \sqrt[q]{a^p} \cdot \sqrt[s]{a^r} = \sqrt[qs]{a^{ps}} \cdot \sqrt[qs]{a^{qr}}$$
$$= \sqrt[qs]{a^{ps+qr}} = a^{\frac{ps+qr}{qs}} = a^{\frac{p}{q}+\frac{r}{s}}.$$

2. When $m = -p/q$ and $n = -r/s$, we have, by Case 1,

$$a^{-\frac{p}{q}} \cdot a^{-\frac{r}{s}} = \frac{1}{a^{\frac{p}{q}} \cdot a^{\frac{r}{s}}} = \frac{1}{a^{\frac{p}{q}+\frac{r}{s}}} = a^{-\frac{p}{q}+\left(-\frac{r}{s}\right)}.$$

3. When $m = p/q$ and $n = -r/s$, and $p/q > r/s$, we have

$$a^{\frac{p}{q}} \cdot a^{-\frac{r}{s}} = \sqrt[q]{a^p} / \sqrt[s]{a^r} = \sqrt[qs]{a^{ps}} / \sqrt[qs]{a^{qr}}$$
$$= \sqrt[qs]{a^{ps-qr}} = a^{\frac{ps-qr}{qs}} = a^{\frac{p}{q}+\left(-\frac{r}{s}\right)}.$$

4. When $m = p/q$ and $n = -r/s$, and $p/q < r/s$, we have, by Case 3,

$$a^{\frac{p}{q}} \cdot a^{-\frac{r}{s}} = \frac{1}{a^{-\frac{p}{q}} \cdot a^{\frac{r}{s}}} = \frac{1}{a^{-\frac{p}{q}+\frac{r}{s}}} = a^{\frac{p}{q}+\left(-\frac{r}{s}\right)}.$$

598 **Theorem 2.** *The law* $(a^m)^n = a^{mn}$ *holds good for all rational values of* m *and* n.

For, let m denote *any* rational number. Then

1. When n is a *positive integer*, we have, § 597,

$$(a^m)^n = a^m \cdot a^m \cdots \text{to } n \text{ factors} = a^{m+m+\cdots \text{ to } n \text{ terms}} = a^{mn}.$$

2. When $n = p/q$, where p and q are positive integers, we have, by Case 1,

$$(a^m)^{\frac{p}{q}} = \sqrt[q]{(a^m)^p} = \sqrt[q]{a^{mp}} = a^{\frac{mp}{q}} = a^{m \cdot \frac{p}{q}}.$$

3. When $n = -s$, where s is any positive rational, we have, by Cases 1, 2,

$$(a^m)^{-s} = \frac{1}{(a^m)^s} = \frac{1}{a^{ms}} = a^{-ms} = a^{m(-s)}.$$

599 **Theorem 3.** *The law* $(ab)^n = a^n b^n$ *holds good for all rational values of* n.

1. Let $n = p/q$, where p and q denote positive integers. Then

$$(ab)^{\frac{p}{q}} = \sqrt[q]{(ab)^p} = \sqrt[q]{a^p b^p} = \sqrt[q]{a^p} \cdot \sqrt[q]{b^p} = a^{\frac{p}{q}} b^{\frac{p}{q}}.$$

2. Let $n = -s$, where s denotes any positive rational, whether integral or fractional. Then, by Case 1,

$$(ab)^{-s} = \frac{1}{(ab)^s} = \frac{1}{a^s b^s} = a^{-s} b^{-s}.$$

600 **Applications.** The following examples will illustrate the use of fractional and negative exponents. A complicated piece of reckoning with radicals often becomes less confusing when this notation is employed.

Example 1. Simplify $\sqrt{a/\sqrt[3]{a}}$.

We have $\sqrt{a/\sqrt[3]{a}} = (aa^{-\frac{1}{3}})^{\frac{1}{2}} = (a^{\frac{2}{3}})^{\frac{1}{2}} = a^{\frac{1}{3}} = \sqrt[3]{a}$.

Example 2. Simplify $\sqrt[4]{ab^3} \cdot \sqrt[6]{a^5 b} \div \sqrt[3]{a^2 b^2}$.

We have
$$\begin{aligned}\sqrt[4]{ab^3} \cdot \sqrt[6]{a^5 b} \div \sqrt[3]{a^2 b^2} &= a^{\frac{1}{4}} b^{\frac{3}{4}} \cdot a^{\frac{5}{6}} b^{\frac{1}{6}} \cdot a^{-\frac{2}{3}} b^{-\frac{2}{3}} \\ &= a^{\frac{1}{4}+\frac{5}{6}-\frac{2}{3}} b^{\frac{3}{4}+\frac{1}{6}-\frac{2}{3}} = a^{\frac{5}{12}} b^{\frac{3}{12}} = \sqrt[12]{a^5 b^3}.\end{aligned}$$

Example 3. Expand $(x^{\frac{2}{3}} + y^{-\frac{2}{3}})^3$.

We have
$$\begin{aligned}(x^{\frac{2}{3}} + y^{-\frac{2}{3}})^3 &= (x^{\frac{2}{3}})^3 + 3(x^{\frac{2}{3}})^2 y^{-\frac{2}{3}} + 3x^{\frac{2}{3}}(y^{-\frac{2}{3}})^2 + (y^{-\frac{2}{3}})^3 \\ &= x^2 + 3x^{\frac{4}{3}} y^{-\frac{2}{3}} + 3x^{\frac{2}{3}} y^{-\frac{4}{3}} + y^{-2}.\end{aligned}$$

Example 4. Divide $x - y$ by $x^{\frac{2}{3}} + x^{\frac{1}{3}} y^{\frac{1}{3}} + y^{\frac{2}{3}}$.

Arranging the reckoning as in § 401, we have

$$\begin{array}{l|l} x - y & x^{\frac{2}{3}} + x^{\frac{1}{3}} y^{\frac{1}{3}} + y^{\frac{2}{3}} \\ \hline x + x^{\frac{2}{3}} y^{\frac{1}{3}} + x^{\frac{1}{3}} y^{\frac{2}{3}} & x^{\frac{1}{3}} - y^{\frac{1}{3}} \\ \hline -x^{\frac{2}{3}} y^{\frac{1}{3}} - x^{\frac{1}{3}} y^{\frac{2}{3}} - y & \\ -x^{\frac{2}{3}} y^{\frac{1}{3}} - x^{\frac{1}{3}} y^{\frac{2}{3}} - y & \\ \hline \end{array}$$

Hence the quotient is $x^{\frac{1}{3}} - y^{\frac{1}{3}}$.

EXERCISE XXXV

Express as simply as possible without radical signs

1. $\sqrt[12]{a^8}$.
2. $\sqrt{c^{\frac{4}{3}}}$.
3. $a^{\frac{3}{5}}/\sqrt[3]{a^{\frac{3}{5}}}$.
4. $b\sqrt[3]{b^4}\cdot\sqrt[6]{b^5}$.

Express without negative or fractional exponents

5. $a^{1\frac{1}{4}}$.
6. $c^{-1.5}$.
7. $(d^{\frac{2}{3}})^{-6}$.
8. $(e^{-3\frac{1}{2}})^{-\frac{1}{7}}$.

Express with positive exponents and without radical signs

9. $a^{-1}/b^{-3}c^{-2}$.
10. $x^{-\frac{1}{2}}\sqrt{y^{-3}}$.
11. $(1/\sqrt{x^{-5}})^{-4}$.
12. $x^{-2}\sqrt{y^{-3}}/y^{-2}\sqrt{x^{-3}}$.

Express as simply as possible without denominators

13. $\frac{a}{bc}-\frac{b^{-1}}{c^{-2}}-\frac{a^{-1}(b^{-1}+c^{-1})}{a^{-2}(b+c)}+\frac{b+c}{b^{-}+c^{-1}}$.

Reduce each of the following to its simplest exponential form.

14. $(3\frac{1}{8})^{\frac{2}{3}}$.
15. $81^{\frac{3}{4}}$.
16. $(-27)^{\frac{2}{3}}$.
17. $8^{-\frac{5}{3}}$.
18. $a^{\frac{2}{3}}a^{\frac{3}{4}}a^{\frac{5}{6}}$.
19. $a^{\frac{2}{3}}a^{-\frac{1}{3}}a^{-\frac{1}{15}}$.
20. $(a^{\frac{3}{2}}b)^{\frac{1}{2}}a^{\frac{1}{2}}b^{\frac{3}{4}}$.
21. $ab^{-2}/a^{-3}b$.
22. $(a^{\frac{3}{4}})^{\frac{2}{3}}$.
23. $(a^{-1}b^{-2}c^3)^{-2}$.
24. $(-32a^{10})^{\frac{3}{5}}$.
25. $(-a^6b^{-9})^{-\frac{2}{3}}$.
26. $b^{-\frac{1}{3}}\sqrt[12]{b^{-5}}\div b^{-1}\sqrt{b^{-1}}$.
27. $(a^{-\frac{3}{2}}\sqrt{bc^5})^{\frac{2}{3}}$.
28. $(8a^{-15}/\sqrt{125a^3})^{-\frac{2}{3}}$.
29. $\sqrt{a^{\frac{2}{3}}(bc^{-1})^{-2}}$.
30. $\sqrt[3]{a^{-1}\sqrt[4]{a^3}}$.
31. $\sqrt[3]{a^{\frac{9}{2}}\sqrt{a^{-3}}}/\sqrt{\sqrt[3]{a^{-7}}\cdot\sqrt[3]{a}}$.
32. $[(x^x)^x]^x$.
33. $(x^{x^2+xy}y^{y^2+xy})^{\frac{xy}{x+y}}$.
34. $(x^{\frac{1}{2}}-y^{\frac{1}{2}})/(x^{-\frac{1}{4}}+y^{-\frac{1}{4}})$.
35. Multiply $x^{\frac{1}{4}}+x^{\frac{1}{8}}y^{\frac{1}{8}}+y^{\frac{1}{4}}$ by $x^{\frac{1}{4}}-x^{\frac{1}{8}}y^{\frac{1}{8}}+y^{\frac{1}{4}}$.
36. Divide a^2-b^3 by $a^{\frac{1}{3}}-b^{\frac{1}{2}}$.
37. Expand $(x^{\frac{1}{2}}-y^{\frac{1}{4}}z^{\frac{1}{4}})^4$.
38. Simplify $[(e^x+e^{-x})^2-4)]^{\frac{1}{2}}$.
39. Find square root of $x^2+4x^{\frac{3}{2}}y^{\frac{1}{2}}+4xy+6x^{\frac{1}{2}}y^{\frac{3}{2}}+12y^2+9x^{-1}y^3$.
40. Find cube root of $x^3+3x^2+6x+7+6x^{-1}+3x^{-2}+x^{-3}$.

THE BINOMIAL THEOREM FOR NEGATIVE AND FRACTIONAL EXPONENTS

If in the binomial expansion, § 561,

$$(a+b)^n = a^n + na^{n-1}b + \frac{n(n-1)}{1 \cdot 2} a^{n-2}b^2 + \cdots$$

601 we assign a fractional or negative value to n, we shall have on the right a never-ending, or *infinite*, series; for none of the coefficients n, $n(n-1)/2$, $\cdots$ will then be 0.

It will be shown further on that if $b < a$ the sum of the first m terms of this series will approach the value of $(a+b)^n$ as limit when m is indefinitely increased; in other words, that, by adding a sufficient number of the terms of this series, we may obtain a result *approximating* as closely as we please to the value of $(a+b)^n$.

This is what is meant when it is said that the binomial theorem holds good for $(a+b)^n$ when n is fractional or negative and $b < a$.

Example 1. Expand $(8 + x^{-\frac{1}{2}})^{\frac{1}{3}}$ to four terms.

Putting $n = 1/3$, $a = 8$, $b = x^{-\frac{1}{2}}$ in the formula, we have

$$\begin{aligned}(8 + x^{-\frac{1}{2}})^{\frac{1}{3}} &= 8^{\frac{1}{3}} + \tfrac{1}{3} \cdot 8^{-\frac{2}{3}} x^{-\frac{1}{2}} + \frac{\frac{1}{3}\left(-\frac{2}{3}\right)}{2} 8^{-\frac{5}{3}} (x^{-\frac{1}{2}})^2 \\ &\quad + \frac{\frac{1}{3}\left(-\frac{2}{3}\right)\left(-\frac{5}{3}\right)}{2 \cdot 3} 8^{-\frac{8}{3}} (x^{-\frac{1}{2}})^3 + \cdots \\ &= 2 + \frac{x^{-\frac{1}{2}}}{12} - \frac{x^{-1}}{288} + \frac{5x^{-\frac{3}{2}}}{20736} \cdots.\end{aligned}$$

Example 2. Find the sixth term in the expansion of $1/(a^{\frac{1}{2}} + x^{\frac{2}{3}})^2$ or $(a^{\frac{1}{2}} + x^{\frac{2}{3}})^{-2}$.

Putting $n = -2$, $a = a^{\frac{1}{2}}$, $b = x^{\frac{2}{3}}$, $r = 5$ in the formula for the $(r+1)$th term, § 565, we have

$$\frac{(-2)(-3)(-4)(-5)(-6)}{1 \cdot 2 \cdot 3 \cdot 4 \cdot 5} (a^{\frac{1}{2}})^{-2-5} (x^{\frac{2}{3}})^5 = -6\, a^{-\frac{7}{2}} x^{\frac{10}{3}}.$$

Example 3. Expand $\sqrt{1+x}$ to four terms.

Since $\sqrt{1+x} = (1+x)^{\frac{1}{2}}$, we have $n = \frac{1}{2}$, $a = 1$, $b = x$.

$$\text{Hence}\quad \sqrt{1+x} = 1 + \tfrac{1}{2}x + \frac{\frac{1}{2}\left(-\frac{1}{2}\right)}{2}x^2 + \frac{\frac{1}{2}\left(-\frac{1}{2}\right)\left(-\frac{3}{2}\right)}{2\cdot 3}x^3 + \cdots$$

$$= 1 + \frac{x}{2} - \frac{x^2}{8} + \frac{x^3}{16}\cdots.$$

The result is the same as that obtained in § 572, Ex. 1.

Example 4. Find an approximate value of $\sqrt{10}$.

We have $\sqrt{10} = (3^2+1)^{\frac{1}{2}} = 3\left(1+\frac{1}{9}\right)^{\frac{1}{2}}$,

$$\text{and}\quad 3\left(1+\tfrac{1}{9}\right)^{\frac{1}{2}} = 3\left[1 + \frac{1}{2}\cdot\frac{1}{9} + \frac{\frac{1}{2}\left(-\frac{1}{2}\right)}{2}\left(\frac{1}{9}\right)^2 + \frac{\frac{1}{2}\left(-\frac{1}{2}\right)\left(-\frac{3}{2}\right)}{2\cdot 3}\left(\frac{1}{9}\right)^3 + \cdots\right]$$

$$= 3 + \frac{1}{6} - \frac{1}{216} + \frac{1}{3888} + \cdots$$

$$= 3 + .16666 - .00462 + .00025 + \cdots = 3.1623 \text{ nearly}$$

EXERCISE XXXVI

Expand each of the following to *four* terms.

1. $(1+x)^{\frac{1}{3}}$.
2. $(a^{\frac{2}{3}} + x^{-\frac{2}{3}})^{-\frac{1}{2}}$.
3. $\sqrt[3]{(27-2x)^2}$.
4. $(a^m + x)^{\frac{1}{m}}$.
5. $(a^{-1} - b^{-\frac{1}{2}})^{-4}$.
6. $(\sqrt{x} + \sqrt[3]{y})^{-6}$.
7. $\dfrac{1}{2+3x}$.
8. $\dfrac{1}{\sqrt[5]{(1+x)^2}}$.
9. $\left(\dfrac{1}{\sqrt{1+3\sqrt{x}}}\right)^3$.

10. Find the tenth term in $(1+x)^{-3}$.

11. Find the seventh term in $(x^{-2} - 2y^{\frac{1}{3}})^{\frac{3}{4}}$.

12. Find the term involving $x^{\frac{9}{2}}$ in $(1 - x^{\frac{1}{2}})^{\frac{1}{4}}$.

13. Find the term involving x^{-2} in $x^{-\frac{3}{2}}(2 + x^{-\frac{1}{6}})^{-3}$.

14. By the method illustrated in § 601, Ex. 4, find approximate values of the following.

1. $\sqrt{99}$.   2. $\sqrt[3]{62}$.   3. $\sqrt[5]{31}$.

RATIONALIZING FACTORS

Rationalizing factors. When the product of two given radical expressions is rational, each of these expressions is called a *rationalizing factor* of the other. **602**

Thus, $(\sqrt{a}+\sqrt{b})(\sqrt{a}-\sqrt{b})=a-b$. Hence $\sqrt{a}+\sqrt{b}$ is a rationalizing factor of $\sqrt{a}-\sqrt{b}$, and *vice versa*.

It can be proved that every finite expression which involves *simple* radicals only has a rationalizing factor. The following sections will serve to illustrate this general theorem.

Rationalizing factors of functions of square roots. Every expression which is rational and integral with respect to $\sqrt{x}$ can be reduced to the form $A+B\sqrt{x}$, where A and B are rational and integral with respect to x; and $A+B\sqrt{x}$ has, with respect to x, the rationalizing factor $A-B\sqrt{x}$, obtained by merely changing the sign of $\sqrt{x}$. **603**

Thus, $2(\sqrt{x})^4+3x(\sqrt{x})^3$ may be written $2x^2+3x^2\sqrt{x}$. Hence this expression has the rationalizing factor $2x^2-3x^2\sqrt{x}$.

We may obtain a rationalizing factor of an expression which is rational and integral with respect to any finite number of square roots, as $\sqrt{x}$, $\sqrt{y}$, $\sqrt{z}$, $\cdots$, by repetitions of the process just explained. For we shall obtain a result which is completely rational if we multiply the given expression by its rationalizing factor with respect to $\sqrt{x}$, the product by its rationalizing factor with respect to $\sqrt{y}$, and so on.

Example. Find the rationalizing factor of $1+\sqrt{x}+\sqrt{y}+2\sqrt{xy}$.

We have $$1+\sqrt{y}+\sqrt{x}(1+2\sqrt{y}). \quad (1)$$

Multiply (1) by $$1+\sqrt{y}-\sqrt{x}(1+2\sqrt{y}). \quad (2)$$

We obtain $$(1+\sqrt{y})^2-x(1+2\sqrt{y})^2,$$

or $$1-x+y-4xy+2\sqrt{y}(1-2x). \quad (3)$$

Multiply (3) by $$1-x+y-4xy-2\sqrt{y}(1-2x). \quad (4)$$

We obtain $$(1-x+y-4xy)^2-4y(1-2x)^2. \quad (5)$$

Therefore, since (5) is completely rational, the product of (2) and (4) is the rationalizing factor of (1).

604 **Rationalizing factors of binomial radical expressions.** The rationalizing factor of an expression of the form $\sqrt[m]{a} \pm \sqrt[n]{b}$ may be found as in the following example.

Example. Find the rationalizing factor of $\sqrt[3]{a} + \sqrt{b}$.

We have
$$\sqrt[3]{a} + \sqrt{b} = a^{\frac{1}{3}} + b^{\frac{1}{2}} = (a^2)^{\frac{1}{6}} + (b^3)^{\frac{1}{6}}. \qquad (1)$$

But, § 438, $(a^2)^{\frac{1}{6}} + (a^3)^{\frac{1}{6}}$ will exactly divide the rational expression $a^2 - b^3$, the quotient being $(a^2)^{\frac{5}{6}} - (a^2)^{\frac{4}{6}}(b^3)^{\frac{1}{6}} + \cdots - (b^3)^{\frac{5}{6}}$. (2)
Hence (2) is the rationalizing factor of (1).

605 **On rationalizing the denominator of a fraction.** Any irrational expression of the form A / B, in which B involves simple radicals only, may be reduced to an equivalent expression having a *rational denominator* by multiplying both A and B by the rationalizing factor of B.

Example 1. Rationalize the denominator of $1 / \sqrt[4]{a^3}$.

We have
$$\frac{1}{\sqrt[4]{a^3}} = \frac{1}{a^{\frac{3}{4}}} = \frac{a^{\frac{1}{4}}}{a^{\frac{3}{4}} \cdot a^{\frac{1}{4}}} = \frac{a^{\frac{1}{4}}}{a} = \sqrt[4]{a} / a.$$

Example 2. Rationalize the denominator of $\dfrac{\sqrt{x^2 + a^2} + \sqrt{x^2 - a^2}}{\sqrt{x^2 + a^2} - \sqrt{x^2 - a^2}}$.

We have
$$\frac{\sqrt{x^2 + a^2} + \sqrt{x^2 - a^2}}{\sqrt{x^2 + a^2} - \sqrt{x^2 - a^2}} = \frac{(\sqrt{x^2 + a^2} + \sqrt{x^2 - a^2})^2}{(\sqrt{x^2 + a^2} - \sqrt{x^2 - a^2})(\sqrt{x^2 + a^2} + \sqrt{x^2 - a^2})}$$
$$= \frac{x^2 + \sqrt{x^4 - a^4}}{a^2}.$$

606 In computing an approximate value of a fractional numerical expression which involves radicals, one should begin by rationalizing the denominator. Much unnecessary reckoning is thus avoided.

Example. Find an approximate value of $(1 + \sqrt{8}) / (3 - \sqrt{2})$ which is correct to the third decimal figure.

We have
$$\frac{1 + \sqrt{8}}{3 - \sqrt{2}} = \frac{(1 + 2\sqrt{2})(3 + \sqrt{2})}{(3 - \sqrt{2})(3 + \sqrt{2})} = 1 + \sqrt{2} = 2.414 \cdots.$$

Division of radical expressions. To divide one radical expression by another, we write the quotient in the form of a fraction and then rationalize the denominator of this fraction. **607**

Example. Divide $4 + 2\sqrt{5}$ by $1 - \sqrt{2} + \sqrt{5}$.

We have

$$\frac{4 + 2\sqrt{5}}{1 + \sqrt{2} + \sqrt{5}} = \frac{(4 + 2\sqrt{5})(1 + \sqrt{2} - \sqrt{5})}{(1 + \sqrt{2} + \sqrt{5})(1 + \sqrt{2} - \sqrt{5})} = \frac{-3 + 2\sqrt{2} - \sqrt{5} + \sqrt{10}}{\sqrt{2} - 1}$$

$$= \frac{(-3 + 2\sqrt{2} - \sqrt{5} + \sqrt{10})(\sqrt{2} + 1)}{(\sqrt{2} - 1)(\sqrt{2} + 1)} = 1 - \sqrt{2} + \sqrt{5}.$$

General result. It follows from § 592 and § 607 that every expression which involves simple radicals only can be reduced to an algebraic sum of simple radicals. **608**

EXERCISE XXXVII

Find rationalizing factors of the following.

1. $\sqrt[7]{a^5}$.
2. $\sqrt[3]{a^2}\sqrt{b^3}$.
3. $x^{\frac{3}{2}} + x^{\frac{5}{2}} + x^{\frac{7}{2}}$.
4. $\sqrt{a} + \sqrt{bc}$.
5. $\sqrt{x} + \sqrt{y} + \sqrt{z}$.
6. $\sqrt{xy} + \sqrt{yz} + \sqrt{zx}$.
7. $\sqrt{x} + \sqrt{y} - \sqrt{z} - \sqrt{u}$.
8. $\sqrt{x} + \sqrt[4]{x} + 1$.
9. $x^{\frac{1}{3}} + y^{\frac{1}{3}}$.
10. $\sqrt[3]{a} - \sqrt[3]{b^2}$.
11. $x^{\frac{1}{4}} - y^{\frac{1}{4}}$.
12. $x^{\frac{3}{4}} + y^{\frac{2}{3}}$.
13. $1 + x^{\frac{1}{2}}y^{\frac{1}{3}}$.
14. $x^{\frac{2}{3}} + x^{\frac{1}{3}} + 1$.
15. $3 - \sqrt{5}$.
16. $1 + \sqrt{2} + \sqrt{3}$.
17. $1 + \sqrt[3]{2}$.
18. $\sqrt[3]{9} + \sqrt[3]{3} + 1$.
19. $\sqrt[3]{12} + \sqrt[3]{6} + \sqrt[3]{3}$.

Reduce each of the following to a fraction having a rational number or expression for its denominator.

20. $\dfrac{1}{\sqrt{a}\sqrt[5]{b^2}}$.
21. $\dfrac{a + \sqrt{b}}{a - \sqrt{b}}$.
22. $\dfrac{\sqrt{3} - \sqrt{2}}{2\sqrt{3} + 3\sqrt{3}}$.
23. $\dfrac{1}{b + \sqrt{b^2 - a^2}}$.
24. $\dfrac{\sqrt{x + y} + \sqrt{x - y}}{\sqrt{x + y} - \sqrt{x - y}}$.
25. $\dfrac{1 + \sqrt{2} + \sqrt{3}}{1 - \sqrt{2} + \sqrt{3}}$.
26. $\dfrac{1}{1 + \sqrt{2} + \sqrt{3} + \sqrt{6}}$.

27. $\dfrac{x\sqrt{y}+y\sqrt{x}}{\sqrt{x}+\sqrt{y}+\sqrt{x+y}}$.

28. $\dfrac{1}{\sqrt[3]{3}-1}+\dfrac{1}{\sqrt[3]{3}+1}$.

Find approximate values of the following expressions correct to the third decimal figure.

29. $\dfrac{5}{\sqrt{125}}$.

30. $\dfrac{2+\sqrt{28}}{\sqrt{7}}$.

31. $\dfrac{3+\sqrt{6}}{\sqrt{2}+\sqrt{3}}$.

IRRATIONAL EQUATIONS

609 **On solving an irrational equation.** The general method of solving an irrational equation is described in the following rule.

First, rationalize the equation.

Next, solve the resulting rational equation.

Finally, test all the solutions thus obtained in the given equation and reject those which do not satisfy it.

For, let $P=0$ denote the given equation, and $PR=0$ the rational equation obtained by multiplying both members of $P=0$ by R, the rationalizing factor of P. By § 341, the roots of $PR=0$ are those of $P=0$ and $R=0$ jointly. We discover which of them are the roots of $P=0$ by testing them in this equation.

Example. Solve $x-7-\sqrt{x-5}=0$.

Multiplying both members by the rationalizing factor $x-7+\sqrt{x-5}$, we obtain

$$(x-7)^2-(x-5)=0,$$

or simplifying,

$$x^2-15x+54=0.$$

Solving, by § 455, $\quad x=9$ or 6.

Substituting 9 for x in $x-7-\sqrt{x-5}=0$, we have $9-7-\sqrt{9-5}=0$, which is true. Hence 9 is a root.

But substituting 6, we have $6-7-\sqrt{6-5}=0$, which is false. Hence 6 is not a root.

But observe that 6 is a root of the equation $x-7+\sqrt{x-5}=0$, obtained by equating the rationalizing factor to 0; for $6-7+\sqrt{6-5}=0$ is true.

610 An equation which involves the radical $\sqrt[n]{A}$ may be rationalized with respect to this radical by collecting the terms which involve $\sqrt[n]{A}$ in one member and the remaining terms in the other, and then raising both members to the nth power. By repetitions of this process an equation which involves square roots only may be completely rationalized. It follows from § 345 that this method is equivalent to that described in § 609, but it involves less reckoning.

Example 1. Solve $\sqrt[3]{\sqrt{x} + a} = \sqrt{b}$.

Cubing both members, $\sqrt{x} + a = b^{\frac{3}{2}}$.

Transposing and squaring, $x = (b^{\frac{3}{2}} - a)^2$.

Substituting this result in the given equation, we find it to be a root.

Example 2. Solve $\sqrt{x+5} + \sqrt{x-4} = 9$.

Transposing, $\sqrt{x-4} = 9 - \sqrt{x+5}$.

Squaring, $x - 4 = 81 - 18\sqrt{x+5} + x + 5$.

Simplifying, $\sqrt{x+5} = 5$.

Squaring, $x + 5 = 25$.

Solving, $x = 20$.

Substituting 20 for x in the given equation, we have $\sqrt{25} + \sqrt{16} = 9$, which is true. Hence 20 is a root.

611 **Notes.** 1. Observe, as in Ex. 1, that we rationalize an equation with respect to the *unknown letter only* and make no attempt to rid it of radicals which do not involve this letter.

2. Observe also that *an irrational equation may have no root.*

Thus, the equation $\sqrt{x+5} - \sqrt{x-4} = 9$ has no root. For if we attempt to solve it we shall merely repeat the reckoning in Ex. 2 and shall again obtain the result $x = 20$; and $\sqrt{25} - \sqrt{16} = 9$ is false.

3. We may add that the simplest method of rationalizing an equation of the form $\sqrt{A} + \sqrt{B} + \sqrt{C} + \sqrt{D} = 0$ (or $\sqrt{A} + \sqrt{B} + \sqrt{C} + E = 0$) is to begin by writing it thus:

$$\sqrt{A} + \sqrt{B} = -\sqrt{C} - \sqrt{D} \quad (\text{or } \sqrt{A} + \sqrt{B} = -\sqrt{C} - E)$$

and then to square both members. The resulting equation will involve but two radicals and it may be rationalized as in Ex. 2.

612 **Simultaneous irrational equations.** To solve a system of such equations we may first rationalize each equation, then solve the resulting rational system, and finally test the results thus obtained in the given system.

But if the equations are of the form described in § 379, they should be solved by the method there explained.

Example 1. Solve $\sqrt{x-5}+\sqrt{y+5}=\sqrt{x}+\sqrt{y}$, (1)

$$x+2y=17. \qquad (2)$$

Squaring (1), $x-5+y+5+2\sqrt{xy+5x-5y-25}=x+y+2\sqrt{xy}$,

or $$\sqrt{xy+5x-5y-25}=\sqrt{xy}. \qquad (3)$$

Squaring (3) and simplifying, $x-y=5$. (4)

Solving (4), (2), $x=9,\ y=4$. (5)

Substituting $x=9,\ y=4$ in (1), we have $\sqrt{4}+\sqrt{9}=\sqrt{9}+\sqrt{4}$, which is true. Hence $x=9,\ y=4$ is the solution of (1), (2).

Example 2. Solve $\sqrt{x+6}+2/\sqrt{y}=4$, (1)

$$2\sqrt{x+6}+6/\sqrt{y}=9. \qquad (2)$$

Solving for $\sqrt{x+6}$ and $1/\sqrt{y}$, we find $\sqrt{x+6}=3,\ 1/\sqrt{y}=1/2$. (3)

And from (3) we obtain $x=3,\ y=4$, which is the solution of (1) and (2).

EXERCISE XXXVIII

Solve the following equations for x.

1. $x^{\frac{1}{4}}=4$.
2. $x^{-\frac{1}{2}}=3$.
3. $x^{\frac{3}{4}}=8$.
4. $(\sqrt{2x-1})^{\frac{1}{3}}=\sqrt{3}$.
5. $\sqrt{2+\sqrt{3+\sqrt{x}}}=2$.
6. $\sqrt{ax}+\sqrt{bx}+\sqrt{cx}=d$.
7. $\sqrt{4x^2+x+10}=2x+1$.
8. $\sqrt{x+4}+\sqrt{x+11}=7$.
9. $\sqrt{4x+5}+\sqrt{x+1}-\sqrt{9x+10}=0$.
10. $\sqrt{x+1}+\dfrac{x-6}{\sqrt{x+2}}=0$.
11. $\sqrt{x^2+3x-1}-\sqrt{x^2-x-1}=2$.
12. $\sqrt{x+7}+\sqrt{x-2}=\sqrt{x+2}+\sqrt{x-1}$.
13. $\dfrac{\sqrt{x+3}+\sqrt{x-5}}{\sqrt{x+3}-\sqrt{x-5}}=2$.
14. $\dfrac{1}{\sqrt{x+1}}-\dfrac{1}{\sqrt{x-1}}+\dfrac{1}{\sqrt{x^2-1}}=0$.

Solve the following for x and y.

15. $\begin{cases} \sqrt{x+17}+\sqrt{y-2}=\sqrt{x+5}+\sqrt{y+6}, \\ \sqrt{y-x}=\sqrt{3-x}+\sqrt{y-3}. \end{cases}$

16. $\begin{cases} 3\sqrt{x-2y}-\sqrt{x+y-4}=3, \\ \sqrt{x-2y}+2\sqrt{x+y-4}=8. \end{cases}$

17. Show that $\sqrt{x+a}+\sqrt{x+b}+\sqrt{x+c}+\sqrt{x+d}=0$ will reduce to a rational equation of the first degree.

18. Show that $\sqrt{ax+b}+\sqrt{cx+d}-\sqrt{ex+f}=0$ will reduce to a rational equation of the first degree if $\sqrt{a}+\sqrt{c}-\sqrt{e}=0$.

QUADRATIC SURDS

Surds. Numerical radicals like $\sqrt{2}$ and $\sqrt[3]{5}$, in which the radicand is rational but the radical itself is irrational, are called *surds*. A surd is called *quadratic, cubic*, and so on, according as its index is two, three, and so on. **613**

Theorem 1. *The product of two dissimilar quadratic surds is a quadratic surd.* **614**

Suppose that when the surds have been reduced to their simplest forms, their radical factors are $\sqrt{a}$ and $\sqrt{b}$. The product of $\sqrt{a}$ and $\sqrt{b}$ is $\sqrt{ab}$, and this is a surd unless ab is a perfect square.

But ab cannot be a perfect square, since by hypothesis a and b are integers none of whose factors are square numbers, and at least one of the factors of a is different from every factor of b.

Thus, $\sqrt{2}\cdot\sqrt{3}=\sqrt{6},\ \sqrt{6}\cdot\sqrt{15}=\sqrt{90}=3\sqrt{10}$.

Theorem 2. *The sum and the difference of two unequal quadratic surds are irrational numbers.* **615**

This is obvious when the surds are similar.

Hence let $\sqrt{a}$ and $\sqrt{b}$ denote dissimilar surds.

Suppose, if possible, that $\sqrt{a} + \sqrt{b} = c$, (1)
where c is rational.

Squaring both members of (1) and transposing,

$$2\sqrt{ab} = c^2 - a - b, \qquad (2)$$

which is impossible since $2\sqrt{ab}$ is irrational, § 614, while $c^2 - a - b$ is rational.

616 **Theorem 3.** *If* $a + \sqrt{b} = c + \sqrt{d}$, *where* $\sqrt{b}$ *and* $\sqrt{d}$ *are surds, then* $a = c$ *and* $b = d$.

For, by hypothesis, $\sqrt{b} - \sqrt{d} = c - a$.

But this is impossible unless $\sqrt{b} - \sqrt{d} = 0$ and $c - a = 0$, since otherwise $\sqrt{b} - \sqrt{d}$ would be irrational, § 615, and equal to $c - a$, which is rational.

Hence $b = d$ and $a = c$.

617 **Square roots of binomial surds.** We have

$$(\sqrt{x} \pm \sqrt{y})^2 = x + y \pm 2\sqrt{xy}.$$

Hence if $a + 2\sqrt{b}$ denote a given binomial surd, and we can find two *positive rational* numbers x and y such that

$$x + y = a \text{ and } xy = b,$$

then $\sqrt{x} + \sqrt{y}$ will be a square root of $a + 2\sqrt{b}$ and $\sqrt{x} - \sqrt{y}$ will be a square root of $a - 2\sqrt{b}$, and both these square roots will be binomial surds.

When such numbers x, y exist they may be found by inspection.

Example 1. Find the square root of $37 - 20\sqrt{3}$.

Reducing to the form $a - 2\sqrt{b}$, $37 - 20\sqrt{3} = 37 - 2\sqrt{300}$.

But $300 = 25 \cdot 12$ and $37 = 25 + 12$.

Hence $\sqrt{37 - 2\sqrt{300}} = \sqrt{25} - \sqrt{12} = 5 - 2\sqrt{3}$.

Example 2. Find the square root of $13/12 + \sqrt{5/6}$.

We have $\dfrac{13}{12} + \sqrt{\dfrac{5}{6}} = \dfrac{13}{12} + \dfrac{\sqrt{30}}{6} = \dfrac{13 + 2\sqrt{30}}{12}$.

Since $30 = 10 \cdot 3$ and $13 = 10 + 3$, we have $\sqrt{13 + 2\sqrt{30}} = \sqrt{10} + \sqrt{3}$.

Hence $$\sqrt{\frac{13 + 2\sqrt{30}}{12}} = \frac{\sqrt{10} + \sqrt{3}}{\sqrt{12}} = \frac{\sqrt{120} + \sqrt{36}}{12} = \frac{1}{2} + \frac{\sqrt{30}}{6}.$$

Note. We may obtain formulas for x and y as follows: **618**

By hypothesis $$\sqrt{x} + \sqrt{y} = \sqrt{a + 2\sqrt{b}}, \quad (1)$$

and $$\sqrt{x} - \sqrt{y} = \sqrt{a - 2\sqrt{b}}. \quad (2)$$

Multiplying (1) by (2), $x - y = \sqrt{a^2 - 4b}$. (3)

But $$x + y = a. \quad (4)$$

Solving (3), (4), $$x = \frac{a + \sqrt{a^2 - 4b}}{2}, \quad y = \frac{a - \sqrt{a^2 - 4b}}{2}.$$

Observe that these values are rational only when $a^2 - 4b$ is a perfect square. Hence in this case only is the square root of $a + 2\sqrt{b}$ a *binomial surd.*

EXERCISE XXXIX

Find square roots of the following.

1. $9 + \sqrt{56}$.
2. $20 + 2\sqrt{96}$.
3. $32 - 2\sqrt{175}$.
4. $1 + \frac{2\sqrt{6}}{5}$.
5. $7 - 3\sqrt{5}$.
6. $8\sqrt{2} + 2\sqrt{30}$.
7. $2(a + \sqrt{a^2 - b^2})$.
8. $b - 2\sqrt{ab - a^2}$.

Simplify the following.

9. $\sqrt[4]{17 + 12\sqrt{2}}$.
10. $\sqrt{9 + 4\sqrt{4 + 2\sqrt{3}}}$.

IMAGINARY AND COMPLEX NUMBERS

Complex numbers. Since all even powers of negative numbers are positive, no even root of a negative number can be a real number. Such roots are *imaginary* numbers. **619**

Definitions of the imaginary numbers and of the operations by which they may be combined are given in §§ 217–228, which the student should read in this connection.

According to these definitions,

1. The symbol $i = \sqrt{-1}$ is called the *unit of imaginaries.*

2. Symbols of the form ai, where a is real, are called *pure imaginaries.*

3. Symbols of the form $a + bi$, where a and b are real, are called *complex numbers.*

4. Two complex numbers are *equal* when, and only when, their real parts and their imaginary parts are equal, so that

$$\text{If } a + bi = c + di, \text{ then } a = c \text{ and } b = d.$$

5. The *sum, difference, product,* or *quotient* of two complex numbers is itself a complex number (in special cases a real number or a pure imaginary) which may be found by applying the ordinary rules of reckoning and the relation $i^2 = -1$. The like is true of any positive integral *power* of a complex number, since by definition $(a + bi)^n = (a + bi)(a + bi) \cdots$ to n factors.

Example 1. Add $5 + 3i$ and $2 - 4i$.

We have $5 + 3i + (2 - 4i) = (5 + 2) + (3 - 4)i = 7 - i.$

Example 2. Subtract $6 + 2i$ from $3 + 2i$.

We have $3 + 2i - (6 + 2i) = (3 - 6) + (2 - 2)i = -3.$

Example 3. Multiply $2 + 3i$ by $1 + 4i$.

We have
$$(2 + 3i)(1 + 4i) = 2 + 3i + 8i + 12i^2$$
$$= 2 + 3i + 8i - 12 = -10 + 11i.$$

Example 4. Expand $(1 + i)^2$.

We have $(1 + i)^2 = 1 + 2i + i^2 = 1 + 2i - 1 = 2i.$

Example 5. Find real values of x, y satisfying the equation

$$(x + yi)i - 2 + 4i = (x - yi)(1 + i).$$

Carrying out the indicated operations, we have

$$-(y + 2) + (x + 4)i = (x + y) + (x - y)i.$$

Equating the real and the imaginary parts, § 619, 4,

$$-(y + 2) = x + y \text{ and } x + 4 = x - y,$$

or, solving,
$$x = 6, \ y = -4.$$

In §§ 238–241 we have given a method for representing complex numbers by points called their *graphs,* and rules for obtaining from the graphs of two complex numbers the graphs of their sum and product. Let the student apply these rules to Exs. 1, 3, 4.

Conjugate imaginaries. Two complex numbers like $a + bi$ and $a - bi$, which differ only in the signs connecting their real and imaginary parts, are called *conjugate imaginaries*. **620**

The product of two conjugate imaginaries is a positive real number. **621**

Thus, $$(a + bi)(a - bi) = a^2 - b^2 i^2 = a^2 + b^2.$$

Hence a fraction, as $(a + bi)/(c + di)$, may be reduced to the form of a complex number by multiplying both its terms by the conjugate of its denominator. **622**

Example. Divide $5 + 7i$ by $2 - 4i$.

We have $$\frac{5 + 7i}{2 - 4i} = \frac{(5 + 7i)(2 + 4i)}{(2 - 4i)(2 + 4i)}$$
$$= \frac{-18 + 34i}{20} = -\frac{9}{10} + \frac{17}{10}i.$$

The powers of i. From the equation $i^2 = -1$ it follows that the even powers of i are either -1 or 1, and the odd powers either i or $-i$. **623**

Thus, $i^3 = i^2 \cdot i = -i$; $i^4 = i^3 \cdot i = -i \cdot i = -i^2 = 1$; and so on.

To find the value of i^n for any given value of n, divide n by 4. Then, according as the remainder is 0, 1, 2, 3, the value of i^n is $1, i, -1, -i$.

Thus, $i^{24} = (i^4)^6 = 1$; $i^{25} = i^{24} \cdot i = i$; and so on.

Even roots of negative numbers. The number -4 has the two square roots $2i$ and $-2i$; for $(2i)^2 = 2^2 i^2 = -4$, and $(-2i)^2 = (-2)^2 i^2 = -4$. We select $2i$ as the *principal* square root, and write $\sqrt{-4} = 2i$ and $-\sqrt{-4} = -2i$. **624**

Similarly the *principal square root* of any given negative number $-a$ is $\sqrt{a}i$, that is, $\sqrt{-a} = \sqrt{a}i$.

From this definition of principal square root it follows that if $-a$ and $-b$ are any two negative numbers, then **625**

$$\sqrt{-a}\sqrt{-b} = -\sqrt{ab}.$$

For $$\sqrt{-a} \cdot \sqrt{-b} = \sqrt{a}i \cdot \sqrt{b}i = i^2 \sqrt{a}\sqrt{b} = -\sqrt{ab}.$$

Thus, while the product of the principal square roots of two negative numbers $-a$, $-b$, is *one* of the square roots of their product ab, it is not, as in the case of real numbers, the *principal* square root of this product.

When reckoning with imaginaries, it is important to bear in mind this modification of the rule $\sqrt{a}\sqrt{b} = \sqrt{ab}$. All chance of confusion is avoided if at the outset we replace every symbol $\sqrt{-a}$ by $\sqrt{a}i$.

Example 1. Simplify $\sqrt{-2}\cdot(\sqrt{-3})^5\cdot(\sqrt{-5})^7$.

We have
$$\sqrt{-2}\cdot(\sqrt{-3})^5\cdot(\sqrt{-5})^7 = \sqrt{2}\,i\cdot(\sqrt{3}\,i)^5(\sqrt{5}\,i)^7$$
$$= \sqrt{2}\cdot(\sqrt{3})^5\cdot(\sqrt{5})^7\,i^{13}$$
$$= 1125\sqrt{30}\,i.$$

Example 2. Multiply $2+\sqrt{-9}$ by $1+\sqrt{-1}$.

We have
$$(2+\sqrt{-9})(1+\sqrt{-1}) = (2+3i)(1+i) = -1+5i.$$

626 The *higher* even roots of negative numbers are *complex* numbers. This will be proved subsequently.

Thus, one of the fourth roots of -4 is $1+i$; for
$$(1+i)^4 = 1+4i+6i^2+4i^3+i^4 = 1+4i-6-4i+1 = -4.$$

627 **Square roots of complex numbers.** As will be proved farther on, all roots of complex numbers are themselves complex numbers. We may find their *square* roots as follows.

We have
$$(\sqrt{x}\pm i\sqrt{y})^2 = x-y\pm 2i\sqrt{xy}.$$

Hence, if $a+bi$ denote a given complex number in which b is *positive*, and we can find two positive numbers x and y such that
$$x-y=a, \quad (1) \qquad \text{and} \qquad 2\sqrt{xy}=b, \quad (2)$$
then $\sqrt{x}+i\sqrt{y}$ will be a square root of $a+bi$, and $\sqrt{x}-i\sqrt{y}$ will be a square root of $a-bi$.

We may find such numbers x and y as follows.

By hypothesis
$$\sqrt{x}+i\sqrt{y} = \sqrt{a+bi}, \qquad (3)$$
and
$$\sqrt{x}-i\sqrt{y} = \sqrt{a-bi}. \qquad (4)$$

Multiplying (3) by (4),
$$x+y = \sqrt{a^2+b^2}. \qquad (5)$$

But, by (1), $$x - y = a. \tag{6}$$

Hence, solving (5) and (6), $$x = \frac{a + \sqrt{a^2 + b^2}}{2},$$

and $$y = \frac{-a + \sqrt{a^2 + b^2}}{2}.$$

And both these values are positive since $\sqrt{a^2 + b^2} > a$.

Example. Find a square root of $-1 + 4\sqrt{5}\,i$.

Here $a = -1$ and $\sqrt{a^2 + b^2} = \sqrt{(-1)^2 + (4\sqrt{5})^2} = 9$.

Hence $x = (-1 + 9)/2 = 4$ and $y = (1 + 9)/2 = 5$.

Therefore $\sqrt{-1 + 4\sqrt{5}\,i} = 2 + \sqrt{5}\,i$.

EXERCISE XL

Simplify the following.

1. $\sqrt{-49}$.
2. $\sqrt{-18}$.
3. $\sqrt{-8} \cdot \sqrt{-12}$.
4. $\sqrt{-2^2}$.
5. $(\sqrt{-2})^2$.
6. i^{12}.
7. i^{-7}.
8. i^{15}.
9. $\sqrt{x - y} \cdot \sqrt{y - x}$.
10. $(2 + \sqrt{-3})(1 + \sqrt{-2})$.
11. $(\sqrt{-2})^7 (\sqrt{-3})^9$.
12. $(1 + 2i)^3 + (1 - 2i)^3$.
13. $\dfrac{a}{\sqrt{-a^2}} - \dfrac{b}{i\sqrt{b^2}}$.
14. $\dfrac{4 + 6i}{1 + i} + \dfrac{4 - 6i}{1 - i}$.
15. $(\sqrt{3 + 4i} + \sqrt{3 - 4i})^2$.
16. $(1 + i^3)/(1 + i)$.
17. $\dfrac{a + bi}{a - bi}$.
18. $\dfrac{9 + 3\sqrt{2}\,i}{(3 + \sqrt{2}\,i)(1 + \sqrt{2}\,i)}$.
19. Divide 4 by $1 + \sqrt{-3}$.
20. Find a fourth root of -16.
21. Show that $(-1 + \sqrt{3}\,i)/2$ is a cube root of 1.
22. Show that $(1 + i)/\sqrt{2}$ is a fourth root of -1.
23. Find real values of x and y satisfying the equation
$$3 + 2i + x(i - 1) + 2yi = (3i + 4)(x + y).$$

Find square roots of the following.

24. $5 + 12i$.
25. $2i$.
26. $4ab + 2(a^2 - b^2)i$.

XIII. QUADRATIC EQUATIONS

628 **General form of the equation.** Every quadratic equation in one unknown letter, as x, may be reduced to the form

$$ax^2 + bx + c = 0,$$

where a, b, and c denote known numbers.

If, as may happen, $b = 0$, the equation is called a *pure* quadratic; if $b \neq 0$, it is called an *affected* quadratic.

629 **The roots found by inspection.** The roots of the equation $ax^2 + bx + c = 0$ are those particular values of x for which the polynomial $ax^2 + bx + c$ vanishes, § 332. There are two of these roots.

If the factors of $ax^2 + bx + c$ are known, the roots of $ax^2 + bx + c = 0$ are also known, for they are the values of x for which the factors of $ax^2 + bx + c$ vanish, §§ 253, 341. If the factors are $x - \alpha$ and $x - \beta$, the roots are α and β.

Example 1. Solve the equation $x^2 + x - 6 = 0$.

We have $\quad x^2 + x - 6 = (x + 3)(x - 2)$.

The factor $x + 3$ vanishes when $x = -3$, and the factor $x - 2$ vanishes when $x = 2$. Hence the roots are -3 and 2.

Example 2. Solve $abx^2 - (a^2 + b^2)x + (a^2 - b^2) = 0$.

Factoring, by § 443, $\quad [ax - (a + b)][bx - (a - b)] = 0$.

Hence the roots are $\quad (a + b)/a$ and $(a - b)/b$.

In particular, since $x^2 - q = (x - \sqrt{q})(x + \sqrt{q})$, the roots of the pure quadratic $x^2 - q = 0$ are $\sqrt{q}$ and $-\sqrt{q}$.

Again, since $ax^2 + bx = (ax + b)x$, the roots of a quadratic of the form $ax^2 + bx = 0$ are $-b/a$ and 0.

Thus, the roots of $4x^2 = 9$ are $3/2$ and $-3/2$; the roots of $2x^2 - x = 0$ are 0 and $1/2$; the roots of $5x^2 = 0$ are 0 and 0.

630 Conversely, to obtain the quadratic whose roots are two given numbers, as α and β, we form the product $(x - \alpha)(x - \beta)$ and equate this product to 0.

Thus, the quadratic whose roots are -2 and $1/3$ is $(x + 2)(x - 1/3) = 0$, or $3x^2 + 5x - 2 = 0$.

Example 1. Solve the following quadratics.

1. $x^2 + 2x - 8 = 0$. 2. $2x^2 - 7x + 3 = 0$.
3. $(2x - 1)(x - 2) = x^2 + 2$. 4. $(x - 1)(x - 3) = (2x - 1)^2$.

Example 2. Find the quadratics whose roots are

1. $-2/3,\ -3/2$. 2. $a,\ -a$. 3. $1/4,\ 0$.

General formula for the roots. But $ax^2 + bx + c$ may always be factored; for, as was shown in § 444, **631**

$$ax^2 + bx + c$$
$$= a\left[x - \frac{-b + \sqrt{b^2 - 4ac}}{2a}\right]\left[x - \frac{-b - \sqrt{b^2 - 4ac}}{2a}\right].$$

Therefore, since the roots of

$$ax^2 + bx + c = 0 \qquad (1)$$

are the values of x for which the factors of $ax^2 + bx + c$ vanish, these roots are

$$x = \frac{-b + \sqrt{b^2 - 4ac}}{2a} \text{ and } x = \frac{-b - \sqrt{b^2 - 4ac}}{2a},$$

or, as we usually write them,

$$x = \frac{-b \pm \sqrt{b^2 - 4ac}}{2a}. \qquad (2)$$

This formula (2) should be carefully memorized, for it enables one by mere substitution to obtain the roots of any given quadratic which has been reduced to the form (1).

Example. Solve $4x^2 + 105x = 81$.

Reducing to the form (1), $4x^2 + 105x - 81 = 0$.

Here $a = 4$, $b = 105$, and $c = -81$.

Hence $x = \dfrac{-105 \pm \sqrt{105^2 + 4 \cdot 4 \cdot 81}}{8}$, that is, $\dfrac{3}{4}$ or -27.

When b is an *even* integer, a and c also being integers, it is more convenient to use the formula **632**

$$x = \frac{-b/2 \pm \sqrt{(b/2)^2 - ac}}{a}, \qquad (3)$$

which is obtained by dividing the numerator and denominator of (2) by 2.

Example. Solve $3x^2 + 56x - 220 = 0$.

Here $b/2 = 28$, and substituting in (3) we have

$$x = \frac{-28 \pm \sqrt{28^2 + 3.220}}{3}, \text{ that is, } \frac{10}{3} \text{ or } -22.$$

633 Any given quadratic may also be solved by applying directly to it the process of *completing the square*, § 444, as in the following example. But since this method involves needless reckoning, it is not to be recommended — except when the formula of § 631 has been forgotten.

Example. Solve $3x^2 - 6x + 2 = 0$.

Transposing the known term and dividing by the coefficient of x^2,

$$x^2 - 2x = -2/3.$$

Completing the square of the first member,

$$x^2 - 2x + 1 = 1/3.$$

Extracting the square root of both members,

$$x - 1 = \pm\sqrt{3}/3, \text{ whence } x = (3 \pm \sqrt{3})/3.$$

634 The methods just explained enable one to solve any *fractional equation* which yields a quadratic when cleared of fractions. But see §§ 524–527.

Example 1. Solve $\dfrac{1}{x+1} + \dfrac{1}{x+2} = \dfrac{1}{x+3} + \dfrac{1}{x+4}$.

Clearing of fractions and simplifying, $2x^2 + 10x + 11 = 0$.

Solving, $x = \dfrac{-5 \pm \sqrt{3}}{2}$.

Both of these values of x are roots of the given equation, for they cause none of its denominators to vanish.

Example 2. Solve $\dfrac{x+3}{x^2-1} + \dfrac{x-3}{x^2-x} + \dfrac{x+2}{x^2+x} = 0$.

Clearing of fractions by multiplying by the lowest common denominator $x(x^2-1)$, and simplifying, we obtain

$$3x^2 + 2x - 5 = 0, \text{ whence } x = 1 \text{ or } -5/3.$$

But 1 cannot be a root of the given equation, since its first two denominators vanish when $x = 1$. Hence $-5/3$ is the only root of this equation.

EXERCISE XLI

Solve the following equations.

1. $x^2 + 2x = 35.$

2. $4x^2 - 4x = 3.$

3. $x^2 = 10x - 18.$

4. $9x^2 + 6x + 5 = 0.$

5. $2x^2 + 3x - 4 = 0.$

6. $(2x - 3)^2 = 8x.$

7. $x^2 + 9x - 252 = 0.$

8. $12x^2 + 56x - 255 = 0.$

9. $8x^2 - 82x + 207 = 0.$

10. $15x^2 - 86x - 64 = 0.$

11. $x^2 - 3x - 1 + \sqrt{3} = 0.$

12. $x^2 - (6 + i)x + 8 + 2i = 0.$

13. $(x - 2)^2(x - 7) = (x + 2)(x - 3)(x - 6).$

14. $\dfrac{2x}{x + 2} + \dfrac{x + 2}{2x} = 2.$

15. $\dfrac{x + 1}{x} + 1 = \dfrac{x}{x - 1}.$

16. $\dfrac{3}{2(x^2 - 1)} + \dfrac{x}{4x + 4} = \dfrac{3}{8}.$

17. $\dfrac{3}{2x+1} - \dfrac{1}{4x-2} - \dfrac{2x}{1-4x^2} = \dfrac{7}{8}$

18. $\dfrac{2x - 1}{x - 2} + \dfrac{3x + 1}{x - 3} = \dfrac{5x - 14}{x - 4}.$

19. $\dfrac{x + 1}{x(x - 2)} - \dfrac{1}{2x - 2} + \dfrac{1}{2x} = 0.$

20. $\dfrac{4}{x - 1} - \dfrac{1}{4 - x} = \dfrac{3}{x - 2} - \dfrac{2}{3 - x}.$

21. $\dfrac{x + 3}{4(x + 2)(3x - 1)} + \dfrac{2x + 1}{3(3x - 1)(x + 4)} - \dfrac{17x + 7}{6(x + 4)(x + 2)} = 0.$

22. $\dfrac{x + 7}{2x^2 - 7x + 3} + \dfrac{x}{x^2 - 2x - 3} + \dfrac{x + 3}{2x^2 + x - 1} = 0.$

23. $3x^2 + (9a - 1)x - 3a = 0.$

24. $x^2 - 2ax + a^2 - b^2 = 0.$

25. $c^2x^2 + c(a - b)x - ab = 0.$

26. $x^2 - 4ax + 4a^2 - b^2 = 0.$

27. $x^2 - 6acx + a^2(9c^2 - 4b^2) = 0.$

28. $(a^2 - b^2)x^2 - 2(a^2 + b^2)x + a^2 - b^2 = 0.$

29. $1/(x - a) + 1/(x - b) + 1/(x - c) = 0.$

30. $\dfrac{(x - a)^2 - (x - b)^2}{(x - a)(x - b)} + \dfrac{4ab}{a^2 - b^2} = 0.$

EXERCISE XLII

1. Find two consecutive integers whose product is 506.

2. Find two consecutive integers the sum of whose squares is 481.

3. Find two consecutive integers the difference of whose cubes is 91.

4. Find three consecutive integers the sum of whose products by pairs is 587.

5. Find a number of two digits from the following data: the product of the digits is 48, and if the digits be interchanged the number is diminished by 18.

6. The numerator of a certain fraction exceeds its denominator by 2, and the fraction itself exceeds its reciprocal by 24/35. Find the fraction.

7. A cattle dealer bought a certain number of steers for $1260. Having lost 4 of them, he sold the rest for $10 a head more than they cost him, and made $260 by the entire transaction. How many steers did he buy?

8. A man sold some goods for $48, and his gain per cent was equal to one half the cost of the goods in dollars. What was the cost of the goods?

9. If $4000 amounts to $4410 when put at compound interest for two years, interest being compounded annually, what is the rate of interest?

10. A man inherits $25,000, but after a certain percentage has been deducted for the inheritance tax and then a percentage for fees at a rate one greater than that of the inheritance tax, he receives only $22,800. What is the rate of the inheritance tax?

11. A man bought a certain number of $50 shares for $4500 when they were at a certain discount. Later he sold all but 10 of them for $5850 when the premium was three times the discount at which he bought them. How many shares did he buy?

12. The circumference of a hind wheel of a wagon exceeds that of a fore wheel by 8 inches, and in traveling 1 mile this wheel makes 88 less revolutions than a fore wheel. Find the circumference of each wheel.

13. A square is surrounded by a border whose width lacks 1 inch of being one fourth of the length of a side of the square, and whose area in square inches exceeds the length of the perimeter of the square in inches by 64. Find the area of the square and that of the border.

14. The corners of a square the length of whose side is 2 are cut off in such a way that a regular octagon remains. What is the length of a side of this octagon?

15. A vintner draws a certain quantity of wine from a full cask containing 63 gallons. Having filled up the cask with water, he draws the same quantity as before and then finds that only 28 gallons of pure wine remain in the cask. How many gallons did he draw each time?

16. A man travels 50 miles by the train A, and then after a wait of 5 minutes returns by the train B, which runs 5 miles an hour faster than the train A. The entire journey occupies $2\frac{4}{9}$ hours. What are the rates of the two trains?

17. A pedestrian walked 6 miles in a certain interval of time. Had the time been $1/2$ hour less, the rate would have been 2 miles per hour greater. Required the time and rate.

18. A pedestrian walked 12 miles at a certain rate and then 6 miles farther at a rate $1/2$ mile per hour greater. Had he walked the entire distance at the greater rate, his time would have been 20 minutes less. How long did it take him to walk the 18 miles?

19. From the point of intersection of two straight roads which cross at right angles, two men, A and B, set out simultaneously, A on the one road at the rate of 3 miles per hour, B on the other at the rate of 4 miles per hour. After how many hours will they be 30 miles apart?

20. If A and B walk on the roads just described, but at the rates of 2 and 3 miles per hour respectively, and A starts 2 hours before B, how long after B starts will they be 10 miles apart?

21. If from a height of a feet a body be thrown vertically upward with an initial velocity of b feet per second, its height at the end of t seconds is given by the formula $h = a + bt - 16t^2$. The corresponding formula when the body is thrown vertically downward is $h = a - bt - 16t^2$.

(1) If a body be thrown vertically upward from the ground with an initial velocity of 32 feet per second, when will it be at a height of 7 feet? of 16 feet? Will it ever reach a height of 17 feet?

(2) A body is thrown from a height of 64 feet vertically downward with an initial velocity of 48 feet per second. When will it reach the height of 36 feet?

(3) If a body be dropped from a height of 36 feet, when will it reach the ground?

XIV. A DISCUSSION OF THE QUADRATIC EQUATION. MAXIMA AND MINIMA

635 **Character of the roots. The discriminant.** Let α and β denote the roots of $ax^2 + bx + c = 0$, so that, § 631,

$$\alpha = \frac{-b + \sqrt{b^2 - 4ac}}{2a}, \quad \beta = \frac{-b - \sqrt{b^2 - 4ac}}{2a}.$$

The radicand $b^2 - 4ac$ is called the *discriminant* of

$$ax^2 + bx + c = 0.$$

When the coefficients a, b, c are real, the character of the roots α, β is indicated by the *sign* of the discriminant. Thus:

1. *When* $\mathrm{b}^2 - 4\,\mathrm{ac}$ *is positive, the roots are real and distinct.*
2. *When* $\mathrm{b}^2 - 4\,\mathrm{ac}$ *is* 0, *the roots are real and equal.*
3. *When* $\mathrm{b}^2 - 4\,\mathrm{ac}$ *is negative, the roots are conjugate imaginaries.*

It should also be observed that

1. When $b^2 - 4ac = 0$, then $ax^2 + bx + c$ is a perfect square.
2. When a is positive and c is negative, the roots are always real, since $b^2 - 4ac$ is then positive.
3. If a, b, c are rational, the roots are rational when, and only when, $b^2 - 4ac$ is a perfect square.

Example 1. Show that the roots of $x^2 - 6x + 10 = 0$ are imaginary.

They are imaginary since $b^2 - 4ac = (-6)^2 - 4 \cdot 1 \cdot 10 = -4$.

Example 2. For what value of m are the roots of $mx^2 + 3x + 2 = 0$ equal?

We must have $3^2 - 4 \cdot m \cdot 2 = 0$, that is, $m = 9/8$.

Example 3. If possible, factor $y^2 + xy - 2x^2 + 11x + y - 12$.

Arranging the polynomial according to powers of y and equating it to 0, we have $y^2 + (x + 1)y - (2x^2 - 11x + 12) = 0$.

Solving, $$y = \frac{-(x+1) \pm \sqrt{9x^2 - 42x + 49}}{2},$$

that is, $$y = x - 4, \text{ or } y = -2x + 3.$$

Hence, § 631, $y^2 + xy - 2x^2 + 11x + y - 12 = (y - x + 4)(y + 2x - 3)$.

Observe that the factorization is possible only because the radicand $9x^2 - 42x + 49$ is a perfect square.

Relations between roots and coefficients. If α and β denote the roots of $ax^2 + bx + c = 0$, we have, § 631, **636**

$$ax^2 + bx + c \equiv a(x - \alpha)(x - \beta).$$

Dividing both members of this identity by a and carrying out the multiplication in the second member, we have

$$x^2 + \frac{b}{a}x + \frac{c}{a} \equiv x^2 - (\alpha + \beta)x + \alpha\beta.$$

Since this is an identity, the coefficients of like powers of x in its two members are equal, § 264, that is,

$$\alpha + \beta = -b/a \text{ and } \alpha\beta = c/a.$$

This may also be proved by adding and multiplying the values of α and β given in § 631. Therefore, since α, β are the roots of $x^2 + bx/a + c/a = 0$, we have the theorem:

In any quadratic of the form $x^2 + px + q = 0$ *the coefficient of* x *with its sign changed is equal to the sum of the roots, and the constant term is equal to the product of the roots.*

Thus, in the quadratic $6x^2 + x = 2$, that is, $x^2 + x/6 - 1/3 = 0$, the sum of the roots is $-1/6$, and their product is $-1/3$.

Example 1. Solve $9x^2 - 10x + 1 = 0$.

Obviously one of the roots is 1, for $9 - 10 + 1 = 0$. Therefore, since the product of the roots is $1/9$, the other root is $1/9 \div 1$ or $1/9$.

Example 2. Find the equation whose roots are *three* times those of $3x^2 + 8x + 5 = 0$.

Let α and β denote the roots of $3x^2 + 8x + 5 = 0$.

Then $\alpha + \beta = -8/3$ and $\alpha\beta = 5/3$.

Hence the required equation is

$$x^2 - (3\alpha + 3\beta)x + 3\alpha \cdot 3\beta \equiv x^2 - 3(\alpha + \beta)x + 9\alpha\beta \equiv x^2 + 8x + 15 = 0.$$

Symmetric functions of the roots. The expressions $\alpha + \beta$ and $\alpha\beta$ are symmetric functions of the roots α, β, § 540. All other rational symmetric functions of α and β can be expressed **637**

rationally in terms of these two functions, $\alpha+\beta$ and $\alpha\beta$, and therefore rationally in terms of the coefficients of the equation.

For every such function can be reduced to the form of an *integral* symmetric function or to that of a quotient of two such functions. If an integral symmetric function contains a term of the type $k\alpha^p\beta^{p+q}$, it must also contain the term $k\alpha^{p+q}\beta^p$, § 542, and therefore $k\alpha^p\beta^p(\alpha^q+\beta^q)$. But $\alpha^p\beta^p=(\alpha\beta)^p$, and it may readily be shown by successive applications of the binomial theorem that $\alpha^q+\beta^q$ can be expressed in terms of powers of $\alpha+\beta$ and $\alpha\beta$.

Thus, since $(\alpha+\beta)^2=\alpha^2+2\,\alpha\beta+\beta^2$, we have $\alpha^2+\beta^2=(\alpha+\beta)^2-2\,\alpha\beta$. Similarly we find $\alpha^3+\beta^3=(\alpha+\beta)^3-3\,\alpha\beta(\alpha+\beta)$.

Example. The roots of $x^2+px+q=0$ being α, β, express $1/\alpha+1/\beta$ and $\alpha^3\beta+\alpha\beta^3$ in terms of p and q.

We have $1/\alpha+1/\beta=(\alpha+\beta)/\alpha\beta=-p/q$,

and $\alpha^3\beta+\alpha\beta^3=\alpha\beta(\alpha^2+\beta^2)=\alpha\beta[(\alpha+\beta)^2-2\,\alpha\beta]=q(p^2-2q)$.

638 Infinite roots. Suppose that, instead of being constants, the coefficients of $ax^2+bx+c=0$ are variables. We can then show that if a approaches 0 as limit, one of the roots will approach ∞; and if both a and b (but not c) approach 0, both roots will approach ∞.

For the formulas for the roots are

$$\alpha=\frac{-b+\sqrt{b^2-4\,ac}}{2\,a},\quad \beta=\frac{-b-\sqrt{b^2-4\,ac}}{2\,a}.$$

Multiply both terms of the fraction α by $-b-\sqrt{b^2-4\,ac}$ and both terms of the fraction β by $-b+\sqrt{b^2-4\,ac}$. We obtain

$$\alpha=-\frac{2\,c}{b+\sqrt{b^2-4\,ac}},\quad \beta=-\frac{2\,c}{b-\sqrt{b^2-4\,ac}}.$$

By §§ 203, 205, if $a \doteq 0$, then $\sqrt{b^2-4\,ac}\doteq b$.

Therefore if $a\doteq 0$, then $\alpha\doteq -c/b$ and $\beta\doteq\infty$,

and if $a\doteq 0$ and $b\doteq 0$, then $\alpha\doteq\infty$ and $\beta\doteq\infty$.

It is customary to state these conclusions as follows, § 519:

One root of $ax^2+bx+c=0$ *becomes infinite when* a *vanishes, and both roots become infinite when* a *and* b *(but not* c*) vanish simultaneously.*

Maxima and minima. Let y be a function of x, § 278. It may happen that, as x increases, y will increase to a certain value, m, and then begin to decrease, or that y will decrease to a certain value, m', and then begin to increase. We then call m a *maximum* value of y and m' a *minimum* value. **639**

Thus, $y = (x - 1)^2 - 4$ has a minimum value when $x = 1$, this value being -4. For if x start from a value less than 1 and increase, $(x - 1)^2$ will first decrease to 0 and then increase.

Similarly $y = 4 - (x - 1)^2$ has a maximum value, 4, when $x = 1$.

Every quadratic trinomial $ax^2 + bx + c$ with real coefficients has either a maximum or a minimum value, which may be found as in the following examples. **640**

Example 1. Find the maximum or minimum value of $y = x^2 + 6x - 7$.

By completing the square, $x^2 + 6x - 7 = (x + 3)^2 - 16$.

Hence when $x = -3$, y has a minimum value, namely -16.

Example 2. Divide a given line segment into two parts whose rectangle shall have the greatest possible area.

Let $2a$ denote the length of the given segment, x and $2a - x$ the lengths of the parts, y the area of their rectangle.

Then $$y = x(2a - x) = 2ax - x^2 = a^2 - (a - x)^2.$$

Hence y has a maximum value when $x = a$, that is, when the given segment is bisected and the rectangle is a square whose area is a^2.

The maximum and minimum values of quadratic trinomials and of certain more complex functions may also be found by the following method. **641**

Example. Find the maximum and minimum values, if any, of $y = (4x^2 - 2)/(4x - 3)$.

Clearing of fractions and solving for x, we have

$$x = \frac{y \pm \sqrt{y^2 - 3y + 2}}{2} = \frac{y \pm \sqrt{(y - 1)(y - 2)}}{2}.$$

By hypothesis, x is restricted to *real* values. Hence y can only take values for which the radicand $(y - 1)(y - 2)$ is *positive* (or 0), that is, the value 1 and *lesser* values and the value 2 and *greater* values.

It follows from this that 1 is a maximum and 2 a minimum value of y.

For observe that as y increases to 1, the two values of x, namely $(y-\sqrt{y^2-3y+2})/2$ and $(y+\sqrt{y^2-3y+2})/2$, respectively increase and decrease to $1/2$. Hence, conversely, as x increases through $1/2$, y first increases to 1 and then decreases.

642 Variation of a quadratic trinomial. Given $y = ax^2 + bx + c$, where a is positive. By completing the square, we obtain

$$y = a\left[\left(x + \frac{b}{2a}\right)^2 + \frac{4ac - b^2}{4a^2}\right].$$

Hence y has a minimum value when $x = -b/2a$, this minimum value being $(4ac - b^2)/4a$.

As x increases from $-\infty$ to $+\infty$, y will first decrease from $+\infty$ to $(4ac - b^2)/4a$ and then increase to $+\infty$.

Thus, let $y = x^2 - 2x - 3 = (x-1)^2 - 4$.

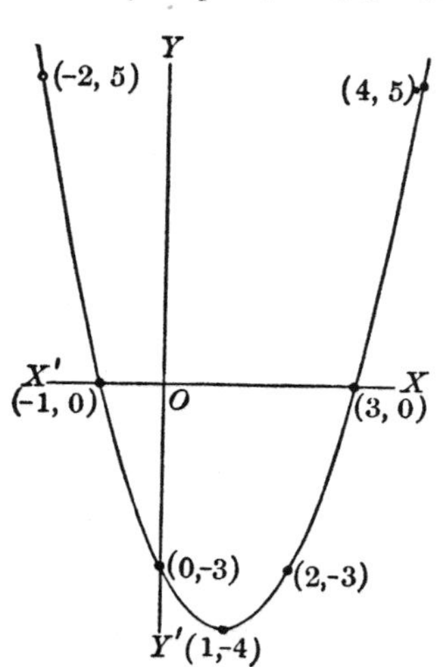

As x increases from $-\infty$ to $+\infty$, y first decreases from ∞ to -4 and then increases from -4 to ∞.

Moreover $y = 0$ when $x^2 - 2x - 3 = 0$, that is, when $x = -1$ or 3.

Until x reaches the value -1, y is positive; it then remains negative until $x = 3$, when it again becomes positive.

When

$x = \cdots -3, -2, -1, \quad 0, \quad 1, \quad 2, 3, 4, \quad 5, \cdots$

we have

$y = \cdots \quad 12, \quad 5, \quad 0, -3, -4, -3, 0, 5, 12, \cdots.$

We may obtain the graph of $y = x^2 - 2x - 3$ by plotting these pairs of values and passing a curve through them, as in § 389.

Observe that to the zero values of y there correspond the points where the graph cuts the x-axis, and that to the minimum value of y there corresponds the lowermost point of the graph, which is also a *turning point* of this curve.

EXERCISE XLIII

1. For what values of m are the roots of $(m+2)x^2 - 2mx + 1 = 0$ equal?

2. What are the roots of $(m^2 + m)x^2 + 3mx - 2 = 0$ when $m = -1$? when $m = 0$?

3. If possible, factor $3x^2 + 5xy - 2y^2 - 5x + 4y - 2$.

4. For what values of m can $x^2 - y^2 + mx + 5y - 6$ be factored?

5. The roots of $x^2 + px + q = 0$ being α and β, express $(\alpha - \beta)^2$, $\alpha^4 + \beta^4$, and $\alpha / \beta + \beta / \alpha$ in terms of p and q.

6. The roots of $2x^2 - 3x + 4 = 0$ being α and β, find the values of $\alpha / \beta^2 + \beta / \alpha^2$ and $\alpha^3\beta + \alpha\beta^3$.

7. The roots of $x^2 + x + 2 = 0$ being α and β, find the equations whose roots are $-\alpha$, $-\beta$; $1/\alpha$, $1/\beta$; 2α, 2β; $\alpha + 1$, $\beta + 1$.

8. Find the maximum and minimum values of the following.

1. $x^2 - 8x + 3$. 2. $2x^2 - x + 4$. 3. $1 + 4x - x^2$.
4. $x/(x^2 + 1)$. 5. $1/x + 1/(1 - x)$. 6. $(x + 1)/(2x^2 - 1)$.

9. Find the greatest rectangle that can be inscribed in a given circle; also the rectangle of greatest perimeter.

10. A man who is in a boat 2 miles from the nearest point on the shore wishes to reach as quickly as he can a point on the shore distant 6 miles from that nearest point. If he can row 4 miles an hour and walk 5 miles an hour, toward what point should he row?

11. What height will a body reach if thrown vertically upward from the ground with an initial velocity of 48 feet per second, and when will it reach this height? See p. 303, Ex. 21.

XV. EQUATIONS OF HIGHER DEGREE WHICH CAN BE SOLVED BY MEANS OF QUADRATICS

Equations which can be factored. Given an integral equation in the form $A = 0$. If we can resolve A into factors of the first or second degrees, we can find all the roots of $A = 0$ by equating the several factors of A to zero and solving the resulting equations. For if $A \equiv BC \cdots$, then $A = 0$ is equivalent to $B = 0$, $C = 0$, $\cdots$, jointly, § 341. **643**

Example 1. Solve $x^4 + x^2 + 1 = 0$.

By § 436, $$x^4 + x^2 + 1 = (x^2 + x + 1)(x^2 - x + 1).$$

Hence $x^4 + x^2 + 1 = 0$ is equivalent to the two equations

$$x^2 + x + 1 = 0 \text{ and } x^2 - x + 1 = 0.$$

Solving these equations, $x = \dfrac{-1 \pm i\sqrt{3}}{2}$ or $\dfrac{1 \pm i\sqrt{3}}{2}$.

Example 2. Solve $x^4 - x^3 - 5x^2 - 7x + 12 = 0$.

Factoring by the method of § 451, we find that

$$x^4 - x^3 - 5x^2 - 7x + 12 = (x - 1)(x - 3)(x^2 + 3x + 4).$$

Hence $x^4 - x^3 - 5x^2 - 7x + 12 = 0$ is equivalent to the three equations

$$x - 1 = 0,\ x - 3 = 0, \text{ and } x^2 + 3x + 4 = 0,$$

whose roots are $1,\ 3, \text{ and } (-3 \pm i\sqrt{7})/2$.

Example 3. Solve the following equations.

1. $6x^3 - 11x^2 + 8x - 2 = 0$.
2. $x^4 - 5x^3 + x^2 + 11x + 4 = 0$.

644 **Equations of the type $au^2 + bu + c = 0$, where u denotes some function of x.** If the roots of $au^2 + bu + c = 0$ when solved for u are α and β, this equation is equivalent to the two equations $u = \alpha$ and $u = \beta$, for, § 631,

$$au^2 + bu + c \equiv a(u - \alpha)(u - \beta).$$

Hence to solve $au^2 + bu + c = 0$ for x, we have only to solve the two equations $u = \alpha$ and $u = \beta$ for x.

Example 1. Solve $3x^4 + 10x^2 - 8 = 0$.

Solving for x^2, $x^2 = 2/3 \text{ or } -4$.

Hence $x = \pm\sqrt{6}/3 \text{ or } \pm 2i$.

Example 2. Solve $x^{\frac{2}{3}} + 3 - 10x^{-\frac{2}{3}} = 0$.

Multiplying by $x^{\frac{2}{3}}$, $x^{\frac{4}{3}} + 3x^{\frac{2}{3}} - 10 = 0$.

Solving for $x^{\frac{2}{3}}$, $x^{\frac{2}{3}} = 2 \text{ or } -5$.

Hence $x = \pm 2\sqrt{2} \text{ or } \pm 5i\sqrt{5}$.

Example 3. Solve $(x^2 + 3x + 4)(x^2 + 3x + 5) = 6$.

We may reduce this equation to the form

$$(x^2 + 3x)^2 + 9(x^2 + 3x) + 14 = 0.$$

Solving for $x^2 + 3x$, we obtain the two equations

$$x^2 + 3x = -2, \text{ and } x^2 + 3x = -7,$$

whose roots are $-1,\ -2, \text{ and } (-3 \pm i\sqrt{19})/2$.

Example 4. Solve $(x+1)(x+2)(x+3)(x+4)=120$.

By multiplying together the first and fourth factors, and the second and third, we reduce the equation to the form

$$(x^2+5x+4)(x^2+5x+6)=120,$$

which may be solved in the same way as the equation in Ex. 3.

Example 5. Solve $x^4+10x^3+31x^2+30x+5=0$.

By completing the square of the first two terms, we obtain

$$(x^2+5x)^2+6(x^2+5x)+5=0.$$

Solving for x^2+5x, we obtain the two equations

$$x^2+5x=-5 \text{ and } x^2+5x=-1,$$

whose roots are $(-5\pm\sqrt{5})/2$ and $(-5\pm\sqrt{21})/2$.

Example 6. Solve $8\dfrac{x^2+2x}{x^2-1}+3\dfrac{x^2-1}{x^2+2x}-11=0$.

Observing that the second fraction is the reciprocal of the first, we multiply both members of the equation by the first fraction, thus obtaining

$$8\left(\frac{x^2+2x}{x^2-1}\right)^2-11\left(\frac{x^2+2x}{x^2-1}\right)+3=0.$$

Solving for $(x^2+2x)/(x^2-1)$, we obtain the two equations

$$\frac{x^2+2x}{x^2-1}=1 \text{ and } \frac{x^2+2x}{x^2-1}=\frac{3}{8},$$

whose roots are $\quad -1/2$ and $-3,\ -1/5$.

All the values of x thus found are roots of the given equation since they cause none of its denominators to vanish.

Example 7. Solve the following equations.

1. $3x^4-29x^2+18=0$.
2. $x^4-6x^3+8x^2+3x=2$.
3. $(x-a)(x+2a)(x-3a)(x+4a)=24a^4$.
4. $(4x^2+2x)/(x^2+6)+(x^2+6)/(2x^2+x)-3=0$.

645 **Reciprocal equations.** These are equations which remain unchanged when we replace x by $1/x$ and clear of fractions. If we arrange the terms of such an equation in descending powers of x, the first and last coefficients will be the same, also the second and next to last, and so on; or each of these pairs of coefficients will have the same absolute values but contrary signs.

Thus, $$2x^4+3x^3+4x^2+3x+2=0$$
and $$x^5-2x^4+4x^3-4x^2+2x-1=0$$
are reciprocal equations.

Reciprocal equations of the *fourth* degree may be reduced to the quadratic form and solved as follows.

Example 1. Solve $2x^4-3x^3+4x^2-3x+2=0$.

Grouping the terms which have like coefficients and dividing by x^2, we reduce the given equation to the form

$$2\left(x^2+\frac{1}{x^2}\right)-3\left(x+\frac{1}{x}\right)+4=0.$$

Since $x^2+1/x^2=(x+1/x)^2-2$, we may reduce this equation to the form

$$2\left(x+\frac{1}{x}\right)^2-3\left(x+\frac{1}{x}\right)=0.$$

Solving for $x+1/x$, we obtain the two equations

$$x+\frac{1}{x}=0 \text{ and } x+\frac{1}{x}=\frac{3}{2},$$

whose roots are $\quad i,\ -i,$ and $(3\pm i\sqrt{7})/4$.

Every reciprocal equation of odd degree has the root 1 or -1; and if the corresponding factor $x-1$ or $x+1$ be separated, the "depressed" equation will also be reciprocal. Hence reciprocal equations of the *third* and *fifth* degrees can be solved by aid of quadratics.

Example 2. Solve $2x^3-3x^2-3x+2=0$.

Grouping terms, $$2(x^3+1)-3(x^2+x)=0.$$

Since both terms of this equation are divisible by $x+1$, it is equivalent to the two equations

$$x+1=0 \text{ and } 2x^2-5x+2=0,$$

whose roots are $\quad -1$, and $2,\ 1/2$.

Example 3. Solve $x^5-5x^4+9x^3-9x^2+5x-1=0$.

Grouping terms, $(x^5-1)-5x(x^3-1)+9x^2(x-1)=0$.

Dividing by $x-1$, we find that this equation is equivalent to

$$x-1=0 \text{ and } x^4-4x^3+5x^2-4x+1=0,$$

whose roots are $\quad 1$, and $(1\pm i\sqrt{3})/2,\ (3\pm\sqrt{5})/2$.

Example 4. Solve the following equations.

1. $x^3 - 2x^2 + 2x - 1 = 0.$ 2. $x^4 - 4x^3 + 5x^2 - 4x + 1 = 0.$

3. $x^5 + x^4 + x^3 + x^2 + x + 1 = 0.$

Binomial equations. This name is given to equations of the form $x^n + a = 0$. They can be solved by methods already given when $x^n + a$ can be resolved into factors of the first or second degrees. **646**

Example 1. Solve $x^3 - 1 = 0$.

Since $x^3 - 1 = (x-1)(x^2 + x + 1)$, the equation $x^3 - 1 = 0$ is equivalent to the two equations

$$x - 1 = 0 \text{ and } x^2 + x + 1 = 0.$$

Solving, $$x = 1 \text{ or } (-1 \pm i\sqrt{3})/2.$$

Example 2. Solve $x^5 - 32 = 0$.

From $x^5 - 32 = 0$, by setting $x = \sqrt[5]{32}\,y = 2y$, we obtain $y^5 - 1 = 0$.

By §§ 438, 643, $y^5 - 1 = 0$ is equivalent to the two equations

$$y - 1 = 0 \text{ and } y^4 + y^3 + y^2 + y + 1 = 0.$$

Solving, $$y = 1,\ (-1 \pm \sqrt{5} + i\sqrt{10 \pm 2\sqrt{5}})/4,$$

or $$(-1 \pm \sqrt{5} - i\sqrt{10 \pm 2\sqrt{5}})/4.$$

Hence $$x = 2y = 2,\ (-1 \pm \sqrt{5} + i\sqrt{10 \pm 2\sqrt{5}})/2,$$

or $$(-1 \pm \sqrt{5} - i\sqrt{10 \pm 2\sqrt{5}})/2.$$

By the method here employed every binomial equation $x^n \pm a = 0$ can be reduced to the reciprocal form $y^n \pm 1 = 0$.

Example 3. Solve the following equations.

1. $x^3 + 8 = 0.$ 2. $x^4 + 1 = 0.$ 3. $x^6 + 1 = 0.$

These examples illustrate the theorem: *Every number has* n n*th roots.* Thus, a cube root of 1 is any number which satisfies the equation $x^3 = 1$; and in Ex. 1, we found *three* such numbers, namely 1, $(-1 + i\sqrt{3})/2$, and $(-1 - i\sqrt{3})/2$. **647**

Irrational equations. If asked to solve an irrational equation, we ordinarily begin by rationalizing it, § 609. But, as will be illustrated below, certain equations admit of a simpler treatment than this. Whatever method is used, care must be taken to test the values obtained for the unknown letter before accepting them as roots of the given equation. **648**

Example 1. Solve $\sqrt{2x-3}-\sqrt{5x-6}+\sqrt{3x-5}=0$.

Transposing, $\sqrt{2x-3}+\sqrt{3x-5}=\sqrt{5x-6}$.

Squaring and simplifying, $\sqrt{(2x-3)(3x-5)}=1$.

Squaring and simplifying, $6x^2-19x+14=0$.

Solving, $x=2$ or $7/6$.

Testing these values of x in the given equation, we find that 2 is a root but that $7/6$ is not a root.

We may also rationalize the given equation by the method of § 603. We thus discover that $-4(6x^2-19x+14)$ is identically equal to

$$(\sqrt{2x-3}-\sqrt{5x-6}+\sqrt{3x-5})(\sqrt{2x-3}+\sqrt{5x-6}-\sqrt{3x-5}).$$
$$(\sqrt{2x-3}+\sqrt{5x-6}+\sqrt{3x-5})(\sqrt{2x-3}-\sqrt{5x-6}-\sqrt{3x-5}).$$

There are but two values of x for which the product $6x^2-19x+14$ can vanish. The first factor on the right vanishes for one of these values, 2; the second, for the other, $7/6$. Hence there is no value of x for which the third or fourth factor can vanish.

Example 2. Solve $\sqrt{4x+3}+\sqrt{12x+1}=\sqrt{24x+10}$.

Some equations which involve a single radical can be reduced to the form of a quadratic with respect to this radical. We then begin by solving for the radical.

Example 3. Solve $2x^2-6x-5\sqrt{x^2-3x-1}-5=0$.

Observing that the x terms outside the radical are twice those under the radical, we are led to write the equation in the form

$$2(x^2-3x-1)-5\sqrt{x^2-3x-1}-3=0.$$

Solving for $\sqrt{x^2-3x-1}$, we obtain the two equations

$$\sqrt{x^2-3x-1}=3, \text{ and } \sqrt{x^2-3x-1}=-1/2.$$

The second of these equations must be rejected since according to the convention made in § 579 a radical of the form $\sqrt{a}$ cannot have a negative value.

Squaring the first equation, we obtain

$$x^2-3x-1=9,$$

whose roots are 5 and -2.

Testing 5 and -2 in the given equation, we find that both of them are roots.

Example 4. Solve $2x^2-14x-3\sqrt{x^2-7x+10}+18=0$.

Sometimes an equation may be reduced to a form in which both members are perfect squares or one member is a perfect square and the other is a constant.

Example 5. Solve $4x^2 + x + 2x\sqrt{3x^2 + x} = 9$.

We may write this equation in the form

$$3x^2 + x + 2x\sqrt{3x^2 + x} + x^2 = 9.$$

The first member is a perfect square, and extracting the square root of both members, we obtain the two equations

$$\sqrt{3x^2 + x} + x = 3, \text{ and } \sqrt{3x^2 + x} + x = -3.$$

Solving these equations, $x = 1, \ -9/2, \text{ or } (5 \pm \sqrt{97})/4$.

Testing these results, we find that only 1 and $-9/2$ are roots of the given equation.

Sometimes all the terms, when properly grouped, have a common irrational factor.

Example 6. Solve $\sqrt{x^2 - 7ax + 10a^2} - \sqrt{x^2 + ax - 6a^2} = x - 2a$.

Here the first two terms and also $x - 2a$ have the factor $\sqrt{x - 2a}$. Separating this factor, we find that the equation is equivalent to

$$\sqrt{x - 2a} = 0 \text{ and } \sqrt{x - 5a} - \sqrt{x + 3a} = \sqrt{x - 2a}.$$

Solving these equations, $\quad x = 2a, \ -10a/3, \text{ or } 6a$.

Testing, we find that only $2a$ and $-10a/3$ are roots of the given equation.

Example 7. Solve $\sqrt{3x^2 - 5x - 12} - \sqrt{2x^2 - 11x + 15} = x - 3$.

If one or more of the terms of the equation are fractions with irrational denominators, it is often best to rationalize these denominators at the outset.

Example 8. Solve $(\sqrt{x} + \sqrt{x - 3})/(\sqrt{x} - \sqrt{x - 3}) = 2x - 5$.

Rationalizing the denominator in the first member and simplifying, we have

$$\sqrt{x^2 - 3x} = 2x - 6.$$

Solving, $\quad x = 3 \text{ or } 4.$

Testing, we find that both 3 and 4 are roots of the given equation.

Example 9. Solve $(\sqrt{x - 1} - \sqrt{x + 1})/(\sqrt{x - 1} + \sqrt{x + 1}) = x - 3$.

EXERCISE XLIV

Solve the following equations.

1. $4x^4 - 17x^2 + 18 = 0.$

2. $3x^{\frac{3}{2}} - 4x^{\frac{3}{4}} = 7.$

3. $(x^2 - 4)(x^2 - 9) = 7x^2.$

4. $(2x^2 - x - 3)(3x^2 + x - 2)^2 = 0$

5. $x^4 + x^3 + x^2 + 3x - 6 = 0.$

6. $x^4 - 2x^3 + x^2 + 2x - 2 = 0.$

7. $(3x^2 - 2x + 1)(3x^2 - 2x - 7) + 12 = 0.$

8. $x^4 - 12x^3 + 33x^2 + 18x - 28 = 0.$

9. $4x^4 + 4x^3 - x^2 - x - 2 = 0.$

10. $x^4 - 2x^3 + 2x^2 - 2x + 1 = 0.$

11. $x^4 + x^3 + 2x^2 + x + 1 = 0.$

12. $x^5 - 11x^4 + 36x^3 - 36x^2 + 11x - 1 = 0.$

13. $x^5 - 243 = 0.$

14. $(2x - 1)^8 = 1.$

15. $(1 + x)^3 = (1 - x)^3.$

16. $(x - 2)^4 - 81 = 0.$

17. $(a + x)^3 + (b + x)^3 = (a + b + 2x)^3.$

18. $(a - x)^4 - (b - x)^4 = (a - b)(a + b - 2x).$

19. $\dfrac{x^2 + 3x + 1}{4x^2 + 6x - 1} - 3\dfrac{4x^2 + 6x - 1}{x^2 + 3x + 1} - 2 = 0.$

20. $x^2 + \dfrac{1}{x^2} = a^2 + \dfrac{1}{a^2}.$

21. $3x^2 - 2x - 5\sqrt{3x^2 - 2x + 3} + 9 = 0.$

22. $4x^2 - 2x - 1 = \sqrt{2x^2 - x}.$

23. $\sqrt{3 - x} + \sqrt{2 - x} = \sqrt{5 - 2x}.$

24. $\sqrt{2x + 3} + \sqrt{3x - 5} - \sqrt{x + 1} - \sqrt{4x - 3} = 0.$

25. $\dfrac{x^2 - x + 1}{x - 1} - x = \sqrt{\dfrac{6}{x}}.$

26. $\sqrt{x} + \sqrt{x - \sqrt{1 - x}} = 1.$

27. $\sqrt{x + 3} - \sqrt{x^2 + 3x} = 0.$

28. $\sqrt[4]{x^3} - 5\sqrt{x} + 6\sqrt[4]{x} = 0.$

29. $\sqrt{\dfrac{2x - 5}{x - 2}} - 3\sqrt{\dfrac{x - 2}{2x - 5}} + 2 = 0.$

30. $\dfrac{\sqrt{x - 1} - \sqrt{x + 1}}{\sqrt{x - 1} + \sqrt{x + 1}} = x - 3.$

31. $\sqrt{5x^2 - 6x + 1} - \sqrt{5x^2 + 9x - 2} = 5x - 1.$

32. $\dfrac{\sqrt{2x-1}+\sqrt{3x}}{\sqrt{2x-1}-\sqrt{3x}}+3=0.$ 33. $\sqrt[3]{x}+\sqrt[3]{2-x}=2.$

34. $(x+a)^{\frac{1}{3}}+(x+b)^{\frac{1}{3}}+(x+c)^{\frac{1}{3}}=0.$

35. $x(x-1)(x-2)(x-3)=6\cdot 5\cdot 4\cdot 3.$

36. $(x+a)^2+4(x+a)\sqrt{x}=a^2-4a\sqrt{x}.$

37. $\sqrt[3]{1+\left(\dfrac{2x}{x^2-1}\right)^2}+\sqrt[3]{1+\dfrac{2}{x^2-1}}=6.$

XVI. SIMULTANEOUS EQUATIONS WHICH CAN BE SOLVED BY MEANS OF QUADRATICS

A PAIR OF EQUATIONS IN X, Y, ONE OF THE FIRST DEGREE, THE OTHER OF THE SECOND

649 A pair of simultaneous equations of the form

$$f(x, y) = ax^2 + bxy + cy^2 + dx + ey + f = 0,$$
$$\phi(x, y) = a'x + b'y + c' = 0$$

may be solved as in the following example.

Example. Solve $y^2 - x^2 + 2x + 2y + 4 = 0,$ (1)

$2x - y - 7 = 0.$ (2)

From (2), $y = 2x - 7.$ (3)

Substituting (3) in (1), $3x^2 - 22x + 39 = 0.$ (4)

Solving (4), $x = 13/3$ or $3.$ (5)

Substituting in (3), $y = 5/3$ or $-1.$ (6)

The solutions of (1), (2) are the pairs of values

$$x = 13/3,\ y = 5/3;\ x = 3,\ y = -1. \quad (7)$$

For, §§ 368, 371, the following pairs of equations are equivalent: (1), (2) to (4), (2); (4), (2) to (5), (2); (5), (2) to (7).

We may indicate the solutions (7) thus: $13/3,\ 5/3;\ 3,\ -1.$ Care must always be taken to group *corresponding* values of x and y.

650 Ordinarily such a pair of equations will have *two finite solutions.* But if the group of first-degree terms in $\phi(x, y)$, namely

$a'x + b'y$, is a factor of the group of second-degree terms in $f(x, y)$, namely $ax^2 + bxy + cy^2$, while $\phi(x, y)$ itself is not a factor of $f(x, y)$, there will be only one finite solution or no such solution. And if $\phi(x, y)$ is a factor of $f(x, y)$, there will be infinitely many solutions.

Example 1. Solve $y^2 - x^2 + 2x + 2y + 4 = 0$, (1)

$y - mx = 0$. (2)

Eliminating y, $(m^2 - 1)x^2 + 2(m + 1)x + 4 = 0$. (3)

If $m^2 - 1 \neq 0$, (3) has two finite roots, and (1), (2) two finite solutions.

But if $y - mx$ is a factor of $y^2 - x^2$, that is, if $m = \pm 1$, then $m^2 - 1 = 0$ and (3) does not have two finite roots.

Thus, if $m = 1$, (3) reduces to $x + 1 = 0$, which has but one finite root, the other being infinite, § 638. And if $m = -1$, (3) reduces to $4 = 0$, which has no finite root, both roots being infinite, § 638.

Hence if (2) has the form $y - x = 0$, the pair (1), (2) has but one finite solution, the other being infinite. And if (2) has the form $y + x = 0$, the pair (1), (2) has no finite solution, both solutions being infinite.

Example 2. Solve $y^2 - x^2 + 2x + 2y = 0$, (1)

$y + x = 0$. (2)

Eliminating y, $x^2 - x^2 + 2x - 2x = 0$. (3)

But (3) is an identity and is satisfied by *every* value of x.

Hence every pair of numbers $x = a$, $y = -a$ is a solution of (1), (2).

The reason for this result is that $y + x$ is a factor of $y^2 - x^2 + 2x + 2y$.

651 When A, B, C are integral functions, the pair of equations $AB = 0$, $C = 0$ is equivalent to the two pairs $A = 0$, $C = 0$ and $B = 0$, $C = 0$, § 371. This principle and § 649 enable us to solve two integral equations $f(x, y) = 0$, $\phi(x, y) = 0$ whenever $f(x, y)$ can be resolved into factors of the first or second degrees and $\phi(x, y)$ into factors of the first degree.

Example. Solve $x^3 + xy^2 - 5x = 0$, (1)

$(2x - y)(x + y - 1) = 0$. (2)

This pair of equations is equivalent to the four pairs

$x = 0, \ 2x - y = 0$, (3)

$x = 0, \ x + y - 1 = 0$, (4)

$x^2 + y^2 - 5 = 0, \ 2x - y = 0$, (5)

$x^2 + y^2 - 5 = 0, \ x + y - 1 = 0$. (6)

Solving the pairs (3), (4), (5), (6), we find the solutions of (1), (2) to be $0, 0;\ 0, 1;\ 1, 2;\ -1, -2;\ 2, -1;\ -1, 2$.

652 A pair of integral equations in x, y can be solved by means of quadratics only when it has one of the forms described in §§ 649, 651 or when an equivalent pair which has one of these forms can be derived from it.

Thus, the pair of equations of the *second* degree, $y^2 - x + 1 = 0$, $y = x^2$, cannot be solved by quadratics. For there is no simpler method of solving this pair than to eliminate y, which gives $x^4 - x + 1 = 0$, an equation of the *fourth* degree which cannot be solved by quadratics.

653 The preceding sections illustrate the truth of the following important theorem:

A pair of integral equations $f(x, y) = 0$, $\phi(x, y) = 0$, *whose degrees are* m *and* n *respectively, has* mn *solutions.*

Thus, the pair $x^3 + xy^2 - 5x = 0$, $(2x - y)(x + y - 1) = 0$ has $3 \cdot 2$, or 6, solutions, § 651. See § 381 also.

654 It should be added, however, that if the groups of terms of highest degree in $f(x, y)$ and $\phi(x, y)$, but not $f(x, y)$ and $\phi(x, y)$ themselves, have a common factor, there are less than mn *finite* solutions. Thus, for every factor of the first degree which is common to the groups of terms of highest degree in $f(x, y)$ and $\phi(x, y)$ there is at least one infinite solution; for every such factor which is also common to the groups of terms of next highest degree there are at least two infinite solutions; and so on. If $f(x, y)$ and $\phi(x, y)$ themselves have a common factor, there are infinitely many solutions.

Thus, the pair $x^3 - xy^2 + xy - y^2 - y = 0$ (1), $x^2 - y^2 - 1 = 0$ (2) cannot have more than three finite solutions; for there are $3 \cdot 2$, or 6, solutions all told, and at least one of these is infinite since $x + y$ is a common factor of the groups of terms of highest degree in (1) and (2), namely $x^3 - xy^2$ and $x^2 - y^2$, and at least two others are infinite since $x - y$ is a common factor of the groups of highest and next highest degree in (1), (2), namely $x^3 - xy^2$, $x^2 - y^2$ and $xy - y^2$, $0(x - y)$.

EXERCISE XLV

Solve the following pairs of equations.

1. $\begin{cases} 7x^2 - 6xy = 8, \\ 2x - 3y = 5. \end{cases}$
2. $\begin{cases} xy = 1, \\ 3x - 5y = 2. \end{cases}$
3. $\begin{cases} x^2 + x = 4y^2, \\ 3x + 6y = 1. \end{cases}$
4. $\begin{cases} 3x^2 - 3xy - y^2 - 4x - 8y + 3 = 0, \\ 3x - y - 8 = 0. \end{cases}$
5. $\begin{cases} x^2 + 5y^2 - 8x - 7y = 0, \\ x + 3y = 0. \end{cases}$
6. $\begin{cases} 2x^2 - xy - 3y = 0, \\ 7x - 6y - 4 = 0. \end{cases}$
7. $\begin{cases} x^2 + 3xy + 2y^2 - 1 = 0, \\ x + y = 0. \end{cases}$
8. $\begin{cases} 2x + 3y = 37, \\ 1/x + 1/y = 14/45. \end{cases}$
9. $\begin{cases} 1/y - 3/x = 1, \\ 7/xy - 1/y^2 = 12. \end{cases}$
10. $\begin{cases} x^2 + xy + 2 = 0, \\ (3x + y)(2x + y - 1) = 0. \end{cases}$
11. $\begin{cases} x^2 + y^2 - 8 = 0, \\ (x + 1)^2 = (y - 1)^2. \end{cases}$
12. $\begin{cases} x^2 - xy - 2y^2 + y = 0, \\ (x - 2y)(x + y - 3) = 0. \end{cases}$

13. Determine m so that the two solutions of the pair $y^2 + 4x + 4 = 0$, $y = mx$ shall be equal.

14. Determine m and c so that both solutions of the pair

$$x^2 + xy - 2y^2 + x = 0, \quad y = mx + c$$

shall be infinite.

15. By the method of § 650, Ex. 2, show that $2x - y + 4$ is a factor of $2x^2 + xy - y^2 + 10x + y + 12$.

16. Show that the pair $xy = 1$, $xy + x + y = 0$ has not more than two finite solutions, and that the pair $x^2y + xy = 1$, $x^2y + y^2 = 2$ has not more than four finite solutions.

PAIRS OF EQUATIONS WHICH CAN BE SOLVED BY FACTORIZATION, ADDITION, OR SUBTRACTION

655 **When both equations are linear with respect to some pair of functions of x and y.** We begin by solving the equations for this pair of functions by the methods of §§ 374–376.

Example 1. Solve

$$2x^2 - 3y^2 = -58, \quad (1)$$
$$3x^2 + y^2 = 111. \quad (2)$$

Solving for x^2, y^2, we obtain $x^2 = 25$, $y^2 = 36$,

whence, $$x = \pm 5,\ y = \pm 6.$$

By §§ 367–372, the pair (1), (2) is equivalent to the four pairs $x = 5$, $y = 6$; $x = -5$, $y = 6$; $x = 5$, $y = -6$; $x = -5$, $y = -6$.

Hence the solutions of (1), (2) are 5, 6; -5, 6; 5, -6; -5, -6.

Example 2. Solve the following pairs.

1. $\begin{cases} ax^2 + by^2 = a, \\ bx^2 - ay^2 = b. \end{cases}$ 2. $\begin{cases} 3x^2 - 1/y^2 = 2, \\ 5x^2 + 3/y^2 = 120. \end{cases}$

656 **When one of the equations can be factored.** This is always possible when the equation in question has the form

$$ax^2 + bxy + cy^2 = 0,$$

and, in general, when it is reducible to the form

$$au^2 + bu + c = 0,$$

where u denotes a function of x, y.

Example 1. Solve $x^2 + y^2 + x - 11y - 2 = 0,$ (1)

$$x^2 - 5xy + 6y^2 = 0. \quad (2)$$

Factoring (2) by solving for x in terms of y,

$$x = 2y, \quad (3)$$

or $$x = 3y. \quad (4)$$

Solving (1), (3) and (1), (4), we obtain all the solutions of (1), (2), namely 4, 2; $-2/5$, $-1/5$; 3, 1; $-3/5$, $-1/5$.

Example 2. Solve $2x^2 + 4xy + 2y^2 + 3x + 3y - 2 = 0,$ (1)

$$3x^2 - 32y^2 + 5 = 0. \quad (2)$$

We may write (1) thus: $2(x + y)^2 + 3(x + y) - 2 = 0.$

Solving, $$x + y = 1/2, \quad (3)$$

or $$x + y = -2. \quad (4)$$

Solving (2), (3) and (2), (4), we obtain all the solutions of (1), (2), namely 1, $-1/2$; 3/29, 23/58; -3, 1; $-41/29$, $-17/29$.

Example 3. Solve the following pairs.

1. $\begin{cases} x^2 + xy - 6 = 0, \\ x^2 - 5x + 6 = 0. \end{cases}$ 2. $\begin{cases} \dfrac{x + y}{x - y} + \dfrac{x - y}{x + y} = \dfrac{26}{5}, \\ y^2 - 2x^2 = 1. \end{cases}$

657 **When the given equations may be combined by addition or subtraction so as to yield an equation which can be factored.** This is always possible when both the given equations are of the form $ax^2 + bxy + cy^2 = d$.

Example 1. Solve
$$6x^2 - xy - 2y^2 = 56, \quad (1)$$
$$5x^2 - xy - y^2 = 49. \quad (2)$$

We combine (1) and (2) so as to eliminate the constant terms.

Multiply (1) by 7, $42x^2 - 7xy - 14y^2 = 392.$ (3)

Multiply (2) by 8, $40x^2 - 8xy - 8y^2 = 392.$ (4)

Subtract (4) from (3), $2x^2 + xy - 6y^2 = 0.$ (5)

Solve (5) for x, $x = 3y/2,$ (6)

or $x = -2y.$ (7)

Solving (2), (6) and (2), (7), we obtain all the solutions of (1), (2), namely $\pm 3\sqrt{35}/10, \pm\sqrt{35}/5; \pm 2\sqrt{21}/3, \mp\sqrt{21}/3$.

And, in general, we obtain an equation which can be factored when the given equations are of the second degree, and can be combined by addition or subtraction so as to eliminate (1) all terms of the second degree; (2) all terms except those of the second degree; (3) all terms which involve x (or y); or (4) all terms which do not involve x (or y).

Example 2. Solve
$$2x^2 + 4xy - 2x - y + 2 = 0, \quad (1)$$
$$3x^2 + 6xy - x + 3y = 0. \quad (2)$$

Here we can eliminate all terms of the second degree by multiplying (1) by 3, and (2) by 2, and subtracting. We thus obtain
$$4x + 9y - 6 = 0. \quad (3)$$

Solving (2), (3), we obtain $-3, 2; -2, 14/9$, and these are all the finite solutions of (1), (2). See § 654.

Example 3. Solve
$$x^2 - 3xy + 2y^2 + 4x + 3y - 1 = 0, \quad (1)$$
$$2x^2 - 6xy + y^2 + 8x + 2y - 3 = 0. \quad (2)$$

Here all the terms which involve x can be eliminated by multiplying (1) by 2 and then subtracting (2). We thus obtain
$$3y^2 + 4y + 1 = 0. \quad (3)$$

Solving (1), (3), we obtain all the four solutions of (1), (2), namely
$$1/3, -1/3; -16/3, -1/3; (-7 \pm \sqrt{57})/2, -1.$$

Consider the following example also.

Example 4. Solve $x^2 + 2xy + 2y^2 + 3x = 0,$ (1)

$$xy + y^2 + 3y + 1 = 0. \quad (2)$$

Multiply (2) by 2 and add to (1). We obtain

$$(x + 2y)^2 + 3(x + 2y) + 2 = 0. \quad (3)$$

Solving (3), $x + 2y = -1,$ (4)

or $x + 2y = -2.$ (5)

Solving (2), (4) and (2), (5), we obtain all the solutions of (1), (2), namely $-3 \pm 2\sqrt{2},\ 1 \mp \sqrt{2};\ -3 \pm \sqrt{5},\ (1 \mp \sqrt{5})/2.$

Example 5. Solve the following pairs of equations.

1. $\begin{cases} 2x^2 + xy + 5y = 0, \\ x^2 + y^2 + 10y = 0. \end{cases}$ 2. $\begin{cases} x^2 + y^2 - 13 = 0, \\ xy + y - x = -1. \end{cases}$

When the equation obtained by eliminating x or y can be factored. **658** From any pair of equations of the second degree we can eliminate x or y by the following method. The resulting equation will ordinarily be of the fourth degree and not solvable by means of quadratics. But if we can resolve it into factors of the first or second degrees, we can solve it and so obtain the solutions of the given pair.

Example 1. Solve $10x^2 + 5y^2 - 27x - 4y + 5 = 0,$ (1)

$$x^2 + y^2 - 3x - y = 0. \quad (2)$$

First eliminate y^2 by multiplying (2) by 5 and subtracting the result from (1). We obtain

$$5x^2 - 12x + y + 5 = 0, \text{ or } y = -5x^2 + 12x - 5. \quad (3)$$

Substituting (3) in (2),

$$5x^4 - 24x^3 + 40x^2 - 27x + 6 = 0. \quad (4)$$

Factoring, by § 451,

$$(x - 1)(x - 2)(5x^2 - 9x + 3) = 0. \quad (5)$$

Solving (5), § 643, $x = 1,\ 2,\ (9 \pm \sqrt{21})/10.$ (6)

Substituting (6) in (3), $y = 2,\ -1,\ (7 \pm 3\sqrt{21})/10.$ (7)

The pairs of corresponding values (6), (7) are the solutions of (1), (2).

Example 2. Solve $x^2 - 3xy + 2y^2 + 3x - 3y = 0,$

$$2x^2 + xy - y^2 + x - 2y + 3 = 0.$$

EXERCISE XLVI

Solve the following pairs of equations.

1. $\begin{cases} x^2 + 3y^2 = 31, \\ 7x^2 - 2y^2 = 10. \end{cases}$

2. $\begin{cases} 36/x^2 + 1/y^2 = 18, \\ 1/y^2 - 4/x^2 = 8. \end{cases}$

3. $\begin{cases} y^2 + xy + 6 = 0, \\ y^2 - y - 2 = 0. \end{cases}$

4. $\begin{cases} x^2 + y^2 - 3x + 2y - 39 = 0, \\ 3x^2 - 17xy + 10y^2 = 0. \end{cases}$

5. $\begin{cases} y^2 - x^2 - 5 = 0, \\ 4x^2 + 4xy + y^2 + 4x + 2y = 3. \end{cases}$

6. $\begin{cases} x^2 + 5xy - 2x + 3y + 1 = 0, \\ 3x^2 + 15xy - 7x + 8y + 4 = 0. \end{cases}$

7. $\begin{cases} x^2 - 15xy - 3y^2 + 2x + 9y = 98, \\ 5xy + y^2 - 3y = -21. \end{cases}$

8. $\begin{cases} 2x^2 + 3xy - 4y^2 = 25, \\ 15x^2 + 24xy - 31y^2 = 200. \end{cases}$

9. $\begin{cases} x(x + 3y) = 18, \\ x^2 - 5y^2 = 4. \end{cases}$

10. $\begin{cases} x^2 - 3xy + 3y^2 = x^2y^2, \\ 7x^2 - 10xy + 4y^2 = 12x^2y^2. \end{cases}$

11. $\begin{cases} x^2 + xy + y^2 = 38, \\ x^2 - xy + y^2 = 14. \end{cases}$

12. $\begin{cases} x^2 - xy + y^2 = 21(x - y), \\ xy = 20. \end{cases}$

13. $\begin{cases} x^2 + y - 8 = 0, \\ y^2 + 15x - 46 = 0. \end{cases}$

PAIRS WHICH CAN BE SOLVED BY DIVISION

659 In solving a pair of equations it is sometimes advantageous to combine them by multiplication or division; but care must then be taken not to introduce extraneous solutions nor to lose actual ones (see §§ 362, 342).

660 If given a pair of the form $AB = CD$ (1), $B = D$ (2), where A, B, C, D denote integral functions of x, y, we may replace B by D in (1), thus obtaining the pair $AD = CD$, $B = D$ which is evidently equivalent to the two pairs $A = C$, $B = D$ and $D = 0$, $B = 0$. We may obtain the pair $A = C$, $B = D$ by dividing each member of (1) by the corresponding member of (2); but if we then merely solve this pair $A = C$, $B = D$, we lose some of the solutions of (1), (2), except, of course, when either B or D is a constant, so that the pair $B = 0$, $D = 0$ has no solution.

Example 1. Solve $x^4 + x^2y^2 + y^4 = 21,$ (1)

$x^2 + xy + y^2 = 7.$ (2)

Dividing (1) by (2), $x^2 - xy + y^2 = 3.$ (3)

Solving (2), (3), we obtain all the finite solutions of (1), (2), namely 2, 1; -2, -1; 1, 2; -1, -2. See § 654.

Example 2. Solve $x^3 - y^3 = -3(x + 1)y,$ (1)

$x^2 + xy + y^2 = x + 1.$ (2)

Dividing (1) by (2), $x - y = -3y.$ (3)

The pair (1), (2) has the same finite solutions as the pair (2), (3) and the pair $x^2 + xy + y^2 = 0$ (4), $x + 1 = 0$ (5) jointly. And the solutions of (2), (3) and (4), (5) are 2, -1; $-2/3$, $1/3$; -1, $(1 \pm i\sqrt{3})/2$.

Example 3. Solve $(x + y)^2 = x,$ (1)

$x^2 - y^2 = -6y.$ (2)

Dividing (1) by (2), $(x + y)/(x - y) = -x/6y.$ (3)

Clearing of fractions, $x^2 + 5xy + 6y^2 = 0.$ (4)

The pair (2), (4) has the four solutions 0, 0; 0, 0; 4, -2; $9/4$, $-3/4$.

The process by which (4) was derived from (1), (2) is reversible when x, y have the values 4, -2 or $9/4$, $-3/4$, but not when they have the values 0, 0. Hence this reckoning only proves that 4, -2 and $9/4$, $-3/4$ are solutions of (1), (2), § 362.

It is obvious by inspection that 0, 0 *is* a solution of (1), (2); but it should be counted only *once* as a solution, not twice as in the case of (2), (4). This follows from the fact that (1), (2) can have but three finite solutions, § 654. It may also be shown thus: In (1), (2) make the substitution $y = tx$ (5). We obtain $(1 + t)^2x^2 = x$ (1′), $(1 - t^2)x^2 = -6tx$ (2′). And (5), (1′), (2′) yield $x = 0$, $y = 0$ once and but once.

EXERCISE XLVII

1. $\begin{cases} x^3 - y^3 = 63, \\ x - y = 3. \end{cases}$

2. $\begin{cases} x + y = 98, \\ \sqrt[3]{x} + \sqrt[3]{y} = 2. \end{cases}$

3. $\begin{cases} x^4 + x^2y^2 + y^4 = 931, \\ x^2 + xy + y^2 = 49. \end{cases}$

4. $\begin{cases} (x + y)(x^2 - 2y^2) = -70, \\ (x - y)(x^2 - 2y^2) = 14. \end{cases}$

5. $\begin{cases} (x + y)^2(x - y) = 3xy + 6y, \\ x^2 - y^2 = x + 2. \end{cases}$

6. $\begin{cases} x^2 - 3xy + 2y^2 = 6x, \\ x^2 - y^2 = -5y. \end{cases}$

SYMMETRIC PAIRS OF EQUATIONS

661 A pair of equations in x, y is said to be **symmetric** if it remains unchanged when x and y are interchanged.

Thus, the following pairs, (a) and (b), are symmetric.

$$(a)\ \begin{cases} 2x^2 + 2y^2 + 3x + 3y = 0, \\ x^2y^2 + xy + 1 = 0. \end{cases} \qquad (b)\ \begin{cases} x^2 = 2x + 3y, \\ y^2 = 2y + 3x. \end{cases}$$

Symmetric pairs are of two types, those like (a) in which the individual equations remain unchanged when x and y are interchanged, and those like (b) in which the two equations change places when x and y are interchanged.

662 **Symmetric pairs of the first type.** The simplest pair of symmetric equations is $x + y = a$, $xy = b$. This pair may be solved as in § 649, but the following is a more symmetric method.

Example. Solve
$$x + y = 5, \tag{1}$$
$$xy = 6. \tag{2}$$
Square (1),
$$x^2 + 2xy + y^2 = 25. \tag{3}$$
Multiply (2) by 4,
$$4xy = 24. \tag{4}$$
Subtract (4) from (3),
$$x^2 - 2xy + y^2 = 1. \tag{5}$$
Hence
$$x - y = 1, \tag{6}$$
or
$$x - y = -1. \tag{7}$$
From (1), (6), $x = 3$, $y = 2$; and from (1), (7), $x = 2$, $y = 3$.

663 If given a more complicated pair of symmetric equations, we may transform each equation into an equation in $x + y$ and xy, § 637, and then solve for these functions; or in the given equations we may set $x = u + v$, $y = u - v$ and then solve for u and v. The second method is essentially the same as solving the given equations for $x + y$ and $x - y$; for, since $x = u + v$, $y = u - v$, we have $u = (x + y)/2$, $v = (x - y)/2$.

Example 1. Solve
$$2x^2 + 5xy + 2y^2 + x + y + 1 = 0, \tag{1}$$
$$x^2 + 4xy + y^2 + 12x + 12y + 10 = 0. \tag{2}$$
In (1) and (2) for $x^2 + y^2$ substitute $(x + y)^2 - 2xy$.

Collecting terms, $2(x+y)^2 + xy + (x+y) + 1 = 0,$ (3)

$$(x+y)^2 + 2xy + 12(x+y) + 10 = 0. \quad (4)$$

Eliminating xy, $3(x+y)^2 - 10(x+y) - 8 = 0.$ (5)

Solving, $x + y = 4,$ (6)

or $x + y = -2/3.$ (7)

Hence, from (3), $xy = -37,$ (8)

or $xy = -11/9.$ (9)

And solving the pairs (6), (8) and (7), (9) for x, y, we have

$$x, y = 2 \pm \sqrt{41},\ 2 \mp \sqrt{41};\ (-1 \pm 2\sqrt{3})/3,\ (-1 \mp 2\sqrt{3})/3.$$

Example 2. Solve $x^4 + y^4 = 97,$ (1)

$$x + y = 5. \quad (2)$$

In (1) and (2) set $x = u + v,\ y = u - v.$

We obtain $(u+v)^4 + (u-v)^4 = 97,$ (3)

and $2u = 5.$ (4)

Eliminating u, $16v^4 + 600v^2 - 151 = 0.$ (5)

Solving, $v = \pm 1/2$ or $\pm i\sqrt{151}/2.$ (6)

Substituting $u = 5/2$ (4) and the four values (6) of v in the formulas $x = u + v$, $y = u - v$, we obtain

$$x, y = 2, 3;\ 3, 2;\ (5 \pm i\sqrt{151})/2,\ (5 \mp i\sqrt{151})/2.$$

Evidently if $x = \alpha$, $y = \beta$ is one solution of a symmetric pair, $x = \beta$, $y = \alpha$ is another solution. Unless $\alpha = \beta$, these two solutions are different; but xy and $x + y$ have the same values for $x = \alpha$, $y = \beta$ as for $x = \beta$, $y = \alpha$, and the corresponding values of $x - y$, namely $\alpha - \beta$ and $\beta - \alpha$, differ only in sign.

Hence the values of xy or $x + y$ derived from a symmetric pair will be less numerous than the values of x or y, that is the degree of the equation in xy or $x + y$ derived from the pair by elimination, as in Ex. 1, will be less than the degree of an equation in x or y similarly derived would be. As for the equation in $x - y$, if c is one of its roots, $-c$ must be another root. Hence this equation will involve only even powers of $x - y$, as in Ex. 2, or only odd powers with no constant term.

664 **Note.** The methods just given are applicable to pairs of equations which are symmetric with respect to x and $-y$ or some other pair of functions of x and y. Thus, $x^4 + y^4 = a$, $x - y = b$ may be written $x^4 + (-y)^4 = a$, $x + (-y) = b$.

665 **Symmetric pairs of the second type.** Such a pair may sometimes be solved as follows.

Example. Solve
$$x^3 = 7x + 3y, \quad (1)$$
$$y^3 = 7y + 3x. \quad (2)$$

Adding (1) and (2), $$x^3 + y^3 = 10(x + y). \quad (3)$$

Subtracting (2) from (1), $$x^3 - y^3 = 4(x - y). \quad (4)$$

By § 341, (3) is equivalent to the two equations
$$x + y = 0, \quad (5)$$
and
$$x^2 - xy + y^2 = 10. \quad (6)$$

Similarly (4) is equivalent to the two equations
$$x - y = 0, \quad (7)$$
and
$$x^2 + xy + y^2 = 4. \quad (8)$$

And solving (5), (7); (5), (8); (6), (7); (6), (8), we obtain $0, 0$; $2, -2$; $-2, 2$; $\pm\sqrt{10}, \pm\sqrt{10}$; $(1 \pm \sqrt{13})/2, (1 \mp \sqrt{13})/2$; $(-1 \pm \sqrt{13})/2, (-1 \mp \sqrt{13})/2$.

EXERCISE XLVIII

Solve the following pairs of equations.

1. $\begin{cases} x + y = 5, \\ xy + 36 = 0. \end{cases}$
2. $\begin{cases} x^2 + y^2 = 200, \\ x + y = 12. \end{cases}$
3. $\begin{cases} x^2 + y^2 = 293, \\ xy = 34. \end{cases}$
4. $\begin{cases} x^2 + y^2 = 85, \\ x - y = 7. \end{cases}$
5. $\begin{cases} x^3 + y^3 = 513, \\ x + y = 9. \end{cases}$
6. $\begin{cases} x^3 + y^3 = 468, \\ x^2y + xy^2 = 420 \end{cases}$
7. $\begin{cases} 27x^3 + 64y^3 = 65, \\ 3x + 4y = 5. \end{cases}$
8. $\begin{cases} x^4 + y^4 = 82, \\ x - y = 2. \end{cases}$
9. $\begin{cases} x^5 + y^5 = 32, \\ x + y = 2. \end{cases}$
10. $\begin{cases} x + y = 1/2, \\ 56\left(\frac{x}{y} + \frac{y}{x}\right) + 113 = 0. \end{cases}$
11. $\begin{cases} xy + x + y + 19 = 0, \\ x^2y + xy^2 + 20 = 0. \end{cases}$
12. $\begin{cases} x^4 + y^4 - (x^2 + y^2) = 72, \\ x^2 + x^2y^2 + y^2 = 19. \end{cases}$
13. $\begin{cases} x^2y + xy^2 = 30, \\ 1/x + 1/y = 3/10. \end{cases}$
14. $\begin{cases} x^2 + 3xy + y^2 + 2x + 2y = 8, \\ 2x^2 + 2y^2 + 3x + 3y = 14. \end{cases}$
15. $\begin{cases} x^3 = 5y, \\ y^3 = 5x. \end{cases}$

SYSTEMS INVOLVING MORE THAN TWO UNKNOWN LETTERS

666 A system of three equations in three unknown letters can be solved by means of quadratics when one of the equations is of the second degree and the other two of the first degree; also when it is possible to reduce the system to one or more equivalent systems each consisting of one equation of the second degree and the rest of the first degree. The like is true of a system of four equations in four unknown letters, and so on.

If A, B, C are integral functions of degrees m, n, p in x, y, z, and no two of them have a common factor, the equations $A = 0$, $B = 0$, $C = 0$ will have mnp solutions. But some of these solutions may be infinite.

Example 1. Solve
$$z^2 - xy - 7 = 0, \quad (1)$$
$$x + y + z = 0, \quad (2)$$
$$3x - 2y + 2z + 2 = 0. \quad (3)$$

Solving (2), (3) for x and y in terms of z,
$$x = -(4z + 2)/5, \quad (4)$$
$$y = (-z + 2)/5. \quad (5)$$

Substituting in (1) and simplifying,
$$7z^2 + 2z - 57 = 0. \quad (6)$$

Solving (6), $\quad z = -3$ or $19/7$.

Hence, from (4), (5), $\quad x, y, z = 2, 1, -3; \; -\frac{18}{7}, \; -\frac{1}{7}, \; \frac{19}{7}.$

Example 2. Solve the system
$$xy = 6, \quad (1)$$
$$yz = 12, \quad (2)$$
$$zx = 8. \quad (3)$$

Dividing (2) by (1), $\quad z/x = 2$ or $z = 2x$. (4)

Substituting (4) in (3), $\quad x^2 = 4$, whence $x = \pm 2$. (5)

From (5), (4), (1) we obtain $\quad x, y, z = 2, 3, 4; \; -2, -3, -4.$

EXERCISE XLIX

Solve the following systems of equations.

1. $\begin{cases} x + y = 3, \\ y + z = 2, \\ x^2 - yz = 19. \end{cases}$

2. $\begin{cases} x(y + z) = 12, \\ y(z + x) = 6, \\ z(x + y) = 10. \end{cases}$

3. $\begin{cases} (y + b)(z + c) = a^2, \\ (z + c)(x + a) = b^2, \\ (x + a)(y + b) = c^2. \end{cases}$

EXERCISE L

Solve the following systems of equations by any of the methods of the present chapter.

1. $\begin{cases} 7x^2 - 6xy = 8, \\ 2x - 3y = 5. \end{cases}$
2. $\begin{cases} x^2 + y^2 = 25, \\ x - y = 1. \end{cases}$
3. $\begin{cases} x - y = a, \\ xy = (b^2 - a^2)/4. \end{cases}$

4. $\begin{cases} \dfrac{a}{x^2} + \dfrac{b}{y^2} = a^2 + b^2, \\ x^2 + y^2 = 0. \end{cases}$
5. $\begin{cases} \dfrac{1}{y} - \dfrac{3}{x} = 1, \\ \dfrac{7}{xy} - \dfrac{1}{y^2} = 12. \end{cases}$
6. $\begin{cases} x + y = a + b, \\ \dfrac{a}{x+b} + \dfrac{b}{y+a} = 1. \end{cases}$

7. $\begin{cases} \dfrac{1}{x^3} + \dfrac{1}{y^3} = \dfrac{1001}{125}, \\ \dfrac{1}{x} + \dfrac{1}{y} = \dfrac{11}{5}. \end{cases}$
8. $\begin{cases} \dfrac{a^2}{x^2} + \dfrac{b^2}{y^2} = 5, \\ \dfrac{ab}{xy} = 2. \end{cases}$
9. $\begin{cases} x^2 + y^2 = \dfrac{17}{4}xy, \\ x - y = \dfrac{3}{4}xy. \end{cases}$

10. $a(x + y) = b(x - y) = xy.$
11. $40xy = 21(x^2 - y^2) = 210(x + y).$

12. $\begin{cases} 4x^2 - 25y^2 = 0, \\ 2x^2 - 10y^2 - 3y = 4. \end{cases}$
13. $\begin{cases} x^2 + 3xy - 9y^2 = 9, \\ x^2 - 13xy + 21y^2 = -9. \end{cases}$

14. $\begin{cases} x^2 - 7y^2 - 29 = 0, \\ x^2 - 6xy + 9y^2 - 2x + 6y = 3. \end{cases}$
15. $\begin{cases} x/y + y/x = 65/28, \\ 2(x^2 + y^2) + (x - y) = 34. \end{cases}$

16. $\begin{cases} x^2y = a, \\ xy^2 = b. \end{cases}$
17. $\begin{cases} x^2y + xy^2 = a, \\ x^2y - xy^2 = b. \end{cases}$
18. $\begin{cases} x = a(x^2 + y^2), \\ y = b(x^2 + y^2). \end{cases}$

19. $\begin{cases} (x + y)/(x - y) = 5/3, \\ (2x + 3y)(3x - 2y) = 110a^2. \end{cases}$
20. $\begin{cases} 3(x^3 - y^3) = 13xy, \\ x - y = 1. \end{cases}$

21. $\begin{cases} x^4 + y^4 = a^4, \\ x + y = a. \end{cases}$
22. $\begin{cases} 21(x + y) = 10xy, \\ x + y + x^2 + y^2 = 68. \end{cases}$

23. $x^2 + y^2 = xy = x + y.$
24. $x^2 - xy + y^2 = 3a^2 = x^2 - y^2.$

25. $\begin{cases} x^2 + xy + y^2 = 21, \\ x + \sqrt{xy} + y = 7. \end{cases}$
26. $\begin{cases} 4x^2 - 3y^2 = 12(x - y), \\ xy = 0. \end{cases}$

27. $\begin{cases} x^2 + y^2 = x + y + 20, \\ xy + 10 = 2(x + y). \end{cases}$
28. $\begin{cases} x^2 + 4x - 3y = 0, \\ y^2 + 10x - 9y = 0. \end{cases}$

29. $\begin{cases} 28(x^5 + y^5) = 61(x^3 + y^3), \\ x + y = 2. \end{cases}$
30. $\begin{cases} xy - x/y = a, \\ xy - y/x = 1/a. \end{cases}$

31. $\begin{cases} (x+1)^3 + (y-2)^3 = 19, \\ x + y = 2. \end{cases}$

32. $\begin{cases} x^2 + y = 8/3, \\ x + y^2 = 34/9. \end{cases}$

33. $\begin{cases} y^2 - xy - yz = 3, \\ x + 4y + z = 14, \\ x - y + 2z = 0. \end{cases}$

34. $\begin{cases} x + y + z + u = 0, \\ 3x + z + u = 0, \\ 3y + 2z = 0, \\ x^2 + y^2 + zu = 5. \end{cases}$

35. $\begin{cases} (y+z)(x+y+z) = 10, \\ (z+x)(x+y+z) = 20, \\ (x+y)(x+y+z) = 20. \end{cases}$

36. $\begin{cases} x^2 + y^2 + z^2 = 6, \\ xy + yz + zx = -1, \\ 2x + y - 2z = -3. \end{cases}$

EXERCISE LI

1. The difference of the cubes of two numbers is 218 and the cube of their difference is 8. Find the numbers.

2. The square of the sum of two numbers less their product is 63, and the difference of their cubes is 189. What are they?

3. The sum of the terms of a certain fraction is 11, and the product of this fraction by one whose numerator and denominator exceed its numerator and denominator by 3 and 4 respectively is 2/3. Find the fraction.

4. Separate 37 into three parts whose product is 1440 and such that the product of two of them exceeds three times the third by 12.

5. The diagonal of a rectangle is 13 feet long. If each side were 2 feet longer than it is, the area would be 38 square feet greater than it is. What are the sides?

6. The perimeter of a right-angled triangle is 36 inches long and the area of the triangle is 54 square inches. Find the lengths of the sides.

7. The hypotenuse of a right-angled triangle is longer than the two perpendicular sides by 3 and 24 inches respectively. Find the sides of the triangle.

8. Find the dimensions of a room from the following data: its floor is a rectangle whose area is 224 square feet, and the areas of two of its side walls are 126 and 144 square feet respectively.

9. A rectangle is surrounded by a border whose width is 5 inches. The area of the rectangle is 168 square inches, that of the border 360 square inches. Find the length and breadth of the rectangle.

10. In buying coal A gets 3 tons more for \$135 than B does and pays \$7 less for 4 tons than B pays for 5. Required the price each pays per ton.

11. A certain principal at a certain rate amounts to \$1248 in one year at simple interest. Were the principal \$100 greater and the rate $1\frac{1}{2}$ times as great, the amount at the end of 2 years would be \$1456. What is the principal and what is the rate?

12. A man leaves \$60,000 to his children and grandchildren, seven in all. The children receive $\frac{1}{3}$ of it, which is \$2000 more apiece than the grandchildren get. How many children are there and how many grandchildren, and what does each receive?

13. At his usual rate a man can row 15 miles downstream in 5 hours less time than it takes him to return. Could he double his rate, his time downstream would be only 1 hour less than his time upstream. What is his usual rate in dead water and what is the rate of the current?

14. Three men A, B, C together can do a piece of work in 1 hour, 20 minutes. To do the work alone it would take C twice as long as A and 2 hours longer than B. How long would it take each man to do the work alone?

15. Two bodies A and B are moving at constant rates and in the same direction around the circumference of a circle whose length is 20 feet. A makes one circuit in 2 seconds less time than B, and A and B are together once every minute. What are their rates?

16. On two straight lines which meet at right angles at O the points A and B are moving toward O at constant rates. A is now 28 inches from O and B 9 inches; 2 seconds hence A and B will be 13 inches apart, and 3 seconds hence they will be 5 inches apart. At what rates are A and B moving?

17. Three men A, B, and C set out at the same time to walk a certain distance. A walks $4\frac{1}{2}$ miles an hour and finishes the journey 2 hours before B. B walks 1 mile an hour faster than C and finishes the journey in 3 hours less time. What is the distance?

18. Two couriers A and B start simultaneously from P and Q respectively and travel toward each other. When they meet A has traveled 12 miles farther than B. After their meeting A continues toward Q at the same rate as before, arriving in $4\frac{2}{3}$ hours. Similarly B arrives at P in $7\frac{1}{7}$ hours after the meeting. What is the distance from P to Q?

GRAPHS OF EQUATIONS OF THE SECOND DEGREE IN X, Y

667 **Examples of such graphs.** The graph of any given equation of the second degree in x, y may be obtained by the method illustrated in the following examples.

Example 1. Find the graph of $y^2 = 4x$. (1)

Solving for y, $y = \pm 2\sqrt{x}$. (2)

From (2) it follows that when x is negative, y is imaginary; when x is 0, y is 0; when x is positive, y has two real values which are equal numerically but of opposite signs. Hence the graph of (1) lies entirely to the right of the y-axis, passes through the origin, and is symmetric with respect to the x-axis.

When $x = 0,\ 1/4,\ 1/2,\ 1,\ 2,\ 3,\ 4, \cdots,$

we have $y = 0,\ \pm 1,\ \pm\sqrt{2},\ \pm 2,\ \pm 2\sqrt{2},\ \pm 2\sqrt{3},\ \pm 4, \cdots.$

We obtain the part of the graph given in the figure by plotting these solutions $(0,\ 0)$, $(\frac{1}{4},\ 1)$, $(\frac{1}{4},\ -1)$, $\cdots$ and passing a curve through the points thus found. Compare § 389. It touches the y-axis.

This curve is called a **parabola**. It consists of one "infinite branch," here extending indefinitely to the right.

Example 2. In what points is the graph of $y^2 = 4x$ (1) met by the graphs of $y = x - 3$ (2), $y = x + 1$ (3), $y = x + 3$ (4)?

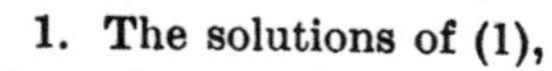

1. The solutions of (1), (2) are $1,\ -2$; $9,\ 6$. Hence, § 386, the graphs of (1), (2) intersect in the points $(1,\ -2)$, $(9,\ 6)$, as is shown in the preceding figure.

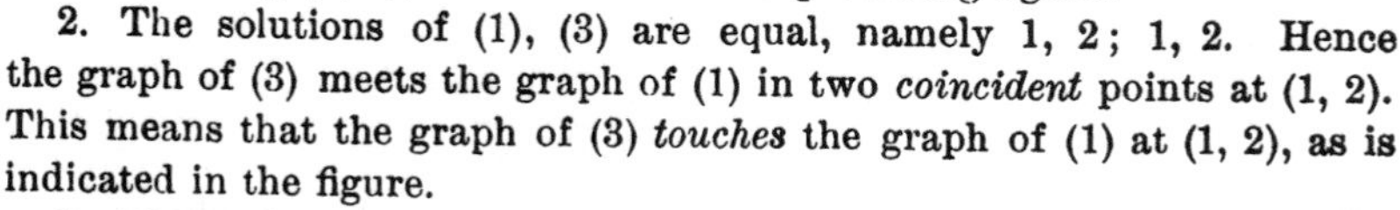

2. The solutions of (1), (3) are equal, namely $1,\ 2$; $1,\ 2$. Hence the graph of (3) meets the graph of (1) in two *coincident* points at $(1,\ 2)$. This means that the graph of (3) *touches* the graph of (1) at $(1,\ 2)$, as is indicated in the figure.

3. The solutions of (1), (4) are *imaginary*, namely $-1 \pm 2\sqrt{2}\,i$, $2 \pm 2\sqrt{2}\,i$. Hence, as the figure shows, the graphs of (1), (4) do not meet.

Example 3. Find the graph of

$$y^2 - 2xy + 2x^2 - 6x + 2y + 1 = 0. \qquad (1)$$

Solving for y, we have $\quad y = x - 1 \pm \sqrt{4x - x^2}. \qquad (2)$

The values of y given by (2) are real when the radicand $4x - x^2$, or $x(4 - x)$, is positive (or 0), that is, when x lies between 0 and 4 (or is 0 or 4). Hence the graph of (1) lies between the lines $x = 0$ and $x = 4$.

When $x = 0$ and when $x = 4$ the values of y given by (2) are equal: namely $-1, -1$ when $x = 0$, and $3, 3$, when $x = 4$. This means that the graph of (1) *touches* the line $x = 0$ at the point $(0, -1)$ and the line $x = 4$ at the point $(4, 3)$. See Ex. 2, 2. The line $y = x - 1$ passes through these points of tangency, for when $4x - x^2$ vanishes, (2) gives the same values for y that $y = x - 1$ gives.

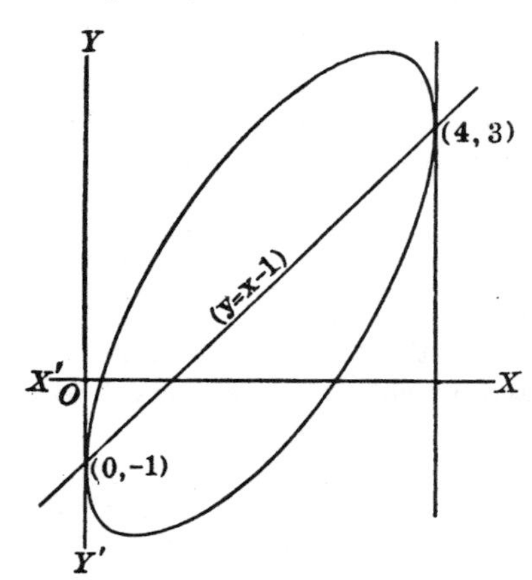

For each value of x between 0 and 4 the equation (2) gives two real and distinct values of y, obtained by increasing and diminishing the value of $x - 1$ by that of $\sqrt{4x - x^2}$. Hence for each of these values of x there are two points of the graph of (1). They are most readily obtained by drawing the line $y = x - 1$ and then increasing and diminishing its ordinate for the value of x in question by the value of $\sqrt{4x - x^2}$.

Thus, when	$x = 0,$	$1,$	$2,$	$3,$	$4,$
we have for the line	$y = -1,$	$0,$	$1,$	$2,$	$3,$
and for the graph of (1)	$y = -1,$	$\pm\sqrt{3},$	$1 \pm 2,$	$2 \pm \sqrt{3},$	$3.$

The figure shows the oval-shaped curve which the points thus found determine. It is called an **ellipse**.

By solving (1) for x and applying the method of § 641, we may show that the highest and lowest points of the curve are $(2 + \sqrt{2}, 1 + 2\sqrt{2})$ and $(2 - \sqrt{2}, 1 - 2\sqrt{2})$.

Example 4. Find the graph of $y^2 - x^2 + 2x + 2y + 4 = 0. \qquad (1)$

Solving for y and factoring the radicand in the result,

$$y = -1 \pm \sqrt{(x + 1)(x - 3)}. \qquad (2)$$

The radicand $(x + 1)(x - 3)$ vanishes when $x = -1$ and when $x = 3$, and in both cases (2) gives equal values of y, namely $-1, -1$. This means that the graph of (1) touches the line $x=-1$ at the point $(-1, -1)$ and the line $x=3$ at the point $(3, -1)$. The line $y=-1$ passes through these points of tangency.

The radicand $(x+1)(x-3)$ is positive when and only when $x<-1$ or $x>3$. For every such value of x the equation (2) gives two real and distinct values of y and therefore two points of the graph of (1). These points may be obtained by drawing the line $y=-1$ and then increasing and diminishing its constant ordinate -1 by the value of $\sqrt{(x+1)(x-3)}$.

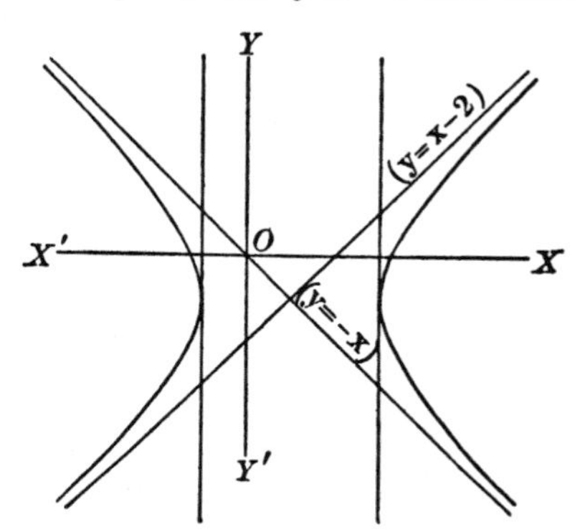

Hence, as is indicated in the figure, the graph of (1) consists of two infinite branches, the one touching the line $x=-1$ and extending indefinitely to the left, the other touching the line $x=3$ and extending indefinitely to the right.

This curve is called an **hyperbola.**

There are two straight lines called *asymptotes*, which the infinite branches of this hyperbola tend to touch, and which they are said to touch at infinity. These lines are the graphs of the equations $y=x-2$ and $y=-x$, which we obtain as follows. Compare § 650, Ex. 1.

Eliminating y between (1) and the equation $y=mx+c$, (3)

we obtain $(m^2-1)x^2+2(mc+m+1)x+(c^2+2c+4)=0$. (4)

Both roots of (4) are infinite, § 638,

if $$m^2-1=0 \text{ and } mc+m+1=0,$$

that is, if $$m=1,\ c=-2, \text{ or if } m=-1,\ c=0.$$

Hence both solutions of (1), (3) are infinite if (3) has one of the forms

$$y=x-2 \quad (3') \qquad \text{or} \qquad y=-x. \quad (3'')$$

Therefore the graphs of (3′) and (3″) each meet the graph of (1) in *two infinitely distant coincident* points.

Example 5. Find the graph of $y^2-4xy+3x^2+6x-2y=0$. (1)

Solving for y, we have $y=2x+1\pm\sqrt{x^2-2x+1}$, (2)

that is, $$y=3x \text{ or } y=x+2.$$

Hence the graph of (1) consists of the pair of right lines $y=3x$ and $y=x+2$.

Except when the radicand vanishes, that is, when $x-1=0$, the equation (2) gives two real and distinct values of y. But when $x-1=0$ it gives two equal values of y, namely 3, 3. Hence the line $x-1=0$ meets the graph of (1) in two coincident points at (1, 3). Of course this cannot mean that the line $x-1=0$ touches the graph of (1) at (1, 3). It means that the points coincide in which the line $x-1=0$ meets the two lines, $y=3x$ and $y=x+2$, which together constitute the graph of (1).

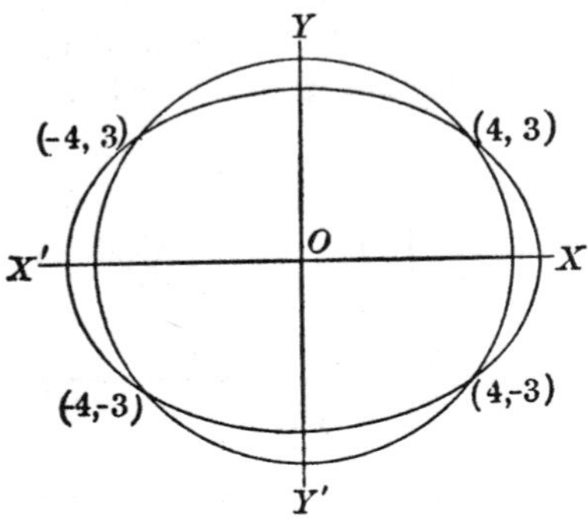

Example 6. Find the graphs, and their intersections, of

$$x^2+y^2=25, \qquad (1)$$

$$x^2/16+y^2/9=2. \qquad (2)$$

The graph of (1) is a circle whose center is at the origin, O, and whose radius is 5. The graph of (2) is an ellipse.

These curves intersect at the four points (4, 3), (− 4, 3), (− 4, − 3), (4, − 3), these points being the graphs of the solutions of (1), (2).

Example 7. Find the graphs, and their intersections, of

$$xy-3y-2=0, \qquad (1)$$

$$xy+2y+3=0. \qquad (2)$$

From (1) we obtain $\qquad y=2/(x-3). \qquad (3)$

Here there is *one* real value of y, and hence one point of the graph, for each real value of x.

When $x=-\infty,\ -1,\ 0,\ 1,\ 2,\ 2\frac{3}{4},\ 3,\ 3\frac{1}{4},\ 4,\ \infty,$
we have $y=0,\ -\frac{1}{2},\ -\frac{2}{3},\ -1,\ -2,\ -8,\ \pm\infty,\ 8,\ 2,\ 0.$

And plotting these solutions, we obtain an hyperbola whose two infinite branches are indicated by the unbroken curved lines in the figure, and whose asymptotes are $y=0$ (found as in Ex. 4), and $x-3=0$ (since, when $x=3$, then $y=\infty$).

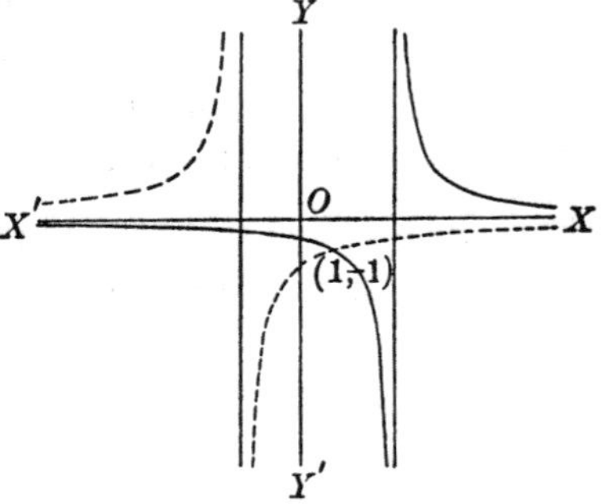

In a similar manner, we find the graph of (2) to be the hyperbola indicated by the dotted curve and having the asymptotes $y=0$ and $x+2=0$.

The equations (1), (2) have the single finite solution $x=1$, $y=-1$, the remaining three solutions being infinite, § 654.

The hyperbolas which are the graphs of (1), (2) meet in the single finite point (1, − 1). But since they have the common asymptote $y = 0$, they are regarded as having *two* infinitely distant coincident intersections at $(\infty, 0)$; and since they have the parallel asymptotes $x - 3 = 0$ and $x + 2 = 0$, they are regarded as having *one* infinitely distant intersection at $(0, \infty)$.

General discussion of such graphs. Generalizing the results obtained in the preceding examples, we reach the following conclusions. **668**

Suppose any equation of the second degree in x, y given, with real coefficients, as

$$ax^2 + 2hxy + by^2 + 2gx + 2fy + c = 0. \qquad (1)$$

If b is not 0, and we solve for y, we have

$$by = -(hx + f) \pm \sqrt{R}, \qquad (2)$$

where $\quad R = (h^2 - ab)x^2 + 2(hf - bg)x + (f^2 - bc).$

From (2) we obtain two real values of y for each value of x for which the radicand R is *positive*. Corresponding to these two values of y there are two points of the graph which may be found by drawing the line

$$by = -(hx + f)$$

and then increasing and decreasing its ordinate for the value of x in question by the value of $\sqrt{R}/b$. See § 667, Exs. 1, 3, 4.

The form of the graph depends on the character of the factors of R.

1. When $\quad (hf - bg)^2 - (h^2 - ab)(f^2 - bc) = 0.$

In this case R is a perfect square, § 635, and the first member of (1) can be resolved into factors of the first degree, § 635, Ex. 3. If these factors have real coefficients, the graph of (1) is a pair of right lines. See § 667, Ex. 5.

2. When $\quad (hf - bg)^2 - (h^2 - ab)(f^2 - bc) > 0.$

In this case, unless $h^2 - ab = 0$, the radicand R can be reduced to the form $R = (h^2 - ab)(x - \alpha)(x - \beta)$ (3) where α and β are *real* and $\alpha < \beta$, § 635.

If $h^2 - ab < 0$, the product (3) is positive when and only when x lies between α and β. Hence the graph of (1) will be a closed curve lying between the lines $x - \alpha = 0$ and $x - \beta = 0$, which it touches. It is therefore an *ellipse* (or circle). See § 667, Ex. 3.

If $h^2 - ab > 0$, the product (3) is positive when and only when $x < \alpha$ or $x > \beta$. Hence the graph will consist of two infinite branches, the one touching the line $x - \alpha = 0$ and extending to its left, the other touching the line $x - \beta = 0$ and extending to its right. It is therefore an *hyperbola*. See § 667, Ex. 4.

If $h^2 - ab = 0$, we have $R = 2(hf - bg)x + (f^2 - bc)$, where $hf - bg \neq 0$, and this is positive when and only when we have $x > -(f^2 - bc)/2(hf - bg)$. Hence the graph will consist of one infinite branch lying entirely to one side of the line $2(hf - bg)x + (f^2 - bc) = 0$, which it touches. It is therefore a *parabola*. See § 667, Ex. 1.

3. When $(hf - bg)^2 - (h^2 - ab)(f^2 - bc) < 0$.

In this case the roots of $R = 0$ are conjugate imaginaries, § 635, and if we call them $\lambda + \mu i$ and $\lambda - \mu i$, we can reduce R to the form $R = (h^2 - ab)[(x - \lambda)^2 + \mu^2]$, (4).

If $h^2 - ab > 0$, the product (4) is positive for *all* values of x. Hence the graph of (1) will consist of two infinite branches which lie on opposite sides of the line $by = -(hx + f)$. It is therefore an *hyperbola*. Thus, $y^2 - x^2 = 1$.

If $h^2 - ab < 0$, the product (4) is negative for all values of x. Hence the graph of (1) will be wholly imaginary. Thus, $x^2 + y^2 + 1 = 0$.

In the preceding discussion it is assumed that $b \neq 0$. But if $b = 0$, while $a \neq 0$, and we solve (1) for x instead of y, we arrive at similar conclusions. If both $a = 0$ and $b = 0$, the graph of (1) is an hyperbola, as in § 667, Ex. 7, or a pair of straight lines of which one is parallel to the x-axis, the other to the y-axis.

EXERCISE LII

Find the graphs of the following equations.

1. $y^2 = -8x$.
2. $x^2 + y^2 = 9$.
3. $(y - x)^2 = x$.
4. $x^2 + 2xy + 2y^2 = 8$.
5. $y^2 - 4xy + 3x^2 + 4x = 0$.
6. $y^2 - 2xy + 1 = 0$.
7. $y^2 - 2xy - 1 = 0$.
8. $2x^2 + 3y^2 - 4x + 6y = 0$.
9. $y^2 - x^2 - 3x + y - 2 = 0$.
10. $2x^2 + 4xy + 4y^2 + x + 4y - 5 = 0$.
11. $4x^2 - 12xy + 9y^2 + 3x - 6y = 0$.

Find the graphs of the following pairs of equations and their points of intersection.

12. $\begin{cases} xy = 1, \\ 3x - 5y = 2. \end{cases}$
13. $\begin{cases} x^2 - y^2 = 1, \\ x^2 - xy + x = 0. \end{cases}$
14. $\begin{cases} x^2 + y^2 = 3, \\ y^2 = 2x. \end{cases}$
15. $\begin{cases} y^2 - xy - 2x^2 - 2x - 2y - 2 = 0, \\ y^2 - xy - 2x^2 + 2 = 0. \end{cases}$
16. $\begin{cases} (x - 2y)(x + y) + x - 3y = 0, \\ (x - 2y)(x - y) + 2x - 6y = 0. \end{cases}$

17. Find the graph of $x^2 + y^2 - 6x - 2y + 1 = 0$ and its points of intersection with the axes of x and y.

18. Show that the graph of $(x - y)^2 - 2(x + y) + 1 = 0$ touches the x and y axes.

19. Show that the line $y = 3x + 5$ touches the graph of $16x^2 + y^2 - 16 = 0$ at the point $(-3/5, 16/5)$.

20. For what values of m will the line $y = mx + 3$ touch the graph of $x^2 + 2y^2 = 6$?

21. For what values of c will the line $7x - 4y + c = 0$ touch the graph of $3x^2 - y^2 + x = 0$?

22. Show that the lines $y = 0$ and $x - 2y + 1 = 0$ are the asymptotes of the graph of $xy - 2y^2 + y + 6 = 0$.

23. Find the asymptotes of the graph of the equation

$$2x^2 + 3xy - 2y^2 + x + 2y + 2 = 0.$$

24. For what values of λ is the graph of $x^2 + \lambda xy + y^2 = x$ an ellipse? a parabola? an hyperbola?

XVII. INEQUALITIES

669 **Single inequalities.** An *absolute inequality* is one like $x^2 + y^2 + 1 > 0$ which holds good for all real values of the letters involved; a *conditional inequality* is one like $x - 1 > 0$ which does not hold good for all real values of these letters, but, on the contrary, imposes a restriction upon them.

670 The principles which control the reckoning with inequalities are given in § 261. From these principles it follows that the sign $>$ or $<$ connecting the two members of an inequality will remain unchanged if we transpose a term, with its sign changed, from one member to the other, or if we multiply both members by the same *positive* number; but that the sign $>$ will be changed to $<$, or *vice versa,* if we multiply both members by the same *negative* number.

Example 1. Prove that $a^2 + b^2 > 2ab$.

We have $(a - b)^2 > 0$,

that is, $a^2 - 2ab + b^2 > 0$,

and therefore $a^2 + b^2 > 2ab$.

Example 2. Prove that $a^2 + b^2 + c^2 > ab + bc + ca$.

We have $a^2 + b^2 > 2ab$, $b^2 + c^2 > 2bc$, $c^2 + a^2 > 2ca$.

Adding the corresponding members of these three inequalities and dividing the result by 2, we have $a^2 + b^2 + c^2 > ab + bc + ca$.

Example 3. Solve the inequality $3x + 5 > x + 11$, that is, find what restriction it imposes on the value of x.

Transposing terms, $2x > 6$,

whence $x > 3$.

Example 4. Solve $x^2 - 2x - 3 < 0$.

Factoring, $(x + 1)(x - 3) < 0$.

To satisfy this inequality one factor must be positive and the other negative. Hence we must have $x > -1$ and < 3, that is, $-1 < x < 3$.

671 **Simultaneous inequalities.** A system of one or more inequalities of the form

$$ax + by + c > 0$$

may be solved for the variables x, y by a simple graphical method which is based upon the following consideration:

Draw the straight line which is the graph of $ax + by + c = 0$, § 385. Then for all pairs of values of x, y whose graphs lie on one side of this line we shall have $ax + by + c > 0$, and for all pairs whose graphs lie on the other side of the line we shall have $ax + by + c < 0$.

Thus, let (x_1, y_1) be a point on the graph of $y - (mx + c) = 0$ so that $y_1 - (mx_1 + c) \equiv 0$. Then, if $y_2 < y_1$ so that the point (x_1, y_2) lies *below* the line, we have $y_2 - (mx_1 + c) < 0$, and if $y_3 > y_1$ so that the point (x_1, y_3) lies *above* the line, we have $y_3 - (mx_1 + c) > 0$.

Example. Solve the simultaneous inequalities

$$k_1 = x - 2y + 1 < 0,\ k_2 = x + y - 5 < 0,\ k_3 = 2x - y - 1 > 0.$$

Find the graphs of $k_1 = 0$, $k_2 = 0$, $k_3 = 0$, as indicated in the figure.

The inequality $k_2 < 0$ is satisfied by those pairs of values of x, y whose graphs lie on the side of the line $k_2 = 0$ toward the origin; for when $x = 0$, $y = 0$, we have $k_2 = -5$, that is < 0. It may be shown in a similar manner that the inequalities $k_1 < 0$ and $k_3 > 0$ are satisfied by the pairs of values of x, y whose graphs lie on the sides of the lines $k_1 = 0$, $k_3 = 0$ remote from the origin.

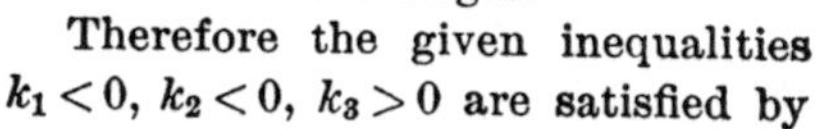

Therefore the given inequalities $k_1 < 0$, $k_2 < 0$, $k_3 > 0$ are satisfied by the pairs of values of x, y whose graphs lie within the triangle formed by the three lines.

EXERCISE LIII

In the following examples the letters a, b, c are supposed to denote unequal positive numbers.

1. Prove that $a/b + b/a > 2$.

2. Prove that $(a + b)(a^3 + b^3) > (a^2 + b^2)^2$.

3. Prove that $a^3 + b^3 > a^2b + ab^2$.

4. Prove that $a^2b + b^2a + b^2c + c^2b + c^2a + a^2c > 6abc$.

5. Prove that $a^3 + b^3 + c^3 > 3\,abc$.

6. Solve $x + 7 > 3\,x/2 - 8$.

7. Solve $2\,x^2 + 4\,x > x^2 + 6\,x + 8$.

8. Solve $(x+1)(x-3)(x-6) > 0$.

9. Solve $y - x - 2 < 0$, $x - 3 < 0$, $y + 1 > 0$ by the graphical method.

10. Also $y - x > 0$, $y - 2\,x < 0$.

11. Also $x + y + 3 > 0$, $y - 2\,x - 4 < 0$, $y + 2\,x + 4 > 0$.

12. Prove that $x^2 + 2\,x + 5 > 0$ is true for all values of x.

13. Solve $x^2 + y^2 - 1 < 0$, $y^2 - 4\,x < 0$ by a graphical method.

XVIII. INDETERMINATE EQUATIONS OF THE FIRST DEGREE

672 **Single equations in two variables.** Given any equation of the form

$$ax + by = c$$

where a, b, c denote integers, of which a and b have no common factor. We seek an expression for all pairs of *integral* values of x and y which will satisfy this equation; also such of these pairs as are *positive* as well as integral.

673 **Theorem 1.** *All equations* $ax + by = c$ *of the kind just described have integral solutions.*

For since a and b are prime to one another, by the method explained in § 491 we can find two integers p and q, positive or negative, such that $ap + bq = 1$ and therefore $a(pc) + b(qc) = c$, and this proves that $x = pc$, $y = qc$ is a solution of $ax + by = c$.

674 **Theorem 2.** *If* $x = x_0$, $y = y_0$ *be one integral solution of such an equation* $ax + by = c$, *all of its integral solutions are given by the formulas*

$$x = x_0 + bt,\quad y = y_0 - at$$

when all possible integral values are assigned to t.

First, $x = x_0 + bt$, $y = y_0 - at$ is always a solution of $ax + by = c$. (1)

For, substituting in (1), $a(x_0 + bt) + b(y_0 - at) = c$,

or, simplifying, $ax_0 + by_0 = c$,

which is true since, by hypothesis, $x = x_0$, $y = y_0$ is a solution of (1).

Second, every integral solution of (1) is given by $x = x_0 + bt$, $y = y_0 - at$.

For let $x = x_1$, $y = y_1$ denote any second integral solution.

Then $ax_1 + by_1 = c$ and $ax_0 + by_0 = c$,

whence, subtracting, $b(y_1 - y_0) = -a(x_1 - x_0)$. (2)

From (2) it follows that b is a factor of the product of the integers a and $x_1 - x_0$. Therefore, since b is prime to a, it must be exactly contained in $x_1 - x_0$, § 492, 1, and if we call the quotient t', we have

$$x_1 - x_0 = bt', \text{ or } x_1 = x_0 + bt'. \quad (3)$$

And substituting (3) in (2) and simplifying, we also have

$$y_1 = y_0 - at'. \quad (4)$$

675 From §§ 673, 674 it follows that every equation $ax + by = c$ of the kind just described has *infinitely many integral solutions.* When a and b have contrary signs there are also infinitely many *positive* solutions; but when a and b have the same sign there is but a limited number of such solutions or no such solution.

Thus, one solution of $2x + 3y = 18$ is $x = 3$, $y = 4$.

Hence the general solution is $x = 3 + 3t$, $y = 4 - 2t$.

The positive solutions correspond to $t = -1, 0, 1, 2$ and are $x, y = 0, 6$; $3, 4$; $6, 2$; $9, 0$.

676 The theorem of § 674 enables one to write down the general integral solution of an equation of the kind under consideration as soon as a particular solution is known. A particular solution may often be found by inspection. Thus, one solution of $10x + 3y = 12$ is $x = 0$, $y = 4$. A particular solution may always be found by the method indicated in § 673; also by the method illustrated in the following example.

Example. Find the integral solutions of $7x + 19y = 213$. (1)

Solving for the variable with the smaller coefficient, here x, and reducing, we have

$$x = \frac{213 - 19y}{7} = 30 - 2y + \frac{3 - 5y}{7}. \quad (2)$$

Hence if x is to be an integer when y is one, $(3-5y)/7$ must be an integer. Call this integer u, so that $(3-5y)/7=u$.

Then $$5y+7u=3. \tag{3}$$

Treating (3) as we have just treated (1), we have

$$y=\frac{3-7u}{5}=-u+\frac{3-2u}{5}. \tag{4}$$

Set $(3-2u)/5$, which must be integral, equal to v.

Then $$2u+5v=3. \tag{5}$$

Treating (5) as we have already treated (1) and (3), we have

$$u=\frac{3-5v}{2}=1-2v+\frac{1-v}{2}. \tag{6}$$

When $v=1$ the fractional term $(1-v)/2$ vanishes and u has the integral value -1.

Substituting $u=-1$ in (4), we obtain $y=2$.

Substituting $y=2$ in (2), we obtain $x=25$.

Hence the general integral solution of (1) is

$$x=25+19t,\ y=2-7t.$$

There are two positive solutions corresponding to $t=-1$ and $t=0$ respectively, namely: $x, y=6, 9$; $25, 2$.

Observe that in the fractional terms of (2), (4), (6) the numerical values of the coefficients of y, u, v, namely 5, 2, 1, are merely the successive remainders occurring in the process of finding the greatest common divisor of the given coefficients 7 and 19. We finally obtain the remainder, or coefficient, 1, because 7 and 19 have no common factor. The like will be true if we apply the method to any equation $ax+by=c$ in which a and b have no common factor. Hence the method will always yield a solution of such an equation.

But in practice it is seldom necessary to complete the reckoning above indicated. Thus, having obtained (4), we might have observed that $u=-1$ will make $(3-2u)/5$ integral, which would have at once given us $y=2$ and therefore $x=25$, by (2).

677 Observe that an equation $ax+by=c$ with integral coefficients of which a and b have a common factor, as d, can have no integral solution unless d is also a factor of c. For if x and y were integers, d would be a factor of $ax+by$ and therefore of c. Thus, $4x+6y=7$ has no integral solution.

Simultaneous equations. The following example will illustrate a method for finding the integral solutions, if there be any, of a pair of simultaneous equations in three variables with integral coefficients. **678**

Example. Find the integral solutions of

$$3x + 6y - 2z = 22, \qquad (1)$$

$$5x + 8y - 6z = 28. \qquad (2)$$

First eliminate z and simplify the resulting equation.

We obtain $$2x + 5y = 19. \qquad (3)$$

Next find the general solution of (3), as in § 676.

We obtain $$x = 7 + 5t, \quad y = 1 - 2t. \qquad (4)$$

Next substitute (4) in (1) and simplify the result.

We obtain $$2z - 3t = 5. \qquad (5)$$

Next find the general solution of (5).

We obtain $$z = 1 - 3u, \quad t = -1 - 2u, \qquad (6)$$

where u denotes any integer whatsoever.

Finally substitute $t = -1 - 2u$ in (4) and simplify.

We obtain $$x = 2 - 10u, \quad y = 3 + 4u, \quad z = 1 - 3u, \qquad (7)$$

which is the general solution required.

The only *positive* solution is that corresponding to $u = 0$, namely $x = 2$, $y = 3$, $z = 1$.

Observe that the given equations will have no integral solution if either of the derived equations in two variables has none, § 677.

We proceed in a similar manner if given *three* equations in *four* variables, and so on.

Single equations in more than two variables. The following example illustrates a method of obtaining formulas for the integral solutions of a single equation in more than two variables with integral coefficients. **679**

Example. Find the integral solutions of $5x + 8y + 19z = 50$. (1)

Solving for x, $$x = 10 - y - 3z - \frac{3y + 4z}{5}. \qquad (2)$$

Set $(3y + 4z)/5$, which must be integral, equal to u.

Then $$3y + 4z = 5u. \qquad (3)$$

Solving for y, $$y = u - z + \frac{2u - z}{3}. \qquad (4)$$

Set $(2u - z)/3$, which must be integral, equal to v.

Then $$z = 2u - 3v. \qquad (5)$$

Substituting (5) in (4), $$y = -u + 4v. \qquad (6)$$

Substituting (5) and (6) in (2),

$$x = 10 - 6u + 5v. \qquad (7)$$

The formulas (5), (6), (7), in which u, v may have any integral values whatsoever, constitute the general solution required.

Substituting $u = 2$, $v = 1$ in the formulas (5), (6), (7), we obtain a positive solution of (1), namely $x = 3$, $y = 2$, $z = 1$.

EXERCISE LIV

Find the general integral solutions of the following; also the positive integral solutions.

1. $6x - 17y = 18$.
2. $43x - 12y = 158$.
3. $16x + 39y = 1$.
4. $72x + 23y = 845$.
5. $49x - 27y = 28$.
6. $47x - 97y = 501$.
7. $\begin{cases} 2x + 5y - 8z = 27, \\ 3x + 2y + z = 11. \end{cases}$
8. $\begin{cases} 5x + 2y = 42, \\ 3y - 7z = 2. \end{cases}$
9. $4x + 3y = 2z + 3$.
10. $2x + 3y + 4z = 17$.

11. Find the number of the positive integral solutions of the equation $3x + 7y = 1043$.

12. Reduce the fraction $41/35$ to a sum of two positive fractions whose denominators are 5 and 7.

13. A man buys calves at $7 a head and lambs at $6 a head. He spends in all $110. How many does he buy of each?

14. Separate 23 into three parts such that the sum of three times the first part, twice the second part, and five times the third part will be 79.

15. Find the smallest number which when divided by 5, 7, 9 will give the remainders 4, 6, 8.

16. Two rods of equal length are divided into 250 and 253 equal divisions respectively. If one rod is laid along the other so that their ends coincide, which divisions will be nearest together?

XIX. RATIO AND PROPORTION. VARIATION

RATIO AND PROPORTION

Ratio. In arithmetic and algebra it is customary to extend the use of the word *ratio*, § 215, to numbers; and, if a and b denote any two numbers, to define the ratio of a to b as the quotient a/b. (Compare § 216.) **680**

The ratio of a to b is denoted by a/b or by $a:b$.

In the ratio $a:b$ we call a the *antecedent* and b the *consequent*.

Properties of ratios. Since ratios such as $a:b$ are fractions, their properties are the properties of fractions. Hence **681**

The value of a ratio is not changed when both of its terms are multiplied or divided by the same number.

Thus, $a:b = ma:mb = a/n:b/n$.

On the other hand, except when $a = b$, the value of $a:b$ *is* changed when both terms are raised to the same power, or when the same number is added to both. In particular,

If a, b, *and* m *are positive, the ratio* a:b *is increased by adding* m *to both* a *and* b *when* $a < b$; *decreased, when* $a > b$.

For
$$\frac{a+m}{b+m} - \frac{a}{b} = \frac{m(b-a)}{b(b+m)},$$

and $m(b-a)/b(b+m)$ is positive or negative according as $a < b$ or $a > b$.

Proportion. When the ratios $a:b$ and $c:d$ are equal, the four numbers a, b, c, d are said to be *in proportion*, or to be *proportional*. **682**

This proportion may be written in any of the ways

$$a/b = c/d, \text{ or } a:b = c:d, \text{ or } a:b::c:d.$$

It is read "a is to b as c is to d."

In the proportion $a:b = c:d$, the terms a and d are called the *extremes*, and b and c the *means*. Again, d is called the *fourth proportional* to a, b, and c.

683 **Theorem.** *In any proportion the product of the extremes is equal to that of the means; that is,*

If $\quad a : b = c : d,$ *then* $ad = bc.$

For from $a/b = c/d$ we obtain $ad = bc$ by merely clearing of fractions.

Example. The first, second, and fourth terms of a proportion are $1/2$, -3, and 5 respectively; find the third term.

Calling the third term x, $\quad 1/2 : -3 = x : 5.$

Hence $\quad 5 \cdot 1/2 = -3 \cdot x,$

or, solving for x, $\quad x = -5/6.$

684 **Conversely,** *if the product of a first pair of numbers be equal to that of a second pair, the four numbers will be in proportion when arranged in any order which makes one of the pairs means and the other extremes.*

For, let $\quad ad = bc.$

Dividing both members by bd, we have $a/b = c/d$. Hence

$$a : b = c : d \quad (1) \qquad \text{and} \qquad c : d = a : b. \quad (2)$$

Similarly by dividing both members of $ad = bc$ by cd, ab, and ac in turn, we obtain

$$a : c = b : d \quad (3) \qquad \text{and} \qquad b : d = a : c, \quad (4)$$

$$d : b = c : a \quad (5) \qquad \text{and} \qquad c : a = d : b, \quad (6)$$

$$d : c = b : a \quad (7) \qquad \text{and} \qquad b : a = d : c. \quad (8)$$

685 **Allowable rearrangements of the terms of a proportion.** From §§ 683, 684 it follows that if a, b, c, d are in proportion when arranged in any one of the orders (1)–(8), they will also be in proportion when arranged in any other of these orders. In particular,

1. *In any proportion the terms of both ratios may be interchanged.*

Thus, if $\quad a : b = c : d,$ then $b : a = d : c.$

2. *In any proportion either the means or the extremes may be interchanged.*

Thus, if $\quad a : b = c : d,$ then $a : c = b : d.$

The transformations 1 and 2 are called *inversion* and *alternation* respectively.

Other allowable transformations of a proportion. **686**

If we know that $a:b=c:d$, we may conclude that

1. $a+b:b=c+d:d$. 2. $a-b:b=c-d:d$.

3. $a+b:a-b=c+d:c-d$.

4. $ma:mb=nc:nd$. 5. $ma:nb=mc:nd$.

6. $a^n:b^n=c^n:d^n$.

For in 1 take the product of the means and extremes and we have $ad+bd=bc+bd$, that is $ad=bc$, which is true since $a:b=c:d$. Hence 1 is true, § 288. The truth of 2–6 may be proved in a similar manner.

The transformations of $a:b=c:d$ into the forms 1, 2, 3 are called *composition*, *division*, and *composition and division* respectively.

Example. Solve $x^2+2x+3:x^2-2x-3=2x^2+x-1:2x^2-x+1$.

By 3, $$2x^2:2(2x+3)=4x^2:2(x-1).$$

Hence $$x^2=0, \quad (1)$$

or by 4, 5, $$1:2x+3=2:x-1. \quad (2)$$

Solving (1) and (2), $$x=0,\ 0,\ \text{or}\ -7/3.$$

Theorem. *In a series of equal ratios any antecedent is to its consequent as the sum of all the antecedents is to the sum of all the consequents.* **687**

Thus, if $$a_1:b_1=a_2:b_2=a_3:b_3,$$

then $$a_1:b_1=a_1+a_2+a_3:b_1+b_2+b_3.$$

For let r denote the common value of the equal ratios. We then have

$$a_1/b_1=r,\ a_2/b_2=r,\ a_3/b_3=r.$$

Hence $$a_1=rb_1,\ a_2=rb_2,\ a_3=rb_3,$$

or, adding, $$a_1+a_2+a_3=r(b_1+b_2+b_3).$$

Therefore $$\frac{a_1+a_2+a_3}{b_1+b_2+b_3}=r=\frac{a_1}{b_1}.$$

Example 1. If $x:(b-c)yz = y:(c-a)zx = z:(a-b)xy,$
then $x^2+y^2+z^2=0.$

For multiplying the terms of the first ratio by x, those of the second by y, and those of the third by z, and then applying our theorem, we have

$$\frac{x^2}{(b-c)xyz} = \frac{y^2}{(c-a)xyz} = \frac{z^2}{(a-b)xyz} = \frac{x^2+y^2+z^2}{0},$$

which evidently requires that $x^2+y^2+z^2=0$.

The device employed in the proof just given will be found useful in dealing with complicated problems in proportion.

Example 2. Prove that if $a:b = x:y,$
then $a^3+2b^3:ab^2 = x^3+2y^3:xy^2.$

Set $a/b = x/y = r$, so that $a = rb$ and $x = ry$.

Then $(a^3+2b^3)/ab^2 = (r^3b^3+2b^3)/rb^3 = (r^3+2)/r,$
and $(x^3+2y^3)/xy^2 = (r^3y^3+2y^3)/ry^3 = (r^3+2)/r.$

688 Continued proportion. The numbers $a, b, c, d, \cdots$ are said to be in continued proportion if $a:b = b:c = c:d = \cdots$.

If three numbers a, b, c are in continued proportion, so that $a:b=b:c$, then b is called a *mean proportional* to a and c, and c is called a *third proportional* to a and b.

If a, b, c *are in continued proportion, then* $b^2 = ac$.

For since $a:b=b:c$, we have $b^2 = ac$, § 683.

EXERCISE LV

1. Find a fourth proportional to 15, 24, and 20; a third proportional to 15 and 24; a mean proportional between $5a^3b^2$ and $20ab^2$; a mean proportional between $\sqrt{12}$ and $\sqrt{75}$.

2. If $3x-2y = x-5y$, find $x:y$; also $x+y:x-y$.

3. If $2x^2-5xy-3y^2=0$, find $x:y$; also $y:x$.

4. If $ax+by+cz=0$ and $a'x+b'y+c'z=0,$
then $x:y:z = bc'-b'c : ca'-c'a : ab'-a'b.$

5. If $a:b=c:d$, then $ab+cd$ is a mean proportional between a^2+c^2 and b^2+d^2.

6. If $(a^2+b^2)cd = (c^2+d^2)ab$, then either $a:b=c:d$ or $a:b=d:c$.

7. If $a : b = c : d$, then $\sqrt{a} + \sqrt{b} : \sqrt{a+b} = \sqrt{c} + \sqrt{d} : \sqrt{c+d}$.

8. If $\frac{x}{a} = \frac{y}{b} = \frac{z}{c}$, then $\frac{x^3}{a^3} + \frac{y^3}{b^3} + \frac{z^3}{c^3} = 3\frac{(x+y+z)^3}{(a+b+c)^3}$.

9. If the numbers $a_1, a_2, \cdots, a_n$; $b_1, b_2, \cdots, b_n$; $l_1, l_2, \cdots, l_n$ are all positive, the ratio $l_1a_1 + l_2a_2 + \cdots + l_na_n : l_1b_1 + l_2b_2 + \cdots + l_nb_n$ is intermediate in value to the greatest and least of the ratios $a_1 : b_1$, $a_2 : b_2, \cdots, a_n : b_n$.

10. If $a - b : k = b - c : l = c - a : m$, and a, b, c are unequal, then $k + l + m = 0$.

11. If $x : mz - ny = y : nx - lz = z : ly - mx$, then $lx + my + nz = 0$ and $x^2 + y^2 + z^2 = 0$.

12. If $a_1 : b_1 = a_2 : b_2 = a_3 : b_3$, then each of these ratios is equal to $(l_1a_1^n + l_2a_2^n + l_3a_3^n)^{\frac{1}{n}} : (l_1b_1^n + l_2b_2^n + l_3b_3^n)^{\frac{1}{n}}$.

13. By aid of § 686 solve each of the following equations.

(1) $\frac{x^2 + ax - a}{x^2 - ax + a} = \frac{2x^2 + a}{2x^2 - a}$.

(2) $\frac{2x^3 - 3x^2 + 2x + 2}{2x^3 - 3x^2 - 2x - 2} = \frac{3x^3 - x^2 + 10x - 26}{3x^3 - x^2 - 10x + 26}$.

14. Separate 520 into three parts in the ratios $2 : 3 : 5$.

15. Two casks A and B are filled with two kinds of sherry mixed in A in the ratio $3 : 5$, in B in the ratio $3 : 7$. What amount must be taken from each cask to form a mixture which shall consist of 6 gallons of one kind and 12 gallons of the other kind?

VARIATION

One independent variable. If two variables y and x are so related that however their values may change their ratio remains constant, then y is said to *vary as* x, or y and x are said to vary proportionally. **689**

More briefly, y is said to vary as x when $y/x = c$, or $y = cx$, where c denotes a constant.

The notation $y \propto x$ means "y varies as x."

If given that y varies as x, we may at once write $y = cx$; and if also given one pair of corresponding values of x and **690**

y, we may find c. The equation connecting y and x is then known, and from it we may compute the value of y which corresponds to any given value of x.

Example. If y varies as x, and $y = 12$ when $x = 2$, what is the value of y when $x = 20$?

We have $$y = cx,$$
and, by hypothesis, this equation is satisfied when $y = 12$, $x = 2$.

Hence $$12 = c \cdot 2, \text{ that is } c = 6.$$

Therefore $$y = 6x.$$

Hence when $x = 20$ we have $y = 6 \cdot 20 = 120$.

691 Instead of varying as x itself, y may vary as some function of x, for example as x^2, or as $x + 1$, or as $1/x$. In particular, if y varies as $1/x$, we say that y varies *inversely* as x.

Example. Given that y is the sum of a constant and a term which varies inversely as x; also that $y = 1$ when $x = -1$, and $y = 5$ when $x = 1$. Find the equation connecting x and y.

By hypothesis, $y = a + b/x$, where a and b are constants.

Since this equation is satisfied by $x = -1$, $y = 1$, and by $x = 1$, $y = 5$, we have $$1 = a - b \text{ and } 5 = a + b.$$

Hence $a = 3$, $b = 2$, and the required equation is $y = 3 + 2/x$.

692 **More than one independent variable.** Let x and y denote variables which are independent of one another. If a third variable z varies as the product xy, so that $z = cxy$, we say that z *varies as* x *and* y *jointly*; and if z varies as the quotient x/y, so that $z = c \cdot x/y$, we say that z *varies directly as* x *and inversely as* y.

Thus, the area of a rectangle varies as the lengths of its base and altitude jointly; and the length of the altitude varies directly as the area and inversely as the length of the base.

693 **Theorem.** *If when* x *is constant* z *varies as* y, *and when* y *is constant* z *varies as* x, *then when both* x *and* y *vary*, z *varies as the product* xy.

For select any three pairs of values of x and y, such as x_1, y_1; x_2, y_2; x_1, y_2; and let z_1, z_2, z_3 denote the corresponding values of z, so that

$$x_1,\ y_1,\ z_1, \tag{1}$$

$$x_1,\ y_2,\ z_3, \tag{2}$$

$$x_2,\ y_2,\ z_2 \tag{3}$$

are sets of corresponding values of the three variables.

Then since the value of x is the same in (1) as in (2), and, by hypothesis, for any *given* value of x, z varies as y, we have, § 689,

$$z_1 / y_1 = z_3 / y_2. \tag{4}$$

Similarly since y_2 is common to (2) and (3), we have

$$z_3 / x_1 = z_2 / x_2. \tag{5}$$

Multiplying together the corresponding members of (4) and (5),

$$z_1 / x_1 y_1 = z_2 / x_2 y_2. \tag{6}$$

Therefore corresponding values of z and xy are proportional; that is, z varies as xy, § 689.

EXERCISE LVI

1. If y varies as x, and $y = -2$ when $x = 5$, what is the value of y when $x = 7$?

2. If y varies inversely as x^2, and $y = 1$ when $x = 2$, for what values of x will $y = 3$?

3. Given that y is the sum of a constant and a term which varies as x^2; also that $y = 1$ when $x = 1$, and $y = 0$ when $x = 2$. Find the equation connecting x and y.

4. If y varies directly as x^2 and inversely as z^3, and $y = 1$ when $x = -1$ and $z = 2$, what is the value of y when $x = 3$ and $z = -1$?

5. If y varies as x, show that $x^2 - y^2$ varies as xy.

6. If the square of y varies as the cube of z, and z varies inversely as x, show that xy varies inversely as the square root of x.

7. The wages of 3 men for 4 weeks being $108, how many weeks will 5 men work for $135?

8. The volume of a circular disc varies as its thickness and the square of the radius of its face jointly. Two metallic discs having the thicknesses 3 and 2 and the radii 24 and 36 respectively are melted and recast in a single disc having the radius 48. What is its thickness?

9. A right-circular cone whose altitude is a is cut by a plane drawn parallel to its base. How far is the plane from the vertex of the cone when the area of the section is half that of the base? How far is the plane from the vertex when it divides the cone into two equivalent parts?

XX. ARITHMETICAL PROGRESSION

694 Arithmetical progression. This name is given to a sequence of numbers which may be derived from a given number a by repeatedly adding a given number d, that is, to any sequence which may be written in the form

$$a, \ a + d, \ a + 2d, \ \cdots, \ a + (n-1)d. \qquad \text{(I)}$$

Since d is the difference between every two consecutive terms of (I), it is called the *common difference* of this arithmetical progression.

Thus, 2, 5, 8, 11 is an arithmetical progression in which $d = 3$, and 2, -1, -4, -7 is an arithmetical progression in which $d = -3$.

695 The nth term. Observe that in (I) the coefficient of d in each of the terms a, $a + d$, $a + 2d$, $\cdots$ is one less than the number of the term. Hence the general or mth term is $a + (m-1)d$; and if the entire number of terms is n and we call the last term l, we have the formula

$$l = a + (n-1)d. \qquad \text{(II)}$$

Example. The *seventh* term of an arithmetical progression is 15 and its *tenth* term is 21; find the first term a and the common difference d; and if the entire number of terms is 20, find l.

We have $\quad a + 6d = 15$ and $a + 9d = 21$.

Solving for a and d, $\quad a = 3, \ d = 2$.

Hence $\quad l = 3 + 19.2 = 41$.

696 The sum. Evidently the next to the last term of (I) may be written $l - d$, the term before that, $l - 2d$, $\cdots$, the first term, $l - (n-1)d$.

Hence, if S denote the sum of the terms of (I), we have

$$S = a + (a+d) + (a+2d) + \cdots + [a + (n-1)d],$$
$$S = l + (l-d) + (l-2d) + \cdots + [l - (n-1)d].$$

Adding the corresponding members of these two equations, we obtain $2S = n(a + l)$. Therefore

$$S = \frac{n}{2}(a + l). \qquad \text{(III)}$$

Example. Find the sum of an arithmetical progression of six terms whose first term is 5 and whose common difference is 4.

Since $n = 6$, we have $\quad l = 5 + 5 \cdot 4 = 25.$

Hence $\quad S = \frac{6}{2}(5 + 25) = 90.$

Applications. If in an arithmetical progression any three of the five numbers a, l, d, n, S are given, the formulas (II) and (III) enable us to find the other two. The only restriction on the given numbers is that they be such as will lead to positive integral values of n. **697**

Example. Given $d = 1/2$, $l = 3/2$, $S = -15/2$; find a and n.

Substituting in (II), (III), $\quad \frac{3}{2} = a + \frac{n-1}{2}, \qquad (1)$

$$-\frac{15}{2} = \frac{n}{2}\left(a + \frac{3}{2}\right). \qquad (2)$$

Eliminating a, $\quad n^2 - 7n - 30 = 0. \qquad (3)$

Solving (3), $\quad n = 10 \text{ or } -3.$

The value $n = -3$ is inadmissible. Substituting $n = 10$ in (1), we obtain $a = -3$. Hence $n = 10$, $a = -3$, and the arithmetical progression is $-3, -2\frac{1}{2}, -2, -1\frac{1}{2}, -1, -\frac{1}{2}, 0, \frac{1}{2}, 1, 1\frac{1}{2}$.

Arithmetical means. If three numbers form an arithmetical progression, the middle number is called the *arithmetical mean* of the other two. **698**

The arithmetical mean of any two numbers a *and* b *is one half of their sum.*

For if x be the arithmetical mean of a and b, then the sequence a, x, b is an arithmetical progression.

Hence $\quad x - a = b - x,$

and therefore $\quad x = (a + b)/2.$

In any arithmetical progression all the intermediate terms may be called the arithmetical means of the first and last terms. It is always possible to insert or "interpolate" any number of such means between two given numbers a and b.

Example. Interpolate four arithmetical means between 3 and 5.

We are asked to find the intermediate terms of an arithmetical progression in which $a = 3$, $l = 5$, and $n = 4 + 2$ or 6.

Substituting $l = 5$, $a = 3$, $n = 6$ in (II), we have

$$5 = 3 + 5d, \text{ whence } d = 2/5.$$

Hence the required means are $3\frac{2}{5}$, $3\frac{4}{5}$, $4\frac{1}{5}$, $4\frac{3}{5}$.

EXERCISE LVII

1. Find the twentieth term and the sum of the first twenty terms of 3, 6, 9, $\cdots$; of -3, $-1\frac{1}{2}$, 0, $\cdots$.

2. Find a formula for the sum to n terms of 1, 2, 3, $\cdots$; of 1, 3, 5, $\cdots$; of 2, 4, 6, $\cdots$.

3. Find the sum of the first n numbers of the form $6r + 1$, where r denotes 0 or a positive integer.

4. Find the arithmetical progression of ten terms whose fifth term is 1 and whose eighth term is 2.

5. Insert five arithmetical means between -1 and 2

6. Given $n = 16$, $a = 0$, $d = 4/3$; find l and S.

7. Given $n = 7$, $l = -7$, $d = -5/3$; find a and S.

8. Given $n = 12$, $a = -5/3$, $l = 31\frac{1}{3}$; find d and S.

9. Given $a = 2$, $l = -23\frac{1}{2}$, $S = -559$; find n and d.

10. Given $n = 7$, $a = 3/7$, $S = 45$; find d and l.

11. Given $a = 4$, $d = 1/5$, $l = 9\frac{3}{5}$; find n and S.

12. Given $n = 9$, $d = -4$, $S = 135$; find a and l.

13. Given $n = 10$, $l = -2$, $S = 115$; find a and d.

14. Given $d = 5$, $l = -47$, $S = -357$; find n and a.

15. Given $a = -10$, $d = 7$, $S = 20$; find n and l.

16. Show that if a^2, b^2, c^2 are in arithmetical progression, so also are $1/(b + c)$, $1/(c + a)$, $1/(a + b)$.

17. Show that the sum of any n consecutive integers is divisible by n, if n be odd.

18. Find an arithmetical progression such that the sum of the first three terms is one half the sum of the next four terms, the first term being 1.

19. Three numbers are in arithmetical progression. Their sum is 15 and the sum of their squares is 83. Find these numbers.

20. Find the sum of all positive integers of three digits which are multiples of 9.

21. If a person saves $130 a year and at the end of the year puts this sum at simple interest at 4%, to how much will his savings amount at the end of 11 years?

22. Two men A and B set out at the same time from two places 72 miles apart to walk toward one another. If A walks at the rate of 4 miles an hour, while B walks 2 miles the first hour, $2\frac{1}{2}$ miles the second hour, 3 miles the third hour, and so on, when and where will they meet?

XXI. GEOMETRICAL PROGRESSION

699 **Geometrical progression.** This name is given to a sequence of numbers which may be derived from a given number a by repeatedly multiplying by a given number r, that is, to any sequence which may be written in the form

$$a, ar, ar^2, \cdots, ar^{n-1}. \qquad \text{(I)}$$

We call r the *common ratio* of the geometrical progression (I) and say that the progression is *increasing* or *decreasing* according as r is numerically greater or less than 1.

Thus, 1, 2, 4, 8 and 1, -2, 4, -8 are increasing geometrical progressions in which $r = 2$ and -2 respectively; while 1, 1/2, 1/4, 1/8 is a decreasing geometrical progression in which $r = 1/2$.

700 **The nth term.** Observe that the exponent of r in each term of (I) is one less than the number of the term. Hence in a geometrical progression of n terms whose first term is a and whose ratio is r, the formula for the last term l is

$$l = ar^{n-1}. \qquad \text{(II)}$$

701 **The sum.** Let S denote the sum of the geometrical progression (I).

Then $$S = a + ar + ar^2 + \cdots + ar^{n-2} + ar^{n-1}$$

and $$rS = \quad ar + ar^2 + \cdots + ar^{n-2} + ar^{n-1} + ar^n.$$

Subtracting the second of these equations from the first, we obtain $(1 - r)S = a - ar^n$. Therefore

$$S = \frac{a(1 - r^n)}{1 - r}. \qquad \text{(III)}$$

In applying this formula to an increasing geometrical progression we may more conveniently write it thus:

$$S = a(r^n - 1)/(r - 1).$$

From (II) we obtain $rl = ar^n$. Hence (III) may also be written thus: $S = (a - rl)/(1 - r)$, or $S = (rl - a)/(r - 1)$.

Example. In the geometrical progression 2, -4, 8, -16, $\cdots$ to eight terms, find l and S.

Here $\quad a = 2,\ r = -2,$ and $n = 8.$

Hence, by (II), $\quad l = 2(-2)^7 = -256,$

and, by (III), $$S = 2\frac{1 - (-2)^8}{1 - (-2)} = -\frac{510}{3}.$$

702 **Applications.** If in a geometrical progression any three of the five numbers a, l, r, n, S are given, the formulas (II) and (III) determine the other two. Moreover these two numbers can be actually found by methods already explained, except when the given numbers are a, n, S or l, n, S. If one of the unknown numbers is n, it must be found by inspection; but this is always possible if admissible values have been assigned to the given numbers, since n will then be a positive integer.

Example 1. Given $r = 3$, $n = 6$, $S = 728$; find a and l.

Substituting the given values in (II) and (III), we have

$$l = a \cdot 3^5 = 243a, \text{ and } 728 = a\frac{3^6 - 1}{3 - 1} = 364a.$$

Solving these equations, $\quad a = 2,\ l = 486.$

Example 2. Given $a = 6$, $n = 5$, $l = 2/27$; find r and S.

By (II), $2/27 = 6r^4$, whence $r^4 = 1/81$, or $r = \pm 1/3$.

Therefore, by (III), if $r = 1/3$, then $S = 6\dfrac{1-(1/3)^5}{1-1/3} = \dfrac{242}{27}$,

and if $r = -1/3$, then $S = 6\dfrac{1-(-1/3)^5}{1-(-1/3)} = \dfrac{122}{27}$.

Hence there are *two* geometrical progressions in which $a = 6$, $n = 5$, and $l = 2/27$.

Example 3. Given $a = -3$, $l = -46875$, $S = -39063$; find r and n.

Substituting in the formula $S = (a - rl)/(1 - r)$, § 701, we have

$$-39063 = \frac{-3 + 46875r}{1-r}, \text{ whence } r = -5.$$

Therefore, by (II), $-46875 = -3(-5)^{n-1}$, or $(-5)^{n-1} = 15625$. But by factoring 15625 we find $15625 = 5^6 = (-5)^6$.

Hence $n - 1 = 6$, that is $n = 7$.

Example 4. Given $a = 3$, $n = 5$, $S = 93$; find r and l.

By (III), $93 = 3\dfrac{1-r^5}{1-r} = 3(1 + r + r^2 + r^3 + r^4)$.

Hence $r^4 + r^3 + r^2 + r - 30 = 0$.

Thus this problem involves solving an equation of the fourth degree; and, in general, when a, n, S are the given numbers, to find r we must solve an equation of the degree $n - 1$. In this particular case, however, we may find one value of r by the method of § 455. It is 2.

Substituting $r = 2$ in (II), we have $l = 3 \cdot 2^4 = 48$.

Geometrical means. If three numbers form a geometrical progression, the middle number is called the *geometrical mean* of the other two. **703**

The geometrical mean of any two numbers a *and* b *is a square root of their product.*

For if x be the geometrical mean of a and b, the sequence a, x, b is a geometrical progression.

Hence $x/a = b/x$ and therefore $x = \pm\sqrt{ab}$.

In any geometrical progression all the intermediate terms may be called the geometrical means of the first and last terms.

We may insert any number of such means between two given numbers a and b, as in the following example.

Example. Insert four geometrical means between 18 and 2/27.

It is required to find the intermediate terms of a geometrical progression in which $a = 18$, $l = 2/27$, and $n = 4 + 2 = 6$.

Substituting the given values in (II), we have

$$2/27 = 18r^5, \text{ whence } r = 1/3.$$

Hence the means are 6, 2, 2/3, 2/9.

704 Infinite decreasing geometric series. We call an expression of the form

$$a + ar + ar^2 + \cdots + ar^{n-1} + \cdots, \tag{1}$$

supposed continued without end, an *infinite geometric series.*

By the formula (III), the sum of the first n terms of (1) is $a(1 - r^n)/(1 - r)$.

Suppose that $r < 1$ numerically. Then, as n is indefinitely increased, r^n will approach 0 as limit, § 724, and therefore $a(1 - r^n)/(1 - r)$ will approach $a/(1 - r)$ as limit. We call this limit the *sum* of the infinite series (1). Hence, if S denote the sum of (1), we have

$$S = \frac{a}{1 - r}. \tag{2}$$

Example 1. Find the sum of $1 + 1/2 + 1/4 + 1/8 + \cdots$.

Here $a = 1$ and $r = 1/2$.

Hence $S = 1/[1 - 1/2] = 2$.

Example 2. Find the value of the recurring decimal $.72323\cdots$.

The part which recurs may be written $\frac{23}{1000} + \frac{23}{100000} + \cdots$, and, by (2), the sum of this infinite series is $\frac{.023}{1 - .01}$ or $\frac{23}{990}$. Adding .7, the part which does not recur, we obtain for the value of the given decimal $\frac{358}{495}$.

EXERCISE LVIII

1. Find the fifth term and the sum of the first five terms of the geometrical progression $2, -6, 18, \cdots$.

2. Find the fourth term and the sum of the first four terms of the geometrical progression $4, 6, 9, \cdots$.

3. Find the sums of the following infinite geometric series:

$$12 - 6 + 3 - \cdots;\ 1 - \tfrac{1}{2} + \tfrac{1}{4} - \cdots;\ \tfrac{5}{3} + \tfrac{1}{3} + \tfrac{1}{15} + \cdots.$$

4. Find the values of the following recurring decimals:

$.341341\cdots,\qquad .0567272\cdots,\qquad 8.45164516\cdots.$

5. Given $a = -.03$, $r = 10$, $n = 6$; find l and S.

6. Given $n = 7$, $a = 48$, $l = 3/4$; find r and S.

7. Given $a = 1/16$, $r = 2$, $l = 8$; find n and S.

8. Given $n = 5$, $r = -3$, $l = 81$; find a and S.

9. Given $a = 54$, $r = 1/3$, $S = 80\tfrac{2}{3}$; find n and l.

10. Given $n = 4$, $a = -3$, $S = -468$; find r and l.

11. Given $a = -9/16$, $l = -16/9$, $S = -781/144$; find n and r.

12. Given $n = 6$, $r = -2/3$, $S = 665/216$; find a and l.

13. Given $r = 3/2$, $l = 30\tfrac{3}{8}$, $S = 83\tfrac{1}{8}$; find n and a.

14. Given $n = 4$, $l = 54/25$, $S = 544/25$; find a and r.

15. Given $n = 5$, $l = 48$, $S = 93$; find a and r.

16. Find the positive geometrical mean of a^3/b and b^3/a.

17. Insert three geometrical means between 5 and 405.

18. The third term of a geometrical progression is 3 and the sixth term is $-3/8$. Find the seventh term.

19. Find a geometrical progression of four terms in which the sum of the first and last terms is 133 and the sum of the middle terms is 70.

20. Find three numbers in geometrical progression such that their sum shall be 7 and the sum of their squares 91.

21. Three numbers whose sum is 36 are in arithmetical progression. If 1, 4, 43 be added to them respectively, the results are in geometrical progression. Find the numbers.

22. There are four numbers the first three of which are in arithmetical progression and the last three in geometrical progression. The sum of the first and fourth is 16 and the sum of the second and third is 8. Find the numbers.

23. What distance will an elastic ball traverse before coming to rest if it be dropped from a height of 15 feet and if after each fall it rebounds to $2/3$ the height from which it falls?

XXII. HARMONICAL PROGRESSION

705 **Harmonical progression.** This name is given to a sequence of numbers whose reciprocals form an arithmetical progression, that is, to any sequence which may be written in the form

$$1/a,\ 1/(a+d),\ 1/(a+2d),\ \cdots,\ 1/[a+(n-1)d].$$

Thus, 1, 1/2, 1/3, 1/4 and 3/2, 3/4, 3/6, 3/8 are harmonical progressions.

Example. Prove that if a, b, c are in harmonical progression, then $a : c = a - b : b - c$.

Since $1/a$, $1/b$, $1/c$ is an arithmetical progression, we have

$$1/b - 1/a = 1/c - 1/b.$$

Hence $c(a-b) = a(b-c)$, that is $a : c = a - b : b - c$.

706 To find any particular term of an harmonical progression, we obtain the term which occupies the same position in the corresponding arithmetical progression and invert it.

Example. Find the tenth term of the harmonical progression 3/5, 3/7, 3/9, ⋯.

By § 695, the tenth term of the corresponding arithmetical progression 5/3, 7/3, 9/3, ⋯ is 23/3. Hence the tenth term of 3/5, 3/7, 3/9, ⋯ is 3/23.

707 **Harmonical means.** If three numbers are in harmonical progression, the middle number is called the *harmonical mean* of the other two. Again, in any harmonical progression all the intermediate terms may be called the *harmonical means* of the extreme terms.

Example 1. Find the harmonical mean of a and b.

If this mean be x, then $1/a$, $1/x$, $1/b$ is an arithmetical progression.

Hence $\quad 1/x - 1/a = 1/b - 1/x$, or $2/x = 1/a + 1/b$.

Therefore $\quad x = 2ab/(a+b)$.

Example 2. Prove that the geometrical mean of two numbers a and b is also the geometrical mean of their arithmetical and harmonical means.

Let A, G, and H denote respectively the arithmetical, geometrical, and harmonical means of a and b.

Then $$A = \frac{a+b}{2},\quad G = \sqrt{ab},\quad H = \frac{2ab}{a+b}.$$

Hence $$AH = \frac{a+b}{2} \cdot \frac{2ab}{a+b} = ab = G^2.$$

Example 3. Prove that when a and b are positive, $A > G > H$.

We have $$A - H = \frac{a+b}{2} - \frac{2ab}{a+b} = \frac{(a-b)^2}{2(a+b)}.$$

Therefore, since $(a-b)^2/2(a+b)$ is positive, we have $A > H$.

And since, by Ex. 2, G is intermediate in value to A and H, we have $A > G > H$.

EXERCISE LIX

1. Continue the harmonical progressión $3/5$, $3/7$, $1/3$ for two terms.

2. Find the harmonical mean of $3/4$ and 5.

3. Insert four harmonical means between 10 and 15.

4. The second and fourth terms of an harmonical progression are $4/5$ and -4. Find the third term.

5. The arithmetical mean of two numbers is 4 and their harmonical mean is $15/4$. Find the numbers.

6. The geometrical mean of two numbers is 4 and their harmonical mean is $16/5$. Find the numbers.

7. Show that if a, b, c are in harmonical progression, so also are $a/(b+c)$, $b/(c+a)$, $c/(a+b)$.

8. Three numbers are in harmonical progression. Show that if half of the middle term be subtracted from each, the results will be in geometrical progression.

9. Show that if x is the harmonical mean between a and b, then $1/(x-a) + 1/(x-b) = 1/a + 1/b$.

10. The bisector of the vertical angle C of the triangle ABC meets the base AB at D, and the bisector of the exterior angle at C meets AB produced at E. Show that AD, AB, AE are in harmonical progression.

11. The point P lies outside of a circle whose center is O, and the tangents from P touch the circle at T and T'. If the line PO meets the circle at A and B and TT' at C, show that PC is the harmonical mean between PA and PB.

XXIII. METHOD OF DIFFERENCES. ARITHMETICAL PROGRESSIONS OF HIGHER ORDERS. INTERPOLATION

ARITHMETICAL PROGRESSIONS OF HIGHER ORDERS

708 **Differences of various orders.** If in any given sequence of numbers we subtract each term from the next following term, we obtain a sequence called the *first order of differences* of the given sequence; if we treat this new sequence in a similar manner, we obtain the *second order of differences* of the given sequence; and so on.

Thus, if the given sequence be $1^3, 2^3, 3^3, \cdots$, we have

given sequence	1, 8, 27, 64, 125, 216, ⋯,
first differences	7, 19, 37, 61, 91, ⋯,
second differences	12, 18, 24, 30, ⋯,
third differences	6, 6, 6, ⋯.

The fourth and all subsequent differences are 0.

709 **Arithmetical progression of the rth order.** This name is given to a sequence whose rth differences are equal, and whose subsequent differences are therefore 0.

Thus, $1^3, 2^3, 3^3, 4^3, \cdots$ is an arithmetical progression of the *third* order, for, as just shown, its third differences are equal.

An ordinary arithmetical progression, § 694, is of the *first* order, each of its first differences being the common difference d.

710 **The nth term of an arithmetical progression of the rth order.** Given any arithmetical progression of the rth order

$$a_1, a_2, a_3, a_4, \cdots, a_n, a_{n+1}, \cdots, \quad (1)$$

and let $d_1, d_2, \cdots, d_r$ denote the first terms of its successive orders of differences. We are to obtain a formula for a_n in terms of $a_1, d_1, d_2, \cdots, d_r$, and n.

The first order of differences of (1) is

$$a_2 - a_1, a_3 - a_2, a_4 - a_3, \cdots, a_{n+1} - a_n, \cdots. \quad (2)$$

The first term of (2) is d_1, and the first terms of its first, second, $\cdots$ differences are $d_2, d_3, \cdots$; for the first differences of (2) are the *second* differences of (1), and so on.

Hence when we have found an expression for any particular term of (1), we can derive from it an expression for the corresponding term of (2) by applying the rule:

Replace $\quad a_1, d_1, d_2, \cdots$ by $d_1, d_2, d_3 \cdots$. $\quad$ (3)

Now since $d_1 = a_2 - a_1$, we have $a_2 = a_1 + d_1$. Starting with this formula for a_2, we may reckon out $a_3, a_4, \cdots$ as follows:

We have $\quad a_2 = a_1 + d_1$

Hence, by (3), $\quad a_3 - a_2 = d_1 + d_2$

Adding, $\quad a_3 = a_1 + 2d_1 + d_2$

Hence, by (3), $\quad a_4 - a_3 = d_1 + 2d_2 + d_3$

Adding, $\quad a_4 = a_1 + 3d_1 + 3d_2 + d_3$

and so on indefinitely, the reckoning, so far as coefficients are concerned, being precisely the same as that given in § 311 for finding the coefficients of successive powers of $a + b$. Therefore, by § 561, we have the formula

$$a_n = a_1 + (n-1)d_1 + \frac{(n-1)(n-2)}{1 \cdot 2} d_2 + \cdots + \frac{(n-1)\cdots(n-r)}{1 \cdot 2 \cdots r} d_r. \quad \text{(I)}$$

Example. Compute the fifteenth term of $1^3, 2^3, \cdots$ by this formula.

Here, § 708, $\quad a_1 = 1,\ d_1 = 7,\ d_2 = 12,\ d_3 = d_r = 6.$

Hence $\quad a_{15} = 1 + 14 \cdot 7 + \dfrac{14 \cdot 13}{2} \cdot 12 + \dfrac{14 \cdot 13 \cdot 12}{2 \cdot 3} \cdot 6 = 3375.$

711 **Sum of the first n terms of an arithmetical progression of the rth order.** Let S_n denote this sum, the sequence being

$$a_1,\ a_2,\ a_3,\ \cdots,\ a_n,\ a_{n+1},\ \cdots, \quad (1)$$

and $d_1, d_2, \cdots, d_r$ having the same meanings as in § 710.

Form the sequence of which (1) is the first order of differences, namely:

$$0,\ a_1,\ a_1 + a_2,\ a_1 + a_2 + a_3, \cdots,\ a_1 + a_2 + \cdots + a_n, \cdots. \quad (2)$$

Then S_n is the $(n+1)$th term of (2), and since (2) is an arithmetical progression of the $(n+1)$th order whose first term is 0, and the first terms of its several orders of differences are $a_1, d_1, d_2, \cdots d_r$, we have, by (I),

$$S_n = na_1 + \frac{n(n-1)}{1 \cdot 2} d_1 + \cdots + \frac{n(n-1)\cdots(n-r)}{1 \cdot 2 \cdots (r+1)} d_r. \quad \text{(II)}$$

Example. Find the sum of the first fifteen terms of $1^3, 2^3, 3^3, \cdots$.

Here, § 708, $n = 15$, $a_1 = 1$, $d_1 = 7$, $d_2 = 12$, $d_3 = d_r = 6$.

Hence $S_{15} = 15 + \dfrac{15 \cdot 14}{2} \cdot 7 + \dfrac{15 \cdot 14 \cdot 13}{2 \cdot 3} \cdot 12 + \dfrac{15 \cdot 14 \cdot 13 \cdot 12}{2 \cdot 3 \cdot 4} \cdot 6 = 14400.$

712 **Piles of spherical shot.** 1. To find the number of shot when the pile has the form of a triangular pyramid.

The top course contains 1 shot, the next lower course $1 + 2$ shot, the next $1 + 2 + 3$, and so on.

Hence, if there are n courses, the number of shot is the sum of n terms of the sequence $1, 3, 6, 10, 15, \cdots$.

The first differences of this sequence are $2, 3, 4, 5, \cdots$, and the second differences are $1, 1, 1, \cdots$.

Hence $1, 3, 6, \cdots$ is an arithmetical progression of the second order in which $a_1 = 1$, $d_1 = 2$, $d_2 = 1$.

Therefore, by (II), $S_n = n + \dfrac{n(n-1)}{1 \cdot 2} \cdot 2 + \dfrac{n(n-1)(n-2)}{1 \cdot 2 \cdot 3}$

$$= \frac{n(n+1)(n+2)}{1 \cdot 2 \cdot 3}.$$

Thus, in a pile of twenty courses there are $20 \cdot 21 \cdot 22/6 = 1540$ shot.

2. To find the number of shot when the pile has the form of a pyramid with a square base.

Enumerating the shot by courses as before, we obtain the sequence $1^2, 2^2, 3^2, 4^2, \cdots$.

The first differences are $3, 5, 7, \cdots$, and the second are $2, 2, \cdots$.

Hence $1^2, 2^2, 3^2, \cdots$ is an arithmetical progression of the second order in which $a_1 = 1$, $d_1 = 3$, $d_2 = 2$.

Therefore, by (II), $S_n = n + \frac{n(n-1)}{1 \cdot 2} \cdot 3 + \frac{n(n-1)(n-2)}{1 \cdot 2 \cdot 3} \cdot 2$

$$= \frac{n(n+1)(2n+1)}{1 \cdot 2 \cdot 3}.$$

Thus, when $n = 20$, the pile contains $20 \cdot 21 \cdot 41 / 6 = 2870$ shot.

3. To find the number of shot when the pile has a rectangular base and terminates at the top in a row of p shot.

Again enumerating the shot by courses, we obtain the sequence $p,\ 2(p+1),\ 3(p+2),\ 4(p+3), \cdots$.

The first differences are $p+2,\ p+4,\ p+6, \cdots$, and the second differences are $2, 2, \cdots$.

Hence $p,\ 2(p+1),\ 3(p+2), \cdots$ is an arithmetical progression of the second order in which $a_1 = p$, $d_1 = p+2$, and $d_2 = 2$.

Therefore, by (II), $S_n = np + \frac{n(n-1)}{1 \cdot 2}(p+2) + \frac{n(n-1)(n-2)}{1 \cdot 2 \cdot 3} \cdot 2$

$$= \frac{n(n+1)(3p+2n-2)}{1 \cdot 2 \cdot 3}.$$

Thus, when $n = 20$ and $p = 5$, the number of shot is $20 \cdot 21 \cdot 53 / 6 = 3710$.

713 **A theorem respecting arithmetical progressions.** An examination of the formula for the nth term of an arithmetical progression of the rth order, § 710, (I), will show that if we carry out the indicated multiplications and arrange the result according to descending powers of n, we can reduce it to the form

$$a_n = b_0 n^r + b_1 n^{r-1} + \cdots + b_r,$$

where the coefficients $b_0, b_1, \cdots, b_r$ are independent of n.

Thus, when $r = 2$, we have

$$a_n = a_1 + (n-1)d_1 + \frac{(n-1)(n-2)}{1 \cdot 2} d_2$$

$$= \frac{d_2}{2} n^2 + \left(d_1 - \frac{3}{2} d_2\right) n + (a_1 - d_1 + d_2).$$

Therefore the terms of any arithmetical progression of the rth order, $a_1, a_2, a_3, \cdots$, are the values for $n = 1, 2, 3, \cdots$ of a certain polynomial $b_0 n^r + b_1 n^{r-1} + \cdots + b_r$ whose degree with respect to n is r. We are going to show conversely that

714 **Theorem.** *If* $\phi(\mathrm{x})$ *denote any rational integral function of the* r*th degree, as*

$$\phi(\mathrm{x}) = \mathrm{b}_0\mathrm{x}^{\mathrm{r}} + \mathrm{b}_1\mathrm{x}^{\mathrm{r}-1} + \cdots + \mathrm{b}_{\mathrm{r}},$$

the sequence of numbers $\phi(1), \phi(2), \phi(3), \cdots$, *obtained by setting* $\mathrm{x} = 1, 2, 3, \cdots$ *successively in* $\phi(\mathrm{x})$, *is an arithmetical progression of the* rth *order.*

Here the given sequence of numbers is

$$\phi(1),\ \phi(2),\ \phi(3),\ \phi(4), \cdots \qquad (1)$$

and we are to prove that all of its rth differences are equal.

Evidently the first differences of (1), namely

$$\phi(2) - \phi(1),\ \phi(3) - \phi(2),\ \phi(4) - \phi(3) \cdots, \qquad (2)$$

are the values of $\phi(x+1) - \phi(x)$ for $x = 1, 2, 3, \cdots$.

But $\phi(x+1) - \phi(x)$ may be reduced to the form of a polynomial in x. Call this polynomial $\phi_1(x)$. *Its degree is* $\mathrm{r} - 1$.

For, by the binomial theorem, § 561, we have

$$\begin{aligned}\phi(x+1) - \phi(x) &= b_0(x+1)^r + b_1(x+1)^{r-1} + \cdots - (b_0x^r + b_1x^{r-1} + \cdots) \\ &= b_0x^r + rb_0x^{r-1} + \cdots + b_1x^{r-1} + \cdots - (b_0x^r + b_1x^{r-1} + \cdots) \\ &= rb_0x^{r-1} + \cdots\end{aligned}$$

Similarly, if we write

$$\phi_1(x+1) - \phi_1(x) = \phi_2(x),\ \phi_2(x+1) - \phi_2(x) = \phi_3(x),$$

and so on, the values of $\phi_2(x), \phi_3(x), \cdots, \phi_r(x)$ for $x = 1, 2, 3, \cdots$ will be the second, third, $\cdots$, rth differences of (1).

But $\phi_r(x)$ is a *constant* and the rth differences of (1) are therefore equal. For the degree of $\phi_2(x)$ is $(r-1) - 1$, or $r - 2$; that of $\phi_3(x)$ is $r - 3$; and finally that of $\phi_r(x)$ is $r - r$, or 0.

For example, if $\phi(x) = 2x^3 - x + 1$, we have

$$\phi_1(x) = 2(x+1)^3 - (x+1) + 1 - (2x^3 - x + 1) = 6x^2 + 6x + 1,$$
$$\phi_2(x) = 6(x+1)^2 + 6(x+1) + 1 - (6x^2 + 6x + 1) = 12x + 12,$$
$$\phi_3(x) = 12(x+1) + 12 - (12x + 12) = 12.$$

Hence the values of $6x^2 + 6x + 1$, $12x + 12$, 12 for $x = 1, 2, 3, \cdots$ are the first, second, third differences of the corresponding values of $2x^3 - x + 1$; and the third differences are equal, all being 12.

Thus, for $x = 1, 2, 3, 4, 5, \cdots$, we find

$$2x^3 - x + 1 = 2,\ 15,\ 52,\ 125,\ 246, \cdots, \tag{1}$$

$$6x^2 + 6x + 1 = 13,\ 37,\ 73,\ 121,\ 181, \cdots, \tag{2}$$

$$12x + 12 = 24,\ 36,\ 48,\ 60,\ 72, \cdots, \tag{3}$$

$$12 = 12,\ 12,\ 12,\ 12,\ 12, \cdots. \tag{4}$$

And by comparing (1), (2), (3), (4), we find that (2), (3), (4) actually are, as they should be, the first, second, third differences of (1).

Corollary 1. *The* r*th powers of consecutive integers form an arithmetical progression of the* r*th order.* **715**

For $1^r, 2^r, 3^r, \cdots$ are the values of the rational integral function of the rth degree $\phi(x) = x^r$ for $x = 1, 2, 3, \cdots$.

Corollary 2. *The products of the corresponding terms of two arithmetical progressions, the one of the* r*th order and the other of the* s*th order, form an arithmetical progression of the* (r + s)*th order.* **716**

For the product of a rational integral function of the rth degree by one of the sth degree is a rational integral function of the $(r + s)$th degree.

EXERCISE LX

1. Find the twentieth term and the sum of the first twenty terms of the sequence $1, 2, 4, 7, \cdots$.

2. Find the eightieth term and the sum of the first eighty terms of the sequence $3, 8, 15, 24, 35, \cdots$.

3. Determine the order of each of the following arithmetical progressions.

(1) $3, 0, -1, 0, 3, \cdots$, (2) $10, 38, 88, 166, 278, 430, \cdots$,

(3) $285, 204, 140, 91, 55, \cdots$, (4) $2, 20, 90, 272, 650, 1332, \cdots$.

Also find the eighteenth term of (1), the twentieth term of (2), the twelfth term of (3), and the tenth term of (4).

4. What is the order of $1 \cdot 2 \cdot 3, 2 \cdot 3 \cdot 4, 3 \cdot 4 \cdot 5, \cdots$? What is its nth term? the sum of its first n terms?

What is the order and what the nth term of $1 \cdot 4 \cdot 2^2, 2 \cdot 6 \cdot 3^2, 3 \cdot 8 \cdot 4^2, \cdots$?

5. Find the number of shot in a triangular pile of fourteen courses. How many shot are there in the lowest course?

6. If from a square pile of fifteen courses six courses be removed, how many shot remain?

7. How many shot are there in a rectangular pile of twelve courses if the uppermost course contains 5 shot?

8. How many shot are there in a triangular pile whose lowermost course contains 253 shot?

9. The number of shot in a certain triangular pile is four sevenths of the number in a square pile of the same number of courses. How many shot are there in each pile?

10. How many shot are there in a rectangular pile whose top row contains 9 balls and whose bottom course contains 240 balls?

11. Show that $1^3 + 2^3 + \cdots + n^3 = (1 + 2 + \cdots + n)^2$.

12. Show that $1^4 + 2^4 + \cdots + n^4 = \frac{n}{30}(n+1)(2n+1)(3n^2 + 3n - 1)$.

13. What is the order and what the sum of the first n terms of the progression whose nth term is $n^2 - n + 1$? $n(n+1)(n+2)/6$?

14. If we write down the arithmetical progressions of the first order in which $d = 1, 2, 3, \cdots$ respectively and then sum each progression to one, two, three, four, $\cdots$ terms, we obtain the following sequences of numbers, called respectively the triangular, quadrangular, pentagonal, $\cdots$ numbers:

$$1, 3, 6, 10, \cdots; 1, 4, 9, 16, \cdots; 1, 5, 12, 22, \cdots; \cdots.$$

Show that in the kth of these sequences the nth term and the sum of the first n terms are $n(kn - k + 2)/2$ and $n(n+1)(kn - k + 3)/6$ respectively.

15. Show that the order of an arithmetical progression of any order is not changed by adding to its terms the corresponding terms of an arithmetical progression of a lower order.

16. Show that if in a polynomial of the nth degree, $f(x)$, we substitute for x successive terms of any arithmetical progression of the first order, we obtain an arithmetical progression of the nth order; and, in general, that if we substitute for x successive terms of any arithmetical progression of the rth order, we obtain an arithmetical progression of the nrth order.

INTERPOLATION

Interpolation. Suppose that y is known to depend on x in such a manner that for each value of x between a and b, y has a definite value. Suppose also that the values of y which correspond to certain of these values of x are actually known. Then from these known values it is possible, by a process called *interpolation*, to derive values of y corresponding to other values of x between a and b. **717**

This process is employed when the general expression for y in terms of x is unknown, or if known is too complicated to be conveniently used for reckoning out particular values of y.

Briefly stated, the process is as follows: we set y equal to the simplest *integral* expression in x which will take the given values and then derive the values of y which we seek from this equation. Of course the values thus obtained will ordinarily be only approximately correct.

Method of undetermined coefficients. We may proceed as in the following example. **718**

Example. For $x = 2, 3, 4, 5$ it is known that $y = 5, 4, -7, -34$; find y when $x = 5/2$.

Since the simplest polynomial in x which will take given values for *four* given values of x will ordinarily be one of the *third* degree, we assume that

$$y = b_0 + b_1x + b_2x^2 + b_3x^3,$$

and then find the coefficients b_0, b_1, b_2, b_3 as follows.

Since $y = 5$ when $x = 2$, $\quad 5 = b_0 + 2b_1 + 4b_2 + 8b_3$.

Since $y = 4$ when $x = 3$, $\quad 4 = b_0 + 3b_1 + 9b_2 + 27b_3$.

Since $y = -7$ when $x = 4$, $\quad -7 = b_0 + 4b_1 + 16b_2 + 64b_3$.

Since $y = -34$ when $x = 5$, $\quad -34 = b_0 + 5b_1 + 25b_2 + 125b_3$.

Solving these equations, $b_0 = 1$, $b_1 = -2$, $b_2 = 4$, $b_3 = -1$.

Hence $\quad y = 1 - 2x + 4x^2 - x^3$.

Therefore, when $x = 5/2$ we have $y = 1 - 5 + 25 - 125/8 = 43/8$.

And in general, if $r+1$ values of y are known, say the values $y = y_1, y_2, \cdots, y_{r+1}$, corresponding to $x = x_1, x_2, \cdots, x_{r+1}$ respectively, we assume that

$$y = b_0 + b_1 x + b_2 x^2 + \cdots + b_r x^r, \tag{1}$$

find $b_0, b_1, \cdots, b_r$ by the method just illustrated, and then employ (1) as a formula for computing y for values of x intermediate to $x_1, x_2, \cdots, x_{r+1}$.

719 **Method of differences.** When $x_1, x_2, \cdots, x_{r+1}$ are consecutive integers, the formula (1) of § 718, may be reduced to the form

$$y = y_1 + (x - x_1) d_1 + \frac{(x - x_1)(x - x_2)}{1 \cdot 2} d_2 + \cdots$$
$$+ \frac{(x - x_1) \cdots (x - x_r)}{1 \cdot 2 \cdots r} d_r, \tag{2}$$

where $d_1, d_2, \cdots, d_r$ denote the first terms of the successive orders of differences of $y_1, y_2, \cdots, y_{r+1}$.

For since $x_1, x_2, \cdots, x_{r+1}$ are consecutive integers, the corresponding values of $b_0 + b_1 x + \cdots + b_r x^r$, namely $y_1, y_2, \cdots, y_{r+1}$, form an arithmetical progression of the rth order, § 714. Hence we may also obtain y_1, $y_2, \cdots$ by substituting $n = 1, 2, \cdots$ in the formula, § 710, (I), namely

$$y = y_1 + (n-1) d_1 + \frac{(n-1)(n-2)}{1 \cdot 2} d_2 + \cdots + \frac{(n-1) \cdots (n-r)}{1 \cdot 2 \cdots r} d_r.$$

But setting $n = 1, 2, 3, \cdots$ in this formula will give identically the same results as setting $x = x_1, x_2, x_3, \cdots = x_1, x_1 + 1, x_1 + 2, \cdots$ in (2).

Therefore the second member of (2) and that of (1), § 718, have equal values for $r + 1$ values of x. But both are of the rth degree. Hence they are identically equal, § 421.

Thus, as in § 718, for $x = 2, 3, 4, 5$, let $y = 5, 4, -7, -34$. We have

$$y_1, y_2, y_3, y_4 = 5, 4, -7, -34.$$

First differences	$-1, -11, -27.$
Second differences	$-10, -16.$
Third difference	$-6.$

Substituting in (2), $x_1 = 2$, $x_2 = 3$, $x_3 = 4$, $y_1 = 5$, $d_1 = -1$, $d_2 = -10$, $d_3 = -6$, we have $y = 5 - (x-2) - 5(x-2)(x-3) - (x-2)(x-3)(x-4)$, which may be reduced to $y = 1 - 2x + 4x^2 - x^3$, as in § 718.

Example. Given $\sqrt[3]{30} = 3.1072$, $\sqrt[3]{31} = 3.1414$, $\sqrt[3]{32} = 3.1748$, and $\sqrt[3]{33} = 3.2075$; find $\sqrt[3]{31.6}$.

$y_1, y_2, y_3, y_4 =$	3.1072,	3.1414,	3.1748,	3.2075.
First differences		.0342,	.0334,	.0327.
Second differences		$-.0008$,	$-.0007$.	
Third difference		.0001.		

Substituting in (2) $x_1 = 30$, $x_2 = 31$, $x_3 = 32$, $y_1 = 3.1072$, $d_1 = .0342$, $d_2 = -.0008$, $d_3 = .0001$, and $x = 31.6$, we have

$$3.1072 + (1.6)(.0342) + \frac{(1.6)(.6)}{2}(-.0008) + \frac{(1.6)(.6)(-.4)}{2 \cdot 3}(-.0001)$$
$$= 3.1072 + .05472 - .000384 + .0000064 = 3.1615 +.$$

Lagrange's formula. The formula (1) of § 718 may also be reduced to the following form, due to Lagrange: **720**

$$y = y_1 \frac{(x - x_2)(x - x_3)\cdots(x - x_{r+1})}{(x_1 - x_2)(x_1 - x_3)\cdots(x_1 - x_{r+1})}$$
$$+ y_2 \frac{(x - x_1)(x - x_3)\cdots(x - x_{r+1})}{(x_2 - x_1)(x_2 - x_3)\cdots(x_2 - x_{r+1})}$$
$$+ \cdots + y_{r+1} \frac{(x - x_1)(x - x_2)\cdots(x - x_r)}{(x_{r+1} - x_1)(x_{r+1} - x_2)\cdots(x_{r+1} - x_r)}. \quad (3)$$

For the right member of (3) is an integral function of x of the rth degree and its values for $x = x_1, x_2, \cdots, x_{r+1}$ are $y_1, y_2, \cdots, y_{r+1}$. Thus, if we set $x = x_1$, every term except the first vanishes and the first term reduces to y_1. Hence, § 421, the right member of (3) and that of (1), § 718, are equal for $r + 1$ values of x and are therefore equal identically.

Thus, as in § 718, for $x = 2, 3, 4, 5$, let $y = 5, 4, -7, -34$. Substituting in (3), we obtain

$$y = 5\frac{(x-3)(x-4)(x-5)}{(2-3)(2-4)(2-5)}$$
$$+ 4\frac{(x-2)(x-4)(x-5)}{(3-2)(3-4)(3-5)} - 7\frac{(x-2)(x-3)(x-5)}{(4-2)(4-3)(4-5)} - 34\frac{(x-2)(x-3)(x-4)}{(5-2)(5-3)(5-4)},$$

which will reduce to $y = 1 - 2x + 4x^2 - x^3$, as in § 718.

EXERCISE LXI

1. For $x = -3, -2, -1, 0$ it is known that $y = -20, 6, 0, 4$; find y when $x = -5/2$, also when $x = -1/2$.

2. Given that $f(4) = 10$, $f(6) = -12$, $f(7) = -20$, $f(8) = -18$; find $f(x)$ and then compute $f(12)$.

3. Given that $25^2 = 625$, $26^2 = 676$, $27^2 = 729$; find 26.54^2 by the method of differences.

4. Given that $2^3 = 8$, $3^3 = 27$, $4^3 = 64$, $5^3 = 125$; find 4.8^3 by the method of differences.

5. Given that $1/22 = .04546$, $1/23 = .04348$, $1/24 = .04167$, and $1/25 = .04$; find $1/23.6$ by the method of differences.

6. Given that $\sqrt{432} = 20.7846$, $\sqrt{433} = 20.8087$, $\sqrt{434} = 20.8327$, $\sqrt{435} = 20.8566$, $\sqrt{436} = 20.8806$; find $\sqrt{435.7}$ by the method of differences.

7. By aid of Lagrange's formula find the polynomial of the third degree whose values for $x = -2, 0, 4, 5$ are $5, 3, -2, -4$.

XXIV. LOGARITHMS

PRELIMINARY THEOREMS REGARDING EXPONENTS

721 **Theorem 1.** *If* a *denote any real number greater than* 1, *and* p, q *denote positive integers, then* $a^{\frac{p}{q}} > 1$.

For $a > 1$, $\therefore a^p > 1$, $\therefore \sqrt[q]{a^p} > 1$, $\therefore a^{\frac{p}{q}} > 1$, § 261.

722 **Theorem 2.** *If* a *denote any real number greater than* 1, *and* r, s *any two rationals such that* $r > s$, *then* $a^r > a^s$.

For $r - s > 0$, $\therefore a^{r-s} > 1$, $\therefore a^{r-s} \cdot a^s > a^s$, $\therefore a^r > a^s$, §§ 721, 261.

723 **Theorem 3.** *If* $a > 1$ *and* n *be integral, then* $\lim\limits_{n \doteq \infty} a^n = \infty$.

For since $a > 1$, we may write $a = 1 + d$, where d is positive.
Then $a^n = (1 + d)^n$, and since $(1 + d)^n > 1 + nd$, § 561, we have $a^n > 1 + nd$.
Therefore, since $\lim\limits_{n \doteq \infty} (1 + nd) = \infty$, we have $\lim\limits_{n \doteq \infty} a^n = \infty$.

Theorem 4. *If* $0 < a < 1$, *and* n *be integral*, $\lim_{n \doteq \infty} a^n = 0$. **724**

For let $a = 1/b$, where $b > 1$, since $a < 1$.

Then $\lim_{n \doteq \infty} a^n = 1 \big/ \lim_{n \doteq \infty} b^n = 0$, § 512, since $\lim_{n \doteq \infty} b^n = \infty$, § 723.

Theorem 5. *If* n *be integral*, $\lim_{n \doteq \infty} \sqrt[n]{a} = \lim_{n \doteq \infty} a^{\frac{1}{n}} = 1$. **725**

1. When $a > 1$, we have $a^{\frac{1}{n}} > 1$, § 721, so that $a^{\frac{1}{n}} = 1 + d_n$, where d_n is some positive number dependent on n.

Then $a = (1 + d_n)^n$, $\therefore a > 1 + nd_n$, $\therefore d_n < (a - 1)/n$.

Therefore, since $\lim_{n \doteq \infty} (a - 1)/n = 0$, § 512, we have $\lim_{n \doteq \infty} d_n = 0$.

Hence $\lim_{n \doteq \infty} a^{\frac{1}{n}} = \lim_{n \doteq \infty} (1 + d_n) = 1$.

2. When $0 < a < 1$, let $a = 1/b$, where $b > 1$, since $a < 1$.

Then $\lim_{n = \infty} a^{\frac{1}{n}} = 1 \big/ \lim_{n \doteq \infty} b^{\frac{1}{n}} = 1$, since $\lim_{n \doteq \infty} b^{\frac{1}{n}} = 1$, by 1.

Theorem 6. *If* b *be a rational number and* x *be a variable which approaches* b *through rational values, then* $\lim_{x \doteq b} a^x = a^b$. **726**

1. The theorem holds true when b, the limit of x, is 0.

For in this case we can select a variable n which takes integral values only and such that we shall always have $-1/n < x < 1/n$ and that when $x \doteq 0$ then $n = \infty$.

Then a^x will always lie between $a^{\frac{1}{n}}$ and $a^{-\frac{1}{n}}$, § 722, and since $\lim_{n \doteq \infty} a^{\frac{1}{n}} = \lim_{n \doteq \infty} a^{-\frac{1}{n}} = 1$, § 725, we have $\lim_{x \doteq 0} a^x = 1 = a^0$.

2. The theorem holds true when $b \neq 0$.

For since $a^x = a^b \cdot a^{x-b}$, we have $\lim_{x \doteq b} a^x = a^b \cdot \lim_{x \doteq b} a^{x-b} = a^b$, by 1.

Theorem 7. *If* b *be an irrational number and* x *be a variable which approaches* b *through rational values, then* a^x *will approach a limit as* $x \doteq b$, *and the value of this limit is independent of the values which* x *takes in approaching* b. **727**

The reasoning is the same whether $a > 1$ or $a < 1$, but to fix the ideas we shall suppose that $a > 1$.

There are infinitely many sequences of rational values through which x may run in approaching b as limit. From among them select some particular *increasing* sequence, and represent x by x' when supposed to run through this sequence. Then as $x' \doteq b$, the variable $a^{x'}$ continually increases, § 722, but remains finite — less, for instance, than a^c, if c denote any rational greater than b. Hence $a^{x'}$ approaches a limit, § 192. Call this limit L.

It only remains to prove that a^x will approach this same limit L if x approach b through any other sequence of rationals than that through which x' runs. But $a^x = a^{x'} \cdot a^{x-x'}$ and therefore $\lim a^x = \lim a^{x'} \cdot \lim a^{x-x'} = L$, since $\lim a^{x-x'} = 1$, § 726.

728 **Irrational exponents.** We employ the symbol a^b to denote the limit which a^x will approach when x is made to approach b through any sequence of rational values. Hence by a^b, when b is irrational, we shall mean $\lim\limits_{x \doteq b} a^x$.

729 Having thus assigned a meaning to a^x when x is irrational, we can readily prove that $\lim\limits_{x \doteq b} a^x = a^b$ when x approaches b through a sequence of *irrational* values.

For let x', x, x'' denote variables all of which approach b as limit, x' and x'' through sequences of rational values and x through a sequence of irrational values, and such that $x' < x < x''$. It then follows from §§ 726, 727 that a^x lies between $a^{x'}$ and $a^{x''}$, and therefore since $\lim\limits_{x' \doteq b} a^{x'} = \lim\limits_{x'' \doteq b} a^{x''} = a^b$, that $\lim\limits_{x \doteq b} a^x = a^b$.

730 **Theorem 8.** *The laws of exponents are valid for irrational exponents.*

For let b and c denote irrational numbers, and x and y variables which approach b and c as limits. We suppose x and y to take rational values only.

1. $a^b \cdot a^c = a^{b+c}$.

For since $a^x a^y = a^{x+y}$, we have $\lim a^x a^y = \lim a^{x+y}$.

But $\lim a^x a^y = \lim a^x \cdot \lim a^y = a^b a^c$, §§ 203, 728

and $\lim a^{x+y} = a^{\lim (x+y)} = a^{b+c}$. §§ 203, 728

2. $(a^b)^c = a^{bc}$.

For $(a^x)^y = a^{xy}$.

Hence $\lim\limits_{x \doteq b} (a^x)^y = \lim\limits_{x \doteq b} a^{xy}$, or $(a^b)^y = a^{by}$. § 728

Hence $\lim\limits_{y \doteq c} (a^b)^y = \lim\limits_{y \doteq c} a^{by}$, or $(a^b)^c = a^{bc}$. §§ 728, 729

3. $(ab)^c = a^c b^c$.

For $(ab)^y = a^y b^y$.

Hence $\lim (ab)^y = \lim a^y b^y = \lim a^y \cdot \lim b^y$. § 203

That is, $(ab)^c = a^c b^c$. § 728

LOGARITHMS. THEIR GENERAL PROPERTIES

Logarithms. Take a, any positive number except 1, as a *base* or number of reference. We have shown that every real power of a, as a^μ, denotes some definite positive number, as m. In a subsequent section we shall show conversely that every positive number, m, may be expressed in the form a^μ, where μ is real. **731**

If $a^\mu = m$, we call μ the *logarithm* of m to the base a and represent it by the symbol $\log_a m$. Hence the logarithm of m to the base a is the exponent of the power to which a must be raised to equal m, that is $a^{\log_a m} = m$. **732**

Thus, $3^4 = 81$, $\therefore 4 = \log_3 81$; $2^{-3} = 1/8$, $\therefore -3 = \log_2 1/8$.

Since $a^0 = 1$, we always have $\log_a 1 = 0$; and since $a^1 = a$, we always have $\log_a a = 1$. **733**

When $a > 1$, it follows from $a^\mu = m$, by § 722, that to any increase in the number m there corresponds an increase in its logarithm μ; also that if m is greater than 1, its logarithm μ is positive, and that if m lies between 1 and 0, its logarithm μ is negative. **734**

Again, when $a > 1$, we have, § 723, **735**

$$\lim_{\mu \doteq \infty} a^\mu = \infty, \text{ and } \lim_{\mu \doteq \infty} a^{-\mu} = \lim_{\mu \doteq \infty} 1/a^\mu = 0.$$

We therefore say, when $a > 1$, that $\log_a \infty = \infty$, and $\log_a 0 = -\infty$.

Theorem 1. *The logarithm of a product to any base is the sum of the logarithms of the factors to the same base.* **736**

For let $\quad m = a^\mu$, that is $\mu = \log_a m$,

and $\quad n = a^\nu$, that is $\nu = \log_a n$.

Then $\quad mn = a^\mu a^\nu = a^{\mu+\nu}$,

that is, $\quad \log_a mn = \mu + \nu = \log_a m + \log_a n$.

Theorem 2. *The logarithm of a quotient is the logarithm of the dividend minus the logarithm of the divisor.* **737**

For if $\quad m = a^\mu$ and $n = a^\nu$,

we have $\quad m/n = a^\mu / a^\nu = a^{\mu-\nu}$,

that is, $\quad \log_a m/n = \mu - \nu = \log_a m - \log_a n$.

738 **Theorem 3.** *The logarithm of any power of a number is the logarithm of the number multiplied by the exponent of the power.*

For if $m = a^\mu$,

we have $m^r = (a^\mu)^r = a^{\mu r}$,

that is, $\log_a m^r = r\mu = r \log_a m$.

739 **Theorem 4.** *The logarithm of any root of a number is the logarithm of the number divided by the index of the root.*

For if $m = a^\mu$,

we have $\sqrt[s]{m} = \sqrt[s]{a^\mu} = a^{\frac{\mu}{s}}$,

that is, $\log_a \sqrt[s]{m} = \mu / s = (\log_a m) / s$.

740 The practical usefulness of logarithms is due to the properties established in §§ 736–739. Logarithms of numbers to the base 10 have been computed and arranged in tables. If we avail ourselves of such a table, we can find the value of a product by an addition, of a quotient by a subtraction, of a power by a multiplication, and of a root by a division.

Thus, $\log \frac{\sqrt[7]{5}\sqrt[8]{6}}{3^{25}} = \log \sqrt[7]{5} + \log \sqrt[8]{6} - \log 3^{25}$ §§ 736, 737

$= (\log 5)/7 + (\log 6)/8 - 25 \log 3.$ §§ 738, 739

Hence, to obtain the value of $\sqrt[7]{5}\sqrt[8]{6}/3^{25}$, we have only to look up the values of $\log 5$, $\log 6$, and $\log 3$ given in the table, then to reckon out the value of $(\log 5)/7 + (\log 6)/8 - 25 \log 3$, and finally to look up in the table the number of which this value is the logarithm.

EXERCISE LXII

1. Find $\log_2 4$, $\log_4 2$, $\log_{\sqrt{2}} 8$, $\log_5 625$, $\log_3 729$, $\log_{10} .001$, $\log_2 1/64$, $\log_2 .125$, $\log_a \sqrt[3]{a^{-2}}$, $\log_8 128$, $\log_{a^3} a^3$.

2. If $\log_{10} 2 = .3010$ and $\log_{10} 3 = .4771$, find the logarithms to the base 10 of 12, 9/2, $\sqrt{2}$, $\sqrt[3]{6}$.

3. Express $\log_a 600^{\frac{1}{4}}$ in terms of $\log_a 2$, $\log_a 3$, and $\log_a 5$.

4. Express the logarithms of each of the following expressions to the base a in terms of $\log_a b$, $\log_a c$, $\log_a d$.

(1) $b^{\frac{2}{3}}c^{-\frac{1}{2}}/d^{\frac{4}{5}}$. (2) $\sqrt[3]{a^{-2}\sqrt{b^6}} \div \sqrt{b^3\sqrt{a^{-3}}}$.

5. Prove that $\log_3 \sqrt[3]{81\sqrt[4]{729 \cdot 9^{-\frac{2}{3}}}} = 31/18$.

6. Prove that $\log_a \dfrac{x+\sqrt{x^2-1}}{x-\sqrt{x^2-1}} = 2\log_a(x+\sqrt{x^2-1})$.

COMMON LOGARITHMS

Computation of common logarithms. For the purposes of numerical reckoning we employ logarithms to the base 10. These are called *common logarithms.* In what follows $\log m$ will mean $\log_{10} m$. **741**

We have $10^0 = 1$, $\therefore \log 1 = 0$; $10^1 = 10$, $\therefore \log 10 = 1$; $10^2 = 100$, $\therefore \log 100 = 2, \cdots$; also $10^{-1} = .1$, $\therefore \log .1 = -1$; $10^{-2} = .01$, $\therefore \log .01 = -2, \cdots$. **742**

Hence, for the numbers whose common logarithms are *integers* we have the table:

The numbers	$\cdots .001,$	$.01,$	$.1,$	$1,$	$10,$	$100,$	$1000, \cdots,$
their logarithms	$\cdots -3,$	$-2,$	$-1,$	$0,$	$1,$	$2,$	$3, \cdots.$

Observe that in this table the numbers constitute a geometrical progression in which the common ratio is 10, and the logarithms an arithmetical progression in which the common difference is 1.

The numbers in this table are the only rationals whose common logarithms are rational, for all fractional powers of 10 are irrational. But, as we proceed to show, every positive number has a common logarithm, and the value of this logarithm may be obtained correct to as many places of decimals as may be desired. **743**

If we extract the square root of 10, the square root of the result thus obtained, and so on, continuing the reckoning in each case to the fifth decimal figure, we obtain the following table:

$10^{\frac{1}{2}} = 3.16228$,	$10^{\frac{1}{32}} = 1.07461$,	$10^{\frac{1}{512}} = 1.00451$,
$10^{\frac{1}{4}} = 1.77828$,	$10^{\frac{1}{64}} = 1.03663$,	$10^{\frac{1}{1024}} = 1.00225$,
$10^{\frac{1}{8}} = 1.33352$,	$10^{\frac{1}{128}} = 1.01815$,	$10^{\frac{1}{2048}} = 1.00112$,
$10^{\frac{1}{16}} = 1.15478$,	$10^{\frac{1}{256}} = 1.00904$,	$10^{\frac{1}{4096}} = 1.00056$,

and so on, the results obtained approaching 1 as limit as we proceed (compare § 725). The exponents $1/2, 1/4, \cdots$ on the left are the logarithms of the corresponding numbers on the right.

744 By aid of this table we may compute the common logarithm of any number between 1 and 10 as in the following example.

Example. Find the common logarithm of 4.26.

Divide 4.26 by the next smaller number in the table, 3.16228.

The quotient is 1.34719. Hence $4.26 = 3.16228 \times 1.34719$.

Divide 1.34719 by the next smaller number in the table, 1.33352.

The quotient is 1.0102. Hence $4.26 = 3.16228 \times 1.33352 \times 1.0102$.

Continue thus, always dividing the quotient last obtained by the next smaller number in the table.

If q_n denote the quotient in the nth division, we shall obtain by this method an expression for 4.26 in the form of a product of n numbers taken from the table and q_n, the result being

$$\begin{aligned} 4.26 &= 3.16228 \times 1.33352 \times 1.00904 \times \cdots \times q_n \\ &= 10^{\frac{1}{2}} \cdot 10^{\frac{1}{8}} \cdot 10^{\frac{1}{256}} \cdots q_n = 10^{\frac{1}{2} + \frac{1}{8} + \frac{1}{256} \text{ to } n \text{ terms}}\, q_n. \end{aligned}$$

As n increases, the exponent $1/2 + 1/8 + 1/256 + \cdots$ to n terms also increases. But it remains less than 1, since it is always a part of the infinite series $1/2 + 1/4 + 1/8 + \cdots$ whose sum is 1, § 704, Ex. 1. Hence it approaches a limit which is some number less than 1, § 192. Represent this limit thus: $1/2 + 1/8 + 1/256 + \cdots$.

Again, as n increases, q_n approaches 1 as limit. For each quotient lies between the divisor used in obtaining it and 1, and as the process is continued the divisors approach 1 as limit.

Hence $4.26 = \lim\limits_{n \doteq \infty} 10^{\frac{1}{2} + \frac{1}{8} + \frac{1}{256} + \cdots \text{ to } n \text{ terms}}\, q_n = 10^{\frac{1}{2} + \frac{1}{8} + \frac{1}{256} + \cdots}$, and therefore $\log 4.26 = 1/2 + 1/8 + 1/256 + \cdots = .6294 \cdots$.

From the common logarithms of the numbers between 1 and 10 we may derive the common logarithms of all other positive numbers by the addition of positive or negative integers. **745**

Example. Find the common logarithms of 42.6 and .426.

1. We have $42.6 = 10 \times 4.26.$

Hence $\log 42.6 = \log 10 + \log 4.26$

$= 1 + \log 4.26 = 1.6294.$

2. Again, $.426 = 4.26/10 = 10^{-1} \times 4.26.$

Hence $\log .426 = \log 10^{-1} + \log 4.26$

$= -1 + \log 4.26 = \bar{1}.6294.$

Similarly $\log 426 = 2.6294$, $\log .0426 = \bar{2}.6294$, and so on; that is, we may obtain the logarithm of any number which has the same sequence of figures as 4.26 by adding a positive or negative integer to log 4.26.

Characteristic and mantissa. In a logarithm expressed as in the preceding example we call the decimal part the *mantissa* and the integral part the *characteristic.* **746**

Thus, $\log 42.6 = 1.6294$ and $\log .426 = \bar{1}.6294$ have the same mantissa .6294 and the characteristics 1 and -1 respectively.

As has been shown in § 745, if n denote a positive number expressed as an integer or decimal, *the mantissa of log* n *depends solely on the sequence of figures in* n, *and the characteristic on the position of the decimal point in* n. **747**

If n lies between 1 and 10, that is, if it has but one figure in its integral part, the characteristic of $\log n$ is 0, § 744. If we shift the decimal point in such a number n one place to the *right,* that is, if we multiply n by 10, we add 1 to the 0 characteristic of $\log n$. Similarly if we shift the decimal point two places to the right, we add 2 to the characteristic of $\log n$, and so on, § 745. Hence the rule: **748**

If $n > 1$, *the characteristic of log* n *is one less than the number of figures in the integral part of* n.

Thus, $\log 426000 = 5.6294$, $\log 42600000 = 7.6294$, and so on.

749 In like manner, if in a number n which has but one figure in its integral part we shift the decimal point μ places to the *left*, that is, if we multiply n by $10^{-\mu}$, we add $-\mu$ to the 0 characteristic of $\log n$. Thus, $\log .426 = -1 + \log 4.26 = \bar{1}.6294$, $\log .0426 = \bar{2}.6294$, and so on, § 745.

In practice we find it convenient to write these negative characteristics $\bar{1}, \bar{2}, \cdots$ in the form $9 - 10, 8 - 10, \cdots$, and to place the positive part $9, 8, \cdots$ before the mantissa, and the -10 after it. Thus, instead of $\bar{1}.6294$ we write $9.6294 - 10$. Hence the rule:

If n < 1, *the characteristic of log* n *is negative. To obtain it, subtract from* 9 *the number of* 0's *immediately to the right of the decimal point in* n, *then write the result before the mantissa, and* -10 *after it.*

Thus, $\log .00426 = 7.6294 - 10$, $\log .000000426 = 3.6294 - 10$.

If more than nine but less than nineteen 0's immediately follow the decimal point, subtract their number from 19 and write the result before the mantissa and -20 after it; and so on.

Example. Given $\log 2 = .3010$, find the number of digits in 2^{25}.

750 **A table of logarithms.** The accompanying table, pp. 384, 385, contains the mantissas of the logarithms of all numbers of three figures computed to the fourth place of decimals and arranged in rows in the order of their magnitude, the decimal points before the mantissas being omitted.

From this table we may also derive mantissas for numbers of more than three figures by aid of the principle:

When a number is changed by an amount which is very small in comparison with the number itself, the change in the logarithm of the number is nearly proportional to the change in the number.

Numerical results obtained by aid of this table are not to be trusted beyond the fourth figure. When greater accuracy is required we must use tables in which the mantissas are given to more than four places of decimals. The student will find it easy to procure a five-, six-, or seven-place table.

751 **To find the logarithm of a number from this table.** We proceed as in the following examples.

Example 1. Find the logarithm of .00589.

We look up the first two significant figures, 58, in the column headed N in the table, then run along the row to the right of 58 until the column is reached which is headed by the third figure, 9. We there find 7701. This (with a decimal point before it) is the mantissa sought. The characteristic is $7 - 10$, § 749. Hence

$$\log .00589 = 7.7701 - 10.$$

Example 2. Find the logarithms of 8 and 46.

The mantissas of these logarithms are the same as those of 800 and 460 respectively. Hence, proceeding as in Ex. 1, we find

$$\log 8 = .9031, \qquad \log 46 = 1.6628.$$

Example 3. Find the logarithm of 4673.

The mantissa is the same as that of log 467.3. It must therefore lie between mant. log 467 and mant. log 468.

From the table we find mant. log 467 = 6693 and mant. log 468 = 6702, and the difference between these mantissas is 9.

Thus if we add 1 to 467, we add 9 to mant. log 467. Hence if we add .3 of 1 to 467, we should add .3 of 9, or 3 approximately, to mant. log 467.

Hence mant. log 467.3 = 6693 + 3 = 6696, and therefore

$$\log .4673 = 9.6696 - 10.$$

Observe that until the characteristic is introduced we omit the decimal point which properly belongs before the mantissa.

The method illustrated in Ex. 3 for finding the mantissa of the logarithm of a number of more than three figures may be described as follows:

From the table obtain m, *the mantissa corresponding to the first three figures, also* d, *the difference between* m *and the next greater mantissa.*

Multiply d *by the remaining part of the number with a decimal point before it, and add the integral part of the product (increased by* 1 *if the decimal part is .5 or more) to* m.

N	0	1	2	3	4	5	6	7	8	9
10	0000	0043	0086	0128	0170	0212	0253	0294	0334	0374
11	0414	0453	0492	0531	0569	0607	0645	0682	0719	0755
12	0792	0828	0864	0899	0934	0969	1004	1038	1072	1106
13	1139	1173	1206	1239	1271	1303	1335	1367	1399	1430
14	1461	1492	1523	1553	1584	1614	1644	1673	1703	1732
15	1761	1790	1818	1847	1875	1903	1931	1959	1987	2014
16	2041	2068	2095	2122	2148	2175	2201	2227	2253	2279
17	2304	2330	2355	2380	2405	2430	2455	2480	2504	2529
18	2553	2577	2601	2625	2648	2672	2695	2718	2742	2765
19	2788	2810	2833	2856	2878	2900	2923	2945	2967	2989
20	3010	3032	3054	3075	3096	3118	3139	3160	3181	3201
21	3222	3243	3263	3284	3304	3324	3345	3365	3385	3404
22	3424	3444	3464	3483	3502	3522	3541	3560	3579	3598
23	3617	3636	3655	3674	3692	3711	3729	3747	3766	3784
24	3802	3820	3838	3856	3874	3892	3909	3927	3945	3962
25	3979	3997	4014	4031	4048	4065	4082	4099	4116	4133
26	4150	4166	4183	4200	4216	4232	4249	4265	4281	4298
27	4314	4330	4346	4362	4378	4393	4409	4425	4440	4456
28	4472	4487	4502	4518	4533	4548	4564	4579	4594	4609
29	4624	4639	4654	4669	4683	4698	4713	4728	4742	4757
30	4771	4786	4800	4814	4829	4843	4857	4871	4886	4900
31	4914	4928	4942	4955	4969	4983	4997	5011	5024	5038
32	5051	5065	5079	5092	5105	5119	5132	5145	5159	5172
33	5185	5198	5211	5224	5237	5250	5263	5276	5289	5302
34	5315	5328	5340	5353	5366	5378	5391	5403	5416	5428
35	5441	5453	5465	5478	5490	5502	5514	5527	5539	5551
36	5563	5575	5587	5599	5611	5623	5635	5647	5658	5670
37	5682	5694	5705	5717	5729	5740	5752	5763	5775	5786
38	5798	5809	5821	5832	5843	5855	5866	5877	5888	5899
39	5911	5922	5933	5944	5955	5966	5977	5988	5999	6010
40	6021	6031	6042	6053	6064	6075	6085	6096	6107	6117
41	6128	6138	6149	6160	6170	6180	6191	6201	6212	6222
42	6232	6243	6253	6263	6274	6284	6294	6304	6314	6325
43	6335	6345	6355	6365	6375	6385	6395	6405	6415	6425
44	6435	6444	6454	6464	6474	6484	6493	6503	6513	6522
45	6532	6542	6551	6561	6571	6580	6590	6599	6609	6618
46	6628	6637	6646	6656	6665	6675	6684	6693	6702	6712
47	6721	6730	6739	6749	6758	6767	6776	6785	6794	6803
48	6812	6821	6830	6839	6848	6857	6866	6875	6884	6893
49	6902	6911	6920	6928	6937	6946	6955	6964	6972	6981
50	6990	6998	7007	7016	7024	7033	7042	7050	7059	7067
51	7076	7084	7093	7101	7110	7118	7126	7135	7143	7152
52	7160	7168	7177	7185	7193	7202	7210	7218	7226	7235
53	7243	7251	7259	7267	7275	7284	7292	7300	7308	7316
54	7324	7332	7340	7348	7356	7364	7372	7380	7388	7396

N	0	1	2	3	4	5	6	7	8	9
55	7404	7412	7419	7427	7435	7443	7451	7459	7466	7474
56	7482	7490	7497	7505	7513	7520	7528	7536	7543	7551
57	7559	7566	7574	7582	7589	7597	7604	7612	7619	7627
58	7634	7642	7649	7657	7664	7672	7679	7686	7694	7701
59	7709	7716	7723	7731	7738	7745	7752	7760	7767	7774
60	7782	7789	7796	7803	7810	7818	7825	7832	7839	7846
61	7853	7860	7868	7875	7882	7889	7896	7903	7910	7917
62	7924	7931	7938	7945	7952	7959	7966	7973	7980	7987
63	7993	8000	8007	8014	8021	8028	8035	8041	8048	8055
64	8062	8069	8075	8082	8089	8096	8102	8109	8116	8122
65	8129	8136	8142	8149	8156	8162	8169	8176	8182	8189
66	8195	8202	8209	8215	8222	8228	8235	8241	8248	8254
67	8261	8267	8274	8280	8287	8293	8299	8306	8312	8319
68	8325	8331	8338	8344	8351	8357	8363	8370	8376	8382
69	8388	8395	8401	8407	8414	8420	8426	8432	8439	8445
70	8451	8457	8463	8470	8476	8482	8488	8494	8500	8506
71	8513	8519	8525	8531	8537	8543	8549	8555	8561	8567
72	8573	8579	8585	8591	8597	8603	8609	8615	8621	8627
73	8633	8639	8645	8651	8657	8663	8669	8675	8681	8686
74	8692	8698	8704	8710	8716	8722	8727	8733	8739	8745
75	8751	8756	8762	8768	8774	8779	8785	8791	8797	8802
76	8808	8814	8820	8825	8831	8837	8842	8848	8854	8859
77	8865	8871	8876	8882	8887	8893	8899	8904	8910	8915
78	8921	8927	8932	8938	8943	8949	8954	8960	8965	8971
79	8976	8982	8987	8993	8998	9004	9009	9015	9020	9025
80	9031	9036	9042	9047	9053	9058	9063	9069	9074	9079
81	9085	9090	9096	9101	9106	9112	9117	9122	9128	9133
82	9138	9143	9149	9154	9159	9165	9170	9175	9180	9186
83	9191	9196	9201	9206	9212	9217	9222	9227	9232	9238
84	9243	9248	9253	9258	9263	9269	9274	9279	9284	9289
85	9294	9299	9304	9309	9315	9320	9325	9330	9335	9340
86	9345	9350	9355	9360	9365	9370	9375	9380	9385	9390
87	9395	9400	9405	9410	9415	9420	9425	9430	9435	9440
88	9445	9450	9455	9460	9465	9469	9474	9479	9484	9489
89	9494	9499	9504	9509	9513	9518	9523	9528	9533	9538
90	9542	9547	9552	9557	9562	9566	9571	9576	9581	9586
91	9590	9595	9600	9605	9609	9614	9619	9624	9628	9633
92	9638	9643	9647	9652	9657	9661	9666	9671	9675	9680
93	9685	9689	9694	9699	9703	9708	9713	9717	9722	9727
94	9731	9736	9741	9745	9750	9754	9759	9763	9768	9773
95	9777	9782	9786	9791	9795	9800	9805	9809	9814	9818
96	9823	9827	9832	9836	9841	9845	9850	9854	9859	9863
97	9868	9872	9877	9881	9886	9890	9894	9899	9903	9908
98	9912	9917	9921	9926	9930	9934	9939	9943	9948	9952
99	9956	9961	9965	9969	9974	9978	9983	9987	9991	9996

752 **To find a number when its logarithm is given.** We have merely to reverse the process described in the preceding section.

Example 1. Find the number whose logarithm is 5.9552 – 10.

We find the mantissa 9552 in the table in the row marked 90 and in the column marked 2. Hence the required *sequence of figures* is 902.

But since the characteristic is 5 – 10, the number is a decimal with 9 – 5, or 4, 0's at the right of the decimal point, § 749. Hence the required *number* is .0000902.

Example 2. Find the number whose logarithm is 7.5520.

Looking in the table we find that the given mantissa 5520 lies between the mantissas 5514 and 5527 corresponding to 356 and 357 respectively. The lesser of these mantissas, 5514, differs from the greater, 5527, by 13 and from the given mantissa, 5520, by 6.

Thus, if we add 13 to the mantissa 5514, we add 1 to the number 356. Hence if we add 6 to the mantissa 5514, we should add 6/13 of 1, or .5 approximately, to 356.

Hence the required sequence of figures is 3565, and therefore by the rule for characteristic, § 748, the required number is 35650000.

We therefore have the following rule for finding the sequence of figures corresponding to a given mantissa which is not in the table:

Find from the table the next lesser mantissa m, *the three corresponding figures, and* d, *the difference between* m *and the next greater mantissa.*

Subtract m *from the given mantissa and divide the remainder by* d, *annexing the resulting figure to the three figures already obtained.*

753 **Cologarithms.** The *cologarithm* of a number is the logarithm of the reciprocal of the number.

Since $\text{colog}\, m = \log 1/m = \log 1 - \log m = -\log m$, §§ 733, 737, we can find the cologarithm of a number by merely changing the sign of its logarithm. But to avail ourselves of the table we must keep the decimal parts of all logarithms positive. We therefore proceed as follows:

Example 1. Find colog 89.2.

We have	$\log 1 =$	$10 \quad\quad - 10$
and	$\log 89.2 =$	1.9504
Hence	$\text{colog}\, 89.2 =$	$8.0496 - 10$

Example 2. Find colog .929.

We have	$\log 1 =$	$10 \quad\quad - 10$
and	$\log .929 =$	$9.9680 - 10$
Hence	$\text{colog}\, .929 =$	$.0320$

Hence we may find the cologarithm of a number from its logarithm by beginning at the characteristic and subtracting each figure from 9 until the last significant figure is reached, which figure must be subtracted from 10. To this result we do or do not affix -10 according as -10 is not or is affixed to the logarithm. In this way when the number has not more than three figures we may obtain its cologarithm directly from the table.

754 **Computation by logarithms.** The following examples will serve to show how expeditiously approximate values of products, quotients, powers, and roots of numbers may be obtained by aid of logarithms (compare § 740).

Example 1. Find the value of $.0325 \times .6425 \times 5.26$.

$\text{Log}\,(.0325 \times .6425 \times 5.26) = \log .0325 + \log .6425 + \log 5.26.$

But	$\log .0325 =$	$8.5119 - 10$
	$\log .6425 =$	$9.8079 - 10$
	$\log 5.26 =$	$.7210$
Hence log of product		$= 19.0408 - 20 = 9.0408 - 10$

Therefore the product is .1099.

Example 2. Find the value of $46.72 / .0998$.

$\text{Log}\,(46.72 / .0998) = \log 46.72 - \log .0998.$

But	$\log 46.72 =$	$11.6695 - 10$
	$\log .0998 =$	$8.9991 - 10$
Hence log of quotient	$=$	2.6704

Therefore the quotient is 468.2.

We write log 46.72, that is 1.6695, in the form 11.6695 – 10 in order to make its positive part greater than that of 8.9991 – 10 which is to be subtracted from it.

Example 3. Find the value of 295 × .05631 ÷ 806.

Log (295 × .05631 ÷ 806) = log 295 + log .05631 + colog 806.

$$
\begin{aligned}
\text{But} \qquad \log 295 &= 2.4698 \\
\log .05631 &= 8.7506 - 10 \\
\operatorname{colog} 806 &= \underline{7.0937 - 10} \\
\text{Hence log of required result} \quad &= 18.3141 - 20 = 8.3141 - 10
\end{aligned}
$$

Therefore the required result is .02061.

Example 4. Find the sixth power of .7929.

Log $(.7929)^6 = 6 \times \log .7929$.

$$
\begin{aligned}
\text{But} \qquad \log .7929 &= 9.8992 - 10 \\
&\quad \underline{\qquad\qquad\quad 6} \\
\text{Hence} \qquad \log (.7929)^6 &= 59.3952 - 60 = 9.3952 - 10
\end{aligned}
$$

Therefore $(.7929)^6 = .2484$.

Example 5. Find the seventh root of .00898.

Log $\sqrt[7]{.00898} = (\log .00898) \div 7$.

$$
\begin{aligned}
\text{But} \qquad \log .00898 &= 7.9533 - 10 \\
&\quad 7\overline{)67.9533 - 70} \\
\text{Hence} \qquad \log \sqrt[7]{.00898} &= 9.7076 - 10
\end{aligned}
$$

Therefore $\sqrt[7]{.00898} = .510$.

Observe that when as here we have occasion to divide a negative logarithm by some number, we add to its positive and negative parts such a multiple of 10 that the quotient of the negative part will be – 10.

Negative numbers do not have real logarithms to the base 10 since all real powers of 10 are positive numbers. If asked to find the value of an expression which involves negative factors, we may first find the absolute value of the expression by logarithms and then attach the appropriate sign to the result.

Thus, if the given expression were 456 × (– 85.96), we should first find the value of 456 × 85.96 by logarithms and then attach the – sign to the result.

EXERCISE LXIII

Find approximate values of the following by aid of logarithms.

1. $79 \times 470 \times .982$.
2. $(-9503) \times (-.0086578)$.
3. 1375600×8799000.
4. $.0356 \times (-.00049)$.
5. $\dfrac{8075}{364.9}$.
6. $\dfrac{.00542}{.04708}$.
7. $\dfrac{24617}{-.00054}$.
8. $\dfrac{.643 \times 7095}{67 \times 9 \times .462}$.
9. $\dfrac{9097 \times 5.4086}{-225 \times 593 \times .8665}$.
10. $(2.388)^5$.
11. $(.57)^{-4}$.
12. $(19/11)^9$.
13. $(1.014)^{25}$.
14. $\sqrt{67.54}$.
15. $\sqrt[3]{-.30892}$.
16. $8^{\frac{5}{4}}$.
17. $(.001)^{\frac{2}{3}}$.
18. $(29\frac{9}{11})^{\frac{1}{2}}$.
19. $\sqrt{\frac{5}{6}} \times \sqrt[3]{\frac{79}{45}}$.
20. $\sqrt{.1} \div (.009)^{\frac{3}{8}}$.
21. $(.00068)^{-\frac{5}{4}}$.
22. $(6\frac{2}{3})^{3.4}$.
23. $(-9306)^{\frac{3}{7}}$.
24. $(.0057)^{2.5}$.
25. $(5648)^{\frac{1}{2}} \times (-.94)^{\frac{1}{3}}$.
26. $28927^3 \div (.8)^{\frac{2}{5}}$.
27. $\dfrac{\sqrt[7]{.0476} \times \sqrt[5]{222}}{\sqrt[3]{5059 \times .0088}}$.
28. $\dfrac{\sqrt[6]{943 \times 7298}}{\sqrt[5]{.00006 \times .99}}$.
29. $\sqrt{\dfrac{854 \times \sqrt[3]{.042}}{7.9856 \times \sqrt[4]{.0005}}}$.
30. $\sqrt[3]{\dfrac{7^{\frac{1}{4}} \times 92^{\frac{1}{5}} \times (.01)^{\frac{1}{2}}}{(.00026)^5 \times 5968^{\frac{1}{3}}}}$.

SOME APPLICATIONS OF COMMON LOGARITHMS

Logarithms to other bases than 10. From the logarithm of a number to the base 10 we can derive its logarithm to any other positive base except 1 by aid of the theorem: **755**

The logarithms of a number m *to two different bases,* a *and* b, *are connected by the formula* $\log_b m = \log_a m / \log_a b$.

For let $\quad m = a^{\mu}$, that is $\mu = \log_a m$,

and let $\quad b = a^{\nu}$, that is $\nu = \log_a b$.

Since $\quad a^{\nu} = b$, we have $a = b^{\frac{1}{\nu}}$.

Hence $\quad m = a^{\mu} = (b^{\frac{1}{\nu}})^{\mu} = b^{\frac{\mu}{\nu}}$,

that is, $\quad \log_b m = \mu / \nu = \log_a m / \log_a b$.

Example. Find the logarithm of .586 to the base 7.

$$\text{Log}_7 .586 = \frac{\log_{10} .586}{\log_{10} 7} = \frac{9.7679 - 10}{.8451} = -\frac{2321}{8451} = -.2746.$$

We reduce $9.7679 - 10$ to the form of a single negative number, namely $-.2321$, and perform the final division by logarithms.

756 When $m = a$ the formula gives $\log_b a = 1/\log_a b$.

757 The only base besides 10 of which any actual use is made is a certain irrational number denoted by the letter e whose approximate value is 2.718. Logarithms to this base are called *natural logarithms.* We shall consider them in another connection.

758 **Exponential and logarithmic equations.** Equations in which the unknown letter occurs in an exponent or in a logarithmic expression may sometimes be solved as follows.

Example 1. Solve the equation $13^{2x+5} = 14^{x+7}$.

Taking logarithms of both members, $(2x + 5) \log 13 = (x + 7) \log 14$.

Solving, $$x = \frac{7 \log 14 - 5 \log 13}{2 \log 13 - \log 14} = \frac{2.4532}{1.0817} = 2.268.$$

Example 2. Solve the equation $\log \sqrt{x - 21} + \frac{1}{2} \log x = 1$.

By §§ 736, 739, we can reduce this equation to the form

$$\log \sqrt{x(x - 21)} = 1 = \log 10.$$

Hence $$x^2 - 21x = 100.$$

Solving, $$x = 25 \text{ or } -4.$$

Example 3. Solve the equation $x^{2 \log x} = 10x$.

Taking logarithms, $2(\log x)^2 = \log x + 1$.

Solving for $\log x$, $\log x = 1 \text{ or } -1/2$.

Hence $x = 10 \text{ or } 1/\sqrt{10}$.

759 **Compound interest.** Suppose that a sum of P dollars is put at compound interest for a period of n years, interest being compounded annually and the interest on one dollar for one year being r.

Then the amount at the end of the first year will be $P + Pr$ or $P(1 + r)$, at the end of the second year it will be $P(1+r)\cdot(1+r)$ or $P(1+r)^2$, and so on. Hence, if A denote the amount at the end of the nth year, we have

$$A = P(1+r)^n.$$

If interest be compounded semiannually, $A = P(1+r/2)^{2n}$; if quarterly, $A = P(1+r/4)^{4n}$; and so on.

We call P the *present worth* of A. If A, n, and r be given, we can find P by means of the formula $P = A(1+r)^{-n}$.

Example 1. Find the amount of $2500 in eighteen years at 4% compound interest.

We have $\log A = \log 2500 + 18 \log 1.04 = 3.7039$.

Hence $A = \$5057$, approximately.

Example 2. At the beginning of each of ten successive years a premium of $120 is paid on a certain insurance policy. What is the worth of the sum of these premiums at the end of the tenth year if computed at 4% compound interest?

The required value is $120[1.04 + (1.04)^2 + \cdots + (1.04)^{10}]$,

that is, by § 701, $120 \times 1.04 \times \dfrac{(1.04)^{10} - 1}{1.04 - 1}$.

By logarithms, $(1.04)^{10} = 1.479$.

Hence the required value is $120 \times 1.04 \times .479 \div .04$; and this, computed by logarithms, gives $1494, approximately.

Annuities. A sum of money which is to be paid at fixed intervals, as annually, is called an *annuity*. **760**

It is required to find the present worth of an annuity of A dollars payable annually for n years, beginning a year hence, the interest on one dollar for one year being r.

The present worth of the first payment is $A(1+r)^{-1}$, that of the second payment is $A(1+r)^{-2}$, and so on.

Hence the present worth of the whole is, § 701,

$$A\left[\frac{1}{1+r} + \frac{1}{(1+r)^2} + \cdots + \frac{1}{(1+r)^n}\right] = \frac{A}{r}\left[1 - \frac{1}{(1+r)^n}\right].$$

If the annuity be *perpetual*, that is, if $n = \infty$, then $(1+r)^n = \infty$, and the formula for the present worth reduces to A/r.

Example. What sum should be paid for an annuity of \$1000 payable annually for twenty years, money being supposed to be worth 3% per annum?

The present worth, P, is $\dfrac{1000}{.03}\left[1 - \dfrac{1}{(1.03)^{20}}\right]$.

By logarithms, we find that $(1.03)^{20} = 1.803$.

Hence $P = \dfrac{1000}{.03}\left[1 - \dfrac{1}{1.803}\right] = \dfrac{1000 \times .803}{.03 \times 1.803} = \$14845$, approximately.

EXERCISE LXIV

1. Find $\log_5 555$, $\log_7 .0463$, $\log_{100} 47$.

2. Solve the following exponential equations.

   (1) $3^x = 729$. (2) $a^{x^2+2} = a^{3x}$. (3) $213^x = 516^{-x+4}$.

3. Solve the following logarithmic equations.

   (1) $\log x + \log(x+3) = 1$. (2) $\log x^2 + \log x = 2$.

   (3) $\log(1-2x)^3 - \log(3-x)^3 = 6$. (4) $x^{\log x} = 2$.

4. Find the amount of \$7500 in thirty-five years at 5% compound interest, the interest being compounded annually.

5. Find the amount of \$5500 in twenty years at 3% compound interest, the interest being compounded semiannually.

6. Show that a sum of money will more than double itself in fifteen years and that it will increase more than a hundredfold in ninety-five years at 5% compound interest.

7. What sum will amount to \$1250 if put at compound interest at 4% for fifteen years?

8. A man invests \$200 a year in a savings bank which pays $3\frac{1}{2}$% per annum on all deposits. What will be the total amount due him at the end of twenty-five years?

9. What sum should be paid for an annuity of \$1200 a year to be paid for thirty years, money being supposed to be worth 4% per annum? What sum should be paid were this annuity to be perpetual?

10. If c denote the length of the hypotenuse of a right-angled triangle and a, b the lengths of the other two sides, $b = \sqrt{(c+a)(c-a)}$. Given $c = 586.4$, $a = 312.2$, find b and the area of the triangle, using logarithms.

11. If a, b, c denote the lengths of the sides of a triangle and $s=(a+b+c)/2$, the area of the triangle is $\sqrt{s(s-a)(s-b)(s-c)}$. Find the area of the triangle in which $a=416.8$, $b=424$, $c=25.68$.

12. Find the area of the surface and the volume of a sphere the length of whose radius is 23.6 by aid of the formulas $S=4\pi r^2$, $V=4\pi r^3/3$, assuming that $\pi=3.1416$.

XXV. PERMUTATIONS AND COMBINATIONS

761 **Definitions of permutation and combination.** Suppose a group of n letters to be given, as $a, b, c, \cdots, k$, denoting objects of any kind.

Any set of r of these letters, considered without regard to order, is called a *combination of the* n *letters* r *at a time*, or, more briefly, an r-*combination* of the n letters.

We shall use the symbol C_r^n to denote the number of such combinations.

Thus, the 2-combinations of the four letters a, b, c, d are

$$ab,\ ac,\ ad,\ bc,\ bd,\ cd.$$

There are *six* of these combinations, that is, $C_2^4=6$.

On the other hand, any arrangement of r of these n letters in a definite order in a row is called a *permutation of the* n *letters,* r *at a time*, or, more briefly, an r-*permutation* of the n letters.

We shall use the symbol P_r^n to denote the number of such permutations.

Thus, the 2-permutations of the four letters a, b, c, d are

$$ab,\ ac,\ ad,\ bc,\ bd,\ cd,$$
$$ba,\ ca,\ da,\ cb,\ db,\ dc.$$

There are *twelve* of these permutations, that is $P_2^4=12$.

Observe that while ab and ba denote the same combination, they denote different permutations.

In what has just been said it is assumed that the letters $a, b, \cdots, k$ are all different and that the repetition of a letter within a permutation or combination is not allowed. This will be the understanding throughout the chapter except where the contrary is stated.

762 A preliminary theorem. We have already had occasion to apply the following principle, § 554:

If a certain thing can be done in m *ways, and if, when it has been done, a certain other thing can be done in* n *ways, the entire number of ways in which both things can be done in the order stated is* mn.

We reason thus: Since for each way of doing the first thing there are n ways of doing both things, for m ways of doing the first thing there are mn ways of doing both things.

More generally, if a first thing can be done in m ways, then a second thing in n ways, then a third in p ways, and so on, the entire number of ways in which all the things can be done in the order stated is $m \cdot n \cdot p \cdots$.

Example. How many numbers of three different figures each can be formed with the digits 1, 2, 3, $\cdots$, 9?

We may choose any one of the *nine* digits for the first figure of the number, then any one of the remaining *eight* digits for its second figure, and finally any one of the *seven* digits still remaining for its third figure. Hence we may form $9 \cdot 8 \cdot 7$, or 504, numbers of the kind required.

763 The number of r-permutations of n different letters. By the reasoning employed in the preceding example we readily prove that this number P_r^n is given by the formula

$$P_r^n = n(n-1)(n-2) \cdots \text{to } r \text{ factors.} \quad (1)$$

For in forming an r-permutation of n letters we may choose any one of the n letters for its first letter, then any one of the remaining $n-1$ letters for its second letter, then any one of the $n-2$ letters still remaining for its third letter, and so on.

Hence, § 762, the entire number of ways in which we may choose its first, second, third, $\cdots$, rth letters, in other words the entire number of ways in which we may form an r-permutation with the n letters, is $n(n-1)(n-2)\cdots$ to r factors.

Thus, the numbers of permutations of the letters a, b, c, d, e one, two, three, four, five at a time are

$$P_1^5 = 5,\ P_2^5 = 5\cdot 4,\ P_3^5 = 5\cdot 4\cdot 3,\ P_4^5 = 5\cdot 4\cdot 3\cdot 2,\ P_5^5 = 5\cdot 4\cdot 3\cdot 2\cdot 1.$$

Evidently the rth factor in the product $n(n-1)(n-2)\cdots$ is $n-(r-1)$, or $n-r+1$. Hence the formula (1) may be written

$$P_r^n = n(n-1)(n-2)\cdots(n-r+1). \qquad (2)$$

When $r = n$, the factor $n-r+1$ is $n-n+1$, or 1, and we have $P_n^n = n(n-1)\cdots 2\cdot 1$, or $1\cdot 2\cdots(n-1)n$. The continued product $1\cdot 2\cdots n$ is called *factorial* n and is denoted by the symbol $n!$ or $\underline{|n}$. Hence the entire number of orders in which we can arrange n letters in a row, using all of them in each arrangement, is given by the formula

$$P_n^n = n!. \qquad (3)$$

For a reason which will appear later, § 775, the meaningless symbol $0!$ is assigned the value 1.

Example 1. How many different signals can be made with four flags of different colors displayed singly, or one or more together, one above another?

There will be one signal for each arrangement of the flags taken 1, 2, 3, or 4 at a time. Hence the number is $P_1^4 + P_2^4 + P_3^4 + P_4^4$, or 64.

Example 2. Of the permutations of the letters of the word *fancies* taken all at a time,

(1) How many begin and end with a consonant?

The first place may be filled in 4 ways, then the last place in 3 ways, then the intermediate places in $5!$ ways. Hence the required number is $4\cdot 3\cdot 5!$, or 1440.

(2) How many have the vowels in the even places?

The vowels may be arranged in the even places in $3!$ ways, the consonants in the odd places in $4!$ ways, and each arrangement of vowels may be associated with every arrangement of consonants. Hence the required number is $3!\cdot 4!$, or 144.

(3) How many do not have c as their middle letter?

Evidently c is the middle letter in $6!$ of the permutations, for the remaining letters may be arranged in all possible orders. Therefore the number of the permutations in which c is not the middle letter is $7! - 6!$, or 4320.

Example 3. Show that $P_6^8 = 4 \cdot P_7^7$, and that $P_3^{16} = 2 \cdot P_4^8$.

Example 4. If $P_4^{2n} = 127\, P_3^{2n}$, find n.

Example 5. How many passenger tickets will a railway company need for use on a division on which there are twenty stations?

Example 6. In how many of the permutations of the letters a, e, i, o, u, y, taken all at a time, do the letters a, e, i stand together?

Example 7. With the letters of the word *numerical* how many arrangements of five letters each can be formed in which the odd places are occupied by consonants?

Example 8. Show that with the digits $0, 1, 2, \cdots, 9$ it is possible to form $P_4^{10} - P_3^9$ numbers, each of which has four different figures.

Example 9. How many numbers all told can be formed with the digits 3, 4, 5, 7, 8, all the figures in each number being different?

Example 10. In how many ways can seven boys be arranged in a row if one particular boy is not permitted to stand at either end of the row?

764 Circular permutations. The number of different orders in which n different letters can be arranged about the circumference of a circle or any other closed curve is $(n-1)!$.

For the relative order of the n letters will not be changed if we shift all the letters the same number of places along the curve. Hence we shall take account of all the distinct orders of the n letters if we suppose one of the letters fixed in position and the remaining $n-1$ then arranged in all possible orders. But these $n-1$ letters can be arranged in $(n-1)!$ orders, § 763, (3).

Thus, eight persons can be seated at a round table in $7!$, or 5040, orders.

Example 1. Show that the number of circular r-permutations of n different letters is P_r^n/r.

Example 2. Taking account of the fact that a circular ring will come into coincidence with itself if revolved about a diameter through an angle of 180°, show that $(n-1)!/2$ different necklaces can be formed by stringing together n beads of different colors.

Example 3. In how many ways can a party of four ladies and four gentlemen be arranged at a round table so that the ladies and gentlemen may occupy alternate seats?

Permutations of different letters when repetitions are allowed. **765** With n different letters we can form n^r arrangements or permutations of r letters each, if allowed to repeat a letter within a permutation.

For in forming a permutation of this kind we may choose any one of the n letters for its first letter, and then again, since repetitions are allowed, any one of the n letters for its second letter, and so on. Hence, § 762, the entire number of ways in which we can form the permutation is $n \cdot n \cdot n \cdots$ to r factors, or n^r.

Thus, with the digits 1, 2, 3, $\cdots$, 9 we can form all told 9^3, or 729, numbers of three figures each.

Example 1. How many numbers of one, two, or three figures each can be formed with the characters 1, 2, 3, 5, 7?

Example 2. In how many ways can three prizes be given to seven boys if each boy is eligible for every prize?

The number of n-permutations of n letters which are not all different. **766** Let us inquire how many distinguishable permutations can be formed with the letters a, a, a, b, c (1), three of which are alike, all the letters being used in each permutation.

Compare these permutations with the corresponding permutations of the letters a, a', a'', b, c (2), all of which are different. If we take any one of the permutations of (1), as $abaca$, and, leaving b, c undisturbed, we interchange the a's, we get nothing new. But if we treat the corresponding permutation of (2), namely $aba'ca''$, in a similar manner, we obtain $3!$ distinct permutations, namely $aba'ca''$, $aba''ca'$, $a'ba''ca$, $a'baca''$, $a''baca'$, $a''ba'ca$. Hence to each permutation of (1) there correspond

3! permutations of (2). The number of the permutations of (2) is 5!, § 763, (3). Therefore the number of the permutations of (1) is 5! ÷ 3!.

By the reasoning here employed we can prove in general that the number of distinguishable n-permutations of n letters of which p are alike, q others alike, and so on, is given by the formula

$$N = \frac{n!}{p!\,q!\cdots}.$$

Example 1. In how many different ways can the letters of the word *independence* be arranged?

Of the 12 letters in this word 4 are e's, 3 are n's, 2 are d's.

Hence the required result is $12!/4!\cdot 3!\cdot 2!$, or 1,663,200.

Example 2. In how many ways can the letters of the word *Antioch* be arranged without changing the relative order of the vowels or that of the consonants?

From the proof just given it is evident that the required number of arrangements is the same as it would be if the three vowels were the same and the four consonants were the same. Hence it is $7!/3!\cdot 4!$, or 35.

Example 3. How many terms has each of the following symmetric functions of five variables x, y, z, u, v, namely, Σx^3y^2z, $\Sigma x^3y^2z^2$, Σx^3yzu, and $\Sigma x^3y^3z^2u^2$?

We shall obtain all the terms of Σx^3y^2z once each if, leaving the exponents 3, 2, 1 fixed in position, we write under them every 3-permutation of the letters x, y, z, u, v. Hence the number of the terms is P_3^5, or 60.

If we apply the same method to $\Sigma x^3y^2z^2$, we obtain the term $x^3y^2z^2$ twice, once in the form $x^3y^2z^2$ and once in the form $x^3z^2y^2$. Similarly every term is obtained twice, once for each of the orders in which its letters under the equal exponents can be written. Hence the number of terms in $\Sigma x^3y^2z^2$ is $P_3^5/2$, or 30.

Similarly Σx^3yzu has $P_4^5/3!$, or 20, terms, and $\Sigma x^3y^3z^2u^2$ has $P_4^5/2!2!$, or 30, terms.

Example 4. In how many ways can five pennies, six five-cent pieces, and four dimes be distributed among fifteen children so that each may receive a coin?

Example 5. In a certain district of a town there are ten streets running north and south, and five running east and west. In how many ways can a person walk from the southwest corner of the district to the northeast corner, always taking the shortest course?

767 **The number of r-combinations of n different letters.** This number, C^n_r, is given by the formula

$$C^n_r = P^n_r \div r! = \frac{n(n-1)\cdots(n-r+1)}{r!}. \qquad (1)$$

For evidently if we were to form all the r-combinations and were then to arrange the letters of each combination in turn in all possible orders, we should obtain all the r-permutations.

But since each combination would thus yield $r!$ permutations, § 763, (3), all the combinations, C^n_r in number, would yield $r! \times C^n_r$ permutations.

Hence $r! \times C^n_r = P^n_r$, and therefore $C^n_r = P^n_r \div r!$.

Thus, the numbers of combinations of the letters a, b, c, d, e one, two, three, four, five at a time are

$$C^5_1 = \frac{5}{1}, \quad C^5_2 = \frac{5 \cdot 4}{1 \cdot 2}, \quad C^5_3 = \frac{5 \cdot 4 \cdot 3}{1 \cdot 2 \cdot 3}, \quad C^5_4 = \frac{5 \cdot 4 \cdot 3 \cdot 2}{1 \cdot 2 \cdot 3 \cdot 4}, \quad C^5_5 = \frac{5 \cdot 4 \cdot 3 \cdot 2 \cdot 1}{1 \cdot 2 \cdot 3 \cdot 4 \cdot 5}.$$

This expression for C^n_r is the coefficient of the $(r+1)$th term in the expansion of $(a+b)^n$ by the binomial theorem, § 565. This was shown in § 560 by an argument which is merely another proof of the formula (1).

768 If in the expression just obtained for C^n_r we multiply both numerator and denominator by $(n-r)!$, we obtain the more symmetrical formula

$$C^n_r = \frac{n!}{r!(n-r)!}. \qquad (2)$$

769 From this formula (2) it follows that the number of the r-combinations of n letters is the same as the number of the $(n-r)$-combinations.

For $C^n_{n-r} = \dfrac{n!}{(n-r)![n-(n-r)]!} = \dfrac{n!}{(n-r)!\,r!} = C^n_r$.

This also follows from the fact that for every set of r-things taken, a set of $n-r$ things is left.

Thus, $C^{14}_{12} = C^{14}_2 = 14 \cdot 13/1 \cdot 2 = 91$. Observe how much more readily C^{14}_{12} is found in this way than by a direct application of (1).

Example 1. There are fifteen points in a plane and no three of these points lie in the same straight line. Find the number of triangles which can be formed by joining them.

Evidently there are as many triangles as there are combinations of the points taken three at a time. Hence the number of triangles is C_3^{15}, that is, $15 \cdot 14 \cdot 13 / 1 \cdot 2 \cdot 3$, or 455.

Example 2. In how many ways can a committee of three be selected from ten persons (1) so as always to include a particular person A? (2) so as always to exclude A?

(1) The other two members of the committee can be chosen from the remaining nine persons in C_2^9, that is, $9 \cdot 8 / 1 \cdot 2$, or 36, ways.

(2) The entire committee can be chosen from the remaining nine persons in C_3^9, that is, $9 \cdot 8 \cdot 7 / 1 \cdot 2 \cdot 3$, or 84, ways.

Example 3. With the vowels a, e, i, o and the consonants b, c, d, f, g how many arrangements of letters can be made, each consisting of two vowels and three consonants?

The vowels for the arrangement can be chosen in C_2^4 ways, the consonants in C_3^5 ways; then each selection of vowels can be combined with every selection of consonants and the whole arranged in $5\,!$ ways. Hence the required result is $C_2^4 \cdot C_3^5 \cdot 5\,!$, or 7200.

Example 4. In how many ways can eighteen books be divided equally among three persons A, B, C?

A's books can be selected in C_6^{18} ways, then B's in C_6^{12} ways, then C's in C_6^6, or 1, way. Hence, § 762, the required result is $C_6^{18} \cdot C_6^{12} \cdot C_6^6$, or $18\,!/(6\,!)^3$.

To find in how many ways the 18 books can be distributed into three sets of 6 books each, we must divide the result just obtained by $3\,!$, which gives $18\,!/(6\,!)^3\,3\,!$; for here the *order* in which the three sets may chance to be arranged is immaterial.

Example 5. With the letters of the word *mathematical* how many different selections and how many different arrangements of four letters each can be made?

As the letters are not all different we cannot obtain the required results by single applications of the formulas for C_r^n and P_r^n.

The letters are a, a, a; m, m; t, t; h, e, i, c, l.

Hence we may classify and then enumerate the possible selections and arrangements as follows:

1. Those having three like letters.

Combining the 3 a's with each of the seven other letters in turn, we obtain 7 selections and $7 \cdot 4!/3!$, or 28, arrangements.

2. Those having two pairs of like letters.

There are 3 such selections and $3 \cdot 4!/2! \cdot 2!$, or 18, such arrangements.

3. Those having two letters alike, the other two different.

Of such selections there are $3 \cdot C_2^7$, or 63; of arrangements there are $63 \cdot 4!/2!$, or 756.

4. Those having four different letters.

Of selections there are C_4^8, or 70; of arrangements, $70 \cdot 4!$, or 1680.

Hence the total number of selections is $7 + 3 + 63 + 70$, or 143; of arrangements, $28 + 18 + 756 + 1680$, or 2482.

Example 6. Find the values of C_{15}^{17}, C_5^{10}, and C_{19}^{23}.

Example 7. If $C_8^n = C_7^n$, find n.

Example 8. If $2\,C_4^n = 5 \cdot C_2^n$, find n.

Example 9. How many planes are determined by twelve points, no four of which lie in the same plane?

Example 10. How many parties of five men each can be chosen from a company of twelve men? In how many of these parties will a particular man A be included? From how many will A be excluded?

Example 11. Of the parties described in the preceding example how many will include two particular men A and B? How many will include one but not both of them? How many will include neither of them?

Example 12. From twenty Republicans and eighteen Democrats how many committees can be chosen, each consisting of four Republicans and three Democrats?

Example 13. With five vowels and fourteen consonants how many arrangements of letters can be formed, each consisting of three vowels and four consonants?

Example 14. In how many ways can a pack of fifty-two cards be divided equally among four players A, B, C, D? In how many ways can the cards be distributed into four piles containing thirteen each?

Example 15. How many numbers, each of five figures, can be formed with the characters 2, 3, 4, 2, 5, 2, 3, 6, 7?

770 **Total number of combinations.** If in the formula for $(a+b)^n$, § 561, we set $a = b = 1$ and then subtract 1 from both members, we obtain

$$C_1^n + C_2^n + \cdots + C_n^n = 2^n - 1.$$

Hence the total number of combinations of n different things taken one, two, $\cdots$, n at a time, in other words, the total number of ways in which one or more things may be chosen from n things, is $2^n - 1$.

This may also be proved as follows: Each particular thing can be dealt with in one of two ways, that is, be taken or left. Hence the total number of ways of dealing with all n things is $2 \cdot 2 \cdots$ to n factors, or 2^n, § 762. Therefore, rejecting the case in which all the things are left, we have as before, $2^n - 1$.

Example 1. How many different sums of money can be paid with one dime, one quarter, one half dollar, and one dollar?

Example 2. By the reasoning just illustrated, show that the total number of ways in which one or more things can be chosen from $p + q + \cdots$ things of which p are alike, q others alike but different from the p things, and so on, is $(p+1)(q+1)\cdots - 1$.

Example 3. How many different sums of money can be paid with two dimes, five quarters, and four half dollars?

771 **Greatest value of C_r^n.** In the expression for C_r^n, namely $n(n-1)\cdots(n-r+1)/r!$, the r factors of the numerator decrease while those of the denominator increase. Hence for a given value of n the value of C_r^n will be greatest when the next greater value of r will make

$$(n-r+1)/r < 1.$$

From this it readily follows that if n be even, C_r^n is greatest when $r = n/2$; and if n be odd, C_r^n is greatest when $r = (n-1)/2$ or $r = (n+1)/2$, the value of C_r^n being the same for these two values of r, § 769.

Example. What is the greatest value of C_r^{12}? of C_r^{15}?

Combinations when repetitions are allowed. Let us inquire in **772** how many ways we can select three of the four digits 1, 2, 3, 4 when repetitions are allowed.

As examples of such selections we may take 111, 112, 124, illustrating respectively the cases in which all three, two, none of the digits are the same.

If to the digits in 111, in 112, and in 124 we add 0, 1, 2 respectively, we obtain 123, 124, and 136, which are three of the 3-combinations *without* repetitions of the digits 1, 2, 3, 4, 5, 6. And a little reflection will show that if we make out a complete list of the selections like 111, 112, 124, arranging the digits in each so that no digit is followed by one of less value, and then to the digits in each selection add 0, 1, 2, we shall obtain once and but once every one of the 3-combinations without repetitions of the $4 + (3 - 1)$ or 6 digits 1, 2, 3, 4, 5, 6. The number of the latter combinations is C^6_3. Hence C^6_3 is the number which we are seeking.

The same reasoning may be applied to the general case of r-combinations, with repetitions, of the n numbers $1, 2, \cdots, n$. And since the numbers may correspond to n different things of any kind, we have the theorem:

The number of the r*-combinations with repetitions of* n *different things is the same as the number of the* r*-combinations without repetitions of* $n + r - 1$ *different things, namely,* C^{n+r-1}_r, *or* $n(n+1)\cdots(n+r-1)/r!$.

Example 1. How many different throws can be made with four dice?

As any one of the faces marked 1, 2, 3, 4, 5, 6 may turn uppermost in the case of one, two, three, or four of the dice, the number of possible throws is the number of 4-combinations with repetitions of 1, 2, 3, 4, 5, 6, namely, C^{6+3}_4, or 126.

Example 2. How many terms has a complete homogeneous polynomial of the rth degree in three variables x, y, z?

Evidently it has as many terms as there are products of the rth degree whose factors are x's, y's, or z's. Hence the number is

$$C^{3+r-1}_r = C^{r+2}_r = C^{r+2}_2 = (r+1)(r+2)/2.$$

773 **Formulas connecting numbers of combinations. The corresponding algebraic identities.** The following relations are of special interest and importance.

$$C_r^n = C^{n-1}_r + C^{n-1}_{r-1}. \qquad (1)$$

For we may distribute the r-combinations of n letters into two classes, — those which contain some particular letter, as a, and those which do not contain this letter. We shall obtain all the combinations of the first class, once each, if we form every $(r-1)$-combination of the remaining $n-1$ letters and then add a to each of these combinations; hence their number is C^{n-1}_{r-1}. The combinations of the second class are the r-combinations of the remaining $n-1$ letters; hence their number is C^{n-1}_r.

$$C^{m+n}_r = C^m_r + C^m_{r-1} \cdot C^n_1 + C^m_{r-2} \cdot C^n_2 + \cdots + C^m_1 \cdot C^n_{r-1} + C^n_r. \qquad (2)$$

For take any group of $m+n$ letters and separate it into two groups, one of m letters, the other of n letters. We shall take account of all the r-combinations of the $m+n$ letters, once each, if we classify them as follows. They consist of

(*a*) The r-combinations of the letters of the m-group. The number of these combinations is C^m_r.

(*b*) The combinations which contain $r-1$ letters of the m-group and one letter of the n-group. As we can choose the $r-1$ letters in C^m_{r-1} ways and the one letter in C^n_1 ways, the number of combinations of this kind is $C^m_{r-1} \cdot C^n_1$.

(*c*) The combinations which contain $r-2$ letters of the m-group and two letters of the n-group. As we can choose the $r-2$ letters in C^m_{r-2} ways and the two letters in C^n_2 ways, the number of combinations of this kind is $C^m_{r-2} \cdot C^n_2$. And so on, until last of all we reach the r-combinations of the letters of the n-group, of which there are C^n_r.

Thus, $C^9_3 = 84$ and $C^5_3 + C^5_2 C^4_1 + C^5_1 C^4_2 + C^4_3 = 10 + 40 + 30 + 4 = 84$.

774 If in (1) and (2) we replace the several symbols C by their expressions in terms of m, n, r, § 767, (1), we obtain formulas connecting m, n, r. The proofs just given only show that

these formulas hold good when m, n, r denote positive integers. But in fact, so far as m and n are concerned, they are true *algebraic identities*, holding good for all values of these letters. This may be shown by algebraic reduction.

Thus, in the case of (1), we have

$$C^{n-1}_r + C^{n-1}_{r-1} = \frac{(n-1)(n-2)\cdots(n-r)}{1\cdot 2\cdots r} + \frac{(n-1)(n-2)\cdots(n-r+1)}{1\cdot 2\cdots(r-1)}$$

$$= \frac{(n-1)(n-2)\cdots(n-r+1)}{1\cdot 2\cdots(r-1)}\cdot\left[1+\frac{n-r}{r}\right]$$

$$= \frac{n(n-1)\cdots(n-r+1)}{1\cdot 2\cdots r} = C^n_r.$$

But it is not necessary to make such a reduction to prove that these formulas are true identities. Thus, when expressed in terms of m, n, r, each member of (2) denotes an integral function of m and n whose degree with respect to each of these letters is r. These two functions must be identically equal, since otherwise, were we to assign some particular integral value to m, thus making them functions of n alone, they could not be equal for more than r values of n, § 421, whereas it has already been shown that in reality they would be equal for all integral values of n.

EXERCISE LXV

1. If there are three roads leading from P to Q, two from Q to R, and four from R to S, by how many routes can a person travel from P to S?

2. In how many ways can a company of five persons be arranged in six numbered seats?

3. If eight runners enter a half-mile race, in how many ways can the first, second, and third places be won?

4. In how many ways can a four-oar crew be chosen from ten oarsmen and in how many ways can all these crews be arranged in the boat?

5. From a company of one hundred soldiers how many pickets of three men can be chosen?

6. Five baseball nines wish to arrange a schedule of games in which each nine shall meet every other nine three times. How many games must be scheduled?

7. In how many ways can the digits 1, 2, 1, 3, 2, 1, 5 be arranged, all the digits occurring in each arrangement?

8. Of the permutations of the letters in the word *factoring*, taken all at a time, (1) how many begin with a vowel and end with a consonant? (2) how many do not begin with f? (3) how many have vowels in the first three places?

9. In how many of the permutations just described do the vowels retain the order a, o, i? In how many do the consonants retain the order f, c, t, r, n, g? In how many do both the vowels and the consonants retain these orders?

10. With the letters of the word *resident* how many permutations of five letters each can be formed in which the first, third, and fifth letters are vowels?

11. In how many ways can a baseball nine be selected from fifteen candidates of whom six are qualified to play in the outfield only and nine in the infield only?

12. In how many ways can two numbers whose sum is even be chosen from the numbers 1, 2, 3, 8, 9, 10?

13. How many numbers of one, two, or three figures can be formed with the digits 1, 2, 3, 4, 5, 6, 7 (1) when the digits may be repeated? (2) when they may not be repeated?

14. How many odd numbers, each having five different figures, can be formed with the digits 1, 2, 3, 4, 5, 6?

15. How many odd numbers without repeated digits are there between 3000 and 8000? How many of these are divisible by 5?

16. In how many ways can a person invite one or more of five friends to dinner?

17. In how many ways can fifteen apples be distributed among three boys so that one boy shall receive six, another five, and another four?

18. In how many ways can six positive and five negative signs be written in a row?

19. How many numbers of four figures each can be formed with the characters 1, 2, 3, 2, 3, 4, 2, 4, 5, 3, 6, 7?

20. From fifteen French and twelve German books eight French and seven German books are to be selected and arranged on a shelf. In how many ways can this be done?

21. From a complete suit of thirteen cards five are to be selected which shall include the king or queen, or both. In how many ways can this be done?

22. In how many ways can four men be chosen from five Americans and six Englishmen so as to include (1) only one Englishman? (2) at least one Englishman?

23. How many parallelograms are formed when a set of ten parallel lines is met by another set of twelve parallel lines?

24. Given n points in a plane no three of which lie in the same straight line, except m which all lie in the same straight line. Show that the number of lines obtained by joining these points is $C_2^n - C_2^m + 1$.

25. Find the number of bracelets that can be formed by stringing together five like pearls, six like rubies, and five like diamonds.

26. In how many ways can ten persons be arranged at two round tables, five at each table?

27. In how many ways can six ladies and five gentlemen arrange a game of lawn tennis, each side to consist of one lady and one gentleman?

28. In how many ways can fifteen persons vote to fill a certain office for which there are five candidates? In how many of these ways will the vote be equally divided among the five candidates?

29. A boat crew consists of eight men two of whom are qualified to row on the stroke side only and one on the bow side only. In how many ways can the crew be arranged?

30. How many baseball nines can be chosen from eighteen players of whom ten are qualified to play in the infield only, five in the outfield only, and three in any position?

31. Show that the number of permutations of six different letters taken all at a time, when two of the letters are excluded each from a particular position, is $6! - 2 \cdot 5! + 4!$.

32. How many combinations four at a time can be formed with the letters p, q, r, s, t, v, when repetitions are allowed?

33. How many different throws can be made with five dice?

34. How many terms has each of the symmetric functions $\Sigma x^4y^3z^2u$, $\Sigma x^2y^2z^2u$, $\Sigma x^3y^3z^2u^2v$, the number of the variables being ten?

35. Show that the number of terms in a complete homogeneous function of the nth degree in four variables is $(n+1)(n+2)(n+3)/3!$

XXVI. THE MULTINOMIAL THEOREM

775 **Multinomial theorem.** Let $a + b + \cdots + k$ denote any poly nomial, and n a positive integer. Then

$$(a + b + \cdots + k)^n = \Sigma \frac{n!}{\alpha! \beta! \cdots \kappa!} a^\alpha b^\beta \cdots k^\kappa,$$

where the sum on the right contains one term for each set of values of $\alpha, \beta, \cdots, \kappa$ that can be selected from $0, 1, 2, \cdots, n$, such that $\alpha + \beta + \cdots + \kappa = n$, it being understood that when $\alpha = 0$, $\alpha!$ is to be replaced by 1, and the like for $\beta, \cdots, \kappa$.

For $(a + b + \cdots + k)^n$ denotes the continued product

$$(a + b + \cdots + k)(a + b + \cdots + k) \cdots \text{to } n \text{ factors},$$

and each of the partial products obtained by actually carrying out this multiplication, without collecting like terms, has the form: a letter from the first parenthesis, times a letter from the second, times a letter from the third, and so on.

But since the letter selected from each parenthesis may be any one of the letters $a, b, \cdots, k$, a list of the products as thus written would also be a complete list of the n-permutations of the letters $a, b, \cdots, k$ when repetitions are allowed. And if $\alpha, \beta, \cdots, \kappa$ denote any particular set of numbers $0, 1, \cdots, n$ whose sum is n, there will be in the list as many products in which α of the factors are a's, β of them b's, $\cdots$, κ of them k's, as there are n-permutations of n letters of which α are alike, β others are alike, and so on, namely, $n!/\alpha! \cdot \beta! \cdots \kappa!$, § 766. And since each of these products is equal to $a^\alpha b^\beta \cdots k^\kappa$, their sum is $\frac{n!}{\alpha! \beta! \cdots \kappa!} a^\alpha b^\beta \cdots k^\kappa$.

The binomial theorem is a particular case of this theorem.

Thus, the expansion of $(a + b + c + d + e)^4$ consists of terms of the five types $abcd$, a^2bc, a^2b^2, a^3b, a^4 with the coefficients $4!/1!1!1!1!$, or 24; $4!/2!1!1!$, or 12; $4!/2!2!$, or 6; $4!/3!1!$, or 4; $4!/4!$, or 1, respectively. Hence, uniting terms of the same type, we have

$$(a + b + c + d + e)^4 = \Sigma a^4 + 4\,\Sigma a^3 b + 6\,\Sigma a^2 b^2 + 12\,\Sigma a^2 bc + 24\,\Sigma abcd.$$

Example. Find the coefficient of x^5 in the expansion of $(2+3x+4x^2)^8$. The general form of a term of this expansion is $\frac{8!}{\alpha!\beta!\gamma!}2^\alpha 3^\beta 4^\gamma x^{\beta+2\gamma}$, where $\alpha+\beta+\gamma=8$ (1), and the terms required are those for which $\beta+2\gamma=5$ (2). A complete list of the solutions of (1), (2) in positive integers or 0's is $\alpha, \beta, \gamma = 3, 5, 0;\ 4, 3, 1;\ 5, 1, 2$. Hence the required coefficient is

$$\frac{8!}{3!5!}2^3\cdot 3^5 + \frac{8!}{4!3!}2^4\cdot 3^3\cdot 4 + \frac{8!}{5!2!}2^5\cdot 3\cdot 4^2, \text{ or } 850{,}752.$$

EXERCISE LXVI

1. Give the expansion of $(a+b+c+d)^3$, collecting terms of the same type.

2. Also the expansion of $(a+b+c+d)^5$.

3. Find the coefficients of $a^5b^4c^2d$, $a^4b^4c^4$, and $a^5b^5c^2$ in the expansion of $(a+b+c+d)^{12}$.

4. Find the coefficient of $ab^2c^3d^4$ in the expansion of $(a-b+c-d)^{10}$.

5. Find the coefficient of a^4b^3c in the expansion of $(a+3b+2c)^8$.

6. Find the coefficient of x^6 in the expansion of $(1+x+x^2+x^3)^{10}$.

7. Find the coefficient of x^7 in the expansion of $(1-x+3x^2)^9$.

XXVII. PROBABILITY

SIMPLE EVENTS

Probability. Consider any future event which, if given a *trial*, that is, an opportunity to happen, must happen or fail to happen in one of a limited number of ways *all equally likely*, that is, ways so related that there is no reason for expecting any one of them rather than any other. The turning of the ace uppermost when a die is thrown is such an event. For one of the six faces of the die must turn uppermost, and there is no reason for expecting any one face to turn rather than any other. **776**

Calling all the equally likely ways in which such an event can happen or fail the *possible cases* with respect to the event, the ways in which it can happen the *favorable cases*, and the ways in which it can fail the *unfavorable cases*, we say:

The probability or chance of the event is the ratio of the number of favorable cases to the entire number of possible cases, favorable and unfavorable.

Hence if m denote the number of possible cases, a the number of favorable cases, and p the probability, we have by definition

$$p = a/m.$$

Thus, the probability that the ace will turn up when a die is thrown is $1/6$; for here $m = 6$ and $a = 1$.

Again, the chance of drawing a white ball from a bag known to contain five balls, three white and two black, is $3/5$.

777 **Corollary 1.** *If an event is certain to happen, its probability is* 1; *if it is certain to fail, its probability is* 0; *in every other case its probability is a positive proper fraction.*

For if the event is certain to happen, there are no ways in which it can fail; hence $a = m$ and $a/m = 1$. If the event is certain to fail, there are no ways in which it can happen; hence $a = 0$ and $a/m = 0$. In every other case a is greater than 0 and less than m, so that a/m is a positive proper fraction.

778 **Corollary 2.** *If the probability that an event will happen is* p, *the probability that it will not happen is* $1 - p$.

For if a of the m possible cases favor the occurrence of the event, the remaining $m - a$ cases favor its non-occurrence. Hence the probability that the event will not happen is $(m - a)/m = 1 - a/m = 1 - p$.

779 **Odds.** If the number of favorable cases with respect to a certain event is a and the number of unfavorable cases is b, we say, when $a > b$, that the *odds* are a to b in favor of the event; when $b > a$, that the odds are b to a against the event; when $a = b$, that the odds are even on the event. In the first case the probability of the event, namely, $a/(a+b)$, is greater than $1/2$; in the second it is less than $1/2$; in the third it is equal to $1/2$.

Thus, if a ball is to be drawn from a bag containing five balls, three white and two black, the odds are 3 to 2 in favor of its being white, and 3 to 2 against its being black.

Expectation. If p denote the chance that a person will win a certain sum of money M, the product Mp is called the value of his *expectation* so far as this sum M is concerned. **780**

Thus, the value of the expectation of a gambler who is to win \$12 if he throws an ace with a single die is $\$12 \times 1/6$, or \$2.

Examples of probability. In applying the definition of probability, § 776, care must be taken to reduce the possible cases to such as are equally likely. The following examples will illustrate the need of this precaution. **781**

Example 1. If two coins be tossed simultaneously, what is the chance that the result will be two heads? two tails? one head and one tail?

We might reason thus: There are three possible cases, one favoring the first result, one the second, one the third; hence the chance of each result is $1/3$.

But our conclusion would be false, since the number of *equally likely* possible cases is not three but four. For if we name the coins A and B respectively, the equally likely cases are: A head, B head; A tail, B tail; A head, B tail; A tail, B head. And since one of these cases favors the result two heads, one the result two tails, and two the result one head and one tail, the chances of these results are $1/4$, $1/4$, and $2/4$ respectively.

Example 2. What is the chance of throwing a total of eight with two dice?

Here the number of equally likely possible cases is $6 \cdot 6$, or 36, for any face of one die may turn up with any face of the other die.

We have a total of eight if the faces which turn up read 2, 6 or 3, 5 or 4, 4. But there are *two* ways in which 2, 6 may turn up, namely 2 on the die A and 6 on the die B, or *vice versa*. Similarly there are *two* ways in which 3, 5 may turn up. On the contrary, there is but *one* way in which 4, 4 can turn up. Hence there are five favorable cases. Therefore the chance in question is 5/36.

Example 3. What is the chance of throwing a total of eight with three dice if at least one die turns ace up?

The number of equally likely possible cases is $6 \cdot 6 \cdot 6$, or 216.

We have a total of eight if the faces which turn up read 1, 1, 6 or 1, 2, 5 or 1, 3, 4. But 1, 1, 6 may turn up in $3!/2!$, or 3, ways, § 766; for the numbers 1, 1, 6 may be distributed among the three dice in any of the orders in which 1, 1, 6 can be written. Similarly 1, 2, 5 and 1, 3, 4 may each turn up in $3!$, or 6, ways. Hence there are $3 + 6 + 6$, or 15, favorable cases. Therefore the chance in question is 15/216, or 5/72.

Example 4. An urn contains six white, four red, and two black balls.

(1) If four balls are drawn, what is the chance that all are white?

There are as many ways of drawing four white balls as there are 4-combinations of the six white balls in the urn, namely C_4^6. Similarly, since the urn contains twelve balls all told, the total number of possible drawings is C_4^{12}. Hence the chance in question is C_4^6/C_4^{12}, or 1/33.

(2) If six balls are drawn, what is the chance that three of them are white, two red, and one black?

The three white balls can be chosen in C_3^6 ways, the two red balls in C_2^4 ways, the one black ball in C_1^2 ways. Hence the number of ways in which the required drawing can be made is $C_3^6 \cdot C_2^4 \cdot C_1^2$. The total number of possible drawings is C_6^{12}. Hence the chance in question is $C_3^6 \cdot C_2^4 \cdot C_1^2/C_6^{12}$, or 20/77.

Example 5. Three cards are drawn from a suit of thirteen cards.

(1) What is the chance that neither king nor queen is drawn?

Aside from the king and queen there are eleven cards. Hence there are C_3^{11} sets of three cards which include neither king nor queen. Therefore the probability in question is C_3^{11}/C_3^{13}, or 15/26.

(2) What is the chance that king or queen is drawn, one or both?

This event occurs when the event described in (1) fails to occur. Hence the probability in question is $1 - 15/26$, or 11/26, § 778.

(3) What is the chance that both king and queen are drawn?

We obtain every set of three cards which includes both king and queen if we combine each of the remaining eleven cards in turn with king and queen. Hence the chance in question is $11/C_3^{13}$, or 1/26.

782 **On the various meanings of probability.** 1. The fraction a/m, which we have called the probability of an event, § 776, means nothing so far as the actual outcome of a single trial, or a small number of trials, of the event is concerned. But it does indicate the frequency with which the event would occur in the long run, that is, in the course of an indefinitely long series of trials.

Thus, if one try the experiment of throwing a die a very great number of times, say a thousand times, he will find that as the number of throws increases the ratio of the number of times that ace turns up to the total number of throws approaches the value $1/6$ more and more closely.

2. There are important classes of events — the duration of life is one — to which the definition of § 776 does not apply, it being impossible to enumerate the ways, all equally likely, in which the event can happen or fail. But we may be able to determine the frequency with which events of such a class have occurred in the course of a very great number of *past* trials. If so, we call the fraction which indicates this frequency the probability of an event of the class. Like $1/6$ in the case of the die, it indicates the frequency with which events of the class may reasonably be expected to occur in the course of a very great number of *future* trials.

Thus, if we had learned from the census reports that of 100,000 persons aged sixty in 1880 about $2/3$ were still living in 1890, we should say that the probability that a person now sixty will be alive ten years hence is $2/3$.

3. But we also use the fraction a/m to indicate the strength of our *expectation* that the event in question will occur on a *single* trial. The greater the ratio of the number of favorable cases to the number of possible cases, or the greater the frequency with which, to our knowledge, events of a similar character have occurred in the past, the stronger is our expectation that this particular event will occur on the single trial under consideration.

We may speak of the probability, in this sense, of any kind of future event. Thus, before a game between two football teams, A and B, we hear it said that the odds are 3 to 2 in favor of A's winning, or that the probability that A will win is $3/5$. This means that the general expectation of A's winning is about as strong as one's expectation of drawing a white ball from an urn known to contain five balls three of which are white.

EXERCISE LXVII

1. The probability of a certain event is $3/8$. Are the odds in favor of the event or against it, and what are these odds? What is the probability that the event will not occur?

2. The odds are 10 to 9 in favor of A's winning a certain game. What is the chance of his winning the game? of losing it?

3. The odds are 5 to 3 in favor of A's winning a stake of $60. What is his expectation?

4. The French philosopher D'Alembert said: "There are two possible cases with respect to every future event, one that it will occur, the other that it will not occur. Hence the chance of every event is $1/2$ and the definition of probability is meaningless." How should he be answered?

5. An urn contains sixteen balls of which seven are white, six black, and three red.

(1) If a single ball be drawn, what is the chance that it is white? black? red?

(2) If two balls be drawn, what is the chance that both are black? one white and one red?

(3) If three balls be drawn, what is the chance that all are red? none red? one white, one black, one red?

(4) If four balls be drawn, what is the chance that one is white and the rest not? two white and the other two not?

(5) If ten balls be drawn, what is the chance that five are white, three black, and two red?

6. What is the chance of throwing doublets with two dice? with three dice?

7. What is the chance of throwing a total of seven with two dice? Show that this is the most probable throw.

8. What is the chance of throwing at least one ace in a throw with two dice? of throwing one ace and but one?

9. One letter is taken at random from each of the words *factor* and *banter*. What is the chance that the same letter is taken from each?

10. A box contains nine tickets numbered 1, 2, ···, 9. If two of the tickets be drawn at random, what is the chance that the product of the numbers on them is even? odd?

11. If five tickets be drawn from this box, find the chance (1) that 1, 2, and 3 are drawn; (2) that one and but one of 1, 2, and 3 is drawn; (3) that none of these numbers is drawn.

12. If four cards be drawn from a complete pack of fifty-two cards, what is the chance that they are ace, king, queen, and knave? ace, king, queen, and knave of the same suit?

13. What is the chance that a hand at whist contains four trumps and three cards of each of the remaining suits?

14. What is the chance of a total of five in a single throw with three dice? of a total of less than five?

15. If eight persons be seated at a round table, what is the chance that two particular persons sit together?

COMPOUND EVENTS. MUTUALLY EXCLUSIVE EVENTS

Independent events. Two or more events are said to be *independent* when the occurrence or non-occurrence of any one of them is not affected by the occurrence or non-occurrence of the rest. In the contrary case the events are said to be *interdependent*. **783**

Thus, the results of two drawings of a ball from a bag are independent if the ball is returned after the first drawing, but interdependent if the ball is not returned.

Theorem 1. *The probability that all of a set of independent events will occur is the product of the probabilities of the single events.* **784**

For consider two such events whose probabilities are a_1/m_1 and a_2/m_2 respectively.

The number of equally likely possible cases for and against the first event is m_1, for and against the second m_2, and since the events are independent any one of the m_1 cases may occur with any one of the m_2 cases. Hence the number of equally likely possible cases for and against the occurrence of both events is m_1m_2. And by the same reasoning, a_1a_2 of these cases favor the occurrence of both events. Therefore the probability that both events will occur is $\frac{a_1a_2}{m_1m_2}$, that is $\frac{a_1}{m_1}\cdot\frac{a_2}{m_2}$, as was to be demonstrated.

The proof for the case of more than two events is similar.

The demonstration applies only to events of the kind described in § 776, but for the reasons indicated in § 782 we may apply the theorem itself to any kind of future event, as in Ex. 2 below.

Thus, the chance of throwing ace twice in succession with a single die is $1/6 \times 1/6$, or $1/36$.

Again, the chance of twice drawing a white ball from a bag which contains five white and four black balls, the ball first drawn being returned before the second drawing, is $5/9 \times 5/9$, or $25/81$.

785 **Theorem 2.** *If the probability of a first event is* p_1, *and if after this event has happened the probability of a second event is* p_2, *the probability that both events will occur in the order stated is* p_1p_2. *And similarly for more than two events.*

This theorem may be proved in the same manner as Theorem 1. It evidently includes that theorem.

Thus, after a white ball has been drawn from a bag containing five white and four black balls, and *not* replaced, the chance of drawing a second white ball is $4/8$. Hence the chance of twice drawing a white ball when the one first drawn is not replaced is $5/9 \times 4/8$, or $5/18$.

Example 1. What is the chance that ace will turn up at least once in the course of three throws with a die?

Ace will turn up at least once unless it fails to turn in every throw. The chance of failure in a single throw being $5/6$, the chance of failure in all three throws is $5/6 \times 5/6 \times 5/6$, or $125/216$. Hence the chance of at least one ace is $1 - 125/216$, or $91/216$.

Example 2. The chance that A will solve a certain problem is 3/4. The chance that B will solve it is 2/3. What is the chance that the problem will be solved if both A and B attempt it independently?

The problem will be solved unless both A and B fail. The chance of A's failure is 1/4, of B's failure 1/3. Hence the chance that both fail is $1/4 \times 1/3$, or 1/12. Therefore the chance that the problem will be solved is 11/12.

Example 3. There are two purses, one containing five silver coins and one gold coin, the other three silver coins. If four coins be drawn from the first purse and put into the second, and five coins be then drawn from the second purse and put into the first, what is the chance that the gold coin is in the second purse? in the first purse?

The chance that the gold coin is taken from the first purse and put into the second is C_3^5 / C_4^6, or 2/3, § 781, Ex. 5. The chance that it is then left in the second purse is $6 / C_5^7$, or 2/7. Hence the chance that after both drawings it is in the second purse is $2/3 \times 2/7$, or 4/21. The chance that it is in the first purse is $1 - 4/21$, or 17/21.

Example 4. If eight coins be tossed simultaneously, what is the chance that at least one of them will turn head up?

Example 5. Four men A, B, C, and D are hunting quail. If A gets on the average one quail out of every two that he fires at, B two out of every three, C four out of every five, and D five out of every seven, what is the chance that they get a bird at which all happen to fire simultaneously?

Example 6. An urn A contains five white and four red balls. A second urn B contains six white and two black balls. What is the chance of drawing a white ball from A and then, this ball having been put into B, of drawing a white ball from B also?

Example 7. The chance that A will be alive five years hence is 3/4; B, 5/6. What is the chance that five years hence both A and B will be alive? A alive, B dead? A dead, B alive? both dead?

Mutually exclusive events. If two or more events are so related that but one of them can occur, they are said to be *mutually exclusive.* **786**

Thus, the turning of an ace and the turning of a deuce on the same throw of a single die are mutually exclusive events.

787 **Theorem 3.** *The probability that some one or other of a set of mutually exclusive events will occur is the sum of the probabilities of the single events.*

For consider two mutually exclusive events A and B.

The possible cases with respect to the *two* events are of three kinds, all mutually exclusive, namely, those for which (1) A happens, B fails; (2) A fails, B happens; (3) A fails, B fails.

Let the numbers of equally likely possible cases of these three kinds be l, m, and n respectively. Then

(*a*) The chance that either A or B happens is $\dfrac{l+m}{l+m+n}$.

For there are $l+m+n$ possible and $l+m$ favorable cases.

(*b*) The chance of the single event A is $\dfrac{l}{l+(m+n)}$.

For since A never happens except when B fails, the l cases in which A happens and B fails are *all* the cases in which A happens, and the $m+n$ cases in which A fails and B happens or both A and B fail are all the cases in which A fails.

(*c*) Similarly the chance of the single event B is $\dfrac{m}{m+(l+n)}$.

But
$$\frac{l+m}{l+m+n}=\frac{l}{l+(m+n)}+\frac{m}{m+(l+n)}.$$

Therefore the chance that either A or B happens is the sum of the chances of the single events A and B.

The proof for more than two events is similar.

Thus, if one ball be drawn from a bag containing four white, five black, and seven red balls, since the chance of its being white is $1/4$, and that of its being black is $5/16$, the chance of its being either white or black is $1/4 + 5/16$, or $9/16$. Of course this result may be obtained directly from the definition of probability, § 776. In fact that definition may be regarded as a special case of Theorem 3.

Care must be taken not to apply this theorem to events which are not mutually exclusive.

Thus, if asked, as in § 785, Ex. 2, to find the chance that a problem will be solved if both A and B attempt it, A's chance of success being

3/4 and B's 2/3, we cannot obtain the result by merely adding 3/4 and 2/3, since the two events A succeeds, B succeeds are not mutually exclusive. The mutually exclusive cases in which the problem will be solved are: A succeeds, B fails; A fails, B succeeds; A succeeds, B succeeds. The chances of these cases are, § 784, $3/4 \times 1/3$ or 3/12, $1/4 \times 2/3$ or 2/12, $3/4 \times 2/3$ or 6/12; and the sum of these three chances, or 11/12, is the chance that the problem will be solved.

Example 1. An urn A contains ten balls three of which are white, and a second urn B contains twelve balls four of which are white. If one of the urns be chosen at random and a ball drawn from it, what is the chance that the ball is white?

We are required to find the chance of one of the following mutually exclusive events: (1) choosing A and then drawing a white ball from it; (2) choosing B and then drawing a white ball from it.

The chance of choosing A is 1/2, and the chance when A has been chosen of drawing a white ball is 3/10. Hence the chance of (1) is $1/2 \times 3/10$, or 3/20. Similarly the chance of (2) is $1/2 \times 4/12$, or 1/6.

Therefore the chance in question is $3/20 + 1/6$, or 19/60.

Example 2. What is the value of the expectation of a person who is to have any two coins he may draw at random from a purse which contains five dollar pieces and seven half-dollar pieces?

The value of his expectation so far as it depends on drawing two dollar pieces is $\$2 \times C_2^5 / C_2^{12} = \$2 \times 5/33 = \$.30$; on drawing two half-dollar pieces, $\$1 \times C_2^7 / C_2^{12} = \$1 \times 7/22 = \$.32$; on drawing one dollar piece and one half-dollar piece, $\$1.50 \times 5 \cdot 7 / C_2^{12} = \$1.50 \times 35/66 = \$.80$.

Hence the total value of his expectation is $\$.30 + \$.32 + \$.80$, or \$1.42.

Example 3. Two persons A and B are to draw alternately one ball at a time from a bag containing three white and two black balls, the balls drawn not being replaced. If A begins, what chance has each of being the first to draw a white ball?

The chance that A succeeds in the first drawing is 3/5.

The chance that A fails and B then succeeds is $2/5 \times 3/4$, or 3/10, for when B draws, the bag contains four balls three of which are white.

The chance that A fails, B fails, and A then succeeds is $2/5 \times 1/4 \times 3/3$, or 1/10, for when A draws, the bag contains three balls all white.

Therefore A's total chance is $3/5 + 1/10$, or 7/10, and B's is 3/10.

Example 4. In the drawing described in Ex. 3 what are the respective chances of A and B if the balls are replaced as they are drawn?

On the first round A's chance is 3/5, B's 2/5 × 3/5, or 6/25; and their chances on every later round, of which there may be any number, will be the same as these.

Hence their total chances are in the ratio 3/5 : 6/25, or 5 : 2; that is, A's total chance is 5/7, B's 2/7.

Example 5. In a room there are three tables and on them nine, ten, and eleven books respectively. I wish any one of six books, two of which are on the first table, three on the second, one on the third. If a friend select a book for me at random from those in the room, what is the chance that it is one of those I wish?

Example 6. An owner of running horses enters for a certain race two horses whose chances of winning are 1/2 and 1/3 respectively. What is the chance that he will obtain the stakes?

Example 7. A and B throw alternately with two dice for a stake which is to be won by the one who first throws a doublet. What are their respective chances of winning if A throws first?

788 **Repeated trials of a single event.** The following theorems are concerned with the question of the chance that a certain event will occur a specified number of times in the course of a series of trials, the chance of its occurrence on a single trial being known.

789 **Theorem 4.** *If the probability that an event will occur on a single trial is* p, *the probability that it will occur exactly* r *times in the course of* n *trials is* $\mathrm{C}^{\mathrm{n}}_{\mathrm{r}}\mathrm{p}^{\mathrm{r}}\mathrm{q}^{\mathrm{n-r}}$, *where* $\mathrm{q} = 1 - \mathrm{p}$.

For the probability that it will occur on all of any particular set of r trials and fail on the remaining $n - r$ trials is $p^r(1-p)^{n-r}$, or p^rq^{n-r}, if $q = 1 - p$, § 784.

But since there are n trials all told, we may select this particular set of r trials in C^n_r ways which, of course, are mutually exclusive.

Hence the probability in question is $C^n_r p^r q^{n-r}$, § 787.

Thus, the chance that ace will turn up exactly twice in five throws with a single die, or that out of five dice thrown simultaneously two and but two will turn ace up, is $C^5_2 \cdot (\frac{1}{6})^2 \cdot (\frac{5}{6})^3$, or 625/3888.

Observe that $C_r^n p^r q^{n-r}$ is the term containing p^r in the expansion of $(p+q)^n$ by the binomial theorem; for $C_r^n = C_{n-r}^n$.

Theorem 5. *The probability that such an event will occur at least* r *times in the course of* n *trials is the sum of the first* $n - r + 1$ *terms in the expansion of* $(p+q)^n$, *namely,* **790**

$$p^n + C_1^n p^{n-1} q + C_2^n p^{n-2} q^2 + \cdots + C_{n-r}^n p^r q^{n-r}.$$

For the event will occur at least r times if it occurs exactly r times or exactly any number of times greater than r, and the terms p^n, $C_1^n p^{n-1} q, \cdots, C_{n-r}^n p^r q^{n-r}$ represent the probability of the occurrence of the event exactly $n, n-1, \cdots, r$ times respectively, § 789.

Thus, the chance that ace will turn up at least twice in the course of five throws with a single die is

$$(\tfrac{1}{6})^5 + 5(\tfrac{1}{6})^4 \tfrac{5}{6} + 10(\tfrac{1}{6})^3(\tfrac{5}{6})^2 + 10(\tfrac{1}{6})^2(\tfrac{5}{6})^3, \text{ or } \tfrac{763}{3888}.$$

Example 1. Two persons A and B are playing a game which cannot be drawn and in which A's skill is twice B's. What is the chance that A will win as many as three such games in a set of five?

A's chance of winning a single game is $2/3$, of losing $1/3$. Hence the chance that A will win as many as three of the five games is the sum of the first three terms of $(\frac{2}{3} + \frac{1}{3})^5$, that is, $(\frac{2}{3})^5 + 5(\frac{2}{3})^4 \frac{1}{3} + 10(\frac{2}{3})^3(\frac{1}{3})^2$, or $64/81$.

Example 2. Under the conditions of Ex. 1 what is the chance that A will win three games before B wins two?

The chance in question is that of A's winning at least three of the first four games played; and this chance is $(\frac{2}{3})^4 + 4(\frac{2}{3})^3 \frac{1}{3}$, or $\frac{16}{27}$.

And, in general, the chance of A's winning m games before B wins n is the same as the chance of A's winning at least m of the first $m + n - 1$ games played.

Example 3. Ten coins are tossed simultaneously. What is the chance that exactly six of them turn heads up? that at least six turn heads up?

Example 4. If four dice be thrown simultaneously, what is the chance that exactly three turn ace up? that at least three turn ace up?

Example 5. Under the conditions stated in Ex. 1 what is the chance that A will win at least four of the five games played?

Example 6. Under the same conditions what is the chance that A will win four games before B wins one?

EXERCISE LXVIII

1. A bag contains three white, five black, and seven red balls. On the understanding that one ball is drawn at a time and replaced as soon as drawn, what are the chances of drawing (1) first a white, then a red, then a black ball? (2) a white, red, and black ball in any order whatsoever?

2. What is the chance of obtaining a white ball in the first only of three successive drawings from this bag, balls not being replaced?

3. What is the value of the expectation of a person who is allowed to draw two coins at random from a purse containing five fifty-cent pieces, four dollar pieces, and three five-dollar pieces?

4. The chance that a certain door is locked is $1/2$. The key to the door is one of a bunch of eight keys. If I select three of these keys at random and go to the door, what is the chance of my being able to open it?

5. There are three independent events whose chances are $1/2$, $2/3$, and $3/4$ respectively. What is the chance that none of the events will occur? that one and but one of them will occur? that two and but two will occur? that all three will occur?

6. Find the odds against throwing one of the totals seven or eleven in a single throw with two dice.

7. What are the odds against throwing a total of ten with three dice? What are the odds in favor of throwing a total of more than five?

8. Three tickets are drawn from a case containing eleven tickets numbered $1, 2, \cdots, 11$. What is the chance that the sum of their numbers is twelve? What is the chance that this sum is an odd number?

9. Two gamblers A and B throw two dice under an agreement that if seven is thrown A wins, if ten is thrown B wins, if any other number is thrown the stakes are to be divided equally. Compare their chances.

10. The same two gamblers play under an agreement that A is to win if he throws six before B throws seven, and that B is to win if he throws seven before A throws six. A is to begin and they are to throw alternately. Compare their chances.

11. Three gamblers A, B, and C put four white and eight black balls into a bag and agree that the one who first draws a white ball shall win.

If they draw in the order A, B, C, what are their respective chances when the balls drawn are not replaced? when they are replaced?

12. What is the worth of a ticket in a lottery of one hundred tickets having five prizes of \$100, ten of \$50, and twenty of \$5?

13. A bag A contains five balls one of which is white, and a bag B six balls none of which is white. If three balls be drawn from A and put into B and three balls be then drawn from B and put into A, what is the chance that the white ball is in A?

14. The bag A contains m balls a of which are white, and the bag B contains n balls b of which are white. Is the chance of obtaining a white ball by drawing a single ball from one of these bags chosen at random the same that it would be if all the balls were put into one bag and a single ball then drawn?

15. In a certain town five deaths occurred within ten days including January first. What is the chance that none of the deaths occurred on January first?

16. If on the average two persons out of three aged sixty live to be seventy, what is the chance that out of five persons now sixty at least three will be alive ten years hence?

17. A boy is able to solve on the average three out of five of the problems set him. If eight problems are given in an examination and five are required for passing, what is the chance of his passing?

18. A person is to receive a dollar if he throws seven at the first throw with two dice, a dollar if he throws seven at the second throw, and so on until he throws seven. What is the total value of his expectation?

19. In playing tennis with B, A wins on the average three games out of four. What is the chance that he will win a set from B by the score of six to three? What is the total chance of his winning a set from B, the case of deuce sets being disregarded?

20. Under the conditions described in the preceding example what chance has A of winning a set in which the score is now four to two against him?

21. Two gamblers A and B are playing a game of chance and each player has staked \$32. They are playing for three points, but when A has gained two points and B one they decide to stop playing. How should they divide the \$64?

XXVIII. MATHEMATICAL INDUCTION

791 **Mathematical induction.** A number of the formulas contained in recent chapters may be established by a method of proof called *mathematical induction.* It is illustrated in the following example.

Example. Prove that the sum of the first n odd numbers is n^2.

We are asked to show that

$$1 + 3 + 5 + \cdots + (2n - 1) = n^2. \tag{1}$$

We see by inspection that (1) is true for certain values of n, as 1 or 2. Suppose that we have thus found it true when n has the particular value k, so that

$$1 + 3 + 5 + \cdots + (2k - 1) = k^2 \tag{2}$$

is *known* to be true. Adding the next odd number, namely, $2(k + 1) - 1$, or $2k + 1$, to both members of (2) and replacing $k^2 + 2k + 1$ by $(k + 1)^2$, we obtain

$$1 + 3 + 5 + \cdots + (2k + 1) = (k + 1)^2. \tag{3}$$

But (3) is what we get if in (1) we replace n by $k + 1$. We have therefore shown that if (1) is true when n has any particular value k, it is also true when n has the next greater value $k + 1$.

But we have already found by inspection that (1) is true when k has the particular value 1. Hence it is true when $n = 1 + 1$, or 2; hence when $n = 2 + 1$, or 3; and so on through all positive integral values of n, which is what we were asked to demonstrate.

And, in general, if a formula involving n has been found true for $n = 1$ and we can demonstrate that if true for $n = k$ it is also true for $n = k + 1$, we may conclude that it is true for *all* positive integral values of n. For we may reason: Since it is true when $n = 1$, it is also true when $n = 1 + 1$, or 2; hence when $n = 2 + 1$, or 3; and so on through all positive integral values of n.

As another illustration of this method we add the following proof of the binomial theorem.

For small values of n we find by actual multiplication that

$$(a + b)^n = a^n + C_1^n a^{n-1} b + C_2^n a^{n-2} b^2 + \cdots + C_r^n a^{n-r} b^r + \cdots. \tag{1}$$

Multiplying both members of (1) by $a + b$, we obtain, § 773, 1,

$$
\begin{aligned}
(a+b)^{n+1} &= a^{n+1} + \left.\begin{matrix} C_1^n \\ +1 \end{matrix}\right| a^n b + \left.\begin{matrix} C_2^n \\ + C_1^n \end{matrix}\right| a^{n-1}b^2 \cdots + \left.\begin{matrix} C_r^n \\ + C_{r-1}^n \end{matrix}\right| a^{n-r+1}b^r + \cdots \\
&= a^{n+1} + C^{n+1}_1 a^n b + C^{n+1}_2 a^{n-1}b^2 + \cdots \\
&\quad + C^{n+1}_r a^{(n+1)-r} b^r + \cdots . \qquad (2)
\end{aligned}
$$

But (2) is the same as (1) with n replaced by $n + 1$.

Hence if (1) is true when $n = k$, it is also true when $n = k + 1$. But (1) is known to be true when $n = 1$. It is therefore true when $n = 1 + 1$, or 2; therefore when $n = 2 + 1$, or 3; and so on.

Since the formula $C_r^n + C_{r-1}^n = C^{n+1}_r$ can be proved independently of the doctrine of combinations, § 774, the proof of the binomial theorem here given is independent of that doctrine.

EXERCISE LXIX

Prove the truth of the following formulas, §§ 701, 712, by the method of mathematical induction.

1. $a + ar + ar^2 + \cdots + ar^{n-1} = a(1 - r^n)/(1 - r)$.
2. $1^2 + 2^2 + 3^2 + \cdots + n^2 = n(n+1)(2n+1)/6$.
3. $1^3 + 2^3 + 3^3 + \cdots + n^3 = n^2(n+1)^2/4$.
4. $1 + 3 + 6 + \cdots + n(n+1)/2! = n(n+1)(n+2)/3!$.

XXIX. THEORY OF EQUATIONS

THE FUNDAMENTAL THEOREM. RATIONAL ROOTS

The two standard forms of the general equation of the nth degree in x. Every rational integral equation involving a single unknown letter, as x, and of the nth degree with respect to that letter, can be reduced to the standard form **792**

$$a_0x^n + a_1x^{n-1} + \cdots + a_{n-1}x + a_n = 0. \qquad (1)$$

When the coefficients $a_0, a_1, \cdots, a_n$ are given numbers, (1) is called a *numerical equation*, but when they are left wholly undetermined, (1) is called the *general equation* of the nth degree.

The final coefficient a_n is often called the *absolute term.*

We call an equation of the form (1) *complete* or *incomplete* according as none or some of the coefficients $a_1, a_2, \cdots, a_n$ are 0. Observe that in a complete equation the number of the terms is $n+1$.

In what follows, when all the coefficients $a_0, a_1, \cdots, a_n$ are real numbers, we may and shall suppose that the leading one a_0 is positive, and when they are rational, that they are integers which have no common factor.

By dividing both members of (1) by a_0 we reduce it to the second standard form

$$x^n + b_1x^{n-1} + \cdots + b_{n-1}x + b_n = 0, \qquad (2)$$

in which the leading coefficient is 1, and $b_1 = a_1/a_0$, and so on. For many purposes (2) is the more convenient form of the equation.

In the present chapter it is to be understood that $f(x) = 0$ denotes an equation of the form (1) or (2).

793 **Roots of equations.** The *roots* of the equation $f(x) = 0$ are the values of x for which the polynomial $f(x)$ vanishes, §§ 332, 333. It is sometimes convenient to call the roots of the equation the roots of the polynomial.

794 From the definition of root it follows that when a_n is 0 one of the roots of $f(x) = 0$ is 0; also that an equation $f(x) = 0$ all of whose coefficients are positive can have no positive root, and that a complete equation $f(x) = 0$ whose coefficients are alternately positive and negative can have no negative root.

Thus, $2x^3 + x^2 + 1 = 0$ can have no positive root since the polynomial $2x^3 + x^2 + 1$ cannot vanish when x is positive; and $2x^3 - x^2 + 3x - 1 = 0$ can have no negative root since $2x^3 - x^2 + 3x - 1$ cannot vanish when x is negative.

795 **Theorem 1.** *If* b *is a root of* f(x) = 0, *then* f(x) *is exactly divisible by* x − b; *and conversely, if* f(x) *is exactly divisible by* x − b, *then* b *is a root of* f(x) = 0.

For, by § 413, the remainder in the division of $f(x)$ by $x-b$ is $f(b)$. But when b is a root of $f(x)=0$ this remainder $f(b)$ is 0, § 793, so that $f(x)$ is exactly divisible by $x-b$; and conversely, when $f(x)$ is exactly divisible by $x-b$, the remainder $f(b)$ is 0, so that b is a root of $f(x)=0$.

Example. Prove that 3 is a root of $f(x)=x^3-2x^2-9=0$.

$$\begin{array}{rrrrl} 1 & -2 & +0 & -9 & \lfloor 3 \\ & 3 & 3 & 9 & \\ \hline 1 & 1 & 3, & 0 & = f(3) \end{array}$$

Dividing x^3-2x^2-9 by $x-3$ synthetically, § 411, we find that the remainder $f(3)$ is 0. Hence 3 is a root of $f(x)=0$.

796 If b is a root of $f(x)=0$, so that $f(x)$ is exactly divisible by $x-b$, and we call the quotient $\phi(x)$, we have

$$f(x)=(x-b)\,\phi(x).$$

Hence the remaining roots of $f(x)=0$ are the values of x for which the polynomial $\phi(x)$ vanishes; in other words, they are the roots of the *depressed equation* $\phi(x)=0$, § 341.

Example. Solve the equation $x^3-3x^2+5x-3=0$.

$$\begin{array}{rrrrl} 1 & -3 & +5 & -3 & \lfloor 1 \\ & 1 & -2 & +3 & \\ \hline 1 & -2 & +3, & 0 & \end{array}$$

We see by inspection that 1 is a root, and dividing x^3-3x^2+5x-3 by $x-1$, we obtain the depressed equation $x^2-2x+3=0$. The roots of this quadratic, found by § 631, are $1\pm i\sqrt{2}$. Hence the roots of the given equation are 1, $1+i\sqrt{2}$, and $1-i\sqrt{2}$.

797 We shall assume now and demonstrate later that *every rational integral equation* $f(x)=0$ *has at least one root.*

From this assumption and § 795 we deduce the following theorem, often called the *fundamental theorem of algebra.*

798 **Theorem 2.** *Every equation of the* n*th degree, as*

$$f(x)=a_0x^n+a_1x^{n-1}+\cdots+a_{n-1}x+a_n=0,$$

has n *and but* n *roots.*

By § 797 there is a value of x for which $f(x)$ vanishes. Call it β_1. Then $f(x)$ is exactly divisible by $x-\beta_1$, § 795, the leading term of the quotient being a_0x^{n-1}. Hence

$$f(x)=(x-\beta_1)(a_0x^{n-1}+\cdots). \tag{1}$$

By the same reasoning, since there is a value of x, call it β_2, for which the polynomial $a_0x^{n-1}+\cdots$ vanishes, we have $a_0x^{n-1}+\cdots=(x-\beta_2)(a_0x^{n-2}+\cdots)$. Therefore, by (1),

$$f(x)=(x-\beta_1)(x-\beta_2)(a_0x^{n-2}+\cdots). \qquad (2)$$

Continuing thus, after n divisions we obtain

$$f(x)=a_0(x-\beta_1)(x-\beta_2)\cdots(x-\beta_n). \qquad (3)$$

We have thus shown that n factors of the first degree exist, namely, $x-\beta_1$, $x-\beta_2$, $\cdots$, $x-\beta_n$, of which $f(x)$ is the product; and by § 419, $f(x)$ can have no other factors than these and their products.

But since a product vanishes when one of its factors vanishes and then only, it follows from (3) that $f(x)$ vanishes when $x=\beta_1$, or β_2, $\cdots$, or β_n, and then only. Hence, § 793, the n numbers β_1, β_2, $\cdots$, β_n are roots of the equation $f(x)=0$ and it has no other roots than these.

799 From this theorem it follows that the problem of solving an equation $f(x)=0$ is essentially the same as that of factoring the polynomial $f(x)$. Also that to form an equation which shall have certain given numbers for its roots, we have merely to subtract each of these numbers in turn from x, and then to equate to 0 the product of the binomial factors thus obtained.

Example. Form the equation whose roots are 2, 1/2, -1, 0.

It is $(x-2)(x-1/2)(x+1)(x-0)=0$, or $2x^4-3x^3-3x^2+2x=0$.

800 **Multiple roots.** Observe that two or more of the roots β_1, β_2, $\cdots$, β_n may be equal. If two or more of them are equal to β, we call β a *multiple root.* And according as the number of the roots equal to β is two, three, $\cdots$, in general, r, we call β a *double root*, a *triple root*, in general, a *root of order* r. A simple root may be described as a root whose order r is 1. Evidently it follows from § 798 that

The condition that β be a root of order r *of* $f(x)=0$ ***is that*** $f(x)$ ***be divisible by*** $(x-\beta)^r$ ***but not by*** $(x-\beta)^{r+1}$.

When we say, therefore, that every equation of the nth degree has n roots, the understanding is that each multiple root of order r is to be counted r times. It is of course *not* true that every equation of the nth degree has n *different* roots.

Thus, $x^3 - 3x^2 + 3x - 1 = 0$ is an equation of the third degree; but since $x^3 - 3x^2 + 3x - 1 = (x-1)^3$, each of its roots is 1.

801 **On finding the rational roots of numerical equations.** Let $f(x) = a_0x^n + a_1x^{n-1} + \cdots + a_n = 0$ denote an equation with *integral* coefficients, and let b denote an integer and b/c a rational fraction in its lowest terms. It follows from §§ 451, 795 that if b is a root of $f(x) = 0$, then b is a factor of a_n; and from §§ 452, 795 that if b/c is a root, then b is a factor of a_n and c is a factor of a_0. Hence, in particular, if $a_0 = 1$, b/c cannot be a root unless $c = \pm 1$, that is, unless b/c denotes the integer $\pm b$. Hence the following theorem, § 454:

An equation of the form $x^n + a_1x^{n-1} + \cdots + a_n = 0$, *where* $a_1, \cdots, a_n$ *denote integers, cannot have a rational fractional root.*

802 It follows from what has just been said that all the rational roots of an equation with rational coefficients can be found by a limited number of tests. These tests are readily made by synthetic division.

Example. Find the rational roots, if any, of the equation

$$3x^5 - 8x^4 + x^2 + 12x + 4 = 0.$$

The only possible rational roots are $\pm 1, \pm 2, \pm 4, \pm 1/3, \pm 2/3, \pm 4/3$.

3	−8	+0	+1	+12	+4	2
	6	−4	−8	−14	−4	
3	−2	−4	−7	−2,	0	2
	6	8	8	2		
3	4	4	1,	0	−1/3	
	−1	−1	−1			
3	3	3,	0			

We see by inspection that 1 is not a root. Testing 2, we find that it is a root and obtain the depressed equation $3x^4 - 2x^3 - 4x^2 - 7x - 2 = 0$. We find that 2 is a root of this depressed equation also and obtain the second depressed equation $3x^3 + 4x^2 + 4x + 1 = 0$. This equation can have no positive root since all its terms are positive, § 794. Testing -1, we find that it is not a root. Testing $-1/3$, we find that it is a root and obtain the third depressed equation $x^2 + x + 1 = 0$.

Hence the rational roots of the given equation are $2, 2, -1/3$. Its remaining roots, found by solving $x^2 + x + 1 = 0$, are $(-1 \pm i\sqrt{3})/2$.

803 The reckoning involved in making these tests will be lessened if one bears in mind the remark made in § 453; also the fact that a number known not to be a root of the given equation cannot be a root of one of the depressed equations, § 796; and finally the following theorem:

If b *is positive and the signs of all the coefficients in the result of dividing* f(x) *by* x − b *synthetically are plus,* f(x) = 0 *can have no root greater than* b; *if* b *is negative and the signs just mentioned are alternately plus and minus,* f(x) = 0 *can have no root algebraically less than* b.

For it follows from the nature of synthetic division that in both cases the effect of increasing b numerically will be to increase the numerical values of all coefficients after the first in the result without changing their signs, so that the final coefficient, that is, the remainder, cannot be 0.

Example 1. Show that $2x^3 + 3x^2 - 4x + 5 = 0$ has no root greater than 1.

$$\begin{array}{rrrr|l} 2 & +3 & -4 & +5 & \underline{1} \\ & 2 & 5 & 1 & \\ \hline 2 & +5 & +1, & 6 & \end{array}$$

Dividing by $x - 1$, we obtain positive coefficients only. Hence there is no root greater than 1.

If we divide by $x - 2$, we obtain a result with larger coefficients, all positive, namely, $2 + 7 + 10, 25$.

Example 2. Show that $3x^3 + 4x^2 - 3x + 1 = 0$ has no root less than -2.

$$\begin{array}{rrrr|l} 3 & +4 & -3 & +1 & \underline{-2} \\ & -6 & +4 & -2 & \\ \hline 3 & -2 & +1, & -1 & \end{array}$$

Dividing by $x + 2$, we obtain coefficients which are alternately plus and minus. Hence there is no root less than -2.

If we divide by $x + 3$, we obtain coefficients with the same signs as those just found but numerically greater, namely, $3 - 5 + 12, -35$.

804 We may add that any number which is known to be algebraically greater than all the real roots of $f(x) = 0$ is called a *superior limit* of these roots, and that any number which is known to be algebraically less than all the real roots of $f(x) = 0$ is called an *inferior limit* of these roots.

Thus, we have just proved that 1 is a superior limit of the roots of $2x^3 + 3x^2 - 4x + 5 = 0$ and that -2 is an inferior limit of the roots of $3x^3 + 4x^2 - 3x + 1 = 0$.

EXERCISE LXX

1. Form the equations whose roots are

(1) $a, -b, a+b$. (2) $3, 4, 1/2, -1/3, 0$.

2. Show that -3 is a triple root of the equation

$$x^4 + 8x^3 + 18x^2 - 27 = 0.$$

3. Show that 1 and 1/2 are double roots of the equation

$$4x^5 - 23x^3 + 33x^2 - 17x + 3 = 0.$$

4. By the method of § 803 find superior and inferior limits of the real roots of $x^5 - 5x^4 - 5x^3 + 4x^2 - 7x - 250 = 0$.

5. Show that $2x^4 - 3x^3 + 4x^2 - 10x - 3 = 0$ has no rational root.

Each of the following equations has one or more rational roots. Solve them.

6. $x^3 - x^2 - 14x + 24 = 0$. 7. $x^3 - 2x^2 - 25x + 50 = 0$.

8. $3x^3 - 2x^2 + 2x + 1 = 0$. 9. $2x^4 + 7x^3 - 2x^2 - x = 0$.

10. $x^4 + 4x^3 + 8x^2 + 8x + 3 = 0$.

11. $2x^4 + 7x^3 + 4x^2 - 7x - 6 = 0$.

12. $3x^4 + 11x^3 + 9x^2 + 11x + 6 = 0$.

13. $x^5 - 9x^4 + 2x^3 + 71x^2 + 81x + 70 = 0$.

14. $2x^5 - 8x^4 + 7x^3 + 5x^2 - 8x + 4 = 0$.

15. $x^5 + 3x^4 - 15x^3 - 35x^2 + 54x + 72 = 0$.

16. $12x^4 - 32x^3 + 13x^2 + 8x - 4 = 0$.

17. $x^5 - 7x^4 + 10x^3 + 18x^2 - 27x - 27 = 0$.

18. $2x^4 - 17x^3 + 25x^2 + 74x - 120 = 0$.

19. $4x^5 - 9x^3 + 6x^2 - 13x + 6 = 0$.

20. $x^5 + 8x^4 + 3x^3 - 80x^2 - 52x + 240 = 0$.

21. $2x^5 + 11x^4 + 23x^3 + 25x^2 + 16x + 4 = 0$.

22. $6x^4 - 89x^3 + 359x^2 - 254x + 48 = 0$.

23. $10x^4 + 41x^3 + 46x^2 + 20x + 3 = 0$.

24. $36x^4 - 108x^3 + 107x^2 - 43x + 6 = 0$.

25. $12x^5 + 20x^4 + 29x^3 + 77x^2 + 69x + 18 = 0$.

26. $2x^6 + 7x^5 + 8x^4 + 7x^3 + 2x^2 - 14x - 12 = 0$.

27. $2x^6 + 11x^5 + 24x^4 + 22x^3 - 8x^2 - 33x - 18 = 0$.

28. $5x^6 - 7x^5 - 8x^4 - x^3 + 7x^2 + 8x - 4 = 0$.

RELATIONS BETWEEN ROOTS AND COEFFICIENTS

805 **Relations between roots and coefficients.** When an equation whose roots are $\beta_1, \beta_2, \cdots, \beta_n$ is reduced to the second standard form, § 792, (2), the identity in § 798, (3), becomes

$$x^n + b_1x^{n-1} + b_2x^{n-2} + b_3x^{n-3} + \cdots + b_n$$
$$= (x - \beta_1)(x - \beta_2)(x - \beta_3) \cdots (x - \beta_n).$$

Carry out the multiplications indicated in the second member and arrange the result as a polynomial in x, § 559. Then equate the coefficients of like powers of x in the two members, § 284. We thus obtain the following relations between the coefficients $b_1, b_2, \cdots, b_n$ and the roots $\beta_1, \beta_2, \cdots, \beta_n$:

$$-b_1 = \beta_1 + \beta_2 + \beta_3 + \cdots + \beta_n, \quad (1)$$
$$b_2 = \beta_1\beta_2 + \beta_1\beta_3 + \cdots + \beta_2\beta_3 + \cdots + \beta_{n-1}\beta_n, \quad (2)$$
$$-b_3 = \beta_1\beta_2\beta_3 + \beta_1\beta_2\beta_4 + \cdots + \beta_{n-2}\beta_{n-1}\beta_n, \quad (3)$$
$$\cdots\cdots\cdots$$
$$(-1)^n b_n = \beta_1\beta_2\beta_3 \cdots \beta_n, \quad (n)$$

where the second members of (2), (3), $\cdots$ represent the sum of the products of every two of the roots, of every three, and so on, and the sign before the first member is plus or minus according as the number of the roots in each term of the second member is even or odd. Hence the theorem :

Theorem. *In every equation reduced to the form* **806**

$$x^n + b_1x^{n-1} + b_2x^{n-2} + \cdots + b_n = 0,$$

the coefficient b_1 *of the second term, with its sign changed, is equal to the sum of all the roots; the absolute term* b_n, *with its sign changed or not according as* n *is odd or even, is equal to the product of all the roots; and the coefficient* b_r *of each intermediate term, with its sign changed or not according as* r *is odd or even, is equal to the sum of the products of every* r *of the roots.*

Before applying this theorem to an equation whose leading coefficient is not 1 we must divide the equation by that coefficient. If the equation be incomplete, it must be remembered that the coefficients of the missing terms are 0.

Thus, without solving the equation $3x^3 - 6x + 2 = 0$, we know the following facts regarding its roots β_1, β_2, β_3. Reduced to the proper form for applying the theorem, the equation is $x^3 + 0x^2 - 2x + 2/3 = 0$. Hence

$$\beta_1 + \beta_2 + \beta_3 = 0,\ \beta_1\beta_2 + \beta_1\beta_3 + \beta_2\beta_3 = -2,\ \beta_1\beta_2\beta_3 = -2/3.$$

If all but one of the roots of an equation are known, we can find the remaining root by subtracting the sum of the known roots from $-b_1$, or by dividing b_n, with its sign changed if n is odd, by the product of the known roots. **807**

Example. Two of the roots of $2x^3 + 3x^2 - 23x - 12 = 0$ are 3 and -4. What is the remaining root?

The remaining root is $-3/2 - [3 + (-4)] = -1/2$; or again, it is $6 \div 3(-4) = -1/2$.

When the roots themselves are connected by some given relation, a corresponding relation must exist among the coefficients. To find this relation we apply the theorem of § 806. **808**

Example 1. Find the condition that the roots of $x^3 + px^2 + qx + r = 0$ shall be in geometrical progression.

Representing the roots by α/β, α, $\alpha\beta$, we have

$$\frac{\alpha}{\beta} + \alpha + \alpha\beta = -p,\ \frac{\alpha^2}{\beta} + \alpha^2 + \alpha^2\beta = q,\ \frac{\alpha}{\beta} \cdot \alpha \cdot \alpha\beta = -r.$$

The third equation reduces to $\alpha^3 = -r$, whence $\alpha = \sqrt[3]{-r}$.

Dividing the second equation by the first, substituting $\alpha = \sqrt[3]{-r}$ in the result, and simplifying, we have $q^3 - p^3r = 0$.

Example 2. Solve the equation $x^3 + 8x^2 + 5x - 50 = 0$, having given that it has a double root.

Representing the roots by α, α, β, we have

$$2\alpha + \beta = -8,\ \alpha^2 + 2\alpha\beta = 5,\ \alpha^2\beta = 50.$$

Solving the first and second of these equations for α and β, we obtain $\alpha = -5$, $\beta = 2$ and $\alpha = -1/3$, $\beta = -22/3$.

The values $\alpha = -5$, $\beta = 2$ satisfy the equation $\alpha^2\beta = 50$, but the values $\alpha = -1/3$, $\beta = -22/3$ do not satisfy this equation.

Hence the required roots are -5, -5, 2.

809 Symmetric functions of the roots. The expressions in the roots to which the several coefficients are equal, § 805, are *symmetric functions* of the roots, § 540. It will be proved in § 868 that all other rational symmetric functions of the roots can be expressed rationally in terms of these functions, and therefore rationally in terms of the coefficients of the equation.

Example 1. Find the sum of the squares of the roots of the equation $2x^3 - 3x^2 - 4x - 5 = 0$.

Calling the roots α, β, γ, we have

$$\alpha^2 + \beta^2 + \gamma^2 = (\alpha + \beta + \gamma)^2 - 2(\alpha\beta + \beta\gamma + \gamma\alpha) = (3/2)^2 + 4 = 6\tfrac{1}{4}.$$

Example 2. If the roots of $x^3 + px^2 + qx + r = 0$ are α, β, γ, what is the equation whose roots are $\beta\gamma$, $\gamma\alpha$, $\alpha\beta$?

If p', q', r' denote the coefficients of the required equation, we have

$$\begin{aligned} -p' &= \beta\gamma + \gamma\alpha + \alpha\beta = q, \\ q' &= \beta\gamma \cdot \gamma\alpha + \gamma\alpha \cdot \alpha\beta + \alpha\beta \cdot \beta\gamma \\ &= \alpha\beta\gamma(\alpha + \beta + \gamma) = (-r)(-p) = rp, \\ -r' &= \beta\gamma \cdot \gamma\alpha \cdot \alpha\beta = (\alpha\beta\gamma)^2 = r^2. \end{aligned}$$

Hence the required equation is $x^3 - qx^2 + prx - r^2 = 0$.

EXERCISE LXXI

1. Two of the roots of $2x^3 - 7x^2 + 10x - 6 = 0$ are $1 \pm i$; find the third root.

2. The roots of each of the following equations are in geometrical progression; find them.

(1) $8x^3 - 14x^2 - 21x + 27 = 0$. (2) $x^3 + x^2 + 3x + 27 = 0$.

3. The roots of each of the following equations are in arithmetical progression; find them.

(1) $x^3 + 6x^2 + 7x - 2 = 0$. (2) $x^3 - 9x^2 + 23x - 15 = 0$.

4. Show that if one root of $x^3 + px^2 + qx + r = 0$ be the negative of another root, $pq = r$.

5. Find the condition that one root of $x^3 + px^2 + qx + r = 0$ shall be the reciprocal of another root.

6. Solve $x^4 + 4x^3 + 10x^2 + 12x + 9 = 0$, having given that it has two double roots.

7. Solve the equation $14x^3 - 13x^2 - 18x + 9 = 0$, having given that its roots are in harmonical progression.

8. Solve the equation $x^4 - x^3 - 56x^2 + 36x + 720 = 0$, having given that two of its roots are in the ratio $2:3$ and that the difference between the other two roots is 1.

9. If α, β, γ are the roots of $x^3 + px^2 + qx + r = 0$, find the equations whose roots are

(1) $-\alpha, -\beta, -\gamma$. (2) $k\alpha, k\beta, k\gamma$.

(3) $1/\alpha, 1/\beta, 1/\gamma$. (4) $\alpha + k, \beta + k, \gamma + k$.

(5) $\alpha^2, \beta^2, \gamma^2$. (6) $-1/\alpha^2, -1/\beta^2, -1/\gamma^2$.

10. If α, β, γ are the roots of $2x^3 + x^2 - 4x + 1 = 0$, find the values of

(1) $\alpha^2 + \beta^2 + \gamma^2$. (2) $\alpha^3 + \beta^3 + \gamma^3$.

(3) $1/\beta\gamma + 1/\gamma\alpha + 1/\alpha\beta$. (4) $\alpha\beta^2 + \beta\alpha^2 + \beta\gamma^2 + \gamma\beta^2 + \gamma^2\alpha + \alpha^2\gamma$.

11. If α, β, γ are the roots of $x^3 - 2x^2 + x - 3 = 0$, find the values of

(1) $\alpha/\beta\gamma + \beta/\gamma\alpha + \gamma/\alpha\beta$ (2) $\alpha\beta/\gamma + \beta\gamma/\alpha + \gamma\alpha/\beta$.

(3) $(\beta + \gamma)(\gamma + \alpha)(\alpha + \beta)$. (4) $(\beta^2 + \gamma^2)(\gamma^2 + \alpha^2)(\alpha^2 + \beta^2)$.

(5) $\alpha\left(\frac{1}{\beta} + \frac{1}{\gamma}\right) + \beta\left(\frac{1}{\gamma} + \frac{1}{\alpha}\right) + \gamma\left(\frac{1}{\alpha} + \frac{1}{\beta}\right)$.

TRANSFORMATIONS OF EQUATIONS

810 **Some important transformations.** It is sometimes advan tageous to transform a given equation $f(x)=0$ into another equation whose roots stand in some given relation to the roots of $f(x)=0$. The transformations most frequently used are the following:

811 *To transform a given equation* $f(x)=0$ *into another whose roots are those of* $f(x)=0$ *with their signs changed.*

The required equation is $f(-y)=0$. For substituting any number, as β, for x in $f(x)$ gives the same result as substituting $-\beta$ for y in $f(-y)$. Hence, if $f(x)$ vanishes when $x=\beta$, $f(-y)$ will vanish when $y=-\beta$; that is, if β is a root of $f(x)=0$, $-\beta$ is a root of $f(-y)=0$.

Therefore every root of $f(x)=0$, with its sign changed, is a root of $f(-y)=0$; and $f(-y)=0$ has no other roots than these, since $f(x)=0$ and $f(-y)=0$ are of the same degree.

If the given equation is

$$a_0x^n + a_1x^{n-1} + a_2x^{n-2} + \cdots + a_n = 0,$$

the required equation will be

$$a_0(-y)^n + a_1(-y)^{n-1} + a_2(-y)^{n-2} + \cdots + a_n = 0,$$

which on being simplified becomes

$$a_0y^n - a_1y^{n-1} + a_2y^{n-2} - \cdots + (-1)^n a_n = 0.$$

Hence the required equation may be obtained from the given one by *changing the signs of the terms of odd degree when* n *is even, and by changing the signs of the terms of even degree, including the absolute term, when* n *is odd.*

We may use x instead of y for the unknown letter in the transformed equation, and write $f(-x)=0$ for $f(-y)=0$.

Example. Change the signs of the roots of

$$4x^5 - 9x^3 + 6x^2 - 13x + 6 = 0.$$

Changing the signs of the terms of even degree, we have

$$4x^5 - 9x^3 - 6x^2 - 13x - 6 = 0.$$

In fact, the roots of the given equation are $1/2$, $3/2$, -2, $\pm i$, and those of the transformed equation are $-1/2$, $-3/2$, 2, $\mp i$.

To transform a given equation $f(x) = 0$ *into another whose roots are those of* $f(x) = 0$ *each multiplied by some constant, as* k. **812**

The required equation is $f(y/k) = 0$. For if $f(x)$ vanishes when $x = \beta$, $f(y/k)$ will vanish when $y/k = \beta$, that is, when $y = k\beta$ (compare § 811).

If the given equation is

$$a_0x^n + a_1x^{n-1} + a_2x^{n-2} + \cdots + a_n = 0,$$

the required equation will be

$$a_0\left(\frac{y}{k}\right)^n + a_1\left(\frac{y}{k}\right)^{n-1} + a_2\left(\frac{y}{k}\right)^{n-2} + \cdots + a_n = 0,$$

which when cleared of fractions becomes

$$a_0y^n + ka_1y^{n-1} + k^2a_2y^{n-2} + \cdots + k^na_n = 0.$$

Hence the required equation may be obtained by *multiplying the second term of the given equation by* k, *its third term by* k^2, *and so on, taking account of missing terms if any.*

When $k = -1$ this transformation reduces to that of § 811.

Example. Multiply the roots of $x^4 + 2x^3 - x + 3 = 0$ by 2. Also divide them by 2.

The first of the required equations is $x^4 + 4x^3 - 8x + 48 = 0$, and since dividing by 2 is the same as multiplying by $1/2$, the second is

$$x^4 + x^3 - x/8 + 3/16 = 0, \text{ or } 16x^4 + 16x^3 - 2x + 3 = 0.$$

The following example illustrates an important application of the transformation now under consideration. **813**

Example. Transform the equation $36x^3 + 18x^2 + 2x + 9 = 0$ into another whose leading coefficient is 1 and its remaining coefficients integers.

Dividing by 36, we have

$$x^3 + x^2/2 + x/18 + 1/4 = 0. \qquad (1)$$

Multiplying the roots by k,

$$x^3 + kx^2/2 + k^2x/18 + k^3/4 = 0. \qquad (2)$$

We see by inspection that the smallest value of k which will cancel all the denominators is 6. And substituting 6 for k in (2), we have

$$x^3 + 3x^2 + 2x + 54 = 0, \qquad (3)$$

which has the form required. The roots of (3) each divided by 6 are the roots of the given equation (1).

814 *To transform a given equation* $f(x) = 0$ *into another whose roots are the reciprocals of those of* $f(x) = 0$.

The required equation is $f(1/y) = 0$. For if $f(x)$ vanishes when $x = \beta$, $f(1/y)$ will vanish when $1/y = \beta$, that is, when $y = 1/\beta$.

If the given equation is

$$a_0x^n + a_1x^{n-1} + \cdots + a_{n-1}x + a_n = 0,$$

the required equation will be

$$a_0/y^n + a_1/y^{n-1} + \cdots + a_{n-1}/y + a_n = 0,$$

which when cleared of fractions becomes

$$a_ny^n + a_{n-1}y^{n-1} + \cdots + a_1y + a_0 = 0.$$

Hence the required equation may be obtained by merely *reversing the order of the coefficients of the given equation.*

Example. Replace the roots of $2x^4 - x^2 - 3x + 4 = 0$ by their reciprocals.

Reversing the coefficients, we have $4x^4 - 3x^3 - x^2 + 2 = 0$.

815 An equation like $2x^3 + 3x^2 - 3x - 2 = 0$, which remains unchanged when this transformation is applied to it, that is, when the order of its coefficients is reversed, is called a *reciprocal equation*, § 645. If β is a root of such an equation, $1/\beta$ must also be a root. Hence when the degree of the equation is even, half of the roots are the reciprocals of the other half. The same is true of all the roots but one when the degree is odd; but in this case there must be one root which is its own reciprocal, that is, one root which is either

1 or -1. Thus, one root of $2x^3 + 3x^2 - 3x - 2 = 0$ is 1 and the other roots are -2 and $-1/2$.

816 From the nature of this transformation it follows that *when an equation has variable coefficients and the leading coefficient vanishes, one of the roots becomes infinite; when the two leading coefficients vanish, two of the roots become infinite; and so on.*

Example 1. Show that one of the roots of $mx^3 + 3x^2 - 2x + 1 = 0$ becomes infinite when m vanishes.

Applying § 814 to $\quad mx^3 + 3x^2 - 2x + 1 = 0, \qquad (1)$

we obtain $\quad x^3 - 2x^2 + 3x + m = 0. \qquad (2)$

If the roots of (2) are $\beta_1, \beta_2, \beta_3$, those of (1) are $1/\beta_1, 1/\beta_2, 1/\beta_3$.

By § 806, $\beta_1\beta_2\beta_3 = -m$. Hence, if m approach 0 as limit, one of the roots $\beta_1, \beta_2, \beta_3$ must also approach 0 as limit, and if this root be β_1, the corresponding root of (1), namely, $1/\beta_1$, must approach ∞, § 512.

Example 2. Show that two of the roots of $mx^3 + m^2x^2 + x + 1 = 0$ become infinite when m vanishes.

Applying § 814 to $\quad mx^3 + m^2x^2 + x + 1 = 0, \qquad (1)$

we have $\quad x^3 + x^2 + m^2x + m = 0. \qquad (2)$

If the roots of (2) are $\beta_1, \beta_2, \beta_3$, those of (1) are $1/\beta_1, 1/\beta_2, 1/\beta_3$. Also,

$$\beta_1\beta_2\beta_3 = -m, \quad \beta_1\beta_2 + \beta_1\beta_3 + \beta_2\beta_3 = m^2. \qquad (3)$$

It follows from (3) that if m approach 0, two of the three roots $\beta_1, \beta_2, \beta_3$ must also approach 0, and if these roots be β_1, β_2, the corresponding roots of (1), namely, $1/\beta_1, 1/\beta_2$, must approach ∞.

817 *To transform a given equation* $\mathrm{f(x)} = 0$ *into another whose roots are those of* $\mathrm{f(x)} = 0$, *each diminished by some constant, as* k.

The required equation is $f(y + k) = 0$. For if $f(x)$ vanishes when $x = \beta$, $f(y + k)$ will vanish when $y + k = \beta$, that is, when $y = \beta - k$.

If the given equation is

$$f(x) = a_0x^n + a_1x^{n-1} + \cdots + a_{n-1}x + a_n = 0,$$

the required equation will be

$$f(y + k) = a_0(y + k)^n + a_1(y + k)^{n-1} + \cdots + a_n = 0,$$

which, when its terms are expanded by the binomial theorem and then collected, will reduce to the form

$$\phi(y) = c_0 y^n + c_1 y^{n-1} + \cdots + c_{n-1} y + c_n = 0,$$

where $c_0 = a_0$, $c_1 = nka_0 + a_1$, and so on.

This method of obtaining $\phi(y)$ from $f(x)$ is usually very laborious. The following method is much more expeditious, at least when the coefficients of $f(x)$ are given rational numbers.

If $x = y + k$, then $y = x - k$, and we have

$$f(x) = f(y + k) = \phi(y) = \phi(x - k),$$

that is,

$$c_0(x - k)^n + \cdots + c_{n-1}(x - k) + c_n \equiv a_0 x^n + \cdots + a_{n-1}x + a_n.$$

If both members of this identity be divided by $x - k$, if again the quotients thus obtained be divided by $x - k$, and so on, the successive *remainders* yielded by the first member, namely, $c_n, c_{n-1}, \cdots$, will be the same as those yielded by the second member. Hence we may obtain $\phi(y)$ from $f(x)$ as follows: *Divide* $f(x)$ *by* $x - k$, *divide the quotient thus obtained by* $x - k$, *and so on. The successive remainders will be* c_n, $c_{n-1}, \cdots, c_1$, *and the final quotient will be* c_0 (compare § 423). The divisions should be performed synthetically.

Example 1. Diminish the roots of $2x^3 - 7x^2 - 3x + 1 = 0$ by 4.

First method. Substituting $y + 4$ for x, we have

$$\begin{aligned} 2x^3 - 7x^2 - 3x + 1 &= 2(y + 4)^3 - 7(y + 4)^2 - 3(y + 4) + 1 \\ &= 2y^3 + 17y^2 + 37y + 5. \end{aligned}$$

Second method. Arranging the reckoning as in § 423, we have

$$\begin{array}{rrrrl}
2 & -7 & -3 & +1 & \underline{|4} \\
 & 8 & 4 & 4 & \\
\hline
2 & +1 & +1, & 5 & \quad \therefore c_3 = 5. \\
 & 8 & 36 & & \\
\hline
2 & +9, & 37 & & \quad \therefore c_2 = 37. \\
 & 8 & & & \\
\hline
2, & 17 & & & \quad \therefore c_1 = 17 \text{ and } c_0 = 2.
\end{array}$$

Hence, as before, we find the required equation to be

$$2y^3 + 17y^2 + 37y + 5 = 0.$$

Example 2. Increase the roots of $x^3 + 4x^2 + x + 3 = 0$ by 4.

To increase the roots by 4 is the same thing as to diminish them by -4. Hence the required equation may be obtained either by substituting $y - 4$ for x or by dividing synthetically by -4. It will be found to be $y^3 - 8y^2 + 17y - 1 = 0$.

818 By aid of § 817 we can transform a given equation into another which lacks some particular power of the unknown letter.

Example 1. Transform the equation $x^3 - 3x^2 + 5x + 6 = 0$ into another which lacks the second power of the unknown letter.

Substituting $x = y + k$, we have $y^3 + (3k - 3)y^2 + \cdots$. Hence we must have $3k - 3 = 0$, that is, $k = 1$. And diminishing the roots of $x^3 - 3x^2 + 5x + 6 = 0$ by 1, we obtain $x^3 + 2x + 9 = 0$.

Example 2. Transform the equation $x^3 - 5x^2 + 8x - 1 = 0$ into another which lacks the first power of the unknown letter.

Substituting $x = y + k$, we have

$$y^3 + (3k - 5)y^2 + (3k^2 - 10k + 8)y + \cdots = 0.$$

Hence we must have $3k^2 - 10k + 8 = 0$, that is, $k = 2$ or $4/3$. Diminishing the roots of $x^3 - 5x^2 + 8x - 1 = 0$ by 2, we obtain

$$x^3 + x^2 + 3 = 0.$$

819 If, when we diminish the roots of $f(x) = 0$ by k, we obtain an equation $\phi(x) = 0$ whose terms are all positive, k is a superior limit of the positive roots of $f(x) = 0$, § 804. For in this case $\phi(x) = 0$ has no positive root, § 794. Hence any positive roots that $f(x) = 0$ may have become negative when diminished by k. They are therefore less than k.

The process of synthetic division is such that it is possible by inspection and trial to find the smallest integer k for which all the terms of $\phi(x) = 0$ will be positive. In most cases this can be accomplished with comparatively little labor.

We may obtain an inferior limit of the negative roots of $f(x) = 0$ by finding a superior limit of the roots of $f(-x) = 0$. For if k is a superior limit of the roots of $f(-x) = 0$, then $-k$ is an inferior limit of the roots of $f(x) = 0$, § 811.

Example. Find superior and inferior limits of the roots of the equation $f(x) = x^4 - 6x^3 + 14x^2 + 48x - 121 = 0$.

We find by inspection and trial that neither $k = 1$ nor $k = 2$ will give a transformed equation $\phi(x) = 0$ all of whose terms are positive, but that $k = 3$ will. In fact, if we diminish the roots of $f(x) = 0$ by 3, we obtain $\phi(x) = x^4 + 6x^3 + 14x^2 + 78x + 68 = 0$. Hence 3 is a superior limit of the roots of $f(x) = 0$.

The equation $f(-x) = 0$ is $x^4 + 6x^3 + 14x^2 - 48x - 121 = 0$. We find by inspection and trial that 3 is a superior limit of its positive roots. Hence -3 is an inferior limit of the negative roots of $f(x) = 0$.

820 On rational transformations in general. If we eliminate x between the equations $f(x) = 0$ and $y = -x$, we obtain $f(-y) = 0$. We have shown in § 811 that the roots y of $f(-y) = 0$ are connected with the roots x of $f(x) = 0$ by the relation $y = -x$. This is an illustration of the general theorem that if we properly eliminate x between $f(x) = 0$ and any equation of the form $y = \phi(x)$, where $\phi(x)$ is rational, we shall obtain an equation $F(y) = 0$ whose roots are connected with those of $f(x) = 0$ by the relation $y = \phi(x)$, so that if the roots of $f(x) = 0$ are $\beta_1, \beta_2, \cdots, \beta_n$, those of $F(y) = 0$ are $\phi(\beta_1)$, $\phi(\beta_2), \cdots, \phi(\beta_n)$. The transformations of §§ 812, 814, 817 afford further illustrations of this theorem. In the first of these transformations the equation $y = \phi(x)$ is $y = kx$, in the second it is $y = 1/x$, and in the third it is $y = x - k$. When, as in these cases, $y = \phi(x)$ can be solved for x, the elimination of x is readily effected.

Example 1. Find the equation whose roots are the squares of the roots of $x^3 + px^2 + qx + r = 0$.

In this case the relation $y = \phi(x)$ is $y = x^2$.

Solving $y = x^2$ for x, we have $x = \pm\sqrt{y}$. And substituting $\pm\sqrt{y}$ for x in the given equation and rationalizing, we obtain

$$y^3 + (2q - p^2)y^2 + (q^2 - 2pr)y - r^2 = 0.$$

Example 2. If the roots of $x^3 + px^2 + qx + r = 0$ are α, β, γ, find the equation whose roots are $\beta\gamma, \gamma\alpha, \alpha\beta$.

We first endeavor to express each of the proposed roots $\beta\gamma$, $\gamma\alpha$, $\alpha\beta$ in terms of a single one of the given roots α, β, γ and the given

coefficients p, q, r. This is readily done, for, since $-r = \alpha\beta\gamma$, we have
$\beta\gamma = \alpha\beta\gamma/\alpha = -r/\alpha$, $\gamma\alpha = \alpha\beta\gamma/\beta = -r/\beta$, $\alpha\beta = \alpha\beta\gamma/\gamma = -r/\gamma$.

Hence each root y of the required equation is connected with the corresponding root x of the given equation by the relation $y = -r/x$.

Solving $y = -r/x$ for x, we have $x = -r/y$.

And substituting $-r/y$ for x in $x^3 + px^2 + qx + r = 0$ and simplifying, we have $y^3 - qy^2 + pry - r^2 = 0$, which is the equation required.

EXERCISE LXXII

1. Change the signs of the roots of $x^7 + 3x^4 - 2x^2 + 6x + 7 = 0$.

2. Multiply the roots of $2x^4 + x^3 - 4x^2 - 6x + 8 = 0$ by -2. Also divide them by 3.

3. In $5x^6 - x^4 + 3x^3 + 9x + 10 = 0$ replace each root by its reciprocal.

4. Diminish the roots of $2x^5 + x^4 - 3x^2 + 6 = 0$ by 2. Also increase them by 1.

5. Transform the equation $x^4 - x^3/3 + x^2/4 + x/25 - 1/48 = 0$ into another whose coefficients are integers, the leading one being 1.

6. Transform the equation $3x^4 - 36x^3 + x - 7 = 0$ into another which lacks the term involving x^3.

7. Transform the following into equations which lack the x term.

(1) $x^3 + 6x^2 + 9x + 10 = 0$. (2) $x^3 - x^2 - x - 3 = 0$.

8. If the roots of $x^4 + x^3 - x + 2 = 0$ are $\alpha, \beta, \gamma, \delta$, find the equation whose roots are $\alpha^2, \beta^2, \gamma^2, \delta^2$.

9. If the roots of $x^4 + 3x^3 + 2x^2 - 1 = 0$ are $\alpha, \beta, \gamma, \delta$, find the equation whose roots are $\beta + \gamma + \delta$, $\alpha + \gamma + \delta$, $\alpha + \beta + \delta$, $\alpha + \beta + \gamma$.

10. If the roots of $x^3 + px^2 + qx + r = 0$ are α, β, γ, find the equations whose roots are

(1) $\dfrac{\alpha\beta}{\gamma}, \dfrac{\beta\gamma}{\alpha}, \dfrac{\gamma\alpha}{\beta}$. (2) $\dfrac{\alpha}{\beta+\gamma}, \dfrac{\beta}{\gamma+\alpha}, \dfrac{\gamma}{\alpha+\beta}$.

11. If the roots of $x^3 + 2x^2 + 3x + 4 = 0$ are α, β, γ, find the equations whose roots are

(1) $\beta^2 + \gamma^2, \gamma^2 + \alpha^2, \alpha^2 + \beta^2$. (2) $\alpha(\beta + \gamma), \beta(\gamma + \alpha), \gamma(\alpha + \beta)$.

(3) $\beta\gamma + \dfrac{1}{\alpha}, \gamma\alpha + \dfrac{1}{\beta}, \alpha\beta + \dfrac{1}{\gamma}$. (4) $\dfrac{\alpha}{\beta+\gamma-\alpha}, \dfrac{\beta}{\gamma+\alpha-\beta}, \dfrac{\gamma}{\alpha+\beta-\gamma}$.

(5) $\alpha\left(\dfrac{1}{\beta}+\dfrac{1}{\gamma}\right), \beta\left(\dfrac{1}{\gamma}+\dfrac{1}{\alpha}\right), \gamma\left(\dfrac{1}{\alpha}+\dfrac{1}{\beta}\right)$.

12. Find superior and inferior limits of the real roots of the following equations.

(1) $x^4 + 3x^3 - 13x^2 - 6x + 28 = 0$. (2) $2x^5 - 120x^2 - 38x + 27 = 0$.

(3) $x^4 - 29x^2 + 50x + 12 = 0$. (4) $2x^5 - 26x^3 + 60x^2 - 92 = 0$.

(5) $x^4 - 14x^3 + 44x^2 + 28x - 92 = 0$.

(6) $3x^6 - 35x^3 + 77x^2 - 50x - 110 = 0$.

IMAGINARY ROOTS. DESCARTES'S RULE OF SIGNS

821 **Theorem.** *Let* f(x) = 0 *denote an equation with real coefficients. If it has imaginary roots, these occur in pairs; that is, if* a + ib *is a root,* a − ib *is also a root.*

For if $a + ib$ is a root of $f(x) = 0$, then $f(x)$ is divisible by $x - (a + ib)$, § 795; and we shall prove that $a - ib$ is a root if we can show that $f(x)$ is also divisible by $x - (a - ib)$, or, what comes to the same thing, if we can show that $f(x)$ is divisible by the product $[x - (a + ib)][x - (a - ib)]$.

This product has real coefficients, for, since $i^2 = -1$,

$$[x - (a + ib)][x - (a - ib)] = (x - a)^2 + b^2$$
$$= x^2 - 2ax + (a^2 + b^2).$$

Since the polynomials $f(x)$ and $x^2 - 2ax + (a^2 + b^2)$ have the common factor $x - (a + ib)$, they have a *highest* common factor. This highest common factor must be either $x - (a + ib)$ or $x^2 - 2ax + (a^2 + b^2)$. But it cannot be $x - (a + ib)$, since this has imaginary coefficients, whereas the highest common factor of two polynomials with real coefficients must itself have real coefficients, § 469. Hence the highest common factor of $f(x)$ and $x^2 - 2ax + (a^2 + b^2)$ is $x^2 - 2ax + (a^2 + b^2)$; in other words, $f(x)$ is divisible by $x^2 - 2ax + (a^2 + b^2)$, as was to be demonstrated.

Example. One root of $2x^3 + 5x^2 + 46x - 87 = 0$ is $-2 + 5i$. Solve this equation.

Since $-2 + 5i$ is a root, $-2 - 5i$ is also a root. But the sum of all the roots is $-5/2$, § 806. Hence the third root is

$$-5/2 - (-2 + 5i - 2 - 5i) = 3/2.$$

Corollary 1. *Every polynomial* f(x) *with real coefficients is the product of real factors of the first or second degree.* **822**

For to each real root c of $f(x) = 0$ there corresponds the real factor $x - c$ of $f(x)$, § 795; and to each pair of imaginary roots $a + ib$, $a - ib$ of $f(x) = 0$ there corresponds the real factor $x^2 - 2ax + (a^2 + b^2)$ of $f(x)$, § 821.

Corollary 2. *The product of those factors of* f(x) *which correspond to the imaginary roots of* f(x) = 0 *is a function of* x *which is positive for all real values of* x. **823**

For it may be expressed as a product of factors of the form $(x - a)^2 + b^2$, § 821, and every such factor, being a sum of squares, is positive for all real values of x.

Corollary 3. *Every equation with real coefficients whose degree is odd has at least one real root.* **824**

For the number of its imaginary roots, if it have any, is even, § 821, and the total number of its roots, real and imaginary, is odd, § 798. Hence at least one root must be real.

Thus, the roots of a cubic equation with real coefficients are either all of them real, or one real and two imaginary.

825 By the reasoning employed in § 821 it may be proved that if $a + \sqrt{b}$ is a root of a given equation with *rational* coefficients, $a - \sqrt{b}$ is also a root; it being understood that a and b are themselves rational, but $\sqrt{b}$ irrational.

826 **Irreducible equations.** Let $\phi(x) = 0$ be an equation whose coefficients are both rational and real. We say that this equation is *irreducible* if $\phi(x)$ has no factor whose coefficients are both rational and real (compare § 486).

Thus, $x^2 - 2 = 0$ and $x^2 + x + 1 = 0$ are irreducible equations, but $x^2 - 4 = 0$ is not irreducible.

Theorem. *Let* f(x) = 0 *be any equation whose coefficients are both rational and real, and let* ϕ(x) = 0 *be an irreducible equation of the same or a lower degree.* **827**

If one of the roots of $\phi(\mathrm{x}) = 0$ *be a root of* $\mathrm{f}(\mathrm{x}) = 0$, *then all of the roots of* $\phi(\mathrm{x}) = 0$ *are roots of* $\mathrm{f}(\mathrm{x}) = 0$.

This may be proved by the reasoning of § 821. For if $f(x) = 0$ and $\phi(x) = 0$ have the root c in common, $f(x)$ and $\phi(x)$ have the common factor $x - c$, § 795, and therefore a highest common factor which is either $x - c$, some factor of $\phi(x)$ which contains $x - c$, or $\phi(x)$ itself.

But since, by hypothesis, $\phi(x) = 0$ is an irreducible equation, $\phi(x)$ is the only one of these factors which has real and rational coefficients such as the highest common factor of $f(x)$ and $\phi(x)$ must have, § 469.

Therefore $\phi(x)$ is itself the highest common factor of $f(x)$ and $\phi(x)$; in other words, $f(x)$ is exactly divisible by $\phi(x)$.

Hence $f(x)$ may be expressed in the form $f(x) = Q\phi(x)$, where Q is integral, and from this identity it follows that $f(x)$ vanishes whenever $\phi(x)$ vanishes; in other words, that every root of $\phi(x) = 0$ is a root of $f(x) = 0$.

828 **Permanences and variations.** In any polynomial $f(x)$, or equation $f(x) = 0$, with real coefficients a *permanence* or *continuation* of sign is said to occur wherever a term follows one of like sign, and a *variation* or *change* of sign wherever a term follows one of contrary sign.

Thus, in $x^6 - x^4 - x^3 + 2x^2 + 3x - 1 = 0$ permanences occur at the terms $-x^3$ and $3x$, and variations at the terms $-x^4$, $2x^2$, and -1.

829 **Theorem.** *If* $\mathrm{f}(\mathrm{x})$ *is exactly divisible by* $\mathrm{x} - \mathrm{b}$, *where* b *is positive and the coefficients of* $\mathrm{f}(\mathrm{x})$ *are real, the quotient* $\phi(\mathrm{x})$ *will have at least one less variation than* $\mathrm{f}(\mathrm{x})$ *has.*

For since b is positive, it follows from the rule of synthetic division, § 411, that when $f(x)$ is divided by $x - b$, the coefficients of the quotient are positive until the first negative coefficient of $f(x)$ is reached. If then or later one of them becomes negative or zero, they continue negative until the next positive coefficient of $f(x)$ is reached, and so on. Hence $\phi(x)$ can have no variations except such as occur at the same or earlier terms of $f(x)$. But since the division is, by hypothesis, exact, the last sign in $\phi(x)$ must be contrary to the last sign in $f(x)$, and therefore $\phi(x)$ must lack the last variation in $f(x)$.

$$\begin{array}{rrrrrrrl} 1 & +1 & -2 & -10 & -1 & +12 & -4 & \lfloor 2 \\ - & 2 & +6 & +8 & -4 & -10 & +4 & \\ \hline 1 & +3 & +4 & -2 & -5 & +2, & 0 & \end{array}$$

Thus, $f(x) = x^6 + x^5 - 2x^4 - 10x^3 - x^2 + 12x - 4$ is exactly divisible by $x - 2$, the quotient being $\phi(x) = x^5 + 3x^4 + 4x^3 - 2x^2 - 5x + 2$. Observe that the first two variations of $f(x)$ are reproduced in $\phi(x)$, but not the third.

$$\begin{array}{rrrrrl} 1 & -1 & +1 & -7 & +2 & \lfloor 2 \\ - & 2 & +2 & +6 & -2 & \\ \hline 1 & +1 & +3 & -1, & 0 & \end{array}$$

Again, $f(x) = x^4 - x^3 + x^2 - 7x + 2$ is exactly divisible by $x - 2$, the quotient being $\phi(x) = x^3 + x^2 + 3x - 1$. In this case only one of the four variations of $f(x)$ is reproduced in $\phi(x)$, and we have an illustration of the fact that when *intermediate* variations of $f(x)$ disappear in $\phi(x)$, they disappear in pairs.

Theorem (Descartes's rule of signs). *An equation* $f(x) = 0$ *cannot have a greater number of positive roots than it has variations, nor a greater number of negative roots than the equation* $f(-x) = 0$ *has variations.* **830**

1. For let $\beta_1, \beta_2, \cdots, \beta_r$ denote the positive roots of $f(x) = 0$.

If we divide $f(x)$ by $x - \beta_1$, the quotient thus obtained by $x - \beta_2$, and so on, we obtain a final quotient $\phi(x)$ which has at least r less variations than $f(x)$ has, § 829. Therefore, since $\phi(x)$ cannot have less than no variations, $f(x)$ must have at least r variations, that is, at least as many as $f(x) = 0$ has positive roots.

2. The negative roots of $f(x) = 0$ become the positive roots of $f(-x) = 0$, § 811. And, as just demonstrated, $f(-x) = 0$ cannot have more positive roots than variations. Hence $f(x) = 0$ cannot have more negative roots than $f(-x) = 0$ has variations.

Thus, the equation $f(x) = x^6 - x^5 - x^3 + x - 1 = 0$ cannot have more than three positive roots nor more than one negative root. For $f(x) = 0$ has three variations, and $f(-x) = 0$, that is, $x^6 + x^5 + x^3 - x - 1 = 0$, has one variation.

831 **Corollary.** *A complete equation cannot have a greater number of negative roots than it has permanences.*

For when $f(x) = 0$ is complete its permanences correspond one for one to the variations of $f(-x) = 0$, since of every two consecutive like signs in $f(x) = 0$ one is changed in $f(-x) = 0$, § 811.

Thus, if $f(x) = 0$ is $x^5 + x^4 - 6x^3 - 8x^2 - 7x + 1 = 0$, (1)

then $f(-x) = 0$ is $x^5 - x^4 - 6x^3 + 8x^2 - 7x - 1 = 0$. (2)

In (1) we have permanences at the terms x^4, $-8x^2$, $-7x$, and at the corresponding terms of (2), namely, $-x^4$, $8x^2$, $-7x$, we have variations.

Since (1) has *two* variations and *three* permanences, $f(x) = 0$ cannot have more than two positive roots nor more than three negative roots.

832 **Detection of imaginary roots.** In the case of an incomplete equation we can frequently prove the existence of imaginary roots by aid of Descartes's rule of signs.

Let $f(x) = 0$ *be an equation of the* n*th degree which has no zero roots, and let* v *and* v′ *denote the number of variations in* $f(x) = 0$ *and* $f(-x) = 0$ *respectively. The equation* $f(x) = 0$ *must have at least* $n - (v + v')$ *imaginary roots.*

For $f(x) = 0$ cannot have more than v positive roots nor more than v' negative roots, § 830, and therefore not more than $v + v'$ real roots all told. The rest of its n roots must therefore be imaginary.

This theorem gives no information as to the imaginary roots of a *complete* equation, since $v + v'$ is equal to n in such an equation.

Example. Show that $x^5 + x^2 + 1 = 0$ has four imaginary roots.

In this case $f(x) = 0$ is $x^5 + x^2 + 1 = 0$, and $f(-x) = 0$ is $x^5 - x^2 - 1 = 0$.

Hence $n - (v + v') = 5 - (0 + 1) = 4$, so that there cannot be *less* than four imaginary roots. But since there are five roots all told and one of them is real, the degree of $x^5 + x^2 + 1 = 0$ being odd, § 824, there cannot be *more* than four imaginary roots. Hence $x^5 + x^2 + 1 = 0$ has exactly four imaginary roots.

EXERCISE LXXIII

1. One root of $2x^4 - x^3 + 5x^2 + 13x + 5 = 0$ is $1 - 2i$. Solve this equation.

2. One root of $2x^4 - 11x^3 + 17x^2 - 10x + 2 = 0$ is $2 + \sqrt{2}$. Solve this equation.

3. Find the equation of lowest degree with rational coefficients two of whose roots are $-5 + 2i$ and $-1 + \sqrt{5}$.

4. Find the irreducible equation one of whose roots is $\sqrt{2} + i$.

5. What conclusions regarding the roots of the following equations can be drawn by aid of Descartes's rule and § 832?

(1) $x^4 + 1 = 0$.

(2) $x^4 - x^2 - 1 = 0$.

(3) $x^4 + 2x^3 + x^2 + x + 1 = 0$.

(4) $x^4 - 2x^3 + x^2 - x + 1 = 0$.

(5) $x^7 + x^5 + x^3 - x + 1 = 0$.

(6) $x^7 + x^4 - x^2 - 1 = 0$.

(7) $x^5 - 4x^2 + 3 = 0$.

(8) $x^{3n} - x^{2n} + x^n + x + 1 = 0$.

6. Show that a complete equation all of whose roots are real has as many positive roots as variations, and as many negative roots as permanences.

7. Given that all the roots of $x^5 + 3x^4 - 15x^3 - 35x^2 + 54x + 72 = 0$ are real, state how many are positive and how many negative.

8. Prove by Descartes's rule that $x^{2n} + 1 = 0$ has no real root. What conclusions can be drawn by aid of this rule regarding the roots of $x^{2n+1} + 1 = 0$? $x^{2n} - 1 = 0$? $x^{2n+1} - 1 = 0$?

9. Prove that an equation which involves only even powers of x with positive coefficients cannot have a positive or a negative root.

10. Prove that an equation which involves only odd powers of x with positive coefficients has no real root except 0.

11. Show that the equation $x^3 + px + q = 0$, where p and q are positive, has but one real root, that root being negative.

12. Show that an incomplete equation which has no zero roots must have two or more imaginary roots except when, as in $x^4 - 3x^2 + 1 = 0$, the missing terms occur singly and between terms which have contrary signs.

13. Show that in any equation $f(x) = 0$ with real coefficients there must be an odd number of variations between two non-consecutive contrary signs, and an even number of variations, or none, between two non-consecutive like signs.

14. Prove that in the product of the factors corresponding to the negative and imaginary roots of an equation with real coefficients the final term is always positive, and then show that if this product has any variations their number is even.

15. Prove that when the number of variations exceeds the number of positive roots, the excess is an even number.

16. Show that $x^4 + x^3 - x^2 + x - 1 = 0$ has either one or three positive roots and one negative root.

17. Show that every equation of even degree whose absolute term is negative has at least one positive and one negative root.

LOCATION OF IRRATIONAL ROOTS

833 **Theorem 1.** *If* f(a) *and* f(b) *have contrary signs, a root of* f(x) = 0 *lies between* a *and* b.

This may be proved as in the following example. A general statement of the proof will be given subsequently.

Example. Prove that $f(x) = x^3 - 3x + 1 = 0$ has a root between 1 and 2.

The sign of $f(1) = -1$ is minus and that of $f(2) = 3$ is plus.

By computing the values of $f(x)$ for $x = 1.1, 1.2, 1.3, \cdots$ successively, we find *two consecutive tenths* between 1 and 2, namely, 1.5 and 1.6, for which $f(x)$ has the same signs as for $x = 1$ and $x = 2$ respectively; for $f(1.5) = -.125$ is minus, and $f(1.6) = .296$ is plus.

By the same method we find *two consecutive hundredths* between 1.5 and 1.6, namely, 1.53 and 1.54, for which $f(x)$ has the same signs as for $x = 1$ and $x = 2$; for $f(1.53) = -.008423$ is minus and $f(1.54) = .032264$ is plus.

This process may be continued indefinitely. It determines the two never-ending sequences of numbers:

(a) $1, 1.5, 1.53, 1.532, \cdots$ (b) $2, 1.6, 1.54, 1.533, \cdots,$

the terms of which approach the same limiting value, §§ 192, 193. Call this limiting value c. It is a root of $f(x) = 0$, that is, $f(c) = 0$.

For, by § 509, if x be made to run through either of the sequences of values (a) or (b), $f(x)$ will approach $f(c)$ as limit. But since $f(x)$ is always negative as x runs through the sequence (a), its limit $f(c)$ cannot be positive; and since $f(x)$ is always positive as x runs through the sequence (b), its limit $f(c)$ cannot be negative. Hence $f(c)$ is zero.

Theorem 2. *If neither* a *nor* b *is a root of* f(x) = 0, *and an odd number of the roots of* f(x) = 0 *lie between* a *and* b, f(a) *and* f(b) *have contrary signs; but if no root or an even number of roots lie between* a *and* b, f(a) *and* f(b) *have the same sign.* **834**

Conversely, if f(a) *and* f(b) *have contrary signs, an odd number of the roots of* f(x) = 0 *lie between* a *and* b; *but if* f(a) *and* f(b) *have the same sign, either no root or an even number of roots lie between* a *and* b.

Suppose that $a < b$ and that $\beta_1, \beta_2, \cdots, \beta_r$ is a complete list of the roots of $f(x) = 0$ between a and b. Then $f(x)$ is exactly divisible by $(x - \beta_1)(x - \beta_2) \cdots (x - \beta_r)$, § 418, and if we call the quotient $\phi(x)$, we have

$$f(x) = (x - \beta_1)(x - \beta_2) \cdots (x - \beta_r)\phi(x). \qquad (1)$$

Substituting first a and then b for x in (1) and dividing the first result by the second, we obtain

$$\frac{f(a)}{f(b)} = \frac{a - \beta_1}{b - \beta_1} \cdot \frac{a - \beta_2}{b - \beta_2} \cdots \frac{a - \beta_r}{b - \beta_r} \cdot \frac{\phi(a)}{\phi(b)}. \qquad (2)$$

In the product (2) the factor $\phi(a)/\phi(b)$ is positive. For $\phi(a)$ and $\phi(b)$ have the same sign, since otherwise, by § 833, between a and b there would be a root of $\phi(x) = 0$ and therefore, by (1), a root of $f(x) = 0$ in addition to the roots $\beta_1, \beta_2, \cdots, \beta_r$.

On the other hand, each of the r factors $(a - \beta_1)/(b - \beta_1)$, and so on, is negative, since each of the r roots $\beta_1, \beta_2, \cdots, \beta_r$ is greater than a and less than b.

Therefore, when r is odd, $f(a)/f(b)$ is negative, that is, $f(a)$ and $f(b)$ have contrary signs; but when r is even or zero, $f(a)/f(b)$ is positive, that is, $f(a)$ and $f(b)$ have the same sign.

Conversely, when $f(a)$ and $f(b)$ have contrary signs, so that $f(a)/f(b)$ is negative, it follows from (2) that r is odd; and when $f(a)$ and $f(b)$ have the same sign, it follows that r is even or zero.

835 Observe that in the proofs of the preceding theorems, §§ 833, 834, no use has been made of the assumption that every equation $f(x) = 0$ has a root. Notice also that in applying these theorems a multiple root of order r is to be counted as r simple roots.

From § 834 it follows that as x varies from a to b, $f(x)$ will change its sign as x passes through each simple root or multiple root of odd order of $f(x) = 0$ which lies between a and b, and that $f(x)$ will experience no other changes of sign than these.

Thus, if $f(x) = (x-2)(x-3)^2(x-4)^3$, and x be made to vary from 1 to 5, the sign of $f(x)$ will be plus between $x = 1$ and $x = 2$, minus between $x = 2$ and $x = 4$, and plus between $x = 4$ and $x = 5$.

836 **Location of irrational roots.** By aid of the theorem of § 833 it is usually possible to determine between what pair of consecutive integers each of the fractional and irrational roots of a given numerical equation lies.

Example. Locate the roots of $f(x) = x^4 - 6x^3 + x^2 + 12x - 6 = 0$.

By Descartes's rule of signs, § 830, this equation cannot have more than three positive roots nor more than one negative root.

To locate the positive roots we compute successively $f(0), f(1), f(2), \cdots$ until three roots are accounted for by § 833 or until we reach a value of x which is a superior limit of the roots, § 803.

Thus, using the method of synthetic division, as in § 414, we find $f(0) = -6$, $f(1) = 2$, $f(2) = -10$, $f(3) = -42$, $f(4) = -70$, $f(5) = -46$, $f(6) = 102$.

Hence, § 833, one root lies between 0 and 1, another between 1 and 2, and the third between 5 and 6. There cannot be more than one root in any of these intervals, since there are only three positive roots all told.

Making a similar search for the negative root, we have $f(0) = -6$, $f(-1) = -10$, $f(-2) = 38$. Hence the negative root lies between -1 and -2.

The mere substitution of integers for x in $f(x)$ will of course not lead to the detection of all the real roots when two or more of them lie between a pair of consecutive integers. This case will be considered in § 844 and again in § 864, where a method is given for determining exactly how many roots lie between any given pair of numbers.

EXERCISE LXXIV

Locate the real roots of each of the following equations.

1. $2x^3 - 3x^2 - 9x + 8 = 0.$
2. $x^3 + x^2 - 4x - 2 = 0.$
3. $x^3 - 3x^2 - 2x + 5 = 0.$
4. $2x^3 + 3x^2 - 10x - 15 = 0.$
5. $x^3 - 4x^2 - 4x + 12 = 0.$
6. $x^3 + 13x^2 + 54x + 71 = 0.$
7. $x^3 + 5x + 19 = 0.$
8. $x^4 - 95 = 0.$
9. $x^4 - 8x^3 + 14x^2 + 4x - 8 = 0.$
10. $x^4 + 5x^3 + x^2 - 13x - 7 = 0.$
11. $x^4 - 11x^3 + 32x^2 - 4x - 46 = 0.$
12. $x^5 + 2x^4 - 16x^3 - 24x^2 + 48x + 32 = 0.$

13. Assuming that when x is very large numerically the sign of $f(x)$ is that of its term of highest degree, show that

(1) Every equation $x^n + b_1x^{n-1} + \cdots + b_n = 0$ with real coefficients, in which n is even and b_n is negative, has at least one positive and one negative root.

(2) The four roots of the equation

$$k^2(x-b)(x-c) + l^2(x-c)(x-a) + m^2(x-a)(x-b) - x(x-a)(x-b)(x-c) = 0$$

lie between $-\infty$ and a, a and b, b and c, c and ∞ respectively, it being assumed that a, b, c, k, l, m are real and that $a < b < c$.

14. Show that every equation of the form $x^3 + (x-1)(ax-1) = 0$, where $a \geqq 3$, has two roots between 0 and 1, namely, one between $1/a$ and $1 - 1/a$ and one between $1 - 1/a$ and 1.

15. Show that $x^4 + (x-1)(2x-1)(ax-1) = 0$, where $a \geqq 5$, has roots between 0 and $1/a$, $1/a$ and $1 - 2/a$, $1 - 2/a$ and 1.

COMPUTATION OF IRRATIONAL ROOTS

837 **Horner's method. Positive roots.** There are several methods by which approximate values of the irrational roots of numerical equations can be computed. The most expeditious of these methods is due, in its perfected form, to an English mathematician named Horner. It may best be explained in connection with an example.

Example. Find the positive root of $f(x) = 2x^3 + x^2 - 15x - 59 = 0$.

1. By the method of § 836, we find that the required root lies between 3 and 4. Hence if it be expressed as a decimal number, it will have the form $3.\beta\gamma\delta\cdots$, where $\beta, \gamma, \delta, \cdots$ denote its decimal figures.

$$\begin{array}{rrrrl}
2 & +1 & -15 & -59 & \underline{|3} \\
 & \underline{6} & \underline{21} & \underline{18} & \\
 & 7 & 6 & -41 & \\
 & \underline{6} & \underline{39} & & \\
 & 13 & 45 & & \\
 & \underline{6} & & & \\
 & 19 & & &
\end{array}$$

2. Diminish the roots of $f(x) = 0$ (1) by 3. We obtain the transformed equation

$$\phi(x) = 2x^3 + 19x^2 + 45x - 41 = 0, \quad (2)$$

which has the root $.\beta\gamma\delta\cdots$ lying between 0 and 1.

Testing $x = .1, .2, .3, \cdots$ in $\phi(x)$, we find that $\phi(.6)$ is $-$ and $\phi(.7)$ is $+$. Hence the root of (2) lies between .6 and .7, that is, β is 6, and the root of (1) to the first decimal figure is 3.6.

$$\begin{array}{rrrrl}
2 & +19 & +45 & -41 & \underline{|.6} \\
 & \underline{1.2} & \underline{12.12} & \underline{34.272} & \\
 & 20.2 & 57.12 & -\ 6.728 & \\
 & \underline{1.2} & \underline{12.84} & & \\
 & 21.4 & 69.96 & & \\
 & \underline{1.2} & & & \\
 & 22.6 & & &
\end{array}$$

3. Diminish the roots of (2) by .6. We obtain

$$\psi(x) = 2x^3 + 22.6x^2 + 69.96x - 6.728 = 0, \quad (3)$$

which has the root $.0\gamma\delta\cdots$ lying between 0 and .1.

Testing $x = .01, .02, \cdots$ in $\psi(x)$, we find that $\psi(.09)$ is $-$ and $\psi(.1)$ is $+$. Hence the root lies between .09 and .1, that is, γ is 9, and the root of (1) to the second decimal figure is 3.69.

4. Diminish the roots of (3) by .09, and so on.

838 The reckoning may be conducted more simply than in this example, as will be shown in the following sections. But before turning from the example, observe that the absolute terms of the first and second transformed equations, (2) and (3), namely, -41 and -6.728, have the same sign. This is as it should be, since $-41 = \phi(0)$ and $-6.728 = \phi(.6)$. For were $\phi(0)$ and $\phi(.6)$ to have contrary signs, the root of $\phi(x) = 0$ would lie between 0 and .6, § 833, and therefore .6 would *not* be its first figure. The like is true of the subsequent transformed equations, and, in general,

When the given equation has but one root with the integral part α, the absolute terms of all the transformed equations used in finding this root will have the same sign, if the reckoning is correctly performed.

839 In the example we found the first figure of the root of each transformed equation, that is, the successive decimal figures of the root of the given equation, by the method of substitution. But the first figure of the root of each transformed equation from the *second* on may ordinarily be found by merely dividing the absolute term of the equation, with its sign changed, by the coefficient of x. This is called the *method of trial divisor*.

Thus, consider the second transformed equation in the example

$$\psi(x) = 2x^3 + 22.6x^2 + 69.96x - 6.728 = 0. \quad (3)$$

This equation is known to have a root c which is less than .1. The second and higher powers of such a number c will be much smaller than c itself. Thus, even $(.09)^2$ is but .0081. Hence, were c known and substituted in (3), the first two terms of the resulting numerical identity

$$2c^3 + 22.6c^2 + 69.96c - 6.728 = 0$$

would be very small numbers in comparison with the last two.

Therefore c is not likely to differ in its first figure from the root of the equation

$$69.96x - 6.728 = 0 \quad (3')$$

obtained by discarding the x^3 and x^2 terms in (3).

But solving (3′), we have $x = 6.728/69.96 = .09+$, that is, we find, as above, that the first figure of the root of (3) is 9.

This method cannot be trusted to give the first figure of the root of the *first* transformed equation correctly. But it will usually give at least some indication as to what that figure is and so lessen the number of tests that need to be made in applying the method of substitutions. Occasionally the method fails to give correctly the first figure of the root of even the second transformed equation. But in such a case the error is readily detected in carrying out the next transformation; for if the figure is too large, a change of sign will occur in the absolute term of this next transformed equation, § 838; if it is too small, the first figure of the root of this equation will be of too high a denomination.

840 We may avoid the troublesome decimals which occur in the transformations after the first by multiplying the roots of each transformed equation by ten, § 812, before making the next transformation. This may be done by affixing one zero to the second coefficient of the equation in question, two zeros to its third coefficient, and so on. We then treat the figure of the root employed in the next transformation as if it were an integer.

Thus, the first transformed equation in the example in § 837 was

$$2x^3 + 19x^2 + 45x - 41 = 0, \tag{2}$$

and we found that it had a root of the form $.6+$.

Multiplying the roots of (2) by 10, we obtain

$$2x^3 + 190x^2 + 4500x - 41000 = 0, \tag{2'}$$

which has a root of the form $6+$.

Diminishing the roots of (2′) by 6, the reckoning differing from that above given only in the absence of decimal points, we have

$$2x^3 + 226x^2 + 6996x - 6728 = 0, \tag{3'}$$

whose roots are ten times as great as those of

$$2x^3 + 22.6x^2 + 69.96x - 6.728 = 0. \tag{3}$$

The method of trial divisor gives $.9+$ as the root of (3′) and therefore, as above, $.09+$ as the root of (3).

841 We may now arrange the reckoning involved in computing the root of $2x^3 + x^2 - 15x - 59 = 0$ to the third decimal figure as follows:

2	+ 1	− 15	− 59	3.693	
	6	21	18		
	7	6	− 41		
	6	39			
	13	45			
	6				
2	+ 190	+ 4500	− 41000	6	
	12	1212	34272		
	202	5712	− 6728		
	12	1284			$\frac{6728}{6996} = .9+$
	214	6996			
	12				
2	+ 2260	+ 699600	− 6728000	9	
	18	20502	6480918		
	2278	720102	− 247082		
	18	20664			
	2296	740766			
	18				$\frac{247082}{740766} = .3+$
	2314				

Observe that here each figure obtained by the trial divisor method is a *tenth*, thus, .9, .3. Had the last coefficient of the second transformed equation been -6728, we should have had $672/6996 = .09$ for the next *two* figures. The root as far as computed would then have been 3.609 instead of 3.69, and before performing the next transformation we should have multiplied the roots of this second transformed equation by 100, that is, we should have affixed *two* zeros to its second coefficient, *four* to the third, and *six* to the fourth.

842 This process may be continued indefinitely. But we soon encounter very large numbers, and after a few decimal figures of the root have been obtained we can find as many more as are likely to be required, with much less reckoning, by the following *contracted method*.

The last transformed equation in the reckoning above given is

$$2x^3 + 2314x^2 + 740766x - 247082 = 0. \qquad (4)$$

Instead of affixing zeros to the coefficients in order to multiply the roots of (4) by 10, we may substitute $x/10$ for x in (4), § 812, thus obtaining

$$.002x^3 + 23.14x^2 + 74076.6x - 247082 = 0. \qquad (4')$$

Ignoring the decimal parts thus cut off from the coefficients as being too small to affect the next few figures of the root, but adding 1 to the corresponding integral part when the decimal part is .5 or greater, we have the quadratic

$$23x^2 + 74077x - 247082 = 0. \qquad (4'')$$

We may then continue the reckoning as follows:

$$\begin{array}{rrrl}
23 & +74077 & -247082 & \underline{|.003332} \\
 & \underline{69} & \underline{222438} & \\
 & 74146 & -\quad 24644 & \\
 & 69 & & \\
\hline
\not{2}3 & +7422\not{5} & -\quad 24644 & \\
 & & \underline{22266} & \\
 & 742\not{2} & -\quad 2378 & \\
 & & \underline{2226} & \\
 & 74\not{2} & -\quad 152 & \\
 & & \underline{148} & \\
 & & -\quad 4 &
\end{array}$$

That is, we diminish the roots of the quadratic (4″) by 3 and thus obtain the transformed equation $23x^2 + 74215x - 24644 = 0$ (5). By the method of trial divisor, we find that the next figure of the root is also 3.

Before performing the next transformation we cut off figures as before and thus reduce (5) to the simple equation $7422x - 24644 = 0$. The next two figures of the root, namely, 3, 2, are then obtained by merely dividing 24644 by 7422 by a contracted process which consists in cutting off figures at the end of the divisor instead of affixing zeros at the end of the dividend.

843 **Negative roots.** To find a negative irrational root of $f(x) = 0$, find the corresponding positive root of $f(-x) = 0$ and then change its sign.

Example. Find the negative root of $f(x) = x^3 + x^2 - 10x + 9 = 0$.

Here $f(-x) = 0$ is $x^3 - x^2 - 10x - 9 = 0$. Its positive root, found by Horner's method, is 4.03293 approximately. Hence the negative root of $f(x) = 0$ is -4.03293 approximately.

844 **Roots nearly equal.** If the given equation has two roots lying between a pair of consecutive integers, they may be found as in the following example.

Example. Find the positive roots, if any, of $f(x) = x^3 + x^2 - 10x + 9 = 0$.

We find that $f(0) = 9$, $f(1) = 1$, $f(2) = 1$, $f(3) = 15$, and the reckoning shows that 3 is a superior limit of the roots, § 803. Hence, § 834, either there is no positive root, or there are two such roots both lying between 0 and 1, or between 1 and 2, or between 2 and 3. But $f(1)$ and $f(2)$ differ less from 0 than $f(0)$ and $f(3)$ do. Hence, if two roots exist, we may expect to find them between 1 and 2 rather than between 0 and 1 or between 2 and 3.

We therefore diminish the roots of $f(x) = 0$ by 1, obtaining

$$\phi(x) = x^3 + 4x^2 - 5x + 1 = 0,$$

which has two roots between 0 and 1 if $f(x) = 0$ has two roots between 1 and 2.

Computing the values of $\phi(x)$ for $x = .1, .2, .3, \cdots$, we find that $\phi(.2)$ is $+$ and $\phi(.3)$ is $-$, also that $\phi(.7)$ is $-$ and $\phi(.8)$ is $+$. Hence $\phi(x) = 0$ has a root between .2 and .3 and another between .7 and .8. By Horner's method we find that these roots are .25560 and .77733 approximately.

Hence $f(x) = 0$ has the two positive roots 1.2556 and 1.77733.

845 **On locating large roots.** In case the given equation $f(x) = 0$ has a root which is greater than ten, we may employ the following method for finding the figures of its integral part.

To obtain the first figure, compute the values of $f(x)$ for $x = 10, 20, \cdots$, or, if necessary, for $x = 100, 200, \cdots$, and so on, applying § 833. Thus, if we found that $f(400)$ and $f(500)$ had contrary signs, so that the root lay between 400 and 500, the first figure would be 4. To find the remaining figures, make successive transformations of the equation, as when finding the decimal figures. Thus, in the case just cited we should diminish the roots of $f(x) = 0$ by 400 and so obtain an equation $\phi(x) = 0$ having a root between 0 and 100. If we found that this root lay between 70 and 80, the second figure of the root would be 7. We should then diminish the roots of $\phi(x) = 0$ by 70 and so obtain an equation $\psi(x) = 0$ having a root between 0 and 10. If we found that this root lay between 8 and 9, we should have shown that the integral part of the root of $f(x) = 0$ was 478.

On solving numerical equations. If asked to find all the real roots of a given numerical equation $f(x) = 0$, it is best, at least when the coefficients are rational numbers, to search first for rational roots by the method of § 802. This process will yield a depressed equation $\phi(x) = 0$ whose real roots, if any, are irrational. We locate these roots by the method of §§ 833, 844, 845, and then find their approximate values by Horner's method. **846**

It may be added that a fractional root may also be found by Horner's method, exactly when the denominator involves only the factors 2 and 5, approximately in other cases.

EXERCISE LXXV

Compute the roots indicated below to the sixth decimal figure.

1. $x^3 + x - 3 = 0$; root between 1 and 2.
2. $x^3 + 2x - 20 = 0$; root between 2 and 3.
3. $x^3 + 6x^2 + 10x - 2 = 0$; root between 0 and 1.
4. $3x^3 + 5x - 40 = 0$; root between 2 and 3.
5. $x^3 + 10x^2 + 8x - 120 = 0$; root between 2 and 3.

6. $2x^3 - x^2 - 9x + 1 = 0$; root between -1 and -2.

7. $x^3 + x^2 - 5x - 1 = 0$; root between 1 and 2.

8. $x^3 - 2x^2 - 23x + 70 = 0$; root between -5 and -6.

9. $x^4 - 10x^2 - 4x + 8 = 0$; root between 3 and 4.

10. $x^4 + 6x^3 + 12x^2 - 11x - 41 = 0$; root between -2 and -3.

11. $x^3 - 3x^2 - 4x + 13 = 0$; two roots between 2 and 3.

Find to the third decimal figure all the roots of the following equations

12. $x^3 - 3x^2 - 4x + 10 = 0$.

13. $x^3 + x^2 - 2x - 1 = 0$.

14. $x^3 - 3x + 1 = 0$.

15. $x^4 + 5x^3 + x^2 - 13x - 7 = 0$.

16. By applying Horner's method to the equation $x^3 - 17 = 0$ compute $\sqrt[3]{17}$ to the fourth decimal figure.

17. By the same method compute $2\sqrt[3]{3}$ and $\sqrt[4]{87}$ each to the third decimal figure.

18. By aid of § 845 and Horner's method find to the second decimal figure the real root of $x^3 + x^2 - 2500 = 0$.

19. By aid of § 844 locate the roots of $x^3 + 5x^2 - 6x + 1 = 0$.

20. Find all the roots of $3x^5 + x^4 - 14x^3 - x^2 + 9x - 2 = 0$.

TAYLOR'S THEOREM. MULTIPLE ROOTS

847 Derivatives. Multiply any monomial of the form ax^n by n, the exponent of x, and then diminish that exponent by 1. We obtain nax^{n-1}, which is called the *derivative* of ax^n, or, more precisely, its derivative with respect to x. In particular, the derivative of a constant a, that is ax^0, is 0.

The sum of the derivatives of the terms of a polynomial $f(x)$ is called the derivative of $f(x)$, or, more precisely, its *first derivative*, and is represented by $f'(x)$.

The derivative of $f'(x)$ is called the *second derivative* of $f(x)$, and is represented by $f''(x)$, and so on.

Evidently every polynomial $f(x)$ of the nth degree has a series of n derivatives, the last of which, $f^{(n)}(x)$, is a constant.

Thus, if $$f(x) = 3x^4 - 8x^3 + 4x^2 - x + 4,$$
we have
$$f'(x) = 12x^3 - 24x^2 + 8x - 1,$$
$$f''(x) = 36x^2 - 48x + 8,$$
$$f'''(x) = 72x - 48,$$
$$f''''(x) = 72.$$

All the subsequent derivatives are 0.

Observe that the second, third, $\cdots$ derivatives of $f(x)$ are the first, second, $\cdots$ derivatives of $f'(x)$.

Taylor's theorem. If in $f(x) = a_0x^n + a_1x^{n-1} + \cdots + a_n$ we replace x by $x + h$, we obtain **848**

$$f(x + h) = a_0(x + h)^n + a_1(x + h)^{n-1} + \cdots + a_n.$$

By expanding $(x + h)^n$, $(x + h)^{n-1}$, and so on, by the binomial theorem and then collecting terms, we can reduce this expression to the form of a polynomial in h. We shall show that the result will be

$$f(x + h) = f(x) + f'(x)\frac{h}{1} + f''(x)\frac{h^2}{2!} + \cdots + f^n(x)\frac{h^n}{n!}, \quad \text{(I)}$$

where $f'(x)$, $f''(x)$, $\cdots$ are the successive derivatives of $f(x)$. This identity is called *Taylor's theorem.*

For when the result of expanding $(x + h)^m$ by the binomial theorem, § 561, is multiplied by a constant a and written in the form

$$a(x + h)^m = ax^m + max^{m-1} \cdot h + m(m-1)ax^{m-2} \cdot \frac{h^2}{2!}$$
$$+ m(m-1)(m-2)ax^{m-3} \cdot \frac{h^3}{3!} + \cdots,$$

each of the coefficients

$$max^{m-1},\ m(m-1)ax^{m-2},\ m(m-1)(m-2)ax^{m-3}, \cdots$$

is the *derivative* of the one which immediately precedes it.

Hence, if we arrange the expansion of each term of

$$f(x + h) = a_0(x + h)^n + a_1(x + h)^{n-1} + \cdots + a_n$$

in this form, the sum of the leading terms in these several expansions will be $f(x)$; the sum of the second terms will be

h times the sum of the derivatives of the leading terms, or $f'(x)h$; the sum of the third terms will be $h^2/2!$ times the sum of the derivatives of the second terms, or $f''(x)h^2/2!$; and so on. In other words, we shall have

$$f(x+h) = f(x) + f'(x)h + f''(x)\frac{h^2}{2!} + \cdots + f^{(n)}(x)\frac{h^n}{n!}.$$

Thus, if $\quad f(x) = a_0x^3 + a_1x^2 + a_2x + a^3,$

we have $\quad f(x+h) = a_0(x+h)^3 + a_1(x+h)^2 + a_2(x+h) + a_3$

$$\begin{array}{rl|l|l|l} = & a_0x^3 & + 3a_0x^2 & h + 6a_0x & \dfrac{h^2}{2!} + 6a_0\dfrac{h^3}{3!} \\ & + a_1x^2 & + 2a_1x & + 2a_1 & \\ & + a_2x & + \quad a_2 & & \\ & + \quad a_3 & & & \end{array}$$

$$= f(x) + f'(x)h + f''(x)h^2/2! + f'''(x)h^3/3!.$$

849 Since $x = a + (x - a)$, we have $f(x) = f[a + (x - a)]$. Hence we may obtain the expression for $f(x)$ in powers of $x - a$, § 423, by merely replacing x by a, and h by $x - a$, in the identity (I). The result is

$$f(x) = f(a) + f'(a)(x-a)$$
$$+ f''(a)\frac{(x-a)^2}{2!} + \cdots + f^{(n)}(a)\frac{(x-a)^n}{n!}. \quad \text{(II)}$$

Example. Express $x^3 - 1$ in powers of $x - 1$.

We have $\quad f(x) = x^3 - 1,\ f'(x) = 3x^2,\ f''(x) = 6x,\ f'''(x) = 6.$

Hence $\quad f(1) = 0,\ f'(1) = 3,\ f''(1)/2 = 3,\ f'''(1)/3! = 1.$

Therefore $x^3 - 1 = 3(x-1) + 3(x-1)^2 + (x-1)^3.$

850 **Multiple roots.** The first, second, $\cdots$ derivatives of $f'(x)$ are the second, third, $\cdots$ derivatives of $f(x)$. Hence, § 849, the expressions for a polynomial $f(x)$ and its first derivative $f'(x)$ in terms of $x - a$ are

$$f(x) = f(a) + f'(a)(x-a) + f''(a)(x-a)^2/2! + \cdots, \quad (1)$$
$$f'(x) = f'(a) + f''(a)(x-a) + f'''(a)(x-a)^2/2! + \cdots. \quad (2)$$

If $f(x)$ is divisible by $x - a$ but not by $(x-a)^2$, it follows from (1) that $f(a) = 0$ but $f'(a) \neq 0$, and therefore from (2) that $f'(x)$ is not divisible by $x - a$. Again, if $f(x)$ is divisible by $(x-a)^2$ but not by $(x-a)^3$, it follows from (1) that $f(a) = f'(a) = 0$ but $f''(a) \neq 0$, and therefore from (2) that $f'(x)$ is divisible by $x - a$ but not by $(x-a)^2$. And in general, if $f(x)$ is divisible by $(x-a)^r$ but not by $(x-a)^{r+1}$, it follows from (1) that $f(a) = f'(a) = \cdots f^{(r-1)}(a) = 0$ but $f^{(r)}(a) \neq 0$, and therefore from (2) that $f'(x)$ is divisible by $(x-a)^{r-1}$ but not by $(x-a)^r$.

Therefore, by § 800, we have the following theorem.

Theorem. *A simple root of* $\mathrm{f}(\mathrm{x}) = 0$ *is not a root of* $\mathrm{f}'(\mathrm{x}) = 0$; *but a double root of* $\mathrm{f}(\mathrm{x}) = 0$ *is a simple root of* $\mathrm{f}'(\mathrm{x}) = 0$, *and, in general, a multiple root of order* r *of* $\mathrm{f}(\mathrm{x}) = 0$ *is a root of order* r − 1 *of* $\mathrm{f}'(\mathrm{x}) = 0$. **851**

Thus, the roots of $f(x) = x^3 - x^2 - 8x + 12 = 0$ are 2, 2, − 3, and the roots of $f'(x) = 3x^2 - 2x - 8 = 0$ are 2, − 4/3.

852 We therefore have the following method for discovering the multiple roots of $f(x) = 0$, if there be any. Seek the highest common factor of $f(x)$ and $f'(x)$ by the method of § 465. If we thus find that $f(x)$ and $f'(x)$ are prime to one another, $f(x) = 0$ has simple roots only. But if we find that $f(x)$ and $f'(x)$ have the highest common factor $\phi(x)$, then every simple root of $\phi(x) = 0$ is a double root of $f(x) = 0$, every double root of $\phi(x) = 0$ is a triple root of $f(x) = 0$, and so on. For, § 850, if $\phi(x)$ is divisible by $(x-a)^r$, then $f'(x)$ is divisible by $(x-a)^r$ and $f(x)$ by $(x-a)^{r+1}$.

Observe that if the quotient $f(x)/\phi(x)$ be $F(x)$, the roots of $F(x) = 0$ are those of $f(x) = 0$, each counted once.

Example. Find the multiple roots, if any, of the equation

$$f(x) = x^5 - x^4 - 5x^3 + x^2 + 8x + 4 = 0.$$

Here $f'(x) = 5x^4 - 4x^3 - 15x^2 + 2x + 8$, and by § 465 we find the highest common factor of $f(x)$ and $f'(x)$ to be $\phi(x) = x^3 - 3x - 2$.

The roots of $\phi(x) = 0$ may be found by § 802 and are − 1, − 1, 2. Hence $f(x) = 0$ has the triple root − 1 and the double root 2, that is, its roots are − 1, − 1, − 1, 2, 2.

Observe that $f(x) = (x+1)^3(x-2)^2$,

that $f'(x) = (x+1)^2(x-2)(5x-4)$,

and that $F(x) = f(x)/\phi(x) = (x+1)(x-2)$.

853 We may add that if any two equations $f(x) = 0$ and $\psi(x) = 0$ have a root in common, it may be discovered by finding the highest common factor of $f(x)$ and $\psi(x)$.

Example. Solve $f(x) = x^4 - x^3 - 3x^2 + 4x - 4 = 0$, having given that one of its roots is the negative of another of its roots.

Evidently the two roots mentioned are common to $f(x) = 0$ and the equation $f(-x) = x^4 + x^3 - 3x^2 - 4x - 4 = 0$, and may therefore be obtained by finding the highest common factor of $f(x)$ and $f(-x)$.

By § 465 we find this highest common factor to be $x^2 - 4$. Hence the roots mentioned are $2, -2$. Dividing $f(x)$ by $x^2 - 4$ and solving the resulting depressed equation $x^2 - x + 1 = 0$, we find that the other two roots of $f(x) = 0$ are $(1 \pm i\sqrt{3})/2$.

EXERCISE LXXVI

1. Find the first, second, $\cdots$ derivatives of $2x^5 - 4x^4 + x^2 - 20x$.

2. Given $f(x) = x^4 - 2x^3 + 1$, find $f(x+h)$ by Taylor's theorem.

3. Using the formula § 849, (II), express (1) $x^4 + x^2 + 1$ in powers of $x+1$; (2) $x^5 - 32$ in powers of $x - 2$; (3) $(x^3+1)/(x^2+1)$ in terms of $x - 1$.

4. The following equations have multiple roots. Solve them.

(1) $x^3 - 3x - 2 = 0$. (2) $9x^3 + 12x^2 - 11x + 2 = 0$.

(3) $4x^4 + 12x^2 + 9 = 0$. (4) $x^4 - 4x^3 + 8x + 4 = 0$.

(5) $2x^4 - 12x^3 + 19x^2 - 6x + 9 = 0$.

(6) $x^5 - x^3 - 4x^2 - 3x - 2 = 0$.

(7) $x^4 - 2x^3 - x^2 - 4x + 12 = 0$.

(8) $x^5 - x^4 - 2x^3 + 2x^2 + x - 1 = 0$.

(9) $3x^5 - 2x^4 + 6x^3 - 4x^2 + 3x - 2 = 0$.

5. Show that $x^n - a^n = 0$ cannot have a multiple root.

6. If the equation $x^3 - 12x + a = 0$ has a double root, find a.

7. Determine a and b so that $3x^3 + ax^2 + x + b = 0$ may have a triple root, and find this root.

8. Show that $x^4 + qx^2 + s = 0$ cannot have a triple root.

9. Find the condition that $x^5 - px^2 + r = 0$ may have a double root.

10. What is the form of $f(x)$ if it is exactly divisible by $f'(x)$?

11. The equations $x^4 + x^3 + 2x^2 + x + 1 = 0$ and $x^4 + x^3 - x - 1 = 0$ have roots in common. Solve both equations.

12. The equation $x^3 - 20x - 16 = 0$ has a root which is twice one of the roots of $x^3 - x^2 - 3x - 1 = 0$. Solve both of these equations.

13. Show that if a cubic equation with rational coefficients has a multiple root, this root must be rational.

14. Show that if an equation of the fourth degree $f(x) = 0$ with rational coefficients has a multiple root, this root must be rational unless $f(x)$ is a perfect square.

15. Prove that if a is a root of $f(x) = 0$, of order r, it is a root of all the equations $f'(x) = 0$, $f''(x) = 0$, $\cdots$, $f^{(r-1)}(x) = 0$.

VARIATION OF A RATIONAL INTEGRAL FUNCTION

854 **Theorem.** *Let* f(x) *denote a polynomial arranged in ascending powers of* x, *and let* b *denote the numerical value of its leading coefficient and* g *that of its numerically greatest coefficient. The leading term of* f(x) *will be numerically greater than the sum of the remaining terms for all values of* x *which are numerically less than* b/(b + g).

First, let $f(x) = b_0 + b_1x + b_2x^2 + \cdots$, so that $b = |b_0|$, and let x' denote the numerical value of x.

Then $b_1x + b_2x^2 + \cdots$ is numerically less than (or equal to) $gx' + gx'^2 + \cdots$ or $g(x' + x'^2 + \cdots)$, § 235, and therefore when $x' < 1$, it is less than $gx'/(1 - x')$, § 704.

But $gx'/(1 - x')$ is less than b when $x' < b/(b + g)$.

Second, let $f(x) = b_1x + b_2x^2 + b_3x^3 + \cdots$, so that $b = |b_1|$. We then have $|b_2x^2 + b_3x^3 + \cdots| < |b_1x|$ when $|b_2x + b_3x^2 + \cdots| < |b_1|$, that is, when $x' < b/(b + g)$, and so on.

Thus, if $f(x) = 5x + 3x^2 - 9x^4$, we have $|3x^2 - 9x^4| < |5x|$ when $x' < 5/(5 + 9)$, that is, when $x' < 5/14$.

855 **Theorem.** *Let* f(x) *denote a polynomial arranged in descending powers of* x, *and let* a *denote the numerical value of its leading coefficient and* g *that of its numerically greatest coefficient. The leading term of* f(x) *will be numerically greater than the sum of the remaining terms for all values of* x *which are numerically greater than* (a + g)/a.

For let $f(x) = a_0x^n + a_1x^{n-1} + \cdots + a_n$, so that $a = |a_0|$, and let x' denote the numerical value of x.

We have $a_0x^n + a_1x^{n-1} + \cdots + a_n = x^n(a_0 + a_1/x + \cdots + a^n/x^n)$. Hence $|a_0x^n| > |a_1x^{n-1} + \cdots + a_n|$ when $|a_0| > |a_1/x + \cdots + a_n/x^n|$. But, § 854, $|a_0| > |a_1/x + \cdots + a_n/x^n|$ when $1/x' < a/(a+g)$, that is, when $x' > (a+g)/a$.

Thus, if $f(x) = 3x^3 + x^2 - 7x + 2$, we have $|3x^3| > |x^2 - 7x + 2|$ when $x' > (3+7)/3$, that is, when $x' > 10/3$.

From this theorem it evidently follows that the number $(a+g)/a$ is greater than the absolute or numerical value of any root of the equation $f(x) = 0$, whether the root be real or imaginary.

856 **Theorem.** *If* a *is a root of* f(x) = 0, *the values of* f(x) *and* f'(x) *have contrary signs when* x *is slightly less than* a, *and the same sign when* x *is slightly greater than* a.

For express $f(x)$ and $f'(x)$ in powers of $x - a$, § 849, and then divide the first expression by the second. When a is a simple root, so that $f(a) = 0$ but $f'(a) \neq 0$, the result may be reduced to the form

$$\frac{f(x)}{f'(x)} = (x-a)\frac{f'(a) + f''(a)(x-a)/2! + \cdots}{f'(a) + f''(a)(x-a) + \cdots}.$$

The numerator and denominator of the fraction on the right are polynomials in $x - a$. Hence for all values of $x - a$ which are small enough to meet the requirements of § 854 their signs will be those of their common leading term $f'(a)$, and the fraction itself will be positive. The sign of $f(x)/f'(x)$ will then be the same as that of $x - a$ and therefore minus or plus according as $x < a$ or $x > a$. But when the sign of $f(x)/f'(x)$

is minus, $f(x)$ and $f'(x)$ have contrary signs, and when the sign of $f(x)/f'(x)$ is plus, $f(x)$ and $f'(x)$ have the same sign.

When a is a multiple root of order r we have, § 850,

$$\frac{f(x)}{f'(x)} = (x-a)\frac{f^{(r)}(a)/r! + \text{terms involving } (x-a)}{f^{(r)}(a)/(r-1)! + \text{terms involving } (x-a)},$$

from which the theorem follows by the same reasoning as when a is a simple root.

Rolle's theorem. *Between two consecutive roots of* $f(x) = 0$ *there is always a root of* $f'(x) = 0$. **857**

For let β_1 and β_2 be the roots in question, and let c denote a number slightly greater than β_1 and d a number slightly less than β_2, so that $\beta_1 < c < d < \beta_2$.

Then $f'(c)$ has the same sign as $f(c)$, § 856, and $f(c)$ has the same sign as $f(d)$, § 834; but $f(d)$ has a different sign from that of $f'(d)$, § 856. Hence $f'(c)$ and $f'(d)$ have contrary signs. Therefore a root of $f'(x) = 0$ lies between c and d, that is, between β_1 and β_2, § 833.

Thus, if $f(x) = x^2 - 3x + 2 = 0$, then $f'(x) = 0$ is $2x - 3 = 0$. The roots of $f(x) = 0$ are 1 and 2, the root of $f'(x) = 0$ is $3/2$, and $3/2$ lies between 1 and 2.

Example. Prove that $f(x) = x^3 + x^2 - 10x + 9 = 0$ has two roots between 1 and 2. (Compare § 844, Ex.)

Since $f(1) = 1$ and $f(2) = 1$, there are two roots or none between 1 and 2. If there are two roots, $f'(x) = 0$ must also have a root between 1 and 2, and this root must lie between the two roots of $f(x) = 0$.

But $f'(x) = 3x^2 + 2x - 10 = 0$ has a root between 1 and 2, for $f'(1) = -5$ and $f'(2) = 6$. Solving, we find that this root is 1.5 approximately. Moreover $f(1.5) = -.375$ is minus. Therefore, since both $f(1)$ and $f(2)$ are plus, $f(x) = 0$ has two roots between 1 and 2, namely, one between 1 and 1.5, and another between 1.5 and 2.

Theorem. *If the variable* x *is increasing, then, as it passes through the value* a, *the value of* $f(x)$ *is increasing if* $f'(a) > 0$, *but decreasing if* $f'(a) < 0$. **858**

If $f'(a) = 0$ *but* $f''(a) \neq 0$, $f(a)$ *is a maximum value of* $f(x)$ *when* $f''(a) < 0$, *a minimum value when* $f''(a) > 0$.

For by § 849 we have

$$f(x)-f(a)=f'(a)(x-a)+f''(a)(x-a)^2/2\,!+\cdots.$$

The second member is a polynomial in $x-a$, and for all values of x which make $x-a$ small enough numerically to meet the requirement of § 854, the leading term will control the sign of the entire expression and therefore that of $f(x)-f(a)$. We shall suppose x restricted to such values. Then

1. If $f'(a)>0$, $f'(a)(x-a)$, and therefore $f(x)-f(a)$, has the same sign as $(x-a)$. Therefore, since $x-a$ changes from minus to plus as x passes through a, the same is true of $f(x)-f(a)$, that is, $f(x)$ is then *increasing* from a value less than $f(a)$ to a value greater than $f(a)$.

2. If $f'(a)<0$, $f'(a)(x-a)$ and $(x-a)$ have contrary signs. Hence, reasoning as in 1, we conclude that $f(x)$ is *decreasing* as x passes through a.

3. If $f'(a)=0$ but $f''(a)\neq 0$, the sign of $f(x)-f(a)$ is that of $f''(a)(x-a)^2/2$ and therefore that of $f''(a)$; for $(x-a)^2$ is positive whether $x<a$ or $x>a$. Hence when $f''(a)<0$, we have $f(x)<f(a)$ just before x reaches a and also just after x passes a, which proves that $f(a)$ is a *maximum* value of $f(x)$, § 639.

And in the same manner when $f''(a)>0$, we may show that $f(a)$ is a *minimum* value of $f(x)$.

It may be added that if $f''(a)=0$ but $f'''(a)\neq 0$, $f(a)$ is not a maximum or minimum value of $f(x)$ (see § 859, Ex. 2). And, in general, if all the derivatives from the first to the rth, but not the $(r+1)$th, vanish when $x=a$, $f(a)$ is a maximum or minimum when r is odd, but not when r is even.

Example. Is $f(x)=x^3-6x^2+9x-1$ increasing or decreasing as x, increasing, passes through the value 2? Find the maximum and minimum values of $f(x)$.

We find $f'(x)=3x^2-12x+9=3(x-1)(x-3)$. Hence $f'(2)=-3$ is negative. Therefore $f(x)$ is decreasing as x passes through 2.

We have $f'(x)=0$ when $x=1$ and when $x=3$. Moreover $f''(x)=6x-12$, and therefore $f''(1)=-6$ is negative and $f''(3)=6$ is positive. Hence $f(1)=3$ is a maximum value of $f(x)$, and $f(3)=-1$ is a minimum value.

Variation of f (x). Let us now consider how the value of a polynomial $f(x)$ with real coefficients varies when x varies continuously, § 214, from $-\infty$ to $+\infty$. **859**

Example 1. Discuss the variation of $f(x)=x^3-2x^2-x+2$.

The roots of $f(x)=0$ are $-1, 1, 2$, and $f(x)=(x+1)(x-1)(x-2)$.

Hence, when $x=-\infty$, $f(x)=-\infty$; when x is between $-\infty$ and -1, $f(x)$ is negative; when $x=-1$, $f(x)=0$; when x is between -1 and 1, $f(x)$ is positive; when $x=1$, $f(x)=0$; when x is between 1 and 2, $f(x)$ is negative; when $x=2$, $f(x)=0$; when x is between 2 and ∞, $f(x)$ is positive; when $x=\infty$, $f(x)=\infty$.

The roots of $f'(x)=3x^2-4x-1=0$ are $(2\pm\sqrt{7})/3$, or $-.2$ and 1.5 approximately. When $x<(2-\sqrt{7})/3$ and when $x>(2+\sqrt{7})/3$, $f'(x)$ is positive, but when x is between $(2-\sqrt{7})/3$ and $(2+\sqrt{7})/3$, $f'(x)$ is negative.

Therefore, § 858, $f(x)$ is continually increasing as x varies from $-\infty$ to $(2-\sqrt{7})/3$, is continually decreasing as x varies from $(2-\sqrt{7})/3$ to $(2+\sqrt{7})/3$, and is again continually increasing as x varies from $(2+\sqrt{7})/3$ to ∞.

It follows from this, § 639, that $f(x)$ has a maximum value when $x=(2-\sqrt{7})/3$, and a minimum value when $x=(2+\sqrt{7})/3$. This is in agreement with § 858, for $f''(x)=6x-4$ is negative when $x=(2-\sqrt{7})/3$, and positive when $x=(2+\sqrt{7})/3$.

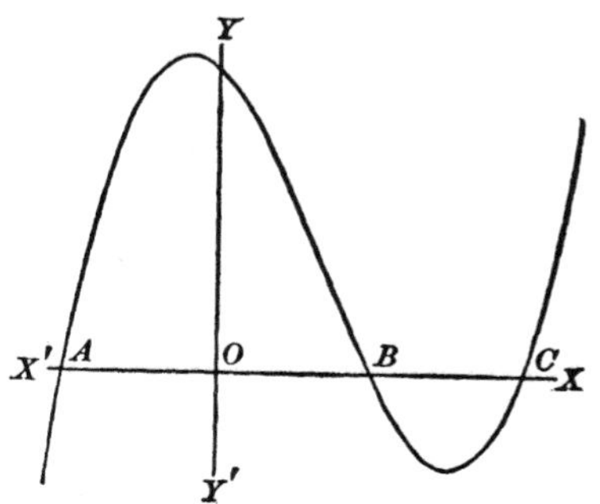

The variation of $f(x)$ will be exhibited to the eye if we put $y=f(x)$ and then construct the graph of this equation by the method of § 389. We thus obtain the curve indicated in the accompanying figure. The points A, B, C at which the curve cuts the x-axis are the graphs of the roots $-1, 1, 2$ of $f(x)=0$. The portions of the curve above the x-axis correspond to positive values of $f(x)$, those below to negative values. The uppermost point on the curve between A and B corresponds to the maximum value of $f(x)$, the lowermost point between B and C to the minimum value.

As x varies from $-\infty$ to ∞ the corresponding point on the curve moves from an infinite distance below the x-axis upward and to the right through A to the maximum point, then downward through B to the minimum point, then upward again through C to an infinite distance above the x-axis.

If we were gradually to increase the absolute term of $f(x)$, the graph of $y = f(x)$ would be shifted vertically upward and the points B and C would at first approach coincidence in a point of tangency and then disappear. The corresponding roots of $f(x) = 0$ would at first become equal and then imaginary.

Example 2. Discuss the variation of $f(x) = x^4 - 2x^3 + 2x - 1$.

The roots of $f(x) = 0$ are $-1, 1, 1, 1$, and $f(x) = (x+1)(x-1)^3$.

Hence when $x = \pm\infty, f(x) = \infty$; when $x = \pm 1, f(x) = 0$; when $x < -1$ and when $x > 1$, $f(x)$ is positive; when x is between -1 and 1, $f(x)$ is negative.

Here $f'(x) = 4x^3 - 6x^2 + 2 = 2(2x+1)(x-1)^2$, and the roots of $f'(x) = 0$ are $-1/2, 1, 1$. When $x < -1/2, f'(x)$ is negative; when x is between $-1/2$ and 1 and also when $x > 1$, $f'(x)$ is positive. Therefore, § 858, $f(x)$ is continually decreasing as x varies from $-\infty$ to $-1/2$, and continually increasing as x varies from $-1/2$ to 1 and from 1 to ∞.

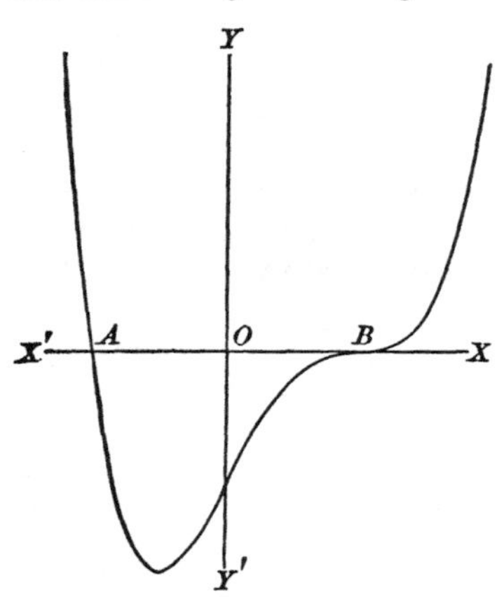

Hence $f(x)$ has a minimum value when $x = -1/2$, but it has neither a maximum nor a minimum value when $x = 1$.

This is in agreement with § 858. For $f''(x) = 12x^2 - 12x = 12x(x-1)$ and $f'''(x) = 24x - 12$. Hence when $x = -1/2, f''(x)$ is positive; but when $x = 1, f''(x) = 0$ and $f'''(x) \neq 0$.

The graph of $y = f(x)$ has the form indicated in the accompanying figure. The point A where the curve merely cuts the x-axis corresponds to the root -1 of $f(x) = 0$, and the point B where the curve both touches and crosses the x-axis corresponds to the triple root 1. The lowermost point of the curve corresponds to the minimum value of $f(x)$. Its coördinates are $-1/2, -27/16$.

860 As in these examples, so in general, if $f(x)$ is of odd degree, its leading coefficient being positive, when x varies from $-\infty$ to $+\infty$, $f(x)$ increases from $-\infty$ to the first maximum value, then decreases to the first minimum value, and so on, and finally increases from the last minimum value to $+\infty$. It is possible, however, that there are no maximum or minimum values, for the equation $f'(x)=0$, being of even degree, may have no real root. The graph of $y=f(x)$ extends from an infinite distance below the x-axis to an infinite distance above the x-axis. It crosses the x-axis an odd number of times, — once at least.

On the other hand, if $f(x)$ is of even degree, $f(x)$ begins by decreasing from $+\infty$ to the first minimum value and ends by increasing from the last minimum value to $+\infty$. In this case the graph of $y=f(x)$ need not cross the x-axis at all. If it does cross the axis, it crosses an even number of times.

In most cases we can obtain a sufficiently accurate representation of the graph of $y=f(x)$ by the method of § 389, which consists in assigning a series of values to x, computing the corresponding values of y, plotting the pairs of values of x, y thus found, and passing a "smooth" curve through all these points. Such a curve will indicate roughly where the true graph crosses the x-axis and where its maximum and minimum points lie. But to obtain the points of crossing with exactness, we must solve the equation $f(x)=0$; and to obtain the actual positions of the maximum and minimum points, we must solve the equation $f'(x)=0$. To every multiple root of $f(x)=0$ there corresponds a point of tangency of the graph with the x-axis. If the order of the multiple root is odd, the graph also crosses the x-axis at this point.

EXERCISE LXXVII

1. Discuss the variation of $f(x)=(x+1)(x-2)^2=x^3-3x^2+4$, finding its maximum and minimum values if any, and draw the graph of $y=f(x)$.

2. Treat in a similar manner each of the following functions.

(1) $2x^2-x+1$.	(2) $(x+1)(x-2)(2x-1)$.
(3) $x^3-12x+14$.	(4) x^3-5x^2+3x+9.
(5) x^3-3x^2+5.	(6) $(x+1)^2(x-2)^2$.
(7) $(x^2+x+1)(x+2)$.	(8) $x(x-1)(x+2)(x+3)$.

3. Find the graphs of each of the following fractional equations by plotting the points corresponding to $x = -1, -1/2, 0, 1/2, 1, \cdots, 4$.

(1) $y = \dfrac{x(x-1)}{(x-2)(x-3)}$. (2) $y = \dfrac{x(x-2)}{(x-1)(x-3)}$.

STURM'S THEOREM

861 **Sturm functions.** Let $f(x) = 0$ be any equation which has no multiple roots, and let $f_1(x)$ be the first derivative of $f(x)$.

Divide $f(x)$ by $f_1(x)$ and call the quotient q_1, and the remainder, with its sign changed, $f_2(x)$.

Again, divide $f_1(x)$ by $f_2(x)$, and call the quotient q_2, and the remainder, with its sign changed, $f_3(x)$.

And so on, modifying the ordinary process of finding the highest common factor of $f(x)$ and $f_1(x)$ in this respect only: *the sign of each remainder is changed, and care is taken to make no other changes of sign than these.*

Since $f(x) = 0$ has no multiple roots and therefore $f(x)$ and $f_1(x)$ have no common factor, § 851, we shall finally obtain a remainder which is a constant different from 0, § 465. Call this remainder, with its sign changed, f_m.

The sequence of functions

$$f(x),\ f_1(x),\ f_2(x),\ f_3(x),\ \cdots,\ f_m,$$

consisting of the given polynomial, its first derivative, and the several remainders in order, each with its sign changed, is called a *sequence of Sturm* or a sequence of *Sturm functions.*

862 **Relations among the Sturm functions.** These functions are, by definition, connected by the series of identical equations

$$f(x) \equiv q_1 f_1(x) - f_2(x), \qquad (1)$$

$$f_1(x) \equiv q_2 f_2(x) - f_3(x), \qquad (2)$$

$$f_2(x) \equiv q_3 f_3(x) - f_4(x), \qquad (3)$$

$$\cdot \quad \cdot \quad \cdot \quad \cdot \quad \cdot \quad \cdot$$

$$f_{m-2}(x) \equiv q_{m-1} f_{m-1}(x) - f_m. \qquad (m-1)$$

From these equations we conclude that

1. *Two consecutive functions cannot vanish for the same value of* x.

Thus, if both $f_1(x)$ and $f_2(x)$ vanish when $x = c$, it follows from (2) that $f_3(x)$ also vanishes; therefore, from (3), that $f_4(x)$ vanishes; and therefore, finally, that f_m is 0. But this is contrary to hypothesis.

2. *When for a certain value of* x *one of the intermediate functions* $f_1(x), f_2(x), \cdots, f_{m-1}(x)$ *vanishes, the functions which immediately precede and follow it have opposite signs.*

Thus, if $$f_2(c) = 0,$$
it follows from (2) that $$f_1(c) = -f_3(c).$$

Sturm's theorem. *Let* a *and* b *be any two real numbers neither of which is a root of* $f(x) = 0$. **863**

The difference between the number of variations of sign in the sequence

$$f(a),\ f_1(a),\ f_2(a),\ \cdots,\ f_m$$

and that in the sequence

$$f(b),\ f_1(b),\ f_2(b),\ \cdots,\ f_m$$

is the number of roots of $f(x) = 0$ *which lie between* a *and* b.

To fix the ideas, suppose $a < b$, and suppose x to vary continuously from a to b, § 214.

As x varies from a to b, the sign of the constant f_m remains unchanged, and the only changes which are possible in the signs of the remaining functions, and therefore in the number of variations of sign in the sequence

$$f(x),\ f_1(x),\ f_2(x),\ \cdots,\ f_m,$$

are such as may occur when x passes through roots of the equations $f(x) = 0$, $f_1(x) = 0$, and so on, § 835. But

1. *The number of variations in the sequence is neither increased nor diminished when any function except the first,* $f(x)$, *changes its sign.*

Suppose, for instance, that c is a root of $f_2(x) = 0$ and that $f_2(x)$ changes from plus to minus as x passes through c.

Since c is a root of $f_2(x) = 0$, it cannot be a root of either $f_1(x) = 0$ or $f_3(x) = 0$, § 862. And if we take a positive number h so small that no root of $f_1(x) = 0$ or of $f_3(x) = 0$ lies between $c - h$ and c, or between c and $c + h$, neither of the functions $f_1(x)$ or $f_3(x)$ will change its sign as x varies from $c - h$ to $c + h$, § 835.

But when $x = c$, $f_1(x)$ and $f_3(x)$ have opposite signs, § 862. Suppose that $f_1(c)$ is plus; then $f_3(c)$ is minus, and we have the following scheme of the signs of $f_1(x)$, $f_2(x)$, $f_3(x)$ for values of x between $c - h$ and $c + h$:

	$f_1(x)$	$f_2(x)$	$f_3(x)$
$x = c - h$	$+$	$+$	$-$
$x = c$	$+$	0	$-$
$x = c + h$	$+$	$-$	$-$

For all these values of x, therefore, the signs of the three functions

$$f_1(x),\ f_2(x),\ f_3(x)$$

present *one* variation. The only effect of the change of the sign of $f_2(x)$, at least in this part of the sequence, is a change in the *position* of a variation.

And there can be no change in the number of variations in the part of the sequence which follows $f_3(x)$ as x passes through c. For either no function after $f_3(x)$ will change its sign as x passes through c, or, if it does, we shall have the case of the function $f_2(x)$ over again, and the only effect will be a change in the position of another variation.

2. *The sequence loses one variation for each root of* $f(x) = 0$ *through which* x *passes in varying from* a *to* b.

For let c be a root of $f(x) = 0$.

The functions $f(x)$ and $f_1(x)$ have opposite signs for values of x which are slightly less than c, and the same sign for values of x which are slightly greater than c, § 856.

In other words, the sequence has a variation between $f(x)$ and $f_1(x)$ just before x reaches c, and this variation is lost as x passes through c.

Hence the sequence of Sturm functions never *gains* a variation as x varies from a to b. But, on the other hand, it *loses* a variation each time that x passes through a root of $f(x) = 0$, and then only.

Therefore the difference between the number of variations

in $\qquad f(a),\ f_1(a),\ f_2(a),\ \cdots,\ f_m$

and in $\qquad f(b),\ f_1(b),\ f_2(b),\ \cdots,\ f_m$

is the number of single variations that are lost as x passes through the roots of $f(x) = 0$ which lie between a and b. In other words, it is the number of these roots, as was to be demonstrated.

If we apply the method of § 861 to an equation $f(x) = 0$ which has *multiple* roots, we obtain a sequence $f(x), f_1(x), \cdots, f_m(x)$, (1) the last term of which is the highest common factor of all the terms, § 465. Divide all the terms of (1) by $f_m(x)$. We thus obtain a sequence of the form $\phi(x)$, $\phi_1(x), \cdots, 1$, (2) which, as is easily shown, possesses all the properties on which Sturm's theorem depends. Hence the number of roots of $\phi(x) = 0$, that is, § 852, the number of *different* roots of $f(x) = 0$, between a and b, is the difference between the number of variations in $\phi(a), \phi_1(a), \cdots, 1$ and in $\phi(b), \phi_1(b), \cdots, 1$. And this difference is the same as that between the number of variations in $f(a), f_1(a), \cdots, f_m(a)$ and in $f(b)$, $f_1(b), \cdots, f_m(b)$; for multiplying the sequences $\phi(a), \cdots, 1$ and $\phi(b), \cdots, 1$ by $f_m(a)$ and $f_m(b)$ respectively will not affect their variations.

Applications of Sturm's theorem. Sturm's theorem enables one to find exactly how many different real roots a given numerical equation has. It also enables one to find how many of these roots lie between any pair of consecutive integers and therefore in every case to solve the problem of locating the roots. But this method of locating roots is very laborious and is used only when the simpler method of § 836 fails. **864**

Example 1. Locate the roots of $x^3 + 3x^2 - 4x + 1 = 0$.

Here $f(x) = x^3 + 3x^2 - 4x + 1$ and $f_1(x) = 3x^2 + 6x - 4$.

Arranging the computation of the remaining functions as in § 468, 3, we have

$3 + 6 - 4$	$1 + 3 - 4 + 1$	$1 + 1$
$6 + 12 - 8$	$3 + 9 - 12 + 3$	
$\underline{6 - 3}$	$\underline{3 + 6 - 4}$	
$15 - 8$	$3 - 8 + 3$	
$30 - 16$	$\underline{3 + 6 - 4}$	
$\underline{30 - 15}$	$-14 + 7$	
-1	$\therefore f_2 = 2 - 1$	$3 + 15$
$\therefore f_3 = 1$		

Hence

$$f(x) = x^3 + 3x^2 - 4x + 1,$$
$$f_1(x) = 3x^2 + 6x - 4,$$
$$f_2(x) = 2x - 1,$$
$$f_3 = 1.$$

Observe that the $f_2(x)$ and f_3 here obtained are not the $f_2(x)$ and f_3 defined in § 861, but these functions multiplied by positive constants.

We should have lessened the reckoning had we divided $f_1(x)$ by $f_2(x) = 2(x - .5)$ synthetically. It is often best to use the synthetic method in the final division.

1. When x is very great numerically, the sign of a polynomial is that of its term of highest degree, § 855. Hence the following table.

	$f(x)$	$f_1(x)$	$f_2(x)$	f_3	
$x = -\infty$	$-$	$+$	$-$	$+$,	three variations.
$x = 0$	$+$	$-$	$-$	$+$,	two variations.
$x = \infty$	$+$	$+$	$+$	$+$,	no variation.

Therefore $f(x) = 0$ has one negative and two positive roots.

2. To locate the positive roots, we substitute $x = 0, 1, \cdots$ and obtain

	$f(x)$	$f_1(x)$	$f_2(x)$	f_3	
$x = 0$	$+$	$-$	$-$	$+$,	two variations.
$x = 1$	$+$	$+$	$+$	$+$,	no variation.

Hence the two positive roots lie between 0 and 1.

Since there is but one negative root, we know that it can be located by the method of § 836. We thus find that it lies between -4 and -5.

Example 2. How many real roots has $f(x) = 2x^3 - x^2 - 2x + 2 = 0$?

Proceeding as in Ex. 1, we find $f_1(x) = 3x^2 - x - 1$ and $f_2(x) = 13x - 17$.

We can find the sign of f_3, which alone concerns us, without dividing $f_1(x)$ by $f_2(x)$.

For since $f_2(x) = 13(x - 17/13)$, the remainder in the division of $f_1(x)$ by $f_2(x)$ is $f_1(17/13)$. But $17/13 > 1$, and $f_1(x)$ is positive when $x > 1$. Hence $f_1(17/13)$ is positive, and therefore f_3 is negative.

For $x = -\infty$ the signs of $f(x)$, $f_1(x)$, $f_2(x)$, f_3 are $-, +, -, -$; for $x = \infty$ they are $+, +, +, -$. Hence $f(x) = 0$ has but one real root.

The only property of the final function f_m of which any use is made in the proof in § 863 is that its *sign* is constant. Hence if, when computing the Sturm functions of $f(x) = 0$ in order to find the number of roots between a and b, we come upon a function $f_p(x)$ which has the same sign for all values of x between a and b, we need not compute the subsequent functions. For it follows from the proof in § 863 that the required number of roots will be the difference between the number of variations in $f(a), \cdots, f_p(a)$ and in $f(b), \cdots, f_p(b)$.

Example 3. How many real roots has $f(x) = x^3 + x^2 + x + 1 = 0$?

Here $f_1(x) = 3x^2 + 2x + 1$, and, since $2^2 < 4 \cdot 3$, this is positive for all real values of x, §§ 635, 823. Hence we need not compute $f_2(x)$ and f_3.

The signs of $f(x)$, $f_1(x)$ for $x = -\infty$ are $-$, $+$; for $x = \infty$ they are $+$, $+$. Hence $f(x) = 0$ has one real root.

EXERCISE LXXVIII

By aid of Sturm's theorem find the situation of the real roots of the following equations.

1. $x^3 - 6x^2 + 5x + 13 = 0$.
2. $x^3 - 4x^2 - 10x + 41 = 0$.
3. $x^3 + 5x + 2 = 0$.
4. $x^3 + 3x^2 + 8x + 8 = 0$.
5. $x^3 - x^2 - 15x + 28 = 0$.
6. $x^4 - 4x^3 - 5x^2 + 18x + 20 = 0$.
7. $2x^4 - 3x^2 + 3x - 1 = 0$.
8. $x^4 - 8x^3 + 19x^2 - 12x + 2 = 0$.
9. $x^4 - 12x^2 + 12x - 3 = 0$.
10. $x^4 + 2x^3 - 6x^2 - 8x + 9 = 0$.

By aid of Sturm's theorem find the number of the real roots of each of the following equations.

11. $4x^3 - 2x - 5 = 0$.
12. $x^4 + x^3 + x^2 + x + 1 = 0$.
13. $x^n + 1 = 0$.
14. $x^4 - 6x^3 + x^2 + 14x - 14 = 0$.

15. Let $f(x) = 0$ be an equation of the nth degree without multiple roots. Show that the condition that all the roots of $f(x) = 0$ be real is that there be $n + 1$ terms in its sequence of Sturm functions $f(x), f_1(x), \cdots, f_m$, and that the leading terms of all these functions have the same sign.

16. By aid of the theorem in Ex. 15 prove that the condition that all the roots of the cubic $x^3 + px + q = 0$ be real and unequal is that $4p^3 + 27q^2$ be negative.

SYMMETRIC FUNCTIONS OF THE ROOTS

865 **Theorem 1.** *If the roots of* $f(x) = x^n + b_1x^{n-1} + \cdots + b_n = 0$ *are* $\beta_1, \beta_2, \cdots, \beta_n$, *so that* $f(x) = (x - \beta_1)(x - \beta_2) \cdots (x - \beta_n)$, *then*

$$f'(x) = \frac{f(x)}{x - \beta_1} + \frac{f(x)}{x - \beta_2} + \cdots + \frac{f(x)}{x - \beta_n}.$$

Thus, suppose that $n = 3$, so that

$$f(x) = (x - \beta_1)(x - \beta_2)(x - \beta_3). \tag{1}$$

Substituting $x + h$ for x in (1), we have

$$f(x + h) = [(x - \beta_1) + h][(x - \beta_2) + h][(x - \beta_3) + h]. \tag{2}$$

We can reduce each member of (2) to the form of a polynomial in h, the first member by Taylor's theorem, § 848, the second by continued multiplication, as in § 558.

Since (2) is an identity, the coefficients of like powers of h in the two polynomials thus obtained must be equal, § 284.

But since $f(x + h) = f(x) + f'(x)h + \cdots$, § 848, the coefficient of h in the first polynomial is $f'(x)$. In the second it is $(x - \beta_2)(x - \beta_3) + (x - \beta_3)(x - \beta_1) + (x - \beta_1)(x - \beta_2)$. Hence

$$\begin{aligned} f'(x) &= (x - \beta_2)(x - \beta_3) + (x - \beta_3)(x - \beta_1) \\ &\quad + (x - \beta_1)(x - \beta_2) \\ &= \frac{f(x)}{x - \beta_1} + \frac{f(x)}{x - \beta_2} + \frac{f(x)}{x - \beta_3}, \end{aligned} \tag{3}$$

since $(x - \beta_2)(x - \beta_3) = f(x)/(x - \beta_1)$, and so on.

This reasoning is applicable to an equation of any degree n. In the general case there are n factors in the second member of (2), and when this member is reduced to the form of a polynomial in h, the coefficient of h is the sum of the products of the binomials $x - \beta_1, x - \beta_2, \cdots, x - \beta_n$, taken $n - 1$ at a time, § 558. That one of these products which lacks the factor $x - \beta_1$ may be written $f(x)/(x - \beta_1)$, and so on.

Thus, if $f(x) = x^3 - 6x^2 + 11x - 6 = (x-1)(x-2)(x-3)$,
we have $f'(x) = 3x^2 - 12x + 11$
and $(x-2)(x-3) + (x-3)(x-1) + (x-1)(x-2) = 3x^2 - 12x + 11$.

Theorem 2. *The sums of like powers of all the roots of an equation* $f(x) = 0$ *can be expressed rationally in terms of its coefficients.* **866**

Thus, suppose that the equation is

$$f(x) = x^3 + b_1x^2 + b_2x + b_3 = 0. \qquad (1)$$

Let α, β, γ denote the roots of (1) and let $s_1, s_2, \cdots, s_r$ have the meanings

$$s_1 = \alpha + \beta + \gamma, \quad s_2 = \alpha^2 + \beta^2 + \gamma^2, \cdots, \; s_r = \alpha^r + \beta^r + \gamma^r.$$

We are to prove that $s_1, s_2, \cdots$ can be expressed rationally in terms of the coefficients b_1, b_2, b_3.

1. By the preceding theorem, § 865, we have

$$f'(x) = \frac{f(x)}{x-\alpha} + \frac{f(x)}{x-\beta} + \frac{f(x)}{x-\gamma}. \qquad (2)$$

Since $f(x)$ is divisible by $x - \alpha$, $x - \beta$, and $x - \gamma$, each of the fractions in (2) represents a polynomial in x which can be found by the rule of § 410. Applying this rule and then adding the results, we have

$$\begin{aligned}
f(x)/(x-\alpha) &= x^2 + (\alpha + b_1)x + (\alpha^2 + b_1\alpha + b_2)\\
f(x)/(x-\beta) &= x^2 + (\beta + b_1)x + (\beta^2 + b_1\beta + b_2)\\
f(x)/(x-\gamma) &= x^2 + (\gamma + b_1)x + (\gamma^2 + b_1\gamma + b_2)\\
\hline
f'(x) &= 3x^2 + (s_1 + 3b_1)x + (s_2 + b_1s_1 + 3b_2) \qquad (3)
\end{aligned}$$

But by definition, § 847, we also have

$$f'(x) = 3x^2 + 2b_1x + b_2. \qquad (4)$$

Equating the coefficients of like powers of x in the two expressions (3) and (4) and solving for s_1, s_2, we have

$$s_1 + 3b_1 = 2b_1, \; \therefore \; s_1 = -b_1, \qquad (5)$$

$$s_2 + b_1s_1 + 3b_2 = b_2, \quad \therefore \; s_2 = b_1^2 - 2b_2. \qquad (6)$$

2. From the values thus found for s_1 and s_2 we can obtain the values of $s_3, s_4, \cdots$ successively as follows:

Since α, β, γ are the roots of (1), we have

$$\alpha^3 + b_1\alpha^2 + b_2\alpha + b_3 = 0, \tag{7}$$

$$\beta^3 + b_1\beta^2 + b_2\beta + b_3 = 0, \tag{8}$$

$$\gamma^3 + b_1\gamma^2 + b_2\gamma + b_3 = 0. \tag{9}$$

Adding these identities, we obtain

$$s_3 + b_1 s_2 + b_2 s_1 + 3\,b_3 = 0, \tag{10}$$

which gives s_3 rationally in terms of b_1, b_2, b_3, s_1, s_2 and therefore, by (5), (6), rationally in terms of b_1, b_2, b_3.

Next multiply the identities (7), (8), (9) by α, β, γ respectively and add the results. We obtain

$$s_4 + b_1 s_3 + b_2 s_2 + b_3 s_1 = 0, \tag{11}$$

which by aid of (5), (6), (10) gives s_4 rationally in terms of b_1, b_2, b_3.

And in like manner, if we multiply (7), (8), (9) respectively by $\alpha^2, \beta^2, \gamma^2$, by $\alpha^3, \beta^3, \gamma^3$, and so on, and after each series of multiplications add results, we obtain identities

$$s_5 + b_1 s_4 + b_2 s_3 + b_3 s_2 = 0, \quad s_6 + b_1 s_5 + b_2 s_4 + b_3 s_3 = 0, \ \cdots,$$

which give $s_5, s_6, \cdots$ rationally in terms of b_1, b_2, b_3.

By similar reasoning the theorem may be proved for an equation $f(x) = 0$ of any degree n.

Example 1. If α, β, γ denote the roots of $x^3 - 2\,x^2 + 4\,x + 2 = 0$, find

$$\Sigma 1/\alpha = 1/\alpha + 1/\beta + 1/\gamma, \ \Sigma 1/\alpha^2 = 1/\alpha^2 + 1/\beta^2 + 1/\gamma^2,$$
$$\Sigma 1/\alpha^3 = 1/\alpha^3 + 1/\beta^3 + 1/\gamma^3.$$

Applying the transformation $x = 1/y$, and dividing the transformed equation by the coefficient of y^3, we have $y^3 + 2\,y^2 - y + 1/2 = 0$.

For this equation, by substituting $b_1 = 2$, $b_2 = -1$, $b_3 = 1/2$ in the formulas (5), (6), (10) above, we obtain $s_1 = -2$, $s_2 = 6$, $s_3 = -31/2$.

Therefore, § 814, $\Sigma 1/\alpha = -2$, $\Sigma 1/\alpha^2 = 6$, $\Sigma 1/\alpha^3 = -31/2$.

Example 2. For the equation $f(x) = x^4 + b_1x^3 + b_2x^2 + b_3x + b_4 = 0$, show that

$s_1 + 4b_1 = 3b_1$, $s_2 + b_1s_1 + 4b_2 = 2b_2$, $s_3 + b_1s_2 + b_2s_1 + 4b_3 = b_3$,
$s_4 + b_1s_3 + b_2s_2 + b_3s_1 + 4b_4 = 0$, $s_5 + b_1s_4 + b_2s_3 + b_3s_2 + b_4s_1 = 0$,

and compute s_1, s_2, s_3, s_4 in terms of b_1, b_2, b_3, b_4.

The preceding formulas also show that $s_1, s_2, s_3, \cdots$ are *integral* functions of the coefficients of the equation when, as in (1), the leading coefficient is 1. **867**

Theorem 2. *Every rational symmetric function of the roots of an equation* f(x) = 0 *can be expressed rationally in terms of its coefficients.* **868**

Let the roots of $f(x) = 0$ be $\alpha, \beta, \gamma, \cdots, \nu$.

Every rational symmetric function of $\alpha, \beta, \cdots, \nu$ can be expressed rationally in terms of functions of the types $\Sigma\alpha^p$, $\Sigma\alpha^p\beta^q$, $\Sigma\alpha^p\beta^q\gamma^r$, and so on, § 544. Hence it is only necessary to prove our theorem for functions of these several types. This was done for the type $\Sigma\alpha^p = s_p$ in § 866, and we shall now show that $\Sigma\alpha^p\beta^q$, and so on, can be expressed rationally in terms of functions of this type s_p.

1. The type $\Sigma\alpha^p\beta^q = \alpha^p\beta^q + \beta^p\alpha^q + \cdots$.

The product $(\alpha^p + \beta^p + \cdots)(\alpha^q + \beta^q + \cdots)$ (1)

is the sum of the two symmetric groups of terms

$$\alpha^{p+q} + \beta^{p+q} + \cdots, \quad (2)$$

$$\alpha^p\beta^q + \beta^p\alpha^q + \cdots. \quad (3)$$

But (1) and (2) are rational functions of the coefficients, § 866. Hence (3), or $\Sigma\alpha^p\beta^q$, which may be obtained by subtracting (2) from (1), is also such a function.

Since (1) is s_ps_q and (2) is s_{p+q}, we have the formula

$$\Sigma\alpha^p\beta^q = s_ps_q - s_{p+q}. \quad (4)$$

When $p = q$, the terms of (3) are equal in pairs and (3) becomes $2\Sigma\alpha^p\beta^p$. We then have $2\Sigma\alpha^p\beta^p = s_p^2 - s_{2p}$.

2. The type $\Sigma\alpha^p\beta^q\gamma^r = \alpha^p\beta^q\gamma^r + \beta^p\alpha^q\gamma^r + \cdots$.

The product $(\alpha^p\beta^q + \beta^p\alpha^q + \cdots)(\alpha^r + \beta^r + \gamma^r + \cdots)$ (5)

is the sum of the three symmetric groups of terms

$$\alpha^{p+r}\beta^q + \beta^{p+r}\alpha^q + \cdots, \quad (6)$$

$$\alpha^p\beta^{q+r} + \beta^p\alpha^{q+r} + \cdots, \quad (7)$$

$$\alpha^p\beta^q\gamma^r + \beta^p\alpha^q\gamma^r + \cdots. \quad (8)$$

But we have already shown that (5), (6), and (7) are rational functions of the coefficients. Hence (8), or $\Sigma\alpha^p\beta^q\gamma^r$, which may be obtained by subtracting the sum of (6) and (7) from (5), is also such a function.

When $p = q$, the group (8) becomes $2\Sigma\alpha^p\beta^p\gamma^r$; when $p = q = r$, it becomes $6\Sigma\alpha^p\beta^p\gamma^p$.

3. The types $\Sigma\alpha^p\beta^q\gamma^r\delta^u$, and so on.

We may prove that these are rational functions of the coefficients by repetitions of the process illustrated in 1 and 2. We begin by multiplying $\Sigma\alpha^p\beta^q\gamma^r$ by $\alpha^u + \beta^u + \gamma^u \cdots$.

Example. Show that

$$\Sigma\alpha^p\beta^q\gamma^r = s_p s_q s_r + 2s_{p+q+r} - s_{p+q}s_r - s_{q+r}s_p - s_{r+p}s_q.$$

EXERCISE LXXIX

1. For the equation $a_0x^3 + a_1x^2 + a_2x + a_3 = 0$ find s_3 and s_4 in terms of a_0, a_1, a_2, a_3.

2. If α, β, γ denote the roots of $x^3 + px^2 + qx + r = 0$, find $\Sigma 1/\alpha^2$, $\Sigma 1/\alpha^3$, and $\Sigma\alpha\beta^2$ in terms of p, q, r.

3. Find the equation whose roots are the cubes of the roots of $x^3 - 2x^2 + 3x - 1 = 0$.

4. If α, β, γ denote the roots of the equation $x^3 - x^2 + 3x + 4 = 0$, find the values of the following symmetric functions of these roots by the methods of §§ 866, 867.

(1) s_1, s_2, s_3, s_4. (2) $\Sigma\alpha^3\beta^2$. (3) $\Sigma\alpha^3\beta\gamma$.

(4) $\Sigma\alpha^3\beta^2\gamma$. (5) $\Sigma 1/\alpha^4$. (6) $\Sigma\alpha^2\beta^2/\gamma$.

XXX. THE GENERAL CUBIC AND BIQUADRATIC EQUATIONS

Algebraic solutions. In the preceding chapter we have shown that the real roots of a *numerical* equation can always be found exactly or approximately, and it is possible to extend the methods there employed to the complex roots of such equations. As hardly need be said, these methods are not applicable to *literal* equations. To solve such an equation we must obtain expressions for its roots in terms of its coefficients. **869**

We say that an equation can be solved *algebraically* when its roots can be expressed in terms of its coefficients by applying a finite number of times the several algebraic operations, namely, addition, subtraction, multiplication, division, involution and evolution.

We have already proved, § 631, that the general quadratic equation has such an algebraic solution, and we are now to prove that the like is true of the general cubic and biquadratic equations. But general equations of a degree higher than four cannot be solved algebraically.

Cube roots of unity. In § 646 we showed that the equation $x^3 = 1$ has the roots 1, $(-1 + i\sqrt{3})/2$, $(-1 - i\sqrt{3})/2$. Hence each of these numbers is a cube root of unity. The third will be found to be the square of the second. Hence if we represent the second by ω, we may represent the third by ω^2. Since $x^3 - 1 = 0$ lacks an x^2 term, we have, § 805, $1 + \omega + \omega^2 = 0$. **870**

Similarly every number a has three cube roots, namely, the three roots of the equation $x^3 = a$. If one of these roots be $\sqrt[3]{a}$, the other two will be $\omega\sqrt[3]{a}$ and $\omega^2\sqrt[3]{a}$.

The general cubic. Cardan's formula. By the method of § 818 every cubic equation can be reduced to the form **871**

$$x^3 + px + q = 0, \tag{1}$$

in which the x^2 term is lacking.

In (1) put $$x = y + z. \qquad (2)$$

We obtain

$$y^3 + 3y^2z + 3yz^2 + z^3 + p(y + z) + q = 0,$$

or $$y^3 + z^3 + (3yz + p)(y + z) + q = 0. \qquad (3)$$

As the variables y and z are subject to the single condition (3), we may impose a second condition upon them.

We suppose $$3yz + p = 0, \qquad (4)$$

and therefore, by (3), $y^3 + z^3 + q = 0. \qquad (5)$

From (5), $$y^3 + z^3 = -q, \qquad (6)$$

and from (4), $$y^3z^3 = -p^3/27. \qquad (7)$$

Therefore, § 636, y^3 and z^3 are the roots of a quadratic equation of the form

$$u^2 + qu - p^3/27 = 0. \qquad (8)$$

Solving (8) and representing the expressions obtained for the roots by A and B respectively, we have

$$y^3 = A = -\frac{q}{2} + \sqrt{\frac{q^2}{4} + \frac{p^3}{27}}, \quad z^3 = B = -\frac{q}{2} - \sqrt{\frac{q^2}{4} + \frac{p^3}{27}}. \qquad (9)$$

These equations (9) give three values for y and three for z, namely, § 870,

$$y = \sqrt[3]{A}, \quad \omega\sqrt[3]{A}, \quad \omega^2\sqrt[3]{A}, \qquad (10)$$

$$z = \sqrt[3]{B}, \quad \omega\sqrt[3]{B}, \quad \omega^2\sqrt[3]{B}. \qquad (11)$$

But by (4), $yz = -p/3$, and the only pairs of the values of y, z in (10), (11) which satisfy this condition are

$$y, z = \sqrt[3]{A}, \ \sqrt[3]{B}; \ \omega\sqrt[3]{A}, \ \omega^2\sqrt[3]{B}; \ \omega^2\sqrt[3]{A}, \ \omega\sqrt[3]{B}.$$

Substituting these pairs in (2), we obtain the three roots of (1), namely,

$$x_1 = \sqrt[3]{A} + \sqrt[3]{B}, \ x_2 = \omega\sqrt[3]{A} + \omega^2\sqrt[3]{B}, \ x_3 = \omega^2\sqrt[3]{A} + \omega\sqrt[3]{B},$$

where $$A = -\frac{q}{2} + \sqrt{\frac{q^2}{4} + \frac{p^3}{27}}, \quad B = -\frac{q}{2} - \sqrt{\frac{q^2}{4} + \frac{p^3}{27}}. \qquad (12)$$

Example 1. Solve $x^3 - 6x^2 + 6x - 2 = 0$.

By § 818 we find that the substitution $x = y + 2$ will transform the given equation to the form $y^3 + py + q = 0$. We thus obtain

$$y^3 - 6y - 6 = 0.$$

The roots of this equation, found by substituting $p = -6$, $q = -6$ in the above formulas, are

$$\sqrt[3]{4} + \sqrt[3]{2},\quad \sqrt[3]{4}\,\omega + \sqrt[3]{2}\,\omega^2,\quad \sqrt[3]{4}\,\omega^2 + \sqrt[3]{2}\,\omega.$$

Hence the roots of the given equation are

$$2 + \sqrt[3]{4} + \sqrt[3]{2},\quad 2 + \sqrt[3]{4}\,\omega + \sqrt[3]{2}\,\omega^2,\quad 2 + \sqrt[3]{4}\,\omega^2 + \sqrt[3]{2}\,\omega.$$

Example 2. Solve the equation $x^3 + 3x^2 + 6x + 5 = 0$.

Discussion of the solution. When p and q are real, the character of the roots depends on the value of $q^2/4 + p^3/27$ as follows: **872**

1. *If* $q^2/4 + p^3/27 > 0$, *one root is real, two are imaginary.*

For in this case A and B are real. Hence x_1 is real, and x_2, x_3 are conjugate imaginaries, § 870.

2. *If* $q^2/4 + p^3/27 = 0$, *all the roots are real and two equal.*

For in this case $A = B = -q/2$. Hence $x_1 = -2\sqrt[3]{q/2}$, and $x_2 = x_3 = -(\omega + \omega^2)\sqrt[3]{q/2} = \sqrt[3]{q/2}$, since $\omega + \omega^2 = -1$, by § 870.

3. *If* $q^2/4 + p^3/27 < 0$, *all the roots are real and unequal.*

This may be proved by Sturm's theorem (see p. 477, Ex. 16). But when $q^2/4 + p^3/27 < 0$, A and B are complex numbers, and though the expressions for x_1, x_2, x_3 denote real numbers, they cannot be reduced to a real form by algebraic transformations. This is therefore called the *irreducible case* of the cubic (see § 885).

The expression $q^2/4 + p^3/27$ is called the *discriminant* of the cubic $x^3 + px + q = 0$, since its vanishing is the condition that two of the roots be equal (compare § 635). **873**

874 **The general biquadratic. Ferrari's solution.** By the method of § 818 every biquadratic equation can be reduced to the form

$$x^4 + ax^2 + bx + c = 0. \qquad (1)$$

With a view to transforming the first member of (1) into a difference of squares, we add and subtract $x^2u + u^2/4$, where u denotes a constant whose value is to be found. We thus obtain

$$x^4 + x^2u + u^2/4 - x^2u - u^2/4 + ax^2 + bx + c = 0,$$

or $$(x^2 + u/2)^2 - [(u - a)x^2 - bx + (u^2/4 - c)] = 0. \qquad (2)$$

To make the second term a perfect square, we must have

$$b^2 = 4(u - a)(u^2/4 - c),$$

or $$u^3 - au^2 - 4cu + (4ac - b^2) = 0. \qquad (3)$$

Let u_1 denote one of the roots of this cubic in u. When u is replaced by u_1 in (2), the second term is the square of $\sqrt{u_1 - a}\,x - b/2\sqrt{u_1 - a}$, and (2) becomes equivalent to the two quadratics

$$x^2 + \sqrt{u_1 - a}\,x + \left(\frac{u_1}{2} - \frac{b}{2\sqrt{u_1 - a}}\right) = 0. \qquad (4)$$

$$x^2 - \sqrt{u_1 - a}\,x + \left(\frac{u_1}{2} + \frac{b}{2\sqrt{u_1 - a}}\right) = 0. \qquad (5)$$

We may therefore obtain the roots of (1) by solving (4) and (5).

Example 1. Solve $x^4 + x^2 + 4x - 3 = 0$.

Here $a = 1$, $b = 4$, $c = -3$, so that the cubic (3) is

$$u^3 - u^2 + 12u - 28 = 0.$$

One of the roots of this cubic is 2, and setting $u_1 = 2$ in (4), (5), we have

$$x^2 + x - 1 = 0 \text{ and } x^2 - x + 3 = 0.$$

Solving these quadratics, we obtain $x = (-1 \pm \sqrt{5})/2$, $(1 \pm i\sqrt{11})/2$.

Example 2. Solve the equation $x^4 - 4x^3 + x^2 + 4x + 1 = 0$.

As the cubic (3) has three roots any one of which may be taken as the u_1 in (4) and (5), it would seem that the method above described might yield $3 \cdot 4$ or 12 values of x, whereas the given equation (1) can have but *four* roots. But it is not difficult to prove that the choice made of u_1 does not affect the values of the four roots of (4), (5) combined, but merely the manner in which these roots are distributed between (4) and (5).

875 **Reciprocal equations.** We may discover by inspection whether or not a given reciprocal equation, § 815, has either of the roots 1 or -1, and if it has, derive from it a depressed equation $\phi(x) = 0$ which has neither of these roots. It follows from § 815 that this depressed equation $\phi(x) = 0$ must be of the form

$$a_0 x^{2m} + a_1 x^{2m-1} + \cdots + a_m x^m + \cdots + a_1 x + a_0 = 0, \quad (1)$$

that is, its degree must be even, and every two coefficients which are equally removed from the beginning and end of $\phi(x)$ must be equal.

We proceed to show that by the substitution $z = x + 1/x$ this equation $\phi(x) = 0$ may be transformed into an equation in z whose degree is one half that of $\phi(x) = 0$. It will then follow that if the degree of $\phi(x) = 0$ be not greater than eight, we may find the roots by aid of §§ 631, 871, 874.

Dividing (1) by x^m and combining terms, we have

$$a_0\left(x^m + \frac{1}{x^m}\right) + a_1\left(x^{m-1} + \frac{1}{x^{m-1}}\right) + \cdots + a_m = 0. \quad (2)$$

But by carrying out the indicated reckoning we find that

$$x^{p+1} + \frac{1}{x^{p+1}} = \left(x^p + \frac{1}{x^p}\right)\left(x + \frac{1}{x}\right) - \left(x^{p-1} + \frac{1}{x^{p-1}}\right); \quad (3)$$

and if $z = x + 1/x$, and in (3) we set $p = 1, 2, 3 \cdots$ successively, we have

$$x^2 + \frac{1}{x^2} = z^2 - 2, \quad x^3 + \frac{1}{x^3} = z^3 - 3z,$$

$$x^4 + \frac{1}{x^4} = z^4 - 4z^2 + 2, \cdots, \quad (4)$$

that is, we obtain for $x^p + 1/x^p$ an expression of the pth degree in z. Substituting these several expressions in (2), we obtain an equation of the mth degree in z, as was to be demonstrated (compare § 645).

Example 1. Solve $2x^8 - x^7 - 12x^6 + 14x^5 - 14x^3 + 12x^2 + x - 2 = 0$.

This is a reciprocal equation having the roots 1 and -1.

Removing the factor $x^2 - 1$,

$$2x^6 - x^5 - 10x^4 + 13x^3 - 10x^2 - x + 2 = 0.$$

Dividing by x^3,

$$2\left(x^3 + \frac{1}{x^3}\right) - \left(x^2 + \frac{1}{x^2}\right) - 10\left(x + \frac{1}{x}\right) + 13 = 0.$$

Hence by (4), $\quad 2(z^3 - 3z) - (z^2 - 2) - 10z + 13 = 0,$

or $\quad 2z^3 - z^2 - 16z + 15 = 0.$

Solving, $\quad z = 1, \ -3, \text{ or } 5/2.$

Hence $\quad x + 1/x = 1, \ -3, \text{ or } 5/2,$

and therefore $x = (1 \pm i\sqrt{3})/2, \ (-3 \pm \sqrt{5})/2, \ 2, \text{ or } 1/2.$

Example 2. Solve $x^6 - x^5 + x^4 + x^2 - x + 1 = 0$.

876 Every binomial equation $x^n + a = 0$ may be reduced to the reciprocal form by aid of the substitution $x = \sqrt[n]{a}y$, the result of this substitution being $y^n + 1 = 0$ (compare § 646, Ex. 2).

877 **Expression of a complex number in terms of absolute value and amplitude.** In the accompanying figure, P is the graph of the complex number $a + bi$, constructed as in § 238.

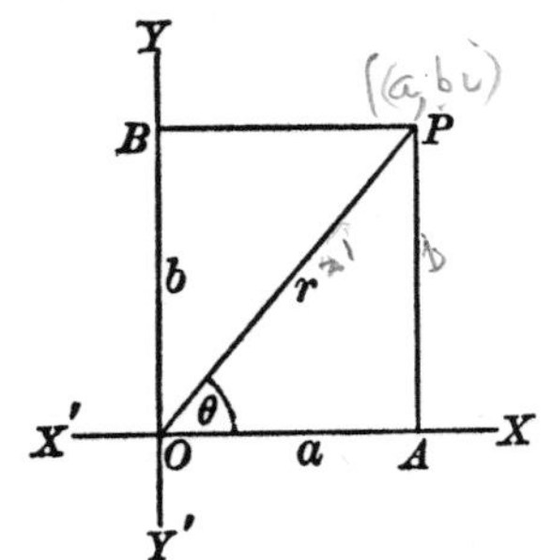

The length of OP is $\sqrt{a^2 + b^2}$, the *absolute value* of $a + bi$, § 239. Represent it by r.

Let θ denote the circular measure of the angle XOP, that is, the length of the arc subtended by this angle on a circle of unit radius described about O as center. We call θ the *amplitude* of $a + bi$.

We call the ratio b/r the *sine* of θ, and write $b/r = \sin\theta$.

We call the ratio a/r the *cosine* of θ, and write $a/r = \cos\theta$.

We thus have $$a = r\cos\theta,\ b = r\sin\theta,$$
and therefore $$a + bi = r(\cos\theta + i\sin\theta).$$

When $\theta = 0$, then $b = 0$ and $a = r$. Hence $\sin 0 = 0$, and $\cos 0 = 1$.

878 The circular measure of $360°$ is 2π, this being the length of a circle of unit radius. Hence a point P given by r, θ is given equally by $r, \theta + 2\pi$; by $r, \theta - 2\pi$; and, in general, by $r, \theta + 2m\pi$, where m denotes any integer. Hence we say that the *general value of the amplitude* of $a + bi$ is $\theta + 2m\pi$.

879 **Theorem.** *The absolute value of the product of two complex numbers is the product of their absolute values; and its amplitude is the sum of their amplitudes.*

For
$$\begin{aligned} & r(\cos\theta + i\sin\theta)\cdot r'(\cos\theta' + i\sin\theta') \\ &= rr'[(\cos\theta\cos\theta' - \sin\theta\sin\theta') \\ &\qquad + i(\sin\theta\cos\theta' + \cos\theta\sin\theta')] \\ &= rr'[\cos(\theta + \theta') + i\sin(\theta + \theta')], \end{aligned}$$

since it is proved in trigonometry that

$$\cos(\theta + \theta') = \cos\theta\cos\theta' - \sin\theta\sin\theta',$$
$$\sin(\theta + \theta') = \sin\theta\cos\theta' + \cos\theta\sin\theta'.$$

The construction given in § 240 is based on this theorem.

880 **Corollary 1.** By repeated applications of § 879 we have

$$\begin{aligned} & r(\cos\theta + i\sin\theta)\cdot r'(\cos\theta' + i\sin\theta')\cdot r''(\cos\theta'' + i\sin\theta'')\cdots \\ &= rr'r''\cdots[\cos(\theta + \theta' + \theta'' + \cdots) \\ &\qquad + i\sin(\theta + \theta' + \theta'' + \cdots)]. \end{aligned}$$

881 **Corollary 2.** Setting $r = r' = r'' = \cdots$, and $\theta = \theta' = \theta'' = \cdots$ in § 880, we obtain the following formula, known as Demoivre's theorem:

$$[r(\cos\theta + i\sin\theta)]^n = r^n(\cos n\theta + i\sin n\theta).$$

882 **Corollary 3.** For a quotient we have the formula

$$\frac{r(\cos\theta + i\sin\theta)}{r'(\cos\theta' + i\sin\theta')} = \frac{r}{r'}[\cos(\theta - \theta') + i\sin(\theta - \theta')].$$

For $\frac{r}{r'}[\cos(\theta - \theta') + i\sin(\theta - \theta')]\cdot r'(\cos\theta' + i\sin\theta')$

$$= r(\cos\theta + i\sin\theta). \qquad \S\ 879$$

883 **Corollary 4.** The n nth roots of a complex number are given by

$$\sqrt[n]{r(\cos\theta + i\sin\theta)} = r^{\frac{1}{n}}\left(\cos\frac{\theta + 2k\pi}{n} + i\sin\frac{\theta + 2k\pi}{n}\right),$$

when k is assigned the n values $0, 1, 2, \cdots, (n-1)$.

For $$\left[r^{\frac{1}{n}}\left(\cos\frac{\theta + 2k\pi}{n} + i\sin\frac{\theta + 2k\pi}{n}\right)\right]^n$$

$$= r[\cos(\theta + 2k\pi) + i(\sin\theta + 2k\pi)] = r(\cos\theta + i\sin\theta).$$

§§ 881, 878

884 **Binomial equations.** The n nth roots of $r(\cos\theta + i\sin\theta)$ are the n roots of the equation $x^n - r(\cos\theta + i\sin\theta) = 0$. Hence, in particular, the n roots of the equation $x^n - r = 0$, where r is real and therefore θ is 0, are

$$\sqrt[n]{r}(\cos 2k\pi/n + i\sin 2k\pi/n),\ k = 0, 1, \cdots, n-1.$$

Thus, the roots of the equation $x^3 - 1 = 0$ are

$$\cos 0 + i\sin 0,\ \cos 2\pi/3 + i\sin 2\pi/3,\ \cos 4\pi/3 + i\sin 4\pi/3,$$

which may be proved equal to

$$1,\ (-1 + i\sqrt{3})/2,\ (-1 - i\sqrt{3})/2.$$

885 **Trigonometric solution of the irreducible case of the cubic.** In the irreducible case of the cubic $x^3 + px + q = 0$, § 872, 3, the expressions A and B, § 871, are conjugate imaginaries. For since in this case $q^2/4 + p^3/27$ is negative, we have

$$A = -\frac{q}{2} + i\sqrt{-\left(\frac{q^2}{4} + \frac{p^3}{27}\right)},\ B = -\frac{q}{2} - i\sqrt{-\left(\frac{q^2}{4} + \frac{p^3}{27}\right)}.$$

Hence the expressions for A and B in terms of absolute value and amplitude, § 877, will be of the form

$$A = r(\cos\theta + i\sin\theta),\ B = r(\cos\theta - i\sin\theta), \qquad (1)$$

where $$r = \left(\frac{q^2}{4} - \frac{q^2}{4} - \frac{p^3}{27}\right)^{\frac{1}{2}} = \left(-\frac{p^3}{27}\right)^{\frac{1}{2}}$$

and $$\cos\theta = \frac{-q/2}{r} = -\frac{q}{2}\left(-\frac{27}{p^3}\right)^{\frac{1}{2}}. \qquad (2)$$

When p and q are given we can find the value of θ from that of $\cos\theta$ by aid of a table of cosines.

In the formulas for the roots of $x^3 + px + q = 0$, § 871, (12), substitute the expressions (1) for A and B, and the expressions for ω and ω^2 given in § 884. The results when simplified are

$$x_1 = 2\,r^{\frac{1}{3}} \cos\frac{\theta}{3}, \quad x_2 = 2\,r^{\frac{1}{3}} \cos\frac{\theta + 2\,\pi}{3}, \quad x_3 = 2\,r^{\frac{1}{3}} \cos\frac{\theta + 4\,\pi}{3}.$$

And, r and θ being known by (2), these formulas enable us to compute the values of the roots by aid of a table of cosines.

Example. Solve $x^3 - x + 1/3 = 0$.

Here $q^2/4 + p^3/27 = -1/108$, so that we have the irreducible case. Substituting in the formulas (2) and simplifying, we find

$$r = 1/\sqrt{27}, \quad \cos\theta = -\sqrt{3}/2, \text{ and therefore } \theta = 150°.$$

Hence by aid of tables of logarithms and cosines we obtain

$$x_1 = \frac{2}{\sqrt[6]{27}} \cos 50° = .7422\,; \quad x_2 = \frac{2}{\sqrt[6]{27}} \cos 170° = -1.1371\,;$$

$$x_3 = \frac{2}{\sqrt[6]{27}} \cos 290° = .3949.$$

EXERCISE LXXX

Solve equations 1 – 10 by the methods of §§ 871 and 874.

1. $x^3 - 9x - 28 = 0$.
2. $x^3 - 9x^2 + 9x - 8 = 0$.
3. $x^3 - 3x - 4 = 0$.
4. $4x^3 - 7x - 6 = 0$.
5. $x^3 + 3x^2 + 9x - 1 = 0$.
6. $3x^3 - 9x^2 + 14x + 7 = 0$.
7. $x^4 + x^2 + 6x + 1 = 0$.
8. $x^4 - 4x^3 + x^2 + 4x + 1 = 0$.
9. $x^4 + 12x - 5 = 0$.
10. $x^4 + 8x^3 + 12x^2 - 11x + 2 = 0$.
11. Solve $3x^6 - 2x^5 + 6x^4 - 2x^3 + 6x^2 - 2x + 3 = 0$.

12. Solve $2x^8 - 9x^7 + 18x^6 - 30x^5 + 32x^4 - 30x^3 + 18x^2 - 9x + 2 = 0$.

13. Solve $6x^7 - x^6 + 2x^5 - 7x^4 - 7x^3 + 2x^2 - x + 6 = 0$.

14. Find the cubic in z on which, by § 875, the solution of $x^7 - 1 = 0$ depends.

15. Find the condition that all the roots of $x^3 + 3ax^2 + 3bx + c = 0$ be real.

16. Write down the trigonometric expressions for the roots of $x^5 - 1 = 0$, and of $x^6 + 1 = 0$.

17. Solve the following irreducible cubics.

(1) $x^3 - 3x - 1 = 0$. (2) $x^3 - 6x - 4 = 0$.

18. In a sphere whose diameter is $3\sqrt{3}$ a right prism with a square base is inscribed. If the volume of the prism is 27, what is its altitude?

19. The volume of a certain right circular cylinder is 50π and its entire superficial area is $105\pi/2$. Find the radius of its base and its altitude.

20. The altitude of a right circular cone is 6 and the radius of its base is 4. In this cone a right circular cylinder is inscribed whose volume is four ninths that of the cone. Find the altitude of the cylinder.

XXXI. DETERMINANTS AND ELIMINATION

DEFINITION OF DETERMINANT

886 **Inversions. Odd and even permutations.** When considering the permutations of a set of objects, as letters or numbers, we may fix upon some particular order of the objects as the *normal* order. Any given permutation is then said to have as many *inversions* as it presents instances in which an object is followed by one which in the normal order precedes it. And the permutation is called *odd* or *even* according as the number of its inversions is odd or even (or 0).

Thus, if the objects in normal order are the numbers 1, 2, 3, 4, 5, the permutation 45312 has the *eight* inversions 43, 41, 42, 53, 51, 52, 31, 32. Hence 45312 is an *even* permutation.

Theorem. *If two of the objects in a permutation are interchanged, the number of inversions is increased or diminished by an odd number.* **887**

For if two *adjacent* objects are interchanged, the number of inversions is increased or diminished by 1. Thus, compare $ApqB$ (1) and $AqpB$ (2), where A and B denote the groups of objects which precede and follow the interchanged objects p and q. Any inversions which may occur in A and B and any which may be due to the fact that A, p and q precede B are common to (1) and (2). Hence the sole difference between (1) and (2), so far as inversions are concerned, is this: If pq is an inversion, qp is not, and (2) has one less inversion than (1); but if pq is not an inversion, qp is, and (2) has one more inversion than (1).

But the interchange of *any* two objects may be brought about by an odd number of interchanges of adjacent objects. Thus, from $pabq$ we may derive $qabp$ by *five* interchanges of adjacent letters. We first interchange p with each following letter in turn, obtaining successively $apbq$, $abpq$, $abqp$, and we then interchange q with each preceding letter, obtaining $aqbp$, $qabp$. There is one less step in the second part of the process than in the first part, because when it is begun q has already been shifted one place in the required direction. Had there been μ letters between p and q, there would have been $\mu + 1$ steps in the first part of the process and μ in the second, and $(\mu + 1) + \mu$, or $2\mu + 1$, is always odd.

Therefore, since each interchange of adjacent objects changes the number of inversions by 1 or -1, and the sum of an odd number of numbers each of which is 1 or -1 is odd, our theorem is demonstrated.

Thus, if in 21457368 (1) we interchange 4 and 6, we get 21657348 (2). It will be found that (1) has *five* inversions and (2) eight, and $8 - 5$ is odd.

Of the $n!$ permutations of n objects taken all at a time, § 763, half are odd and half are even. For from any one of **888** these permutations we can derive all the rest by repeated interchanges of two objects. As thus obtained, the permutations will be alternately odd and even, or *vice versa*, § 887. Therefore, since $n!$ is an even number, half of the permutations are odd and half are even.

889 In what follows we shall have to do with sets of letters with subscripts, as $a_1, a_2, \cdots, b_1, b_2, \cdots$, and so on. Having chosen any set of such symbols in which all the letters and subscripts are different, arrange them in some particular order and then find the sum of the number of inversions of the letters and of inversions of the subscripts. If this sum is even, it will be even when the symbols are arranged in any other order; if odd, odd. For when any two of the symbols are interchanged the inversions of both letters and subscripts are changed by odd numbers, § 887, and therefore their sum by an even number.

In particular, the number of inversions of the subscripts when the letters are in normal order and that of the letters when the subscripts are in normal order are both odd or both even.

Thus, in $a_2b_3c_1$ the number of inversions of the subscripts is *two;* in $c_1a_2b_3$ the number of inversions of the letters is *two;* in $b_3a_2c_1$ the sum of the number of inversions of the letters and that of the subscripts is *four.*

890 **Definition of determinant.** We may arrange any set of 2^2, 3^2, in general n^2 numbers in the form of a *square array*, thus:

$$\begin{matrix} a_1 & a_2 \\ b_1 & b_2 \end{matrix} \qquad \begin{matrix} a_1 & a_2 & a_3 \\ b_1 & b_2 & b_3 \\ c_1 & c_2 & c_3 \end{matrix} \qquad \begin{matrix} a_1 & a_2 & a_3 & a_4 \\ b_1 & b_2 & b_3 & b_4 \\ c_1 & c_2 & c_3 & c_4 \\ d_1 & d_2 & d_3 & d_4 \end{matrix}$$

and so on, where the letter indicates the *row*, and the subscript the *column*, in which any particular number occurs.

We may then call the numbers the *elements* of the array.

With the elements of such an array form all the products that can be formed by taking as factors one element and but one from each row and each column of the array.

In each product arrange the factors so that the row marks (letters) are in normal order, and then count the inversions of the column marks (subscripts). If their number be even (or 0), give the product the plus sign; if odd, the minus sign.

The algebraic sum of all these plus and minus products is called the determinant of the array, and is represented by the array itself with bars written at either side of it.

Thus, $$\begin{vmatrix} a_1 & a_2 \\ b_1 & b_2 \end{vmatrix} = a_1b_2 - a_2b_1$$

and $$\begin{vmatrix} a_1 & a_2 & a_3 \\ b_1 & b_2 & b_3 \\ c_1 & c_2 & c_3 \end{vmatrix} = a_1b_2c_3 - a_1b_3c_2 + a_2b_3c_1 - a_2b_1c_3 + a_3b_1c_2 - a_3b_2c_1.$$

891 It follows from § 889 that in determining the sign of any of the products above described we may arrange the factors so that the *column marks* are in normal order, and then give the product the plus or minus sign according as the number of inversions of the *row marks* is even or odd.

Or we may write the factors in any order whatsoever, and then give the product the plus or minus sign according as the *sum* of the number of inversions of row marks and of column marks is even or odd.

Thus, consider the product $a_3b_2c_1$ to which our first rule gave the minus sign. Writing it so that the subscripts are in normal order, we have $c_1b_2a_3$, and since the letters $c\,b\,a$ now present three inversions, the product must again, by our second rule, be given the minus sign. If the product be written $b_2c_1a_3$, the letters present two inversions and the subscripts one, and $2 + 1$ is odd. Hence again, by the third rule, we must give the product the minus sign.

892 The number of the rows or columns in the array out of which a determinant is formed is called the *order* of the determinant.

893 The products above described, with their proper signs, are called the *terms* of the determinant.

894 To *expand* a determinant is to write out its terms at length.

895 The diagonal of elements $a_1, b_2, c_3, \cdots$ is called the *leading diagonal*, and the product $a_1b_2c_3, \cdots$ is called the *leading term* of the determinant.

The leading term enclosed by bars, thus, $|a_1 \ b_2 \ c_3 \cdots|$, is often used as a symbol for the determinant itself.

896 *The number of the terms of a determinant of the* n*th order is* n!, *and half of these terms have plus signs and half have minus signs.*

For, keeping the letters in normal order, we may form $n!$ permutations of the n subscripts, § 763, and there is one term of the determinant for each of these $n!$ permutations, § 890.

Furthermore half of these $n!$ permutations are even and half are odd, § 888.

Thus, for $n = 3$ we have $3!$ or 6 terms; for $n = 4$, we have $4!$ or 24.

897 **Other notations.** It must be remembered that the letters and subscripts are mere marks of row and column order. Any other symbols which will serve this purpose may be substituted for them.

$$\begin{vmatrix} a_{11} & a_{12} & a_{13} \\ a_{21} & a_{22} & a_{23} \\ a_{31} & a_{32} & a_{33} \end{vmatrix}$$

Thus, the elements of a determinant are often represented by a single letter with two subscripts, as a_{23}, the first indicating the row and the second the column. The symbol a_{23} is read "a two three," and so on.

898 **A rule for expanding a determinant of the third order.** To obtain the three positive terms, start at each element of the first row in turn and, so far as possible, follow the direction of the leading diagonal, thus:

$$\begin{vmatrix} a_1 & a_2 & a_3 \\ b_1 & b_2 & b_3 \\ c_1 & c_2 & c_3 \end{vmatrix}$$

$$a_1b_2c_3, \quad a_2b_3c_1, \quad a_3b_1c_2.$$

To obtain the three negative terms, proceed in a similar manner but follow the direction of the other diagonal, thus:

$$-a_1b_3c_2, \quad -a_2b_1c_3, \quad -a_3b_2c_1.$$

Thus, $\begin{vmatrix} 5 & 3 & 2 \\ -1 & -1 & -3 \\ 2 & 4 & -1 \end{vmatrix} = 5(-1)(-1) + 3(-3)2 + 2(-1)4$
$\qquad -5(-3)4 - 3(-1)(-1) - 2(-1)2 = 40.$

This rule does not apply to determinants of an order higher than the third. Thus, it would give but eight of the twenty-four terms of a determinant of the fourth order.

EXERCISE LXXXI

Expand the following determinants.

1. $\begin{vmatrix} p & q & r \\ q & p & s \\ r & s & p \end{vmatrix}$.

2. $\begin{vmatrix} 1 & x & a \\ 1 & y & b \\ 1 & z & c \end{vmatrix}$.

3. $\begin{vmatrix} p & -q & r \\ q & p & -s \\ -r & s & p \end{vmatrix}$.

4. $\begin{vmatrix} 0 & -q & -r \\ q & 0 & -s \\ r & s & 0 \end{vmatrix}$.

Find the values of the following determinants.

5. $\begin{vmatrix} 1 & 2 & 3 \\ 3 & 1 & 2 \\ 2 & 3 & 1 \end{vmatrix}$.

6. $\begin{vmatrix} 1 & -3 & 4 \\ 2 & 0 & -5 \\ 3 & -1 & 7 \end{vmatrix}$.

7. $\begin{vmatrix} 8 & 9 & 0 & 0 \\ 2 & 3 & 0 & 0 \\ 1 & 1 & 6 & 1 \\ 4 & 3 & 5 & 0 \end{vmatrix}$.

Prove the following relations by expanding the determinants.

8. $\begin{vmatrix} a_1 & a_2 & a_3 \\ b_1 & b_2 & b_3 \\ c_1 & c_2 & c_3 \end{vmatrix} = \begin{vmatrix} a_1 & b_1 & c_1 \\ a_2 & b_2 & c_2 \\ a_3 & b_3 & c_3 \end{vmatrix} = -\begin{vmatrix} b_1 & a_1 & c_1 \\ b_2 & a_2 & c_2 \\ b_3 & a_3 & c_3 \end{vmatrix}$.

9. $\begin{vmatrix} a_1 & a_2 & a_3 \\ b_1 & b_2 & b_3 \\ c_1 & c_2 & c_3 \end{vmatrix} = a_1\begin{vmatrix} b_2 & b_3 \\ c_2 & c_3 \end{vmatrix} - a_2\begin{vmatrix} b_1 & b_3 \\ c_1 & c_3 \end{vmatrix} + a_3\begin{vmatrix} b_1 & b_2 \\ c_1 & c_2 \end{vmatrix}$.

10. In the expansion of the determinant $|a_1\ b_2\ c_3\ d_4|$ collect all the terms which involve as factors (1) c_3d_4, (2) a_1d_4, (3) $a_2b_3d_1$, (4) a_1, (5) c_3.

11. Find the signs of the following terms in the expansion of the determinant $|a_1\ b_2\ c_3\ d_4\ e_5|$.

$a_2b_4c_3d_1e_5$. $\qquad a_4b_2c_1d_5e_3$. $\qquad a_5b_4c_3d_2e_1$.

$c_1d_2a_3e_4b_5$. $\qquad c_1b_2e_3a_4d_5$. $\qquad d_3a_2e_4b_1c_5$.

PROPERTIES OF DETERMINANTS

899 **Theorem 1.** *The value of a determinant is not changed if its rows are made columns, and its columns rows, without changing their relative order.*

Thus,
$$\begin{vmatrix} a_1 & a_2 & a_3 \\ b_1 & b_2 & b_3 \\ c_1 & c_2 & c_3 \end{vmatrix} (1) = \begin{vmatrix} a_1 & b_1 & c_1 \\ a_2 & b_2 & c_2 \\ a_3 & b_3 & c_3 \end{vmatrix} (2).$$

For every term, as $a_2b_3c_1$, in the expansion of the determinant marked (1) contains one element and but one from each row and column of (1). Hence it contains one element and but one from each column and row of (2) and is therefore, except perhaps for sign, a term of (2). Moreover the sign of the term in (1) is the same as its sign in (2). For if the factors of the term be arranged in the order of the letters a, b, c, the inversions of the subscripts will determine this sign in both (1) and (2), in (1) by the rule of § 890, in (2) by the rule of § 891.

Conversely, every term in the expansion of (2) is a term in the expansion of (1).

900 Hence for every theorem respecting the rows of a determinant there is a corresponding theorem respecting its columns.

901 **Theorem 2.** *If all the elements of a row or column of a determinant are* 0, *the value of the determinant is* 0.

For every term of the determinant will contain a factor from the row or column in question, § 890, and will therefore vanish.

902 **Theorem 3.** *If two rows (or columns) of a determinant be interchanged, the determinant merely changes its sign.*

Thus,
$$\begin{vmatrix} a_1 & a_2 & a_3 \\ b_1 & b_2 & b_3 \\ c_1 & c_2 & c_3 \end{vmatrix} (1) = - \begin{vmatrix} c_1 & c_2 & c_3 \\ b_1 & b_2 & b_3 \\ a_1 & a_2 & a_3 \end{vmatrix} (2).$$

For any term of (1) with factors arranged in the order of the rows of (1) may be transformed into a term of (2) with factors arranged in the order of the rows of (2) by interchanging its first and last factors; and conversely. But this interchange will increase or diminish the inversions of the subscripts in the term by an odd number, § 887, and therefore, since the normal order of the subscripts is 123 in both (1) and (2), it will change the sign of the term.

Thus, $a_2b_3c_1$ is a term of (1) and $-c_1b_3a_2$ is the corresponding term of (2). For in $a_2b_3c_1$ the subscripts present two inversions, while in $c_1b_3a_2$ they present one inversion.

Example. Verify the preceding theorem by expanding each of the determinants (1) and (2).

Corollary. *If two of the rows (or columns) of a determinant are identical, the determinant vanishes.* **903**

For let D denote the value of the determinant. An interchange of the identical rows must leave D unchanged; but, by § 902, it will change D into $-D$.

Therefore $D = -D$, that is, $2D = 0$, or $D = 0$.

Thus, $$\begin{vmatrix} a & a & d \\ b & b & e \\ c & c & f \end{vmatrix} = abf + aec + dbc - aec - abf - dbc = 0.$$

Theorem 4. *If all the elements of a row (or column) are multiplied by the same number, as* k, *the determinant is multiplied by* k. **904**

For of the elements thus multiplied by k one and but one occurs as a factor in each term of the determinant, § 890.

The evaluation of a determinant may often be simplified by applying this theorem.

Thus,

$$\begin{vmatrix} -6 & 8 & 2 \\ 15 & -20 & 5 \\ 3 & 4 & -1 \end{vmatrix} = 2 \cdot 5 \cdot \begin{vmatrix} -3 & 4 & 1 \\ 3 & -4 & 1 \\ 3 & 4 & -1 \end{vmatrix} = 2 \cdot 5 \cdot 3 \cdot 4 \cdot \begin{vmatrix} -1 & 1 & 1 \\ 1 & -1 & 1 \\ 1 & 1 & -1 \end{vmatrix} = 480.$$

905 **Corollary.** *If the corresponding elements of two columns (or rows) are proportional, the determinant vanishes.*

Thus, $$\begin{vmatrix} ra & a & d \\ rb & b & e \\ rc & c & f \end{vmatrix} = r \begin{vmatrix} a & a & d \\ b & b & e \\ c & c & f \end{vmatrix} = r \cdot 0 = 0.$$ §§ 903, 904

906 **Theorem 5.** *If one of the columns (or rows) has binomial elements, the determinant may be expressed as a sum of two determinants in the manner illustrated below.*

$$\begin{vmatrix} a_1 + a' & a_2 & a_3 \\ b_1 + b' & b_2 & b_3 \\ c_1 + c' & c_2 & c_3 \end{vmatrix}_{(1)} = \begin{vmatrix} a_1 & a_2 & a_3 \\ b_1 & b_2 & b_3 \\ c_1 & c_2 & c_3 \end{vmatrix}_{(2)} + \begin{vmatrix} a' & a_2 & a_3 \\ b' & b_2 & b_3 \\ c' & c_2 & c_3 \end{vmatrix}_{(3)}.$$

For each term of (1) is the sum of the corresponding terms of (2) and (3). Thus, $(a_1 + a')b_2c_3 = a_1b_2c_3 + a'b_2c_3$.

Observe that any of the numbers a', b', c' may be 0.

By repeated applications of this theorem any determinant with polynomial elements may be resolved into a sum of determinants with simple elements.

Example. Express the determinant $\begin{vmatrix} a_1 + a'_1 & a_2 + a'_2 & a_3 + a'_3 \\ b_1 + b'_1 & b_2 + b'_2 & b_3 + b'_3 \\ c_1 + c'_1 & c_2 + c'_2 & c_3 + c'_3 \end{vmatrix}$ as a sum of eight determinants.

907 **Theorem 6.** *The value of a determinant will not be changed if to the elements of any column (or row) there be added the corresponding elements of any other column (or row) all multiplied by the same number, as* k.

Thus, by the theorems of §§ 905, 906, we have

$$\begin{vmatrix} a_1 + ka_3 & a_2 & a_3 \\ b_1 + kb_3 & b_2 & b_3 \\ c_1 + kc_3 & c_2 & c_3 \end{vmatrix} = \begin{vmatrix} a_1 & a_2 & a_3 \\ b_1 & b_2 & b_3 \\ c_1 & c_2 & c_3 \end{vmatrix} + \begin{vmatrix} ka_3 & a_2 & a_3 \\ kb_3 & b_2 & b_3 \\ kc_3 & c_2 & c_3 \end{vmatrix} = \begin{vmatrix} a_1 & a_2 & a_3 \\ b_1 & b_2 & b_3 \\ c_1 & c_2 & c_3 \end{vmatrix}.$$

And similarly in any other case.

It follows from this theorem that a determinant vanishes if one of its rows may be obtained by adding multiples of any of its remaining rows.

Thus, $\begin{vmatrix} 4 & 7 & 7 \\ 5 & -4 & 2 \\ -2 & 5 & 1 \end{vmatrix} = 0$ since $4, 7, 7 = 2(5, -4, 2) + 3(-2, 5, 1)$.

Theorem 7. *If a determinant whose elements are rational integral functions of some variable, as* x, *vanishes when* x = a, *the determinant is divisible by* x − a. **908**

For the determinant when expanded may be reduced to the form of a polynomial in x. And since this polynomial vanishes when $x = a$ it is divisible by $x - a$, § 415.

The factors of a determinant may often be found by aid of this theorem.

Example. Show that $\begin{vmatrix} 1 & 1 & 1 \\ a & b & c \\ a^2 & b^2 & c^2 \end{vmatrix} = (a-b)(b-c)(c-a)$.

By § 903, this determinant will vanish if $a = b$, if $b = c$, or if $c = a$. Hence it is divisible by $a - b$, $b - c$, and $c - a$, § 416, and therefore by the product $(a-b)(b-c)(c-a)$. But this product and the determinant itself are of the same degree, *three*, with respect to a, b, c. Hence the two will differ by a numerical factor at most.

One term of the determinant is bc^2 and the corresponding term of $(a-b)(b-c)(c-a)$ is bc^2. Hence the numerical factor in question is 1, and the determinant is equal to $(a-b)(b-c)(c-a)$.

EXERCISE LXXXII

1. Evaluate the following determinants.

(1) $\begin{vmatrix} 6 & 42 & 27 \\ 8 & -28 & 36 \\ 20 & 35 & 135 \end{vmatrix}$. (2) $\begin{vmatrix} 10 & 8 & 2 \\ 15 & 12 & 3 \\ 20 & 32 & 12 \end{vmatrix}$. (3) $\begin{vmatrix} -ab & ac & ae \\ bd & -cd & de \\ bf & cf & -ef \end{vmatrix}$.

2. Prove that $\begin{vmatrix} a_1 + ka_2 + la_3 & a_2 + ma_3 & a_3 \\ b_1 + kb_2 + lb_3 & b_2 + mb_3 & b_3 \\ c_1 + kc_2 + lc_3 & c_2 + mc_3 & c_3 \end{vmatrix} = \begin{vmatrix} a_1 & a_2 & a_3 \\ b_1 & b_2 & b_3 \\ c_1 & c_2 & c_3 \end{vmatrix}$.

3. By aid of the theorem of § 907 prove that the value of each of the following determinants is 0.

(1) $\begin{vmatrix} c & a & d & b \\ a & c & d & b \\ a & c & b & d \\ c & a & b & d \end{vmatrix}$. (2) $\begin{vmatrix} 1 & p & q & r+s \\ 1 & q & r & p+s \\ 1 & r & s & p+q \\ 1 & s & p & q+r \end{vmatrix}$. (3) $\begin{vmatrix} 2 & 1 & 4 & -1 \\ 3 & -1 & 2 & -1 \\ 1 & 2 & 3 & -2 \\ 5 & 0 & 6 & -2 \end{vmatrix}$.

4. Prove that $\begin{vmatrix} 1 & p & p^3 \\ 1 & q & q^3 \\ 1 & r & r^3 \end{vmatrix} = (p-q)(q-r)(r-p)(p+q+r).$

5. Prove that $\begin{vmatrix} (b+c)^2 & ab & ac \\ ab & (c+a)^2 & bc \\ ac & bc & (a+b)^2 \end{vmatrix} = 2\,abc\,(a+b+c)^3.$

MINORS. MULTIPLICATION OF DETERMINANTS

909 **Minors.** In any determinant Δ suppress the row and column in which some particular element e lies and then form the determinant of the remaining elements without disturbing their relative positions. This new determinant is called the *complementary minor* of the element e in question and may be denoted by Δ_e.

Thus, in $\Delta = \begin{vmatrix} a_1 & a_2 & a_3 \\ b_1 & b_2 & b_3 \\ c_1 & c_2 & c_3 \end{vmatrix}$ the minor of c_1 is $\Delta_{c_1} = \begin{vmatrix} a_2 & a_3 \\ b_2 & b_3 \end{vmatrix}$.

910 **Theorem.** *In the expansion of any determinant* Δ *the sum of all the terms which involve the leading element* a_1 *is* $a_1\Delta_{a_1}$.

Thus, in $\begin{vmatrix} a_1 & a_2 & a_3 & a_4 \\ b_1 & b_2 & b_3 & b_4 \\ c_1 & c_2 & c_3 & c_4 \\ d_1 & d_2 & d_3 & d_4 \end{vmatrix}$ (1) this sum is $a_1 \begin{vmatrix} b_2 & b_3 & b_4 \\ c_2 & c_3 & c_4 \\ d_2 & d_3 & d_4 \end{vmatrix}$. (2)

For, apart from sign, each term of Δ which involves a_1 is formed by multiplying a_1 by elements chosen from the remaining rows and columns of Δ, one from each, in other words, by multiplying a_1 by a term of Δ_{a_1}. Moreover the sign of the Δ term is that of this Δ_{a_1} term, since writing a_1 before the latter term will not affect the inversions of the subscripts. Thus, the term $-a_1b_4c_3d_2$ of (1) may be formed by multiplying a_1 by the term $-b_4c_3d_2$ of (2). Conversely, the product of a_1 by each term of Δ_{a_1} is a term of Δ.

Corollary. *If* e *denotes the element in the* i*th row and* k*th column of* Δ, *the sum of all the terms of* Δ *which involve* e *is* $(-1)^{i+k}e\Delta_e$. **911**

For we can bring e to the position of leading element without disturbing the relative positions of the elements which lie outside of the row and column in which e stands, namely, by first interchanging the row in which e stands with each preceding row in turn and then interchanging the column in which e stands with each preceding column. In carrying out these successive interchanges of rows and columns we merely change the sign of the determinant $(i-1)+(k-1)$ or $i+k-2$ times, § 902. Hence, if Δ' denote the determinant in its final form, we shall have

$$\Delta' = (-1)^{i+k-2}\Delta = (-1)^{i+k}\Delta.$$

By § 910, the sum of all the terms in Δ' which involve e is $e\Delta'_e$. Hence in Δ this sum is $(-1)^{i+k}e\Delta_e$. For the minor of e in Δ is the same as its minor in Δ'.

Thus, in the case of $\Delta = |a_1\ b_2\ c_3\ d_4|$ the element d_3, for which $i = 4$, $k = 3$, may be brought to the leading position as follows:

$$\begin{vmatrix} a_1 & a_2 & a_3 & a_4 \\ b_1 & b_2 & b_3 & b_4 \\ c_1 & c_2 & c_3 & c_4 \\ d_1 & d_2 & d_3 & d_4 \end{vmatrix} (1) = - \begin{vmatrix} d_1 & d_2 & d_3 & d_4 \\ a_1 & a_2 & a_3 & a_4 \\ b_1 & b_2 & b_3 & b_4 \\ c_1 & c_2 & c_3 & c_4 \end{vmatrix} (2) = - \begin{vmatrix} d_3 & d_1 & d_2 & d_4 \\ a_3 & a_1 & a_2 & a_4 \\ b_3 & b_1 & b_2 & b_4 \\ c_3 & c_1 & c_2 & c_4 \end{vmatrix} (3).$$

By interchanging the fourth row of (1) with the third, second, and first in turn, we obtain (2) before which we place the minus sign because of the *three*, that is, $i - 1$, interchanges of rows.

Then by interchanging the third column of (2) with the second and first in turn, we obtain (3) before which we place the same sign as that before (2) because of the *two*, that is, $k - 1$, interchanges of columns.

The minor of d_3 in (1) is the same as its minor in (3). Hence the sum of all the terms of (1) which involve d_3 is $-d_3 \cdot |a_1\ b_2\ c_4|$.

Theorem. *A determinant may be expressed as the sum of the products of the elements of one of its rows or columns by their* **912**

complementary minors, with signs which are alternately plus and minus, or minus and plus.

Thus, in the case of a determinant of the fourth order $\Delta = |a_1 \; b_2 \; c_3 \; d_4|$ we have

$$\Delta = a_1\Delta_{a_1} - a_2\Delta_{a_2} + a_3\Delta_{a_3} - a_4\Delta_{a_4}.$$

For each term in the expansion of Δ contains one and but one of the elements a_1, a_2, a_3, a_4. And, by §§ 910, 911, the sum of all the terms which involve a_1 is $a_1\Delta_{a_1}$, the sum of all which involve a_2 is $-a_2\Delta_{a_2}$, and so on.

In like manner,

$$\Delta = -b_1\Delta_{b_1} + b_2\Delta_{b_2} - b_3\Delta_{b_3} + b_4\Delta_{b_4}$$
$$= a_1\Delta_{a_1} - b_1\Delta_{b_1} + c_1\Delta_{c_1} - d_1\Delta_{d_1}, \text{ and so on.}$$

913 **Cofactors.** It is sometimes more convenient to write the preceding expressions for Δ in the form

$$\Delta = a_1A_1 + a_2A_2 + a_3A_3 + a_4A_4$$
$$= b_1B_1 + b_2B_2 + b_3B_3 + b_4B_4,$$

and so on, where $A_1 = \Delta_{a_1}$, $A_2 = -\Delta_{a_2}$, and so on. We then call $A_1, A_2, \cdots$ the *cofactors* of $a_1, a_2, \cdots$.

Thus, in

$$\begin{vmatrix} a_1 & a_2 & a_3 \\ b_1 & b_2 & b_3 \\ c_1 & c_2 & c_3 \end{vmatrix} \text{ the cofactors of } a_1, a_2, a_3 \text{ are } \begin{vmatrix} b_2 & b_3 \\ c_2 & c_3 \end{vmatrix}, \; -\begin{vmatrix} b_1 & b_3 \\ c_1 & c_3 \end{vmatrix}, \; \begin{vmatrix} b_1 & b_2 \\ c_1 & c_2 \end{vmatrix}.$$

914 *Any sum like* $b_1A_1 + b_2A_2 + b_3A_3 + b_4A_4$, *obtained by adding the products of the elements of one row by the cofactors of the corresponding elements of another row, is zero.*

For $b_1A_1 + b_2A_2 + b_3A_3 + b_4A_4$ denotes a determinant whose last three rows are the same as those of $\Delta = |a_1 \; b_2 \; c_3 \; d_4|$ but whose first row is b_1, b_2, b_3, b_4. And since both the first and second rows of this determinant are b_1, b_2, b_3, b_4, it vanishes, § 903. And so in general.

Bordering a determinant. Any determinant may be expressed as a determinant of a higher order. For, by § 912, we have **915**

$$\begin{vmatrix} a_1 & a_2 & a_3 \\ b_1 & b_2 & b_3 \\ c_1 & c_2 & c_3 \end{vmatrix} = \begin{vmatrix} 1 & 0 & 0 & 0 \\ 0 & a_1 & a_2 & a_3 \\ 0 & b_1 & b_2 & b_3 \\ 0 & c_1 & c_2 & c_3 \end{vmatrix}, \text{ and so on.}$$

Evaluation of a determinant. The value of a numerical determinant of any order may be obtained by aid of the theorem of § 912 and the rule of § 898 **916**

$$\text{Thus, } \begin{vmatrix} 2 & 3 & 1 & -1 \\ 2 & 0 & 0 & 3 \\ 4 & 1 & 0 & 1 \\ -1 & 2 & -2 & 1 \end{vmatrix} = - \begin{vmatrix} 2 & 0 & 0 & 3 \\ 2 & 3 & 1 & -1 \\ 4 & 1 & 0 & 1 \\ -1 & 2 & -2 & 1 \end{vmatrix}$$

$$= -2 \begin{vmatrix} 3 & 1 & -1 \\ 1 & 0 & 1 \\ 2 & -2 & 1 \end{vmatrix} + 3 \begin{vmatrix} 2 & 3 & 1 \\ 4 & 1 & 0 \\ -1 & 2 & -2 \end{vmatrix}$$

$$= -18 + 87 = 69.$$

But in most cases the value of a numerical determinant may be found with less reckoning by the following method. **917**

It follows from §§ 904, 907, 910 that

$$\begin{vmatrix} a_1 & a_2 & a_3 & \cdots \\ b_1 & b_2 & b_3 & \cdots \\ c_1 & c_2 & c_3 & \cdots \\ \cdot & \cdot & \cdot & \cdots \end{vmatrix} \text{(1)} = \frac{1}{a_1^{n-1}} \begin{vmatrix} a_1 & a_1a_2 & a_1a_3 & \cdots \\ b_1 & a_1b_2 & a_1b_3 & \cdots \\ c_1 & a_1c_2 & a_1c_3 & \cdots \\ \cdot & \cdot & \cdot & \cdots \end{vmatrix} \text{(2)}$$

$$= \frac{1}{a_1^{n-2}} \begin{vmatrix} a_1b_2 - a_2b_1 & a_1b_3 - a_3b_1 & \cdots \\ a_1c_2 - a_2c_1 & a_1c_3 - a_3c_1 & \cdots \\ \cdot & \cdot & \cdots \end{vmatrix}. \text{(3)}$$

Here (1) denotes a determinant of the nth order. By multiplying each column of (1) after the first by a_1 we obtain (2). From the second column of (2) we subtract the first multiplied by a_2; from the third column we subtract the first multiplied by a_3; and so on. The first row of the resulting determinant is $a_1, 0, 0, \cdots$. Hence this determinant is equal to a_1 times its minor, which is the determinant of the $(n-1)$th order (3).

Observe that each element of (3) is obtained from the corresponding element of the minor of a_1 in (1) by multiplying that element by a_1 and from the result subtracting the product of the corresponding elements in the first row and the first column of (1).

Thus, by two reductions of the kind just described, we have

$$\begin{vmatrix} 2 & 2 & -1 & 3 \\ -2 & 1 & 3 & -2 \\ 2 & -1 & 2 & 1 \\ 3 & -2 & -2 & 1 \end{vmatrix} = \frac{1}{2^2}\begin{vmatrix} 6 & 4 & 2 \\ -6 & 6 & -4 \\ -10 & -1 & -7 \end{vmatrix} = \begin{vmatrix} 3 & 2 & 1 \\ 3 & -3 & 2 \\ 10 & 1 & 7 \end{vmatrix}$$

$$= \frac{1}{3}\begin{vmatrix} -15 & 3 \\ -17 & 11 \end{vmatrix} = -38,$$

since $\quad 2\cdot 1 - 2(-2) = 6,\ 2\cdot 3 - (-1)(-2) = 4$, and so on.

When the leading element is 0 we begin by bringing another element into the leading position by the method explained in § 911.

Example. Evaluate each of the following determinants by both of the methods just described.

$$\begin{vmatrix} 1 & -2 & 3 & 4 \\ 2 & 3 & -4 & 2 \\ 1 & -1 & 2 & 3 \\ 0 & 2 & -3 & 1 \end{vmatrix}, \qquad \begin{vmatrix} 1 & -2 & -1 & 3 \\ 2 & 4 & 0 & -1 \\ 3 & 5 & 2 & 3 \\ -1 & 2 & 1 & -3 \end{vmatrix}.$$

918 Multiplication of determinants. The product of two determinants of the same order, Δ and Δ', may be expressed in the form of a third determinant Δ'' obtained as follows:

Multiply the elements of the ith *row of* Δ *by the corresponding elements of the* kth row of Δ'. *The sum of the products thus obtained is the element in the* ith *row and* kth *column of* Δ''.

Thus, $$\begin{vmatrix} a_1 & a_2 \\ b_1 & b_2 \end{vmatrix} \cdot \begin{vmatrix} p_1 & p_2 \\ q_1 & q_2 \end{vmatrix} = \begin{vmatrix} a_1p_1 + a_2p_2 & a_1q_1 + a_2q_2 \\ b_1p_1 + b_2p_2 & b_1q_1 + b_2q_2 \end{vmatrix}.$$

For, by § 906, the third determinant is the sum of

$$\begin{vmatrix} a_1p_1 & a_1q_1 \\ b_1p_1 & b_1q_1 \end{vmatrix}, \ (1) \qquad \begin{vmatrix} a_1p_1 & a_2q_2 \\ b_1p_1 & b_2q_2 \end{vmatrix}, \ (2) \qquad \begin{vmatrix} a_2p_2 & a_1q_1 \\ b_2p_2 & b_1q_1 \end{vmatrix}, \ (3) \qquad \begin{vmatrix} a_2p_2 & a_2q_2 \\ b_2p_2 & b_2q_2 \end{vmatrix}. \ (4)$$

But (1) and (4) vanish, since their columns are proportional, § 905. And simplifying (2) and (3) by aid of §§ 902, 904, and adding them, we have

$$p_1q_2\begin{vmatrix} a_1 & a_2 \\ b_1 & b_2 \end{vmatrix} + p_2q_1\begin{vmatrix} a_2 & a_1 \\ b_2 & b_1 \end{vmatrix} = (p_1q_2 - p_2q_1)\begin{vmatrix} a_1 & a_2 \\ b_1 & b_2 \end{vmatrix}$$

$$= \begin{vmatrix} p_1 & p_2 \\ q_1 & q_2 \end{vmatrix} \cdot \begin{vmatrix} a_1 & a_2 \\ b_1 & b_2 \end{vmatrix}.$$

Again, $$\begin{vmatrix} a_1 & a_2 & a_3 \\ b_1 & b_2 & b_3 \\ c_1 & c_2 & c_3 \end{vmatrix} \cdot \begin{vmatrix} p_1 & p_2 & p_3 \\ q_1 & q_2 & q_3 \\ r_1 & r_2 & r_3 \end{vmatrix}$$

$$= \begin{vmatrix} a_1p_1 + a_2p_2 + a_3p_3 & a_1q_1 + a_2q_2 + a_3q_3 & a_1r_1 + a_2r_2 + a_3r_3 \\ b_1p_1 + b_2p_2 + b_3p_3 & b_1q_1 + b_2q_2 + b_3q_3 & b_1r_1 + b_2r_2 + b_3r_3 \\ c_1p_1 + c_2p_2 + c_3p_3 & c_1q_1 + c_2q_2 + c_3q_3 & c_1r_1 + c_2r_2 + c_3r_3 \end{vmatrix},$$

as may be shown by resolving the third determinant into a sum of determinants with simple columns in the manner just illustrated. There will be twenty-seven such determinants, but twenty-one of them have two or three proportional columns, and therefore vanish. Each of the remaining six is equal to the determinant $|a_1 \ b_2 \ c_3|$ multiplied by a term of $|p_1 \ q_2 \ r_3|$, so that their sum is $|a_1 \ b_2 \ c_3| \cdot |p_1 \ q_2 \ r_3|$.

This proof may easily be generalized.

The rule above given is readily extended to determinants of different orders. We have only to begin by making their orders the same by bordering the one of lower order, § 915.

EXERCISE LXXXIII

Evaluate the following determinants.

1. $\begin{vmatrix} 2 & -1 & 4 & 9 \\ 7 & 5 & -2 & -3 \\ -3 & 2 & 4 & -1 \\ 4 & 7 & 2 & -4 \end{vmatrix}$.

2. $\begin{vmatrix} 1 & 2 & 3 & -1 & -2 \\ 2 & 1 & 3 & -2 & 1 \\ 0 & 1 & 2 & -2 & 1 \\ 0 & -1 & -1 & 2 & 1 \\ 0 & 2 & 3 & 1 & -1 \end{vmatrix}$.

3. $\begin{vmatrix} 3 & 2 & 5 & 10 \\ 6 & 0 & 4 & 0 \\ 9 & 6 & 1 & 30 \\ 12 & 4 & 8 & 20 \end{vmatrix}$. 4. $\begin{vmatrix} 6 & -4 & 10 & 28 \\ 18 & 6 & -30 & 21 \\ 12 & 24 & 40 & 28 \\ 9 & -2 & 20 & 14 \end{vmatrix}$.

Express each of the following products as a determinant.

5. $\begin{vmatrix} a & b & c \\ b & c & a \\ c & a & b \end{vmatrix} \cdot \begin{vmatrix} b & -a & 0 \\ -a & 0 & b \\ 0 & b & -a \end{vmatrix}$. 6. $\begin{vmatrix} p & 0 & r \\ p & q & 0 \\ 0 & q & r \end{vmatrix} \cdot \begin{vmatrix} a & 0 & c \\ a & b & 0 \\ 0 & b & c \end{vmatrix}$.

7. $\begin{vmatrix} a & -a & a & a \\ -b & b & b & b \\ c & c & -c & c \\ d & d & d & -d \end{vmatrix} \cdot \begin{vmatrix} a & b \\ c & d \end{vmatrix}$. 8. $\begin{vmatrix} l & m & n \\ m & n & l \\ n & l & m \end{vmatrix}^2$.

9. Prove that $\begin{vmatrix} a_1 & a_2 & a_3 \\ b_1 & b_2 & b_3 \\ c_1 & c_2 & c_3 \end{vmatrix} \cdot \begin{vmatrix} A_1 & A_2 & A_3 \\ B_1 & B_2 & B_3 \\ C_1 & C_2 & C_3 \end{vmatrix} = \begin{vmatrix} a_1 & a_2 & a_3 \\ b_1 & b_2 & b_3 \\ c_1 & c_2 & c_3 \end{vmatrix}^3$.

10. Prove that a determinant reduces to its leading term when all of the elements at either side of the leading diagonal are zero.

ELIMINATION. LINEAR EQUATIONS

919 **Solution of a system of linear equations.** We are to solve the following system of equations for x_1, x_2, x_3:

$$\left.\begin{aligned} a_1x_1 + a_2x_2 + a_3x_3 &= k \\ b_1x_1 + b_2x_2 + b_3x_3 &= l \\ c_1x_1 + c_2x_2 + c_3x_3 &= m \end{aligned}\right\}. \quad (1)$$

Let Δ denote $|a_1 \ b_2 \ c_3|$, the determinant of the coefficients of x_1, x_2, x_3 arranged as in (1), and, as in § 913, let A_1, A_2, $\cdots$ denote the cofactors of a_1, a_2, $\cdots$ in Δ.

To eliminate x_2 and x_3, we multiply the first equation by A_1, the second by B_1, the third by C_1, and add. We obtain

$$\begin{vmatrix} a_1A_1 \\ b_1B_1 \\ c_1C_1 \end{vmatrix} x_1 + \begin{vmatrix} a_2A_1 \\ +b_2B_1 \\ +c_2C_1 \end{vmatrix} x_2 + \begin{vmatrix} a_3A_1 \\ +b_3B_1 \\ +c_3C_1 \end{vmatrix} x_3 = \begin{vmatrix} kA_1 \\ +lB_1 \\ +mC_1 \end{vmatrix}.$$

But in this equation the coefficients of x_2 and x_3 are 0, § 914; the coefficient of x_1 is Δ, § 913; and the second member denotes a determinant found by replacing the first column of Δ by k, l, m. Hence the equation may be written

$$\begin{vmatrix} a_1 & a_2 & a_3 \\ b_1 & b_2 & b_3 \\ c_1 & c_2 & c_3 \end{vmatrix} x_1 = \begin{vmatrix} k & a_2 & a_3 \\ l & b_2 & b_3 \\ m & c_2 & c_3 \end{vmatrix}. \qquad (2)$$

We may in like manner derive an equation which involves x_2 alone by multiplying the given equations by A_2, B_2, C_2 respectively and then adding them, and an equation which involves x_3 alone by multiplying by A_3, B_3, C_3 and adding. These equations are

$$\begin{vmatrix} a_1 & a_2 & a_3 \\ b_1 & b_2 & b_3 \\ c_1 & c_2 & c_3 \end{vmatrix} x_2 = \begin{vmatrix} a_1 & k & a_3 \\ b_1 & l & b_3 \\ c_1 & m & c_3 \end{vmatrix} \quad (3), \qquad \begin{vmatrix} a_1 & a_2 & a_3 \\ b_1 & b_2 & b_3 \\ c_1 & c_2 & c_3 \end{vmatrix} x_3 = \begin{vmatrix} a_1 & a_2 & k \\ b_1 & b_2 & l \\ c_1 & c_2 & m \end{vmatrix} \quad (4).$$

Hence, if $\Delta \neq 0$, the required solution is

$$x_1 = \frac{|k \; b_2 \; c_3|}{|a_1 \; b_2 \; c_3|}, \quad x_2 = \frac{|a_1 \; l \; c_3|}{|a_1 \; b_2 \; c_3|}, \quad x_3 = \frac{|a_1 \; b_2 \; m|}{|a_1 \; b_2 \; c_3|},$$

that is, the value of each unknown letter x_1, x_2, x_3 may be expressed as a fraction whose denominator is Δ and whose numerator is a determinant differing from Δ only in this, that the coefficients of the unknown letter in question are replaced by the known terms.

It may be proved in the same way that the like is true of a system of n linear equations in n unknown letters.

Example. Solve

$$2x - 3y + z = 4,$$
$$x + y - z = 2,$$
$$4x - y + 3z = 1.$$

We have $$x = \begin{vmatrix} 4 & -3 & 1 \\ 2 & 1 & -1 \\ 1 & -1 & 3 \end{vmatrix} \div \begin{vmatrix} 2 & -3 & 1 \\ 1 & 1 & -1 \\ 4 & -1 & 3 \end{vmatrix} = \frac{26}{20} = \frac{13}{10}.$$

In like manner, we find $y = -21/20$, $z = -35/20 = -7/4$.

920 If Δ is 0 and any one of the determinants $|k \; b_2 \; c_3|$, $|a_1 \; l \; c_3|$, $|a_1 \; b_2 \; m|$ is not 0, it follows from (2), (3), (4) that the given equations (1) have no finite solution (compare § 394).

If Δ and all the determinants $|k\ b_2\ c_3|$, $|a_1\ l\ c_3|$, $|a_1\ b_2\ m|$ vanish, the equations (2), (3), (4) impose no restriction on the values of x_1, x_2, x_3. In this case the given equations (1) are not independent. This follows, by § 394, from the manner in which (2), (3), (4) were derived from (1), unless all the minors A_1, A_2, $\cdots$ vanish. And if all the minors vanish, it may readily be shown that the three equations (1) differ only by constant factors, so that every solution of one of them is a solution of the other two.

These results are readily generalized for a system of n equations in n unknown letters.

921 **Homogeneous linear equations.** When $k = l = m = 0$, the equations (1) of § 919 reduce to a system of homogeneous equations in x_1, x_2, x_3, namely,

$$\left.\begin{aligned} a_1x_1 + a_2x_2 + a_3x_3 &= 0 \\ b_1x_1 + b_2x_2 + b_3x_3 &= 0 \\ c_1x_1 + c_2x_2 + c_3x_3 &= 0 \end{aligned}\right\}, \tag{1}$$

and the equations (2), (3), (4) of § 919 become

$$\Delta x_1 = 0,\ \Delta x_2 = 0,\ \Delta x_3 = 0. \tag{2}$$

Evidently the equations (1) have the solution $x_1 = x_2 = x_3 = 0$, and it follows from (2) that this is the only solution unless $\Delta = 0$.

But if $\Delta = 0$, the equations (1) are satisfied by

$$x_1 = rA_1,\ x_2 = rA_2,\ x_3 = rA_3, \tag{3}$$

where r may denote any constant whatsoever.

For, substituting these values in (1) and simplifying, we have

$$a_1A_1 + a_2A_2 + a_3A_3 = 0,\ b_1A_1 + b_2A_2 + b_3A_3 = 0,$$
$$c_1A_1 + c_2A_2 + c_3A_3 = 0,$$

and these are true identities, the first one because $\Delta = 0$, the other two by § 914. The same thing may be proved as follows: If we solve the second and third of the equations (1) for x_1 and x_2 in terms of x_3, we obtain $x_1/A_1 = x_2/A_2 = x_3/A_3$, or, if

r denote the value of these equal ratios, $x_1 = rA_1$, $x_2 = rA_2$, $x_3 = rA_3$. And as just shown, if $\Delta = 0$, these values will also satisfy the first of the equations (1).

From this second proof it follows that when $\Delta = 0$

$$x_1 : x_2 : x_3 = A_1 : A_2 : A_3 = B_1 : B_2 : B_3 = C_1 : C_2 : C_3,$$

that is, the minors of corresponding elements in the rows of Δ are proportional. It is assumed that these minors are not 0.

922 From the system of three non-homogeneous equations in x, y

$$\left.\begin{array}{l} a_1x + a_2y + a_3 = 0 \\ b_1x + b_2y + b_3 = 0 \\ c_1x + c_2y + c_3 = 0 \end{array}\right\}, \qquad (1')$$

we may derive the homogeneous system (1) of § 921 by substituting $x = x_1/x_3$, $y = x_2/x_3$ and clearing of fractions.

Hence $\Delta = 0$ is the condition that the equations (1′) have a common solution.

EXERCISE LXXXIV

Solve the following systems of equations by determinants.

1. $\begin{cases} 2x + 3y - 5z = 3, \\ x - 2y + z = 0, \\ 3x + y + 3z = 7. \end{cases}$

2. $\begin{cases} 2x + 4y - 3z = 3, \\ 3x - 8y + 6z = 1, \\ 8x - 2y - 9z = 4. \end{cases}$

3. $\begin{cases} ax + by + cz = d, \\ a^2x + b^2y + c^2z = d^2, \\ a^3x + b^3y + c^3z = d^3. \end{cases}$

4. $\begin{cases} 2x - 4y + 3z + 4t = -3, \\ 3x - 2y + 6z + 5t = -1, \\ 5x + 8y + 9z + 3t = 9, \\ x - 10y - 3z - 7t = 2. \end{cases}$

Show that the following systems of equations are consistent, and solve them for the ratios $x : y : z$.

5. $\begin{cases} x + 2y - z = 0, \\ 3x - y + 4z = 0, \\ 4x + y + 3z = 0. \end{cases}$

6. $\begin{cases} a_1x + b_1y + (ka_1 + lb_1)z = 0, \\ a_2x + b_2y + (ka_2 + lb_2)z = 0, \\ a_3x + b_3y + (ka_3 + lb_3)z = 0. \end{cases}$

7. For what values of λ are the following equations consistent?

$$4x + 3y + z = \lambda x,$$
$$3x - 4y + 7z = \lambda y,$$
$$x + 7y - 6z = \lambda z.$$

RESULTANTS

923 **Resultants.** By the *resultant* of two equations $f(x) = 0$ and $\phi(x) = 0$ is meant that integral function of the coefficients of $f(x)$ and $\phi(x)$ whose vanishing is the necessary and sufficient condition that $f(x) = 0$ and $\phi(x) = 0$ have a common root.

Thus, the resultant of $a_0x^2 + a_1x + a_2 = 0$ (1) and $x - b = 0$ (2) is $a_0b^2 + a_1b + a_2$; for when $a_0b^2 + a_1b + a_2 = 0$, the equations (1) and (2) have the common root b.

924 The resultant of any two equations $f(x) = 0$, $\phi(x) = 0$ may be obtained by eliminating x by the following method due to Sylvester.

To fix the ideas, let

$$f(x) = a_0x^3 + a_1x^2 + a_2x + a_3 = 0, \qquad (1)$$

$$\phi(x) = b_0x^2 + b_1x + b_2 = 0. \qquad (2)$$

Multiply (1) by x and 1, and (2) by x^2, x, and 1 successively. We obtain

$$\begin{aligned}
a_0x^4 + a_1x^3 + a_2x^2 + a_3x &= 0,\\
a_0x^3 + a_1x^2 + a_2x + a_3 &= 0,\\
b_0x^4 + b_1x^3 + b_2x^2 &= 0,\\
b_0x^3 + b_1x^2 + b_2x &= 0,\\
b_0x^2 + b_1x + b_2 &= 0.
\end{aligned}$$

These may be regarded as a system of five homogeneous linear equations in the five quantities x^4, x^3, x^2, x, 1. Hence, § 921, they cannot have a common solution unless

$$D = \begin{vmatrix} a_0 & a_1 & a_2 & a_3 & 0 \\ 0 & a_0 & a_1 & a_2 & a_3 \\ b_0 & b_1 & b_2 & 0 & 0 \\ 0 & b_0 & b_1 & b_2 & 0 \\ 0 & 0 & b_0 & b_1 & b_2 \end{vmatrix} = 0. \qquad (3)$$

Hence (3) is the *necessary* condition that (1) and (2) have a common root. It is also the *sufficient* condition. For to the fifth column of D add the first four columns multiplied by x^4,

x^3, x^2, x respectively. We thus transform D into an equivalent determinant, § 907, whose fifth column has the elements $xf(x)$, $f(x), x^2\phi(x), x\phi(x), \phi(x)$. Hence, if $\mu_1, \mu_2, \mu_3, \mu_4, \mu_5$ denote the cofactors of the elements of the fifth column of D, we have, § 913,

$$D = (\mu_1 x + \mu_2) f(x) + (\mu_3 x^2 + \mu_4 x + \mu_5)\,\phi(x) \equiv 0.$$

It follows from this identity that each factor $x - \beta$ of $f(x)$ must be a factor of $(\mu_3 x^2 + \mu_4 x + \mu_5)\,\phi(x)$, and therefore, since $f(x)$ is of the third degree and $\mu_3 x^2 + \mu_4 x + \mu_5$ is of only the second degree, that at least one factor $x - \beta$ of $f(x)$ must be a factor of $\phi(x)$, in other words, that one of the roots of $f(x) = 0$ must be a root of $\phi(x) = 0$, § 795.

It is here assumed that the minors $\mu_1, \mu_2, \cdots, \mu_5$ are not all zero. If the minors of all the elements of D are 0, it can be proved that $f(x) = 0$ and $\phi(x) = 0$ have more than one common root.

925 If $\lambda_1, \lambda_2, \cdots, \lambda_5$ denote the cofactors of the elements of any row of D, it follows from § 921 that when $D = 0$

$$x^4 : x^3 : x^2 : x : 1 = \lambda_1 : \lambda_2 : \lambda_3 : \lambda_4 : \lambda_5,$$

whence $x = \lambda_1/\lambda_2 = \lambda_2/\lambda_3 = \cdots = \lambda_4/\lambda_5$. Therefore when $f(x) = 0$ and $\phi(x) = 0$ have a common root, the value of this root is λ_1/λ_2.

926 In the general case when the degrees of $f(x) = 0$ and $\phi(x) = 0$ are m and n respectively, the resultant D will be a determinant of the $(m + n)$th order whose first n rows consist of the coefficients of $f(x)$ and zeros and whose remaining m rows consist of the coefficients of $\phi(x)$ and zeros, arranged as in § 924, (3). Hence in the terms of D the coefficients of $f(x)$ enter in the degree of $\phi(x)$ and *vice versa*.

Example. By the method just explained show that the equations $x^2 + 3x + 2 = 0$ and $x + 1 = 0$ have a common root and find this root.

Here $D = \begin{vmatrix} 1 & 3 & 2 \\ 1 & 1 & 0 \\ 0 & 1 & 1 \end{vmatrix} = 1 + 2 - 3 = 0$, so that there is a common root.

The values of the cofactors of 1 and 3 in the first row of D are 1 and -1. Hence the common root is $1 : -1$, that is, -1.

927 By the preceding method either of the unknown letters x, y may be eliminated from a pair of algebraic equations of the form $f(x, y) = 0$, $\phi(x, y) = 0$.

Example. Solve

$$x^2 - 2y^2 - x = 0, \tag{1}$$

$$2x^2 - 5y^2 + 3y = 0. \tag{2}$$

We may regard (1) and (2) as quadratics in x, (1) with the coefficients $1, -1, -2y^2$, and (2) with the coefficients $2, 0, -5y^2 + 3y$.

Hence the result of eliminating x is

$$\begin{vmatrix} 1 & -1 & -2y^2 & 0 \\ 0 & 1 & -1 & -2y^2 \\ 2 & 0 & -5y^2+3y & 0 \\ 0 & 2 & 0 & -5y^2+3y \end{vmatrix} = 0, \tag{3}$$

which when expanded and simplified gives

$$y^4 - 6y^3 - y^2 + 6y = 0. \tag{4}$$

Solving (4), we obtain $y = 0, 1, -1, 6$.

It follows from § 924 that when y has any of these values, (1) and (2) are satisfied by the same value of x. And in fact when $y = 0$, (1) and (2) become $x^2 - x = 0$, $2x^2 = 0$, which have the common root $x = 0$; when $y = 1$, (1) and (2) become $x^2 - x - 2 = 0$, $x^2 - 1 = 0$, which have the common root $x = -1$ since $x^2 - x - 2$ and $x^2 - 1$ have the common factor $x + 1$, § 853; and so on. We thus find that the solutions of (1) and (2) are $x, y = 0, 0;\ -1, 1;\ 2, -1;\ 9, 6$. The values of x may also be found from those of y by applying § 925 (compare § 926, Ex.).

This example illustrates the fact that if the result of eliminating x from $f(x, y) = 0$ and $\phi(x, y) = 0$ is $R(y) = 0$, and one of the roots of $R(y) = 0$ is β, the corresponding value or values of x can always be obtained by finding the highest common factor of $f(x, \beta)$ and $\phi(x, \beta)$. Usually this highest common factor will be of the first degree, when but one value of x will correspond to $y = \beta$. But it may be of a higher degree, and then more than one value of x will correspond to $y = \beta$.

928 **Properties of the resultant.** Suppose a pair of equations to be given of the form

$$f(x) = x^m + \cdots + a_m = 0, \quad \phi(x) = x^n + \cdots + b_n = 0.$$

Let α_i denote any root of $f(x)=0$ and β_k any root of $\phi(x)=0$. There will be mn differences of the form $\alpha_i-\beta_k$; let $\Pi(\alpha_i-\beta_k)$ denote their product.

Evidently $\Pi(\alpha_i-\beta_k)=0$ is the necessary and sufficient condition that one of the roots α_i be equal to one of the roots β_k. Moreover, since $\Pi(\alpha_i-\beta_k)$ is a symmetric integral function of the roots α_i and of the roots β_k, it is a rational integral function of the coefficients of $f(x)=0$ and $\phi(x)=0$, §§ 867, 868. Hence, if $R(f, \phi)$ denote the resultant of $f(x)=0$ and $\phi(x)=0$, § 923, we have

$$R(f, \phi)=\Pi(\alpha_i-\beta_k).$$

The product $\Pi(\alpha_i-\beta_k)$ may be written

$$\begin{array}{c}(\alpha_1-\beta_1)(\alpha_1-\beta_2)\cdots(\alpha_1-\beta_n),\\ (\alpha_2-\beta_1)(\alpha_2-\beta_2)\cdots(\alpha_2-\beta_n),\\ \cdot\quad\cdot\quad\cdot\quad\cdot\quad\cdot\quad\cdot\quad\cdot\\ (\alpha_m-\beta_1)(\alpha_m-\beta_2)\cdots(\alpha_m-\beta_n).\end{array}$$

But since $\phi(x)=(x-\beta_1)(x-\beta_2)\cdots(x-\beta_n)$, the product of the factors in the first row is $\phi(\alpha_1)$, in the second row $\phi(\alpha_2)$, and so on. Hence

$$\Pi(\alpha_i-\beta_k)=\phi(\alpha_1)\cdot\phi(\alpha_2)\cdots\phi(\alpha_m).$$

Again, since $f(x)=(x-\alpha_1)(x-\alpha_2)\cdots(x-\alpha_m)$, the product of the factors in the first column is $(-1)^m f(\beta_1)$, in the second column $(-1)^m f(\beta_2)$, and so on. Hence

$$\Pi(\alpha_i-\beta_k)=(-1)^{mn}f(\beta_1)\cdot f(\beta_2)\cdots f(\beta_n).$$

When the given equations have the form

$$f(x)=a_0x^m+\cdots+a_m=0,\quad \phi(x)=b_0x^n+\cdots+b_n=0,$$

that is, when the leading coefficients are not 1, the product of the factors in the first row is $\phi(\alpha_1)/b_0$, and so on; and the product of the factors in the first column is $(-1)^m f(\beta_1)/a_0$, and so on. Hence, in this case, to make $\Pi(\alpha_i-\beta_k)$ an *integral* function of the coefficients of $f(x)=0$ and $\phi(x)=0$ we must multiply it by $a_0^n b_0^m$. We then have

$$\begin{aligned} R(f, \phi) &= a_0^n b_0^m \Pi(\alpha_i - \beta_k) \\ &= a_0^n \phi(\alpha_1) \cdot \phi(\alpha_2) \cdots \phi(\alpha_m) \\ &= (-1)^{mn} b_0^m f(\beta_1) \cdot f(\beta_2) \cdots f(\beta_n). \end{aligned}$$

929 *In the resultant of a pair of equations* $f(x)=0$, $\phi(x)=0$ *the coefficients of* $f(x)=0$ *enter in the degree of* $\phi(x)=0$, *and vice versa.*

For the product $\phi(\alpha_1)\phi(\alpha_2)\cdots\phi(\alpha_m)$ contains m factors, each involving the coefficients of $\phi(x)=0$ to the first degree; and the product $f(\beta_1)\cdot f(\beta_2)\cdots f(\beta_n)$ contains n factors, each involving the coefficients of $f(x)=0$ to the first degree.

We thus have another proof that the determinant D described in §§ 924, 926 is the resultant of $f(x)=0$ and $\phi(x)=0$, that is, that $D=R(f, \phi)$.

930 *The sum of the subscripts of the coefficients of* $f(x)=0$ *and* $\phi(x)=0$ *in each term of* $R(f, \phi)$ *is* mn.

For, by § 812, if we multiply each coefficient of $f(x)$ and $\phi(x)$ by the power of r indicated by its subscript, we obtain two equations

$$\begin{aligned} f_1(x) &= a_0x^m + ra_1x^{m-1} + r^2a_2x^{m-2} + \cdots + r^m a_m = 0, \\ \phi_1(x) &= b_0x^n + rb_1x^{n-1} + r^2b_2x^{n-2} + \cdots + r^n b_n = 0, \end{aligned}$$

whose roots are r times the roots of $f(x)=0$ and $\phi(x)=0$.

Each term of $R(f_1, \phi_1)$ will be equal to the corresponding term of $R(f, \phi)$ multiplied by a power of r whose exponent is the sum of the subscripts of the coefficients of $f(x)$ and $\phi(x)$ which occur in the term. Hence our theorem is demonstrated if we can show that in every term this exponent is mn. But since there are mn factors in the product $\Pi(\alpha_i - \beta_k)$, we have

$$R(f_1, \phi_1) = a_0^n b_0^m \Pi(r\alpha_i - r\beta_k) = r^{mn} \cdot R(f, \phi).$$

Discriminants. The discriminant of $f(x) = a_0x^n + \cdots + a_n = 0$ **931** is that integral function of the coefficients of $f(x)$ whose vanishing is the necessary and sufficient condition that $f(x) = 0$ have a multiple root (compare §§ 635, 873).

If D denote the discriminant of $f(x) = 0$, then

$$D = R(f, f')/a_0.$$

For, by § 851, $f(x) = 0$ has a finite multiple root when and only when $f(x) = 0$ and $f'(x) = 0$ have a finite root in common. By § 928, the condition that $f(x) = 0$ and $f'(x) = 0$ have a root in common is $R(f, f') = 0$. But a_0 is a factor of $R(f, f')$, as may be shown by expressing $R(f, f')$ in the determinant form of § 924. Hence $R(f, f') = 0$ when $a_0 = 0$. But the root which in this case is common to $f(x) = 0$ and $f'(x) = 0$ is ∞, § 816. Therefore, since ∞ is not a multiple root of $f(x) = 0$ unless both a_0 and a_1 vanish, we have $D = R(f, f')/a_0$.

Thus, for $a_0x^2 + a_1x + a_2 = 0$, $D = \begin{vmatrix} a_0 & a_1 & a_2 \\ 2a_0 & a_1 & 0 \\ 0 & 2a_0 & a_1 \end{vmatrix} \div a_0 = -(a_1^2 - 4a_0a_2)$.

The discriminant of f(x) = 0 *is equal to the product of the squares of the differences of every two of the roots of* f(x) = 0 *multiplied by a certain power of the leading coefficient* a_0. **932**

Thus, if $$f(x) = a_0(x - \beta_1)(x - \beta_2)(x - \beta_3), \quad (1)$$
then, § 865, $$f'(x) = a_0[(x - \beta_2)(x - \beta_3) + (x - \beta_3)(x - \beta_1) + (x - \beta_1)(x - \beta_2)]. \quad (2)$$
By § 928, $$R(f, f') = a_0^2 f'(\beta_1)f'(\beta_2)f'(\beta_3). \quad (3)$$
But from (2), $f'(\beta_1) = a_0(\beta_1 - \beta_2)(\beta_1 - \beta_3)$, and so on. (4)

Substituting (4) in (3) and simplifying, we have

$$R(f, f') = -a_0^5(\beta_1 - \beta_2)^2(\beta_2 - \beta_3)^2(\beta_3 - \beta_1)^2,$$
whence $$D = -a_0^4(\beta_1 - \beta_2)^2(\beta_2 - \beta_3)^2(\beta_3 - \beta_1)^2. \quad (5)$$

On the number of solutions of a pair of equations in two unknown letters. Observe that if in the equation $f(x, y) = 0$ we make the substitutions $x = x_1/x_3$, $y = x_2/x_3$, and clear of fractions, we transform $f(x, y) = 0$ into a *homogeneous* equation of the **933**

same degree in x_1, x_2, x_3. Thus, $x^3 + xy + y + 1 = 0$ becomes $x_1^3 + x_1x_2x_3 + x_2x_3^2 + x_3^3 = 0$.

Observe also that a homogeneous equation of the nth degree in x_2, x_3 which is not divisible by x_3 determines n finite values of the ratio x_2/x_3. Thus, from $x_2^2 - 3x_2x_3 + 2x_3^2 = 0$ we obtain $x_2/x_3 = 1$ or 2.

934 Let $$f(x, y) = 0 \text{ and } \phi(x, y) = 0$$

denote two equations whose degrees are m and n respectively. If they involve the terms x^m and x^n, then by substituting $x = x_1/x_3$, $y = x_2/x_3$, clearing of fractions, and collecting terms, we can reduce them to the form

$$F(x_1, x_2, x_3) = a_0x_1^m + a_1x_1^{m-1} + \cdots + a_m = 0, \qquad (1)$$

$$\Phi(x_1, x_2, x_3) = b_0x_1^n + b_1x_1^{n-1} + \cdots + b_n = 0, \qquad (2)$$

where each of the coefficients $a_0, a_1, \cdots, b_0, b_1, \cdots$ denotes a homogeneous function of x_2, x_3 of the degree indicated by its subscript. Hence R, the resultant of (1) and (2) with respect to x_1, is a homogeneous function of x_2, x_3 of the degree mn, § 930.

By § 928 the necessary and sufficient condition that (1) and (2) be satisfied by the same value of x_1 is that

$$R = 0. \qquad (3)$$

If R is not divisible by x_3, then $R = 0$ is satisfied by mn finite values of x_2/x_3 or y, § 933. If β denote any one of these values, the equations $f(x, \beta) = 0$, $\phi(x, \beta) = 0$ have a common root, and if this root be α, then $x = \alpha$, $y = \beta$ is a solution of $f(x, y) = 0$, $\phi(x, y) = 0$ (compare § 927). Moreover it can be shown that to each simple root of $R = 0$ there thus corresponds a single solution of $f(x, y) = 0$, $\phi(x, y) = 0$; and that to a multiple root of order r of $R = 0$ there correspond r solutions of $f(x, y) = 0$, $\phi(x, y) = 0$, all different or some of them equal. Hence $f(x, y) = 0$, $\phi(x, y) = 0$ have mn finite solutions.

If R is divisible by x_3^μ, then $R = 0$ is satisfied by only $mn - \mu$ finite values of x_2/x_3 or y, and therefore $f(x, y) = 0$,

$\phi(x, y) = 0$ have only $mn - \mu$ finite solutions. But since $x = x_1/x_3$ and $y = x_2/x_3$, when $x_3 = 0$ either x or y or both x and y are infinite. We therefore say in this case that $f(x, y) = 0$, $\phi(x, y) = 0$ have μ infinite solutions.

If the given equations $f(x, y) = 0$, $\phi(x, y) = 0$ lack the x^m and x^n terms, we can transform them, by a substitution of the form $y = y' + cx$, into equations of the same degrees which have these terms. By what has just been proved the transformed equations in x, y' will have mn solutions. But if $x = \alpha$, $y' = \beta$ be any one of these solutions, then $x = \alpha$, $y = \beta + c\alpha$ is a solution of $f(x, y) = 0$, $\phi(x, y) = 0$. Hence $f(x, y) = 0$, $\phi(x, y) = 0$ also have mn solutions.

In the preceding discussion it is assumed that R does not vanish identically. If R does thus vanish, $f(x, y)$ and $\phi(x, y)$ have a common factor and therefore $f(x, y) = 0$, $\phi(x, y) = 0$ have infinitely many solutions.

We therefore have the theorem:

If f(x, y) *and* ϕ(x, y) *are of the degrees* m *and* n *respectively and have no common factor, the equations* f(x, y) = 0, $\phi(x, y) = 0$ *have* mn *solutions.*

EXERCISE LXXXV

1. By the method of §§ 924, 925 show that the equations $6x^2 + 5x - 6 = 0$ and $2x^3 + x^2 - 9x - 9 = 0$ have a common root and find this root.

2. Form the resultant of $a_0x^2 + a_1x + a_2 = 0$ and $b_0x^2 + b_1x + b_2 = 0$.

3. Find the resultant of $ax^3 + bx^2 + cx + d = 0$ and $x^3 = 1$.

4. By the method of § 931 find the discriminants of the equations
(1) $x^3 + px + q = 0$. (2) $ax^3 + bx^2 + c = 0$.

5. By aid of § 931 show that $x^3 + x^2 - 8x - 12 = 0$ has a double root and find this root.

6. Solve the following pair of equations by the method of § 927.

$$x^2 - 3xy + 2y^2 - 16x - 28y = 0,$$
$$x^2 - xy - 2y^2 - 5x - 5y = 0.$$

XXXII. CONVERGENCE OF INFINITE SERIES

DEFINITION OF CONVERGENCE

935 **Infinite series.** If $u_1, u_2, \cdots, u_n, \cdots$ denotes any given never-ending sequence of numbers, § 187, the expression

$$u_1 + u_2 + \cdots + u_n + \cdots$$

is called an infinite series (compare § 704).

For $u_1 + u_2 + \cdots$ we may write Σu_n, read "sum of u_n to infinity."

The series Σu_n is called *real* when all its terms $u_1, u_2, \cdots$ are real, *positive* when all its terms are positive. In what follows we shall confine ourselves to real series.

A series is often given by means of a formula for its nth term u_n. Thus, if $u_n = \sqrt{n}/(n+1)$, the series is $\sqrt{1}/2 + \sqrt{2}/3 + \sqrt{3}/4 + \cdots$.

Sometimes such a formula is indicated by writing the first three or four terms of the series. Thus, in $1/2 + 1 \cdot 3/2 \cdot 4 + 1 \cdot 3 \cdot 5/2 \cdot 4 \cdot 6 + \cdots$ we have $u_n = 1 \cdot 3 \cdots (2n-1)/2 \cdot 4 \cdots 2n$.

936 **Convergence and divergence.** Let S_n denote the sum of the first n terms of the series $u_1 + u_2 + \cdots$, so that $S_1 = u_1$, $S_2 = u_1 + u_2$, and in general $S_n = u_1 + u_2 + \cdots + u_n$.

As n increases, S_n will take successively the values u_1, $u_1 + u_2$, $u_1 + u_2 + u_3, \cdots$, and one of the following cases must present itself, namely:

S_n will approach some finite number as limit,

or S_n will approach infinity,

or S_n will be indeterminate.

In the first case the series $u_1 + u_2 + \cdots$ is said to be *convergent*, and $\lim S_n$ is called its *sum*. In the second and third cases the series is said to be *divergent*.

When $u_1 + u_2 + \cdots$ is convergent we may represent its sum, $\lim S_n$, by S and write $S = u_1 + u_2 + \cdots$, that is, we may regard the series as merely another expression for the definite number S.

Thus, the geometric series $1/2 + 1/4 + 1/8 + 1/16 + \cdots$ is convergent and its sum is 1. For here, as n increases S_n takes successively the values $1/2, 3/4, 7/8, 15/16, \cdots$ and, as is proved in § 704, it approaches 1 as limit. Observe that here, as in every convergent series, $\lim u_n = 0$.

The series $1 + 1 + 1 + \cdots$ is divergent. For S_n takes successively the values $1, 2, 3, \cdots$ and therefore approaches ∞.

The series $1 - 1 + 1 - 1 + \cdots$ is divergent. For S_n takes successively the values $1, 0, 1, 0, \cdots$. It is therefore indeterminate.

We therefore have the following definitions: **937**

An infinite series is said to be convergent when the sum of its first n *terms approaches a finite limit as* n *is indefinitely increased. Otherwise it is said to be divergent.*

The limit of the sum of the first n *terms of a convergent series is called the sum of the series.*

This is a new use of the word *sum*. Hitherto *sum* has meant the result of a finite number of additions performed consecutively; here it means the *limit* of such a result. It must therefore not be assumed that the characteristic properties of finite sums, namely, conformity to the commutative and associative laws, always belong to these infinite sums (see §§ 941, 961).

In determining whether a given series is convergent or divergent, a finite number of its terms may be neglected. **938**

For the sum of the neglected terms will have a definite finite value.

If $u_1 + u_2 + \cdots$ (1) is a convergent series having the sum S, and c is any finite number, then $cu_1 + cu_2 + \cdots$ (2) is a convergent series having the sum cS. But if (1) is divergent, so is (2). **939**

For if the sum of the first n terms of (1) be S_n, the sum of the first n terms of (2) is cS_n; and $\lim cS_n = c \lim S_n = cS$.

The sum of a convergent series will not be changed if its terms are combined in groups without changing their order. **940**

Thus, if the given series be $u_1 + u_2 + \cdots$, and $g_1, g_2, \cdots$ denote the sums of its first two terms, its next four terms, and so on, the series $g_1 + g_2 + \cdots$ will have the same sum as $u_1 + u_2 + \cdots$.

For if u_n denote the last term in the group g_m, we have

$$g_1 + g_2 + \cdots + g_m = u_1 + u_2 + \cdots + u_n,$$

and the two members of this equation approach the same limit as m and therefore n is indefinitely increased.

In the same manner it may be shown that a divergent *positive* series remains divergent when its terms are grouped.

941 We may therefore introduce parentheses at will in a convergent series. It is also allowable to remove them unless, as in the following example, the resulting series is divergent.

The convergent series $1/2 + 1/4 + 1/8 + \cdots$, § 936, may be written $(1\frac{1}{2} - 1) + (1\frac{1}{4} - 1) + (1\frac{1}{8} - 1) + \cdots$. But here it is not allowable to remove parentheses since $1\frac{1}{2} - 1 + 1\frac{1}{4} - 1 + 1\frac{1}{8} - 1 + \cdots$ is divergent.

942 It is sometimes possible to find the sum of a series by the removal of parentheses.

Thus, the sum of $1/1 \cdot 2 + 1/2 \cdot 3 + 1/3 \cdot 4 + \cdots$ is 1.

For
$$S_n = \frac{1}{1 \cdot 2} + \frac{1}{2 \cdot 3} + \cdots + \frac{1}{n(n+1)}$$
$$= \left(\frac{1}{1} - \frac{1}{2}\right) + \left(\frac{1}{2} - \frac{1}{3}\right) + \cdots + \left(\frac{1}{n} - \frac{1}{n+1}\right)$$
$$= \frac{1}{1} - \frac{1}{2} + \frac{1}{2} - \frac{1}{3} + \cdots + \frac{1}{n} - \frac{1}{n+1} = 1 - \frac{1}{n+1}.$$

Hence
$$S = \lim S_n = \lim \left(1 - \frac{1}{n+1}\right) = 1.$$

Example. Find the sum of the series whose nth term u_n is $1/n(n+2)$.

943 **Remainder after n terms.** If the series $u_1 + u_2 + \cdots$ (1) is convergent, that portion of the series which follows the nth term, namely, $u_{n+1} + u_{n+2} + \cdots$ (2), will also be convergent, § 938. Let R_n denote the sum of (2). It is called the *remainder after* n *terms* of (1).

Evidently $\lim R_n = 0$.

POSITIVE SERIES

Theorem 1. *A positive series* $u_1 + u_2 + \cdots$ *is convergent if, as* n *increases,* S_n *remains always less than some finite number* c. **944**

For since the series is positive, S_n continually increases as n increases. But it remains less than c. Hence, § 192, it approaches a limit. Therefore, § 937, the series is convergent.

Theorem 2. *Let* $u_1 + u_2 + \cdots$ (1) *denote a given positive series, and let* $a_1 + a_2 + \cdots$ (2) *denote a positive series known to be convergent. The series* (1) *is convergent in any of the cases:* **945**

1. *When each term of* (1) *is less than the corresponding term of* (2).

2. *When the ratio of each term of* (1) *to the corresponding term of* (2) *is less than some finite number* c.

3. *When in* (1) *the ratio of each term to the immediately preceding term is less than the corresponding ratio in* (2).

1. For let S_n denote the sum of the first n terms of $u_1 + u_2 + \cdots$, and let A denote the sum of the series $a_1 + a_2 + \cdots$. If $u_1 < a_1$, $u_2 < a_2$, $\cdots$, we shall always have $S_n < A$. Hence $u_1 + u_2 + \cdots$ is convergent, § 944.

2. For if $\quad \dfrac{u_1}{a_1} < c, \dfrac{u_2}{a_2} < c, \cdots,$ then $u_1 < ca_1, u_2 < ca_2, \cdots.$

Therefore, since $ca_1 + ca_2 + \cdots$ is convergent, § 939, the series $u_1 + u_2 + \cdots$ is convergent, by 1.

3. For if $\quad \dfrac{u_2}{u_1} < \dfrac{a_2}{a_1}, \dfrac{u_3}{u_2} < \dfrac{a_3}{a_2}, \dfrac{u_4}{u_3} < \dfrac{a_4}{a_3}, \cdots,$

then $\quad \dfrac{u_2}{a_2} < \dfrac{u_1}{a_1}, \dfrac{u_3}{a_3} < \dfrac{u_2}{a_2}, \dfrac{u_4}{a_4} < \dfrac{u_3}{a_3}, \cdots.$

It follows from these inequalities that each of the ratios u_2/a_2, u_3/a_3, $\cdots$ is less than the finite number u_1/a_1. Therefore $u_1 + u_2 + \cdots$ is convergent, by 2.

It follows from § 938 that the same conclusions can be drawn if any one of the relations 1, 2, 3 holds good for all but a finite number of the terms of the series (1) and (2).

Example. Prove that

$$1 + 1/2 + 1/2 \cdot 3 + 1/2 \cdot 3 \cdot 4 + \cdots \qquad (1)$$

is convergent by comparing it with the convergent geometric series

$$1 + 1/2 + 1/2 \cdot 2 + 1/2 \cdot 2 \cdot 2 + \cdots \qquad (2)$$

by each of the methods 1, 2, 3.

First. Each term of (1) after the second is less than the corresponding term of (2). Hence (1) is convergent, by 1.

Second. The ratios of the terms of (1) to the corresponding terms of (2), namely, $1, 1, 2/3, 2 \cdot 2/3 \cdot 4, \cdots$, are finite. Hence (1) is convergent, by 2.

Third. The ratios of the terms of (1) to the immediately preceding terms, namely, $1/2, 1/3, 1/4, \cdots$, are less than the corresponding ratios in (2), namely, $1/2, 1/2, 1/2, \cdots$. Hence (1) is convergent, by 3.

946 **Theorem 3.** *Let* $u_1 + u_2 + \cdots$ (1) *denote a given positive series, and let* $b_1 + b_2 + \cdots$ (2) *denote a positive series known to be divergent. The series* (1) *is divergent in any of the cases:*

1. *When each term of* (1) *is greater than the corresponding term of* (2).

2. *When the ratio of each term of* (1) *to the corresponding term of* (2) *is greater than some positive number* c.

3. *When in* (1) *the ratio of each term to the immediately preceding term is greater than the corresponding ratio in* (2).

The proof of this theorem, which is similar to that given in § 945, is left to the student.

947 **Test series.** The practical usefulness of the preceding tests, §§ 945, 946, evidently depends on our possessing *test series* known to be convergent or divergent. The most important of these test series is the geometric series $a + ar + ar^2 + \cdots$, which has been shown, § 704, to be convergent when $r < 1$, and which is obviously divergent when $r \overline{>} 1$. Another very serviceable test series is the following.

948 *The series* $1 + 1/2^p + 1/3^p + \cdots + 1/n^p + \cdots$ *is convergent when* $p > 1$, *divergent when* $p \overline{<} 1$.

1. $p > 1$. Combining the two terms beginning with $1/2^p$, the four terms beginning with $1/4^p$, the eight terms beginning with $1/8^p$, and so on, we obtain the equivalent series, § 940,

$$1 + \left(\frac{1}{2^p} + \frac{1}{3^p}\right) + \left(\frac{1}{4^p} + \frac{1}{5^p} + \frac{1}{6^p} + \frac{1}{7^p}\right) + \cdots. \tag{1}$$

Evidently each term of (1) after the first is less than the corresponding term of the series

$$1 + \left(\frac{1}{2^p} + \frac{1}{2^p}\right) + \left(\frac{1}{4^p} + \frac{1}{4^p} + \frac{1}{4^p} + \frac{1}{4^p}\right) + \cdots, \tag{2}$$

that is, less than the corresponding term of

$$1 + \frac{2}{2^p} + \frac{4}{4^p} + \cdots, \text{ or } 1 + \frac{1}{2^{p-1}} + \frac{1}{(2^{p-1})^2} + \cdots. \tag{3}$$

But since $p > 1$, and therefore $1/2^{p-1} < 1$, the geometric series (3) is convergent. Hence (1) is convergent, § 945, 1.

2. $p = 1$. Combining the two terms ending with $1/4$, the four terms ending with $1/8$, the eight terms ending with $1/16$, and so on, we obtain

$$1 + \frac{1}{2} + \left(\frac{1}{3} + \frac{1}{4}\right) + \left(\frac{1}{5} + \frac{1}{6} + \frac{1}{7} + \frac{1}{8}\right) + \cdots. \tag{4}$$

Evidently each term of (4) after the second is greater than the corresponding term of the series

$$1 + \frac{1}{2} + \left(\frac{1}{4} + \frac{1}{4}\right) + \left(\frac{1}{8} + \frac{1}{8} + \frac{1}{8} + \frac{1}{8}\right) + \cdots, \tag{5}$$

that is, greater than the corresponding term of

$$1 + \frac{1}{2} + \frac{2}{4} + \frac{4}{8} + \cdots, \text{ or } 1 + \frac{1}{2} + \frac{1}{2} + \frac{1}{2} + \cdots. \tag{6}$$

But (6) is divergent. Therefore (4) is divergent, § 946, 1.

3. $p < 1$. In this case the series $1 + 1/2^p + 1/3^p + \cdots$ is divergent since its terms are greater than the corresponding terms of the series $1 + 1/2 + 1/3 + \cdots$, which has just been proved to be divergent, § 946, 1.

949 **Applications of the preceding theorems.** The following examples will serve to illustrate the usefulness of the theorems of §§ 945, 946.

Example 1. Show that $1/1 \cdot 2 + 1/2 \cdot 3 + 1/3 \cdot 4 + \cdots$ is convergent.

It is convergent because its terms after the first are less than the corresponding terms of the convergent series $1/2^2 + 1/3^2 + 1/4^2 + \cdots$, § 945, 1.

Example 2. Show that $1 + 1/3 + 1/5 + 1/7 + \cdots$ is divergent.

The ratios of the terms of this series to the corresponding terms of the divergent series $1 + 1/2 + 1/3 + 1/4 + \cdots$, namely, $1, 2/3, 3/5, 4/7, \cdots, n/(2n-1)$, are all greater than $1/2$. Hence $1 + 1/3 + 1/5 + \cdots$ is divergent, § 946, 2.

Example 3. Is the series in which $u_n = (2n+1)/(n^3+n)$ convergent or is it divergent?

Here
$$u_n = \frac{2n+1}{n^3+n} = \frac{n}{n^3} \cdot \frac{2+1/n}{1+1/n^2} = \frac{1}{n^2} \cdot \frac{2+1/n}{1+1/n^2}.$$

Hence the ratio of u_n to $1/n^2$ is $(2+1/n)/(1+1/n^2)$, an expression which is finite for all values of n, and which approaches the finite limit 2 as n increases. But $1/n^2$ is the nth term of the convergent series $1 + 1/2^2 + 1/3^2 + \cdots$. Therefore the given series is convergent, § 945, 2.

950 By the method employed in Ex. 3, it may be proved that if u_n has the form $u_n = f(n)/\phi(n)$, where $f(n)$ and $\phi(n)$ denote integral functions of n, the series is convergent when the degree of $\phi(n)$ exceeds that of $f(n)$ by more than 1; otherwise, that it is divergent.

Example 1. Show that the following series are convergent.

(1) $\frac{1}{1} + \frac{1}{\sqrt{2^3}} + \frac{1}{\sqrt{3^3}} + \cdots$. (2) $\frac{2}{2 \cdot 3 \cdot 4} + \frac{4}{3 \cdot 4 \cdot 5} + \frac{6}{4 \cdot 5 \cdot 6} + \cdots$.

(3) $\frac{1}{a(a+b)} + \frac{1}{(a+b)(a+2b)} + \frac{1}{(a+2b)(a+3b)} + \cdots$.

Example 2. Show that the following series are divergent.

(1) $\frac{1}{2} + \frac{1}{4} + \frac{1}{6} + \frac{1}{8} + \cdots$. (2) $\frac{1}{a} + \frac{1}{2a+b} + \frac{1}{3a+b} + \cdots$.

(3) $\frac{1}{\sqrt{2}} + \frac{1}{\sqrt{3}} + \frac{1}{\sqrt{4}} + \cdots$. (4) $\frac{2}{1+2\sqrt{2}} + \frac{3}{1+3\sqrt{3}} + \frac{4}{1+4\sqrt{4}} + \cdots$.

Example 3. Write out the first four terms of the series in which u_n has each of the following values and determine which of these series are convergent and which divergent.

(1) $u_n = \dfrac{2n-1}{(n+1)(n+2)}$. (2) $u_n = \dfrac{\sqrt{n}}{n^2+1}$. (3) $u_n = \dfrac{n^2-(n-1)^2}{n^3+(n+1)^3}$.

Theorem 4. *The positive series* $u_1 + u_2 + \cdots$ *is convergent if the ratio of each of its terms to the immediately preceding term is less than some number* r *which itself is less than* 1. **951**

For in $u_1 + u_2 + u_3 + \cdots$ (1) the ratio of each term to the immediately preceding term is less than the corresponding ratio in the geometric series $u_1 + u_1 r + u_1 r^2 + \cdots$ (2), since in (1) the ratio in question is always less than r, while in (2) it is equal to r. But (2) is convergent since $r < 1$. Therefore (1) is convergent, by § 945, 3.

If the ratios above mentioned are equal to 1 or greater than 1, the series is divergent; for in this case $\lim u_n \neq 0$.

Corollary. *If as* n *increases the ratio* u_{n+1}/u_n *approaches a definite limit* λ, *the series is convergent when* $\lambda < 1$, *divergent when* $\lambda > 1$. **952**

1. For if $\lambda < 1$, take any number r such that $\lambda < r < 1$.

Then, since $\lim(u_{n+1}/u_n) = \lambda$, after a certain value of n we shall always have $u_{n+1}/u_n - \lambda < r - \lambda$, § 189, and therefore $u_{n+1}/u_n < r$. Hence the series is convergent, §§ 938, 951.

2. If $\lambda > 1$, after a certain value of n we shall always have $u_{n+1}/u_n > 1$. Hence the series is divergent, § 951.

When $u_{n+1}/u_n > 1$ and $\lim(u_{n+1}/u_n) = 1$, the series is divergent; but when $u_{n+1}/u_n < 1$ and $\lim(u_{n+1}/u_n) = 1$, no conclusion can be drawn from the theorem of § 951.

Example 1. Show that $\dfrac{3}{5} + \dfrac{3 \cdot 5}{5 \cdot 10} + \dfrac{3 \cdot 5 \cdot 7}{5 \cdot 10 \cdot 15} + \cdots$ is convergent.

The nth term of this series is $3 \cdot 5 \cdot 7 \cdots (2n+1)/5 \cdot 10 \cdot 15 \cdots 5n$, and the ratio of this term to the term which precedes it is $(2n+1)/5n$.

But since $(2n+1)/5n = 2/5 + 1/5n$, $\lim(2n+1)/5n = 2/5$, which is < 1. Hence the series is convergent.

Example 2. When is $\frac{1}{1+x}+\frac{1}{1+2x^2}+\frac{1}{1+3x^3}+\cdots$ convergent, x being positive?

Here $$\frac{u_{n+1}}{u_n}=\frac{1+nx^n}{1+(n+1)x^{n+1}}=\frac{x^n+1/n}{x^{n+1}(1+1/n)+1/n},$$

and therefore $$\lim\frac{u_{n+1}}{u_n}=\frac{1}{x}.$$

Hence the series is convergent when $1/x<1$, that is, when $x>1$.

Example 3. Show that $\frac{1}{1}+\frac{1\cdot 3}{1\cdot 4}+\frac{1\cdot 3\cdot 5}{1\cdot 4\cdot 7}+\cdots$ is convergent.

Example 4. Show that $\frac{1}{2}+\frac{1}{2}\cdot\frac{3}{5}+\frac{1}{2}\cdot\frac{3}{5}\cdot\frac{1}{2}+\cdots$ is convergent.

Example 5. When is $\frac{x}{1}+\frac{x^2}{2}+\frac{x^3}{3}+\cdots$ convergent, x being positive?

Example 6. When is $\frac{1}{1+x}+\frac{1}{1+x^2}+\frac{1}{1+x^3}+\cdots$ convergent, x being positive?

953 Series in which $\lim(u_{n+1}/u_n)=1$. In a series of this kind the ratio u_{n+1}/u_n can be reduced to the form

$$u_{n+1}/u_n=1/(1+\alpha_n/n),$$

where $\lim(\alpha_n/n)=0$. We proceed to show that if, as n increases, α_n ultimately becomes and remains greater than some number which is itself greater than 1, the series is convergent; but that if α_n ultimately becomes and remains less than 1, the series is divergent.

1. For suppose that after a certain value of n, which we may call k, we have $\alpha_n>1+\alpha$, where α is positive.

Then $$\frac{u_{n+1}}{u_n}=\frac{1}{1+\alpha_n/n}<\frac{1}{1+(1+\alpha)/n},\text{ when } n\geqq k.$$

But we may reduce this inequality to the form

$$u_{n+1}<\frac{1}{\alpha}[nu_n-(n+1)u_{n+1}],\text{ when } n\geqq k. \tag{1}$$

In (1) set $n=k, k+1, \cdots, k+l-1$ successively, and add the resulting inequalities. We obtain

$$u_{k+1}+u_{k+2}+\cdots+u_{k+l}<\frac{1}{\alpha}[ku_k-(k+l)u_{k+l}]. \tag{2}$$

It follows from (2) that as l increases the sum of the first l terms of the positive series $u_{k+1}+u_{k+2}+\cdots$ remains always less than the finite number ku_k/α, which proves that this series is convergent, § 944. Therefore the complete series $u_1 + u_2 + \cdots$ is convergent, § 938.

2. Suppose that when $n > k$ we have $\alpha_n < 1$.

Then $$\frac{u_{n+1}}{u_n} = \frac{1}{1 + \alpha_n/n} > \frac{1}{1 + 1/n}, \text{ when } n > k.$$

But $1/(1 + 1/n)$ is the ratio of the corresponding terms of the divergent series $1 + 1/2 + 1/3 + \cdots$; for $1/(n+1) \div 1/n = 1/(1 + 1/n)$.

Hence the given series $u_1 + u_2 + \cdots$ is divergent, § 946, 3.

If α_n remains greater than 1 but approaches 1 as limit, the preceding test will not determine whether the series is convergent or divergent. But in this case α_n can be reduced to the form $\alpha_n = 1 + \beta_n/n$, where $\lim \beta_n/n = 0$; and if β_n remains less than some finite number b, the series is divergent.

For since $\beta_n < b$, we have

$$\frac{u_{n+1}}{u_n} = \frac{1}{1 + \alpha_n/n} = \frac{1}{1 + 1/n + \beta_n/n^2} > \frac{1}{1 + 1/n + b/n^2}.$$

But $1/(1 + 1/n + b/n^2)$ in turn is greater than the ratio of the corresponding terms of the divergent series $1/(1-b) + 1/(2-b) + 1/(3-b) + \cdots$.

For $$\frac{1}{(n+1) - b} \div \frac{1}{n - b} = \frac{n - b}{(n - b) + 1} = \frac{1}{1 + 1/(n - b)} < \frac{1}{1 + 1/n + b/n^2},$$

since $$\frac{1}{n - b} = \frac{1}{n}\,\frac{1}{1 - b/n} = \frac{1}{n} + \frac{b}{n^2} + \frac{b^2}{n^3} + \cdots.$$

Therefore the given series $u_1 + u_2 + \cdots$ is divergent, § 946, 3.

954 It follows from the preceding discussion that a series in which u_{n+1}/u_n can be reduced to the form

$$u_{n+1}/u_n = (n^p + an^{p-1} + \cdots)/(n^p + a'n^{p-1} + \cdots)$$

is convergent when $a' - a > 1$, divergent when $a' - a \overset{=}{<} 1$.

For dividing the denominator of this fraction by its numerator, we have

$$\frac{u_{n+1}}{u_n} = \frac{n^p + an^{p-1} + \cdots}{n^p + a'n^{p-1} + \cdots} = \frac{1}{1 + (a' - a)/n + \beta_n/n^2},$$

where β_n is finite.

Example. Prove that the "hypergeometric series"

$$1 + \frac{\alpha \cdot \beta}{1 \cdot \gamma} + \frac{\alpha(\alpha + 1)\beta(\beta + 1)}{1 \cdot 2\,\gamma(\gamma + 1)} + \frac{\alpha(\alpha + 1)(\alpha + 2)\beta(\beta + 1)(\beta + 2)}{1 \cdot 2 \cdot 3\,\gamma(\gamma + 1)(\gamma + 2)} + \cdots$$

is convergent when $\gamma - \alpha - \beta > 0$, divergent when $\gamma - \alpha - \beta \overset{=}{<} 0$.

EXERCISE LXXXVI

Determine whether the following series are convergent or divergent.

1. $\dfrac{1}{2+1}+\dfrac{1}{2^2+1}+\dfrac{1}{2^3+1}+\cdots.$

2. $\dfrac{1}{1}+\dfrac{1\cdot 2}{1\cdot 3}+\dfrac{1\cdot 2\cdot 3}{1\cdot 3\cdot 5}+\cdots.$

3. $\dfrac{2}{2\cdot 3}+\dfrac{4}{3\cdot 4}+\dfrac{6}{4\cdot 5}+\cdots.$

4. $\dfrac{2}{3}+2\left(\dfrac{2}{3}\right)^2+3\left(\dfrac{2}{3}\right)^3+\cdots.$

5. $\dfrac{1}{\sqrt{3}}+\dfrac{1}{\sqrt[3]{3}}+\dfrac{1}{\sqrt[4]{3}}+\cdots.$

6. $\dfrac{1}{a^2}+\dfrac{1}{a^2+1}+\dfrac{1}{a^2+2}+\cdots.$

7. $\dfrac{2}{4}+\dfrac{2\cdot 4}{4\cdot 7}+\dfrac{2\cdot 4\cdot 6}{4\cdot 7\cdot 10}+\cdots+\dfrac{2\cdot 4\cdot 6\cdots 2n}{4\cdot 7\cdot 10\cdots(3n+1)}+\cdots.$

8. $\dfrac{2}{4}+\dfrac{2\cdot 3}{4\cdot 5}+\dfrac{2\cdot 3\cdot 4}{4\cdot 5\cdot 6}+\cdots+\dfrac{2\cdot 3\cdot 4\cdots(n+1)}{4\cdot 5\cdot 6\cdots(n+3)}+\cdots.$

9. $\dfrac{1}{2}+\dfrac{1\cdot 3}{2\cdot 4}+\dfrac{1\cdot 3\cdot 5}{2\cdot 4\cdot 6}+\cdots+\dfrac{1\cdot 3\cdot 5\cdots(2n-1)}{2\cdot 4\cdot 6\cdots 2n}+\cdots.$

Write out the first four terms of the series in which u_n has the following values and determine whether these series are convergent or divergent.

10. $u_n=\dfrac{n+1}{n(n+2)}.$

11. $u_n=\dfrac{\sqrt[3]{n}}{\sqrt[3]{n^3+1}}.$

12. $u_n=\sqrt{n^2+1}-n=\dfrac{1}{\sqrt{n^2+1}+n}.$

Determine whether the series in which u_{n+1}/u_n has the following values are convergent or divergent.

13. $\dfrac{u_{n+1}}{u_n}=\dfrac{2n}{2n+3}.$

14. $\dfrac{u_{n+1}}{u_n}=\dfrac{3n^3-2n^2}{3n^3+n^2+1}.$

For what positive values of x are the following series convergent?

15. $1+\dfrac{3}{5}x+\dfrac{3\cdot 6}{5\cdot 8}x^2+\dfrac{3\cdot 6\cdot 9}{5\cdot 8\cdot 11}x^3+\cdots.$

16. $\dfrac{1}{1+x}+\dfrac{x}{1+x^2}+\dfrac{x^2}{1+x^3}+\dfrac{x^3}{1+x^4}+\cdots.$

17. Show that $\dfrac{a}{1}+\dfrac{a(a+1)}{1\cdot 2}+\dfrac{a(a+1)(a+2)}{1\cdot 2\cdot 3}+\cdots$ is divergent when a is positive.

18. If for all values of n we have $\sqrt[n]{u_n}<r$, where r is positive and less than 1, show that $u_1+u_2+\cdots$ is convergent by comparing it with the convergent series $r+r^2+r^3+\cdots.$

SERIES WHICH HAVE BOTH POSITIVE AND NEGATIVE TERMS

General test of convergence. By definition, § 937, an infinite series of any kind $u_1 + u_2 + \cdots$ is convergent if S_n approaches a finite limit as n is indefinitely increased. **955**

But, §§ 195, 197, S_n will approach a limit if the sequence of values through which it runs as n increases, namely, S_1, S_2, S_3, $\cdots$, possesses the property that for every given positive number δ, however small, a corresponding term S_k can be found which differs numerically from every subsequent term S_{k+p} by less than δ. If this condition is not satisfied, S_n will not approach a limit, § 198.

Since $\quad S_k = u_1 + \cdots + u_k,$

and $\quad S_{k+p} = u_1 + \cdots + u_k + u_{k+1} + \cdots + u_{k+p},$

we have $S_{k+p} - S_k = u_{k+1} + u_{k+2} + \cdots + u_{k+p}.$

Hence the following general test of convergence:

Any infinite series $u_1 + u_2 + \cdots$ *is convergent if for every given positive number* δ, *however small, one can find a term* u_k *such that the sum of any number of the terms after* u_k *is numerically less than* δ; *in other words, such that*

$$|u_{k+1} + u_{k+2} + \cdots + u_{k+p}| < \delta$$

for all values of p. *If the series does not possess this property, it is divergent.*

Hence in particular, a series $u_1 + u_2 + \cdots$ cannot be convergent unless $\lim u_n = 0$. But this single condition is not sufficient for convergence. We must also have $\lim (u_n + u_{n+1}) = 0$, $\lim (u_n + u_{n+1} + u_{n+2}) = 0$, and so on.

Thus, $1 + 1/2 + 1/3 + \cdots$ is divergent although $\lim u_n = \lim 1/n = 0$. For in this series the sum of the k terms which follow the term $1/k$ is always greater than $1/2$. Thus,

$$\frac{1}{k+1} + \frac{1}{k+2} + \cdots + \frac{1}{k+k} > \frac{1}{2k} + \frac{1}{2k} + \cdots \text{ to } k \text{ terms, } i.e. > \frac{1}{2k} \cdot k \text{ or } \frac{1}{2}.$$

Hence k cannot be so chosen that $u_{k+1} + \cdots + u_{k+k}$ is less than every assignable number, and the series is divergent (compare § 948, 2).

956 **Corollary 1.** *A series which has both positive and negative terms is convergent if the corresponding positive series is convergent.*

For let $u_1 + u_2 + \cdots$ (1) be the given series, and let $u'_1 + u'_2 + \cdots$ (2) be the same series with the signs of all its negative terms changed. Then

$$|u_{k+1} + u_{k+2} + \cdots + u_{k+p}| \leqq u'_{k+1} + u'_{k+2} + \cdots + u'_{k+p}.$$

Hence, if by taking k great enough we can make

$$u'_{k+1} + \cdots + u'_{k+p} < \delta,$$

the same will be true of $|u_{k+1} + \cdots + u_{k+p}|$. Therefore (1) is convergent if (2) is, § 955.

957 The preceding demonstration also shows that a series $u_1 + u_2 + \cdots$ with *imaginary* terms is convergent if the series whose terms are the absolute values of $u_1, u_2, \cdots$, § 232, namely, the series $|u_1| + |u_2| + \cdots$, is convergent.

Thus, $i/1 + i^2/2^2 + i^3/3^2 + \cdots$ is convergent since $1 + 1/2^2 + 1/3^2 + \cdots$ is convergent.

958 **Corollary 2.** *A series whose terms are alternately positive and negative is convergent if each term is numerically less than the term which precedes it, and if the limit of the* n*th term is* 0.

For let the series be $a_1 - a_2 + a_3 - \cdots$, where $a_1, a_2, \cdots$ are positive. Using the notation of § 955, we here have

$$|u_{k+1} + u_{k+2} + \cdots + u_{k+p}| = |a_{k+1} - a_{k+2} + \cdots + (-1)^{p-1} a_{k+p}|.$$

We can write $\quad a_{k+1} - a_{k+2} + \cdots + (-1)^{p-1} a_{k+p} \qquad (1)$

in the form $\quad (a_{k+1} - a_{k+2}) + (a_{k+3} - a_{k+4}) + \cdots \qquad (2)$

and in the form $\quad a_{k+1} - (a_{k+2} - a_{k+3}) - \cdots. \qquad (3)$

Since $a_{k+1} > a_{k+2} > a_{k+3} > \cdots$, each of the expressions in parentheses in (2) and (3) is positive. Hence it follows from (2) that (1) is positive, and from (3) that (1) is algebraically less than a_{k+1}, and therefore from (2) and (3) combined that (1) is numerically less than a_{k+1}.

But since $\lim a_n = 0$, we can choose k so that $a_{k+1} < \delta$. Therefore $a_1 - a_2 + a_3 - \cdots$ is convergent, § 955.

959 **Absolute and conditional convergence.** A convergent real series is said to be *absolutely convergent* if it continues to be convergent when the signs of all its negative terms, if any, are changed; *conditionally convergent* if it becomes divergent when these signs are changed.

Thus, $1 - 1/2 + 1/4 - 1/8 + \cdots$ is absolutely convergent since the series $1 + 1/2 + 1/4 + 1/8 + \cdots$ is convergent.

But $1 - 1/2 + 1/3 - 1/4 + \cdots$, which is convergent by § 958, is only conditionally convergent since $1 + 1/2 + 1/3 + 1/4 + \cdots$ is divergent.

960 **Theorem.** *In an absolutely convergent series the positive terms by themselves form a convergent series, and in like manner the negative terms by themselves. And if the sums of these two series be* P *and* − N *respectively, the sum of the entire series is* P − N.

But in a conditionally convergent series both the series of positive terms and the series of negative terms are divergent.

For let $u_1 + u_2 + \cdots$ be a convergent series which has an infinite number of positive and negative terms.

Of the first n terms of this series suppose that p are positive and q negative. Then if S_n denote the sum of all n terms, P_p the sum of the p positive terms, and $-N_q$ the sum of the q negative terms, we shall have $S_n = P_p - N_q$.

When n is indefinitely increased both p and q will increase indefinitely, and since S_n will approach the finite limit S, one of the following cases must present itself, namely, either (1) both P_p and N_q will approach finite limits which we may call P and N, or (2) both P_p and N_q will approach infinity.

In the first case $\lim S_n = \lim (P_p - N_q) = \lim P_p - \lim N_q$, § 203, that is, $S = P - N$. The series is absolutely convergent. In fact, after the change of the signs of the negative terms the sum of the series is $P + N$.

In the second case the series is conditionally convergent. For if S'_n denote the sum of the first n terms of the series obtained by changing the signs of the negative terms, we have $\lim S'_n = \lim (P_p + N_q) = \infty$.

961 **Corollary.** *The terms of a conditionally convergent series may be so arranged that the sum of the series will take any real value that may be assigned.*

For, as just shown, in a conditionally convergent series the positive terms by themselves and the negative terms by themselves each constitute a divergent series the limit of whose nth term is 0.

Hence, for example, if we assign some positive number c, and then, without changing the relative order of the positive terms or that of the negative terms among themselves, form S_n by first adding positive terms until the sum is greater than c, then negative terms until the sum is less than c, and so on indefinitely, the limit of this S_n, as n is indefinitely increased, will be c.

Hence the commutative law of addition does not hold good for a conditionally convergent series.

EXERCISE LXXXVII

1. Determine whether the following series are convergent or divergent.

(1) $\frac{1}{\sqrt{2}} - \frac{1}{\sqrt{3}} + \frac{1}{\sqrt{4}} - \cdots$. (2) $\frac{1}{\sqrt{2}} - \frac{1}{\sqrt[3]{2}} + \frac{1}{\sqrt[4]{2}} - \cdots$.

(3) $\frac{3}{3} - \frac{3 \cdot 5}{3 \cdot 6} + \frac{3 \cdot 5 \cdot 7}{3 \cdot 6 \cdot 9} - \frac{3 \cdot 5 \cdot 7 \cdot 9}{3 \cdot 6 \cdot 9 \cdot 12} + \cdots$.

2. For what real values of x are the following series convergent and for what values are they divergent?

(1) $\frac{1}{1-x} + \frac{1}{1+2x} + \frac{1}{1-3x} + \cdots + \frac{1}{1+(-1)^n nx} + \cdots$.

(2) $\frac{x}{1+x^2} + \frac{x^3}{1+2x^4} + \frac{x^5}{1+3x^6} + \cdots + \frac{x^{2n-1}}{1+nx^{2n}} + \cdots$.

3. If $u_1 + u_2 + u_3 + \cdots$ is absolutely convergent, and $a_1, a_2, a_3, \cdots$ denote any sequence of numbers all of which are numerically less than some finite number c, prove by the method of § 956 that the series $a_1u_1 + a_2u_2 + a_3u_3 + \cdots$ is also convergent.

4. If S denotes the sum of a series of the kind described in § 958, show that the sums $a_1, a_1 - a_2, a_1 - a_2 + a_3, \cdots$ are alternately greater and less than S.

CONVERGENCE OF POWER SERIES

Power series. This name is given to any series which has the form $a_0 + a_1x + a_2x^2 + \cdots + a_nx^n + \cdots$ (1), where x is a variable but $a_0, a_1, \cdots$ are constants. The values of x and $a_0, a_1, \cdots$ may be real or imaginary. **962**

By § 957, the series (1) is convergent if the positive series $|a_0| + |a_1x| + |a_2x^2| + \cdots + |a_nx^n| + \cdots$ (2) is convergent. When (2) is convergent we say that (1) is *absolutely* convergent (compare § 959). Whether (1) is convergent or divergent will depend upon the value of x. Hence the importance of the following theorems.

Theorem 1. *If when* $x = b$ *every term of* $a_0 + a_1x + \cdots$ *is numerically less than some finite positive number* c, *when* $|x| < |b|$ *the series is absolutely convergent.* **963**

For since $$|a_nb^n| < c \text{ for every } n,$$

we have $$|a_nx^n| = |a_nb^n| \cdot \left|\frac{x}{b}\right|^n < c\left|\frac{x}{b}\right|^n \text{ for every } n.$$

Hence each term of $|a_0| + |a_1x| + |a_2x^2| + \cdots$ (1) is less than the corresponding term of $c + c\left|\frac{x}{b}\right| + c\left|\frac{x}{b}\right|^2 + \cdots$ (2).

But (2), being a geometric series, converges when $|x/b| < 1$, that is, when $|x| < |b|$. And when (2) converges, so does (1), § 945, 1.

Thus, $1 + 2x + x^2 + 2x^3 + \cdots$ converges when $|x| < 1$.

Corollary 1. *If* $a_0 + a_1x + \cdots$ *is convergent when* $x = b$, *it is absolutely convergent when* $|x| < |b|$. **964**

This follows immediately from § 963. For since $a_0 + a_1x + \cdots$ is convergent when $x = b$, all its terms have finite values when $x = b$.

Corollary 2. *If* $a_0 + a_1x + \cdots$ *is divergent when* $x = b$, *it is also divergent when* $|x| > |b|$. **965**

For were $a_0 + a_1x + \cdots$ to converge for a value of x which is numerically greater than b, it would also converge for $x = b$, § 964.

966 **Limits of convergence.** It follows from §§ 964, 965 that if we assign to a class A_1 all positive values of x for which $a_0 + a_1x + \cdots$ converges and to a class A_2 all for which it diverges, each number in A_1 will be less than every number in A_2. Hence, § 159, there is either a greatest number in A_1 or a least in A_2. Call this number λ. It represents the *limit of convergence* of $a_0 + a_1x + \cdots$, the series being absolutely convergent when $|x| < \lambda$, divergent when $|x| > \lambda$.

Thus, in both $x + x^2/2 + x^3/3 + \cdots$ (1) and $x + x^2/2^2 + x^3/3^2 + \cdots$ (2) the limit of convergence λ is 1. Observe that (1) diverges and (2) converges when $x = \lambda = 1$. It is possible to construct a series in which $\lambda = 0$; for example, the series $x + 2!\,x^2 + 3!\,x^3 + \cdots$.

What we have called the limit of convergence is more frequently called the *radius of the circle of convergence*. For if we picture complex numbers by points in a plane in the manner described in § 238 and draw a circle whose center is at the origin and whose radius is λ, the series $a_0 + a_1x + \cdots$ will converge for all values of x whose graphs lie within the circle, and it will diverge for all values of x whose graphs lie without the circle, § 239.

967 **Theorem 2.** *If in* $a_0 + a_1x + \cdots$ *the ratio* $|a_n/a_{n+1}|$ *approaches a definite limit* μ, *then* μ *is the limit of convergence.*

For, by § 952, the series $|a_0| + |a_1x| + \cdots$ converges when

$$\lim\left|\frac{a_{n+1}x^{n+1}}{a_nx^n}\right| < 1, \text{ that is, when } |x| < \lim\left|\frac{a_n}{a_{n+1}}\right|.$$

Similarly $|a_0| + |a_1x| + \cdots$ diverges when $|x| > \lim\left|\dfrac{a_n}{a_{n+1}}\right|$.

Example 1. Find the limit of convergence of the series

$$1 + \frac{3}{5}x + \frac{3\cdot 5}{5\cdot 10}x^2 + \cdots + \frac{3\cdot 5\cdots(2n+1)}{5\cdot 10\cdots 5n}x^n + \cdots.$$

Since $\dfrac{a_n}{a_{n+1}} = \dfrac{5(n+1)}{2n+3} = \dfrac{5+5/n}{2+3/n}$, we have $\mu = \lim\dfrac{5+5/n}{2+3/n} = \dfrac{5}{2}$.

Example 2. Find the limits of convergence of the series

$$\cdots + 2^3x^{-3} + 2^2x^{-2} + 2x^{-1} + 1 + x/3 + x^2/3^2 + x^3/3^3 + \cdots.$$

Here $1 + x/3 + x^2/3^2 + \cdots$ is a geometric power series in x which converges when $|x| < 3$, for $a_n/a_{n+1} = 3$.

On the other hand, $2x^{-1} + 2^2x^{-2} + 2^3x^{-3} + \cdots$ is a geometric power series in x^{-1} or $1/x$ which converges when $|x^{-1}| < 1/2$, and therefore when $|x| > 2$.

Hence the given series converges when $2 < |x| < 3$.

Example 3. For what real values of x will the series

$$x/(1+x) + 2x^2/(1+x)^2 + 3x^3/(1+x)^3 + \cdots \text{ converge?}$$

This is a power series in $x/(1+x)$ which converges when $|x/(1+x)| < 1$, for $\lim a_n/a_{n+1} = \lim n/(n+1) = 1$.

But $|x/(1+x)| < 1$ for all positive values of x and for negative values which are greater than $-1/2$. Hence the series converges when $x > -1/2$.

968 **The binomial, exponential, and logarithmic series.** We proceed to apply the preceding theorem to three especially important power series.

1. The *exponential series*, § 990, namely,

$$1 + \frac{x}{1} + \frac{x^2}{2!} + \cdots + \frac{x^n}{n!} + \cdots$$

is convergent for all finite values of x.

For here $$\frac{a_n}{a_{n+1}} = \frac{1}{n!} \div \frac{1}{(n+1)!} = n + 1.$$

Hence $$\lim \frac{a_n}{a_{n+1}} = \lim (n+1) = \infty, \text{ that is, } \mu = \infty.$$

2. The *logarithmic series*, § 992, namely,

$$x - \frac{x^2}{2} + \frac{x^3}{3} - \cdots + (-1)^{n-1}\frac{x^n}{n} + \cdots$$

is convergent when $|x| < 1$, divergent when $|x| > 1$.

For here $$\frac{a_n}{a_{n+1}} = -\frac{1}{n} \div \frac{1}{n+1} = -\frac{n+1}{n}.$$

Hence $$\lim \frac{a_n}{a_{n+1}} = -\lim \frac{n+1}{n} = -\lim \left(1 + \frac{1}{n}\right) = -1, \text{ that is, } \mu = 1.$$

The series converges when $x = 1$, § 958, diverges when $x = -1$, § 948.

3. The *binomial series*, namely,

$$1 + mx + \frac{m(m-1)}{1 \cdot 2} x^2 + \frac{m(m-1)(m-2)}{1 \cdot 2 \cdot 3} x^3 + \cdots,$$

where m is not a positive integer, is convergent when $|x| < 1$, divergent when $|x| > 1$.

For here

$$\frac{a_n}{a_{n+1}} = \frac{m(m-1)\cdots(m-n+1)}{1 \cdot 2 \cdots n} \div \frac{m(m-1)\cdots(m-n)}{1 \cdot 2 \cdots (n+1)} = \frac{n+1}{m-n}.$$

Hence $\lim \frac{a_n}{a_{n+1}} = \lim \frac{n+1}{m-n} = -\lim \frac{1+1/n}{1-m/n} = -1$, that is, $\mu = 1$.

When $x = 1$ the series converges if $m > -1$, diverges if $m \overline{\gtrless} -1$ (see § 1001, Ex. 2).

When $x = -1$ the series converges if $m > 0$, diverges if $m < 0$.

For when $x = -1$, by setting $m = -a$ we may reduce the series to the form

$$1 + a + \frac{a(a+1)}{1 \cdot 2} + \frac{a(a+1)(a+2)}{1 \cdot 2 \cdot 3} + \cdots.$$

Evidently from a certain term on all the terms are of the same sign, so that the test of § 954 is applicable, § 956.

But here $$\frac{u_{n+1}}{u_n} = \frac{a+n-1}{n} = \frac{n+(a-1)}{n}.$$

Hence, § 954, the series converges if $-(a-1) > 1$, that is, if $-a > 0$, or, since $-a = m$, it converges if $m > 0$. But it diverges if $-(a-1) < 1$, that is, if $m < 0$.

EXERCISE LXXXVIII

Determine the limits of convergence of the following series.

1. $1 + mx + m^2x^2/2! + m^3x^3/3! + \cdots$.

2. $2(2x)^2 + 3(2x)^3 + 2(2x)^4 + 3(2x)^5 + \cdots$.

3. $mx + \frac{m(m-2)}{2!} x^2 + \frac{m(m-2)(m-4)}{3!} x^3 + \cdots$.

For what real values of x will the following series converge?

4. $\frac{3x}{x+4} + \frac{1}{2}\left(\frac{3x}{x+4}\right)^2 + \frac{1}{3}\left(\frac{3x}{x+4}\right)^3 + \cdots$.

5. $\frac{x}{x^2+1} + \left(\frac{x}{x^2+1}\right)^2 + \left(\frac{x}{x^2+1}\right)^3 + \cdots$.

6. $\cdots (3x)^{-3} + (3x)^{-2} + (3x)^{-1} + 1 + 2x + (2x)^2 + (2x)^3 + \cdots$.

XXXIII. OPERATIONS WITH INFINITE SERIES

SOME PRELIMINARY THEOREMS

When a given power series $a_0 + a_1x + \cdots$ is convergent, its sum is a definite function of x which we may represent by $f(x)$, writing $f(x) = a_0 + a_1x + \cdots$. In what follows when we write $f(x) = a_0 + a_1x + \cdots$, we assume that $a_0 + a_1x + \cdots$ has a limit of convergence λ which is greater than 0, and suppose that $|x| < \lambda$. **969**

Theorem 1. *Given that* $\phi(x) = a_1x + a_2x^2 + \ldots$, *and that when* x *has the positive value* b *every term of* $\phi(x)$ *is numerically less than some finite positive number* c. **970**

If any positive number δ *be assigned, however small, then* $|\phi(x)| < \delta$, *whenever* $|x| < b\delta/(c + \delta)$.

For, as was shown in the proof in § 963, when $|x| < b$,

$$|\phi(x)| < c\left|\frac{x}{b}\right| + c\left|\frac{x}{b}\right|^2 + \cdots,$$

and therefore $$< c\left|\frac{x}{b}\right|\frac{1}{1 - |x/b|}, \text{ that is, } < \frac{c|x|}{b - |x|}. \quad § 704$$

Hence $$|\phi(x)| < \delta \text{ when } \frac{c|x|}{b - |x|} < \delta,$$

that is, when $$|x| < \frac{b\delta}{c + \delta}.$$

Corollary. *If* $f(x) = a_0 + a_1x + \cdots$, *then* $\lim_{x \doteq 0} f(x) = a_0 = f(0)$. **971**

For, as just shown, $\lim_{x \doteq 0} (a_1x + a_2x^2 + \cdots) = 0$, § 200.

Theorem 2. *If the series* $a_0 + a_1x + \cdots$ *vanishes for every value of* x *for which it converges, then* $a_0 = 0$, $a_1 = 0$, $\cdots$. **972**

For setting $x = 0$, we at once have $a_0 = 0$.

Hence $$a_1x + a_2x^2 + a_3x^3 + \cdots = 0 \quad \textbf{(1)}$$

for every value of x for which it converges.

If $x \neq 0$, we may divide (1) throughout by x.

Hence $$a_1 + a_2x + a_3x^2 + \cdots = 0 \qquad (2)$$

for every value of x for which it converges, except perhaps for $x = 0$. But it follows from this that $a_1 = 0$; for were $a_1 \neq 0$ we could choose x so small (without making it 0) that $|a_2x + a_3x^2 + \cdots| < |a_1|$, § 970, and such a value of x would not satisfy (2), as we have just shown it must.

Hence $a_1 = 0$. And by the same reasoning it may be shown that $a_2 = 0$, $a_3 = 0$, and so on.

The like is true of $a + bx^{\frac{1}{2}} + cx + dx^{\frac{3}{2}} + \cdots$, and of every series in which the exponents of x are positive and different from one another; for the reasoning just given applies to all such series.

The hypothesis that $a_0 + a_1x + \cdots$ vanishes for every value of x for which it converges contains more than is required for the proof that $a_0 = 0$, $a_1 = 0, \cdots$. For the reasoning above given shows that if $\beta_1, \beta_2, \cdots, \beta_n, \cdots$ denote any given never-ending sequence of numbers such that $\lim \beta_n = 0$, and if $a_0 + a_1x + \cdots$ vanishes when $x = \beta_1, \beta_2, \cdots, \beta_n, \cdots$, then $a_0 = 0$, $a_1 = 0, \cdots$. In particular, the numbers $\beta_1, \beta_2, \cdots$ may all be rational.

973 **Theorem 3.** *If* $a_0 + a_1x + a_2x^2 + \cdots = b_0 + b_1x + b_2x^2 + \cdots$ *for every value of* x *for which these series converge, the coefficients of the like powers of* x *are equal; that is,* $a_0 = b_0$, $a_1 = b_1$, $a_2 = b_2$, *and so on.*

For subtracting the second series from both members of the given equation, we have, by § 974,

$$(a_0 - b_0) + (a_1 - b_1)x + (a_2 - b_2)x^2 + \cdots = 0$$

for every value of x for which the given series converge.

Hence, § 972, $a_0 - b_0 = 0$, $a_1 - b_1 = 0$, $a_2 - b_2 = 0, \cdots$,

that is, $$a_0 = b_0,\ a_1 = b_1,\ a_2 = b_2, \cdots.$$

This theorem is called the *theorem of undetermined coefficients.* It asserts that a given function of x cannot be expressed in more than one way as a power series in x (compare § 421).

OPERATIONS WITH POWER SERIES

Since many functions of x can be defined by means of power series only, it is important to establish rules for reckoning with such series. These depend upon the following theorems, §§ 974, 976, which we shall demonstrate for infinite series in general.

Theorem 1. *If the series* $u_1 + u_2 + \cdots$ *and* $v_1 + v_2 + \cdots$ *converge and have the sums* S *and* T *respectively, the series* $(u_1 + v_1) + (u_2 + v_2) + \cdots$ *converges and has the sum* $S + T$. **974**

For, § 203, $\lim\,[(u_1 + v_1) + (u_2 + v_2) + \cdots + (u_n + v_n)]$

$$= \lim\,(u_1 + u_2 + \cdots + u_n) + \lim\,(v_1 + v_2 + \cdots + v_n)$$

$$= S + T.$$

The like is true of the series obtained by adding the corresponding terms of any *finite* number of infinite series. Hence the rule for adding any finite number of functions defined by power series in x is to add the corresponding terms of these series, that is, the terms which involve like powers of x. **975**

Thus, if $f(x) = 1 + x + x^2 + \cdots$ and $\phi(x) = x + 2x^2 + 3x^3 + \cdots$,
then $$f(x) + \phi(x) = 1 + 2x + 3x^2 + 4x^3 + \cdots$$
when the given series converge, that is, when $|x| < 1$.

If there be given an *infinite* number of series whose sums are S, T, $\cdots$, and we add the corresponding terms of these series, we ordinarily obtain a *divergent* series, even when the series $S + T + \cdots$ is convergent. But in the case described in the following theorem we obtain a convergent series by this process, and its sum is $S + T + \cdots$.

Theorem 2. *Let* $U_1 + U_2 + \cdots$ *denote a convergent series each of whose terms is the sum of an absolutely convergent series, namely,* **976**

$$U_1 = u_1^{(1)} + u_2^{(1)} + \cdots \;(1),\quad U_2 = u_1^{(2)} + u_2^{(2)} + \cdots \;(2),\; \cdots.$$

Again, let $U_1', U_2', \cdots$ *denote the sums of the series obtained by replacing the terms of* (1), (2), $\cdots$ *by their absolute values, so that*

$$U_1' = |u_1^{(1)}| + |u_2^{(1)}| + \cdots, \quad U_2' = |u_1^{(2)}| + |u_2^{(2)}| + \cdots,$$

and so on.

If the series $U_1' + U_2' + \cdots$ *is convergent, the several series obtained by adding the corresponding terms of* (1), (2), $\cdots$, *namely, the series* $u_1^{(1)} + u_1^{(2)} + \cdots$, $u_2^{(1)} + u_2^{(2)} + \cdots$, *and so on, are convergent, and if their sums be denoted by* $V_1, V_2, \cdots$, *we shall have*

$$U_1 + U_2 + U_3 + \cdots = V_1 + V_2 + V_3 + \cdots.$$

For let us represent the remainders after n terms in the series (1), (2), $\cdots$ by $R_n^{(1)}, R_n^{(2)}, \cdots$, § 943, so that

$$U_1 = u_1^{(1)} + u_2^{(1)} + \cdots + u_n^{(1)} + R_n^{(1)},$$

$$U_2 = u_1^{(2)} + u_2^{(2)} + \cdots + u_n^{(2)} + R_n^{(2)},$$

$$\cdot \quad \cdot \quad \cdot \quad \cdot \quad \cdot \quad \cdot \quad \cdot,$$

$$U_k = u_1^{(k)} + u_2^{(k)} + \cdots + u_n^{(k)} + R_n^{(k)},$$

$$\cdot \quad \cdot \quad \cdot \quad \cdot \quad \cdot \quad \cdot \quad \cdot.$$

Each of the column series $u_1^{(1)} + u_1^{(2)} + \cdots$, $u_2^{(1)} + u_2^{(2)} + \cdots$, and so on, is convergent since each of its terms is numerically less than the corresponding term of the convergent series $U_1' + U_2' + \cdots$, § 945, 1. Let the sums of these series be denoted by $V_1, V_2, \cdots, V_n, R_n$.

If we add the corresponding terms of these $n + 1$ column series, we obtain the original series $U_1 + U_2 + \cdots$. Therefore, since n is finite, we have, § 975,

$$U_1 + U_2 + \cdots + U_k + \cdots = V_1 + V_2 + \cdots + V_n + R_n.$$

To prove our theorem, therefore, we have only to show that when n is indefinitely increased $\lim R_n = 0$.

But if the remainder after k terms in $R_n^{(1)} + R_n^{(2)} + \cdots$ be denoted by $S_n^{(k)}$, we have $R_n = R_n^{(1)} + R_n^{(2)} + \cdots + R_n^{(k)} + S_n^{(k)}$.

Let δ denote any positive number, it matters not how small. Since each term of $R_n^{(1)} + R_n^{(2)} + \cdots$ (a) is numerically less than the corresponding term of $U_1' + U_2' + \cdots (b)$, the remainder after k terms in (a) is numerically less than the corresponding remainder in (b). But since (b) is convergent we can so choose k that the latter remainder will be less than $\delta/2$. Hence we can so choose k that *whatever the value of* n *may be*, we shall have $S_n^{(k)} < \delta/2$ numerically.

But again, since each of the row series $u_1^{(1)} + u_2^{(1)} + \cdots$, $u_1^{(2)} + u_2^{(2)} + \cdots$, is convergent, as n increases each of the k remainders $R_n^{(1)}, R_n^{(2)}, \cdots R_n^{(k)}$ will ultimately become and remain numerically less than $\delta/2k$, and therefore the sum of these remainders, namely, $R_n^{(1)} + R_n^{(2)} + \cdots + R_n^{(k)}$ will become and remain less than $(\delta/2k)k$, or $\delta/2$.

Therefore, as n increases, $R_n = R_n^{(1)} + R_n^{(2)} + \cdots + R_n^{(k)} + S_n^{(k)}$ will ultimately become and remain numerically less than $\delta/2 + \delta/2$, or δ.

Hence $\lim R_n = 0$, § 200; and therefore

$$U_1 + U_2 + U_3 + \cdots = V_1 + V_2 + V_3 + \cdots,$$

as was to be demonstrated.

A series $U_1 + U_2 + \cdots$ each of whose terms is itself an infinite series is called a *doubly infinite series.*

Thus, consider the series

$$x/(1+x) - x^2/(1+x)^2 + x^3/(1+x)^3 - \cdots \tag{1}$$

which converges for all real values of x which are greater than $-1/2$ (also for imaginary values of x whose real parts are greater than $-1/2$). Is it possible to transform (1) into a power series in x, that is, into a series which will converge for any value of x except 0?

When $|x|<1$, each term of (1) is the sum of a power series which may be obtained by the binomial theorem, § 988. Thus,

$$\left.\begin{array}{rlr} x/(1+x) = & x\,(1+x)^{-1} = x - x^2 + & x^3 - \quad x^4 + \cdots, \\ -x^2/(1+x)^2 = & -x^2(1+x)^{-2} = \quad -x^2 + & 2x^3 - 3x^4 + \cdots, \\ x^3/(1+x)^3 = & x^3(1+x)^{-3} = & x^3 - 3x^4 + \cdots, \\ \cdot \quad \cdot \quad \cdot & \cdot \quad \cdot \quad \cdot \quad \cdot & \cdot \quad \cdot \quad \cdot \end{array}\right\} \tag{2}$$

Replacing each term of the first of these series by its absolute value, we obtain $|x| + |x^2| + |x^3| + \cdots$, whose sum is $|x|/(1-|x|)$.

Treating the remaining series in a similar manner, we obtain series whose sums are $|x^2|/(1-|x|)^2$, $|x^3|/(1-|x|)^3$, and so on.

Hence the series $U_1' + U_2' + U_3' + \cdots$ of our theorem is here

$$|x|/(1-|x|) + |x^2|/(1-|x|)^2 + |x^3|/(1-|x|)^3 + \cdots,$$

which converges when $|x| < 1/2$.

Therefore, when $|x| < 1/2$, the power series obtained by adding the corresponding terms of the series (2), namely, $x - 2x^2 + 4x^3 - 8x^4 + \cdots$, converges and is equal to the given series (1); that is, when $|x| < 1/2$ we have

$$x/(1+x) - x^2/(1+x)^2 + x^3/(1+x)^3 - \cdots = x - 2x^2 + 4x^3 - 8x^4 + \cdots.$$

977 Commutative law valid for absolutely convergent series. We are now in a position to demonstrate that *the terms of an absolutely convergent series may be rearranged at pleasure without changing the sum of the series.*

1. We may rearrange the terms so as to form any other single infinite series out of them.

For let $u_1 + u_2 + \cdots$ (1) denote any absolutely convergent series, and let $u_1' + u_2' + \cdots$ (2) denote the same series with its terms rearranged. Again, let S_n denote the sum of the first n terms of (1), and S_m' the sum of the first m terms of (2).

Assign any value to n; then choose m so that the first n terms of (1) are to be found among the first m terms of (2); and finally choose p so that the first m terms of (2) are to be found among the first $n + p$ terms of (1).

Then $S_m' - S_n$ is made up of terms in $S_{n+p} - S_n$, that is, of terms in the sum $u_{n+1} + u_{n+2} + \cdots + u_{n+p}$.

Hence $|S_m' - S_n| \leqq |u_{n+1}| + |u_{n+2}| + \cdots + |u_{n+p}|$.

But since (1) is absolutely convergent, $\lim (|u_{n+1}| + \cdots + |u_{n+p}|) = 0$.

Therefore $\lim |S_m' - S_n| = 0$, that is, $\lim S_m' = \lim S_n$.

2. We may break the series up into any number (finite or infinite) of series the terms of each of which occur in the same order as in the original series. For we can recover the original series from every such set of series by applying one of the theorems of §§ 974, 976.

Thus, if we form one series out of the terms of $u_1 + u_2 + u_3 + \cdots$ which have odd indices, and another out of those which have even indices, we have, by § 974,

$$u_1 + u_2 + u_3 + u_4 + \cdots = (u_1 + u_3 + u_5 + \cdots) + (u_2 + u_4 + u_6 + \cdots).$$

Or again, arrange the terms of $u_1 + u_2 + u_3 + \cdots$ as follows:

$$\begin{array}{l} u_1 \\ u_2 + u_3 \\ u_4 + u_5 + u_6 \\ u_7 + u_8 + u_9 + u_{10} \\ \cdots\cdots\cdots \end{array}$$

In this scheme there are an infinite number of columns, each forming an infinite series.

The sum of $u_1 + u_2 + u_3 + \cdots$ is equal to the sum of the terms of the scheme added by rows, § 940. And the sum by rows is equal to the sum by columns, § 976.

Hence $u_1 + u_2 + u_3 + \cdots = (u_1 + u_2 + u_4 + \cdots) + (u_3 + u_5 + \cdots) + \cdots$. And similarly in every case.

3. Every possible rearrangement of the terms of

$$u_1 + u_2 + u_3 + \cdots + u_n + \cdots$$

may be had by combining 1 and 2.

978 **Products of power series.** If the functions $f(x)$ and $\phi(x)$ are defined, when $|x| < \lambda$, by the power series $f(x) = a_0 + a_1x + \cdots$ (1), $\phi(x) = b_0 + b_1x + \cdots$ (2), their product $f(x) \cdot \phi(x)$ will be defined, when $|x| < \lambda$, by a power series derived from (1) and (2) by the ordinary rules of multiplication (compare § 314).

Thus,

$$\begin{array}{rlllll} f(x) = & a_0 & + a_1x & + a_2x^2 & + a_3x^3 & + \cdots \quad (1) \\ \phi(x) = & b_0 & + b_1x & + b_2x^2 & + b_3x^3 & + \cdots \quad (2) \\ \hline f(x)\cdot\phi(x) = & a_0b_0 & + a_1b_0 \,|\, x & + a_2b_0 \,|\, x^2 & + a_3b_0 \,|\, x^3 & + \cdots \quad (3) \\ & & + a_0b_1 & + a_1b_1 & + a_2b_1 & \\ & & & + a_0b_2 & + a_1b_2 & \\ & & & & + a_0b_3 & \end{array}$$

For when $|x| < \lambda$ so that (1) and (2) are convergent, we have, § 939, $f(x)\phi(x) = f(x)b_0 + f(x)b_1x + f(x)b_2x^2 + \cdots$.

This is a series of the kind described in § 976, for it remains convergent when all the terms of $f(x)$ and $\phi(x)$ are replaced by their absolute values. We may therefore add the corresponding terms of $f(x)b_0 = a_0b_0 + a_1b_0x + \cdots$, $f(x)b_1x = 0 + a_0b_1x + \cdots$, and so on. The result is the series (3).

Example. Express $(1 + x + x^2 + \cdots)(1 + 2x + 3x^2 + \cdots)$ as a power series.

979 **Transformations of power series.** Suppose that the power series $a_0 + a_1y + a_2y^2 + \cdots$ (1) converges when $|y| < \lambda$.

Suppose also that y may be expressed in terms of x by the power series $y = b_0 + b_1x + \cdots$ (2), where $|b_0| < \lambda$.

By repeatedly multiplying (2) by itself, we obtain expressions for $y, y^2, y^3, \cdots$ in the form of power series in x which converge when (2) converges. If we substitute these expressions in the terms $a_1 y, a_2 y^2, \cdots$ of (1), we obtain a series of the form $a_0 + a_1(b_0 + b_1 x + \cdots) + a_2(b_0^2 + 2\,b_0 b_1 x + \cdots) + \cdots$ (3), and this, when the terms which involve like powers of x are collected, becomes a power series in x of the form $(a_0 + a_1 b_0 + \cdots) + (a_1 b_1 + 2\,a_2 b_0 b_1 + \cdots)x + \cdots$ (4).

This final series will converge and have the same sum as (1) for all values of x such that $|b_0| + |b_1 x| + \cdots < \lambda$. For in this case the condition of § 976 is satisfied by the doubly infinite series (3), the series $U_1' + U_2' + \cdots$ being $|a_0| + |a_1|(|b_0| + |b_1 x| + \cdots) + \cdots$, which by hypothesis converges when $|b_0| + |b_1 x| + \cdots < \lambda$.

980 Quotients of power series. A fraction whose numerator and denominator are power series, as

$$(a_0 + a_1 x + \cdots)/(b_0 + b_1 x + \cdots),$$

where $b_0 \neq 0$, may be transformed into a power series which will converge for all values of x for which $a_0 + a_1 x + \cdots$ converges and $|b_1 x| + |b_2 x^2| + \cdots < |b_0|$.

For let
$$y = b_1 x + b_2 x^2 + \cdots. \tag{1}$$

Then
$$\frac{1}{b_0 + b_1 x + b_2 x^2 + \cdots} = \frac{1}{b_0 + y} = \frac{1}{b_0} \cdot \frac{1}{1 + y/b_0}$$
$$= \frac{1}{b_0}\left(1 - \frac{y}{b_0} + \frac{y^2}{b_0^2} - \cdots\right), \tag{2}$$

since, by hypothesis and § 232, $|y| \overline{\gtrless} |b_1 x| + |b_2 x^2| + \cdots < |b_0|$.

In (2) replace y by its value (1) and then apply § 979. We shall thus transform (2) into a power series in x which converges when

$$|b_1 x| + |b_2 x^2| + \cdots < |b_0|.$$

Multiply this power series by $a_0 + a_1 x + a_2 x^2 + \cdots$, § 978.

The result will be a power series in x which will converge and be equal to the given fraction for all values of x for which $a_0 + a_1 x + \cdots$ converges and $|b_1 x| + |b_2 x^2| + \cdots < |b_0|$.

The quotient series may be obtained to any required term by the process of cancelling leading terms described in § 406, or by the method of undetermined coefficients, § 408.

Example. Expand $(1 + 2x + 2^2x^2 + \cdots)/(1 + x + x^2 + \cdots)$ to four terms.

Using detached coefficients, we have

$$\begin{array}{r|l} 1 + 2 + 4 + 8 + \cdots & 1 + 1 + 1 + 1 + \cdots \\ \underline{1 + 1 + 1 + 1 + \cdots} & 1 + 1 + 2 + 4 + \cdots \\ 1 + 3 + 7 + \cdots & \\ \underline{1 + 1 + 1 + \cdots} & \\ 2 + 6 + \cdots & \\ \underline{2 + 2 + \cdots} & \\ 4 + \cdots & \end{array}$$

Hence the quotient series is $1 + x + 2x^2 + 4x^3 + \cdots$. It converges when $|x| < 1/2$.

981 When instead of being infinite series the numerator and denominator are polynomials in x, so that the fraction has the form $(a_0 + a_1x + \cdots + a_mx^m)/(b_0 + b_1x + \cdots + b_nx^n)$, the quotient series will converge for all values of x which are numerically less than the numerically smallest root of the equation $b_0 + b_1x + \cdots + b_nx^n = 0$. This will be evident from the first of the following examples. The second of these examples illustrates the form of the quotient series when $b_0 = 0$, and the third illustrates the method of expressing the fraction as a power series in $1/x$.

Example 1. Find the limit of convergence of the series which is the expansion of $(3x + 8)/(x^2 + 5x + 6)$.

By the method of partial fractions, § 537, we find

$$\frac{3x + 8}{x^2 + 5x + 6} = \frac{3x + 8}{(x + 2)(x + 3)} = \frac{2}{x + 2} + \frac{1}{x + 3}.$$

But $$\frac{2}{x + 2} = \frac{1}{1 + x/2} = \left(1 + \frac{x}{2}\right)^{-1} = 1 - \frac{x}{2} + \frac{x^2}{2^2} - \cdots \qquad \text{when } |x| < 2,$$

and $$\frac{1}{x + 3} = \frac{1}{3}\frac{1}{1 + x/3} = \frac{1}{3}\left(1 + \frac{x}{3}\right)^{-1} = \frac{1}{3} - \frac{x}{3^2} + \frac{x^2}{3^3} - \cdots \quad \text{when } |x| < 3.$$

Therefore $$\frac{3x + 8}{x^2 + 5x + 6} = \frac{4}{3} - \frac{11x}{18} + \frac{31x^2}{108} - \cdots \qquad \text{when } |x| < 2.$$

Example 2. Expand $(1-x)/(x^2+4x^3)$ in increasing powers of x.

We have
$$\frac{1-x}{x^2+4x^3}=\frac{1}{x^2}\frac{1-x}{1+4x}=\frac{1}{x^2}(1-5x+20x^2-80x^3+\cdots)$$
$$=x^{-2}-5x^{-1}+20-80x+\cdots.$$

Example 3. Expand $(2x^2+x-3)/(x^3+2x+4)$ in powers of $1/x$.
$$\frac{2x^2+x-3}{x^3+2x+4}=\frac{1}{x}\cdot\frac{2+1/x-3/x^2}{1+2/x+4/x^3}=\frac{1}{x}\left(2-\frac{3}{x}-\frac{5}{x^2}\cdots\right)=\frac{2}{x}-\frac{3}{x^2}-\frac{5}{x^3}\cdots$$

982 Reversion of series. From the equation $y=a_1x+a_2x^2+\cdots$, defining y in terms of x, it is possible to derive another of the form $x=b_1y+b_2y^2+\cdots$, defining x in terms of y. The process is called the *reversion* of the given series $a_1x+a_2x^2+\cdots$. It will be observed that this series lacks the constant term a_0, and the understanding is that $a_1\neq 0$. It can be proved that if $a_1x+a_2x^2+\cdots$ has a limit of convergence greater than 0, the like is true of the reverted series $b_1y+b_2y^2+\cdots$.

Example. Revert the series $y=x+2x^2+3x^3+\cdots$.

Assume
$$x=b_1y+b_2y^2+b_3y^3+\cdots. \quad (1)$$

Computing y^2, y^3, $\cdots$ from the given equation by the method of § 978, and substituting the resulting series in (1), we have
$$x=b_1x+\left.\begin{matrix}2b_1\\+\ b_2\end{matrix}\right|x^2+\left.\begin{matrix}3b_1\\+4b_2\\+\ b_3\end{matrix}\right|x^3+\cdots \quad (2)$$

Equating coefficients, $1=b_1$, $0=2b_1+b_2$, $0=3b_1+4b_2+b_3$, $\cdots$ whence $b_1=1$, $b_2=-2$, $b_3=5$, $\cdots$.

Therefore
$$x=y-2y^2+5y^3+\cdots. \quad (3)$$

By the same method, from an equation of the form $y=a_0+a_1x+a_2x^2+\cdots$, or $y-a_0=a_1x+a_2x^2+\cdots$, we can derive another of the form $x=b_1(y-a_0)+b_2(y-a_0)^2+\cdots$. And from an equation of the form $y=a_2x^2+a_3x^3+\cdots$ we can derive two others of the form $x=b_1y^{\frac{1}{2}}+b_2y+b_3y^{\frac{3}{2}}+\cdots$.

983 Expansion of algebraic functions. An algebraic equation of the form $f(x,y)=0$ which lacks a constant term is satisfied when $x=0$ and $y=0$. Hence, if we suppose $f(x,y)=0$

solved for y in terms of x, one or more of the solutions must be expressions in x which vanish when x vanishes. It can be proved that these expressions may be expanded in series in increasing powers of x which have limits of convergence greater than 0. In ordinary cases these series may be obtained to any required term by the method illustrated in the following examples.

Example 1. The equation $y^2 + y - 2x = 0$ lacks a constant term. Find the expansion for the value of y which vanishes when $x = 0$.

When $x = 0$, the equation

$$y^2 + y - 2x = 0 \tag{1}$$

becomes $y^2 + y = 0$. Since one and but one of the roots of this equation is 0, one and but one of the solutions of (1) for y in terms of x vanishes when x vanishes.

Suppose that when this solution is expanded in a series of increasing powers of x its first term is ax^μ, so that

$$y = ax^\mu + \cdots. \tag{2}$$

Substituting (2) in (1), we have

$$a^2x^{2\mu} + \cdots + ax^\mu + \cdots - 2x = 0. \tag{3}$$

Since by hypothesis (3) is an identity, the sums of the coefficients of its terms of like degree must be 0. Hence there must be at least two terms of lowest degree; and since μ is positive, these must be the terms ax^μ and $-2x$. Therefore $\mu = 1$ and $a - 2 = 0$, or $a = 2$.

We therefore assume that

$$y = 2x + bx^2 + cx^3 + \cdots. \tag{2'}$$

Substituting (2′) in (1), we obtain $(4 + b)x^2 + (4b + c)x^3 + \cdots = 0$.

Hence $4 + b = 0$, $4b + c = 0, \cdots$, and therefore $b = -4$, $c = 16, \cdots$.

Therefore the required solution is $y = 2x - 4x^2 + 16x^3 + \cdots$.

Example 2. Find the expansions of the values of y in terms of x which satisfy the equation $y^3 - xy + x^2 = 0$ and vanish when $x = 0$.

When $x = 0$, the equation

$$y^3 - xy + x^2 = 0 \tag{1}$$

becomes $y^3 = 0$, all three of whose roots are 0. Hence we may expect to find three expansions of the kind required.

Let ax^μ denote the leading term in one of these expansions, so that

$$y = ax^\mu + \cdots. \tag{2}$$

Substituting (2) in (1), we have

$$a^3x^{3\mu} + \cdots - ax^{\mu+1} + \cdots + x^2 = 0. \tag{3}$$

By the reasoning of Ex. 1, at least two of the exponents 3μ, $\mu + 1$, and 2 must be equal and less than any other exponent of x in (3).

Setting $3\mu = \mu + 1$, we find $\mu = 1/2$. This is an admissible value of μ, since when $\mu = 1/2$, both 3μ and $\mu + 1$ are less than 2.

Setting $\mu + 1 = 2$, we find $\mu = 1$. This also is an admissible value of μ, since when $\mu = 1$, both $\mu + 1$ and 2 are less than 3μ.

Setting $3\mu = 2$, we find $\mu = 2/3$. But this is not an admissible value of μ, since when $\mu = 2/3$, 3μ and 2 are greater than $\mu + 1$.

Hence μ must have one of the values 1 or 1/2.

When $\mu = 1$, (3) becomes $a^3x^3 + \cdots - ax^2 + \cdots + x^2 = 0$, from which it follows that $-a + 1 = 0$, or $a = 1$.

When $\mu = 1/2$, (3) becomes $a^3x^{\frac{3}{2}} + \cdots - ax^{\frac{3}{2}} + \cdots + x^2 = 0$, from which it follows that $a^3 - a = 0$, or, since $a \neq 0$, that $a = \pm 1$.

We therefore assume that the required solutions are of the form $y = x + bx^2 + cx^3 + \cdots$, $y = x^{\frac{1}{2}} + bx + cx^{\frac{3}{2}} + \cdots$, $y = -x^{\frac{1}{2}} + bx + cx^{\frac{3}{2}} + \cdots$.

And substituting these expressions for y in (1) and determining the coefficients as in Ex. 1, we obtain

$$y = x + x^2 + 3x^3 + \cdots,\ y = x^{\frac{1}{2}} - \frac{x}{2} - \frac{3x^{\frac{3}{2}}}{8} + \cdots,\ y = -x^{\frac{1}{2}} - \frac{x}{2} + \frac{3x^{\frac{3}{2}}}{8} + \cdots.$$

In this method it is assumed that if the leading term of one of the required expansions is $ax^{\frac{p}{q}}$, the expansion will be in powers of $x^{\frac{1}{q}}$. In exceptional cases this is not true and the method fails. But the following method is general. Having found the leading term ax^{μ} of an expansion as in the examples, set $y = x^{\mu}(a + v)$ in the given equation. It becomes an equation in v and x. From this equation find the leading term of the expansion of v in powers of x, and so on.*

Thus, in Ex. 2, setting $y = x^{\frac{1}{2}}(1 + v)$ in

$$y^3 - xy + x^2 = 0 \tag{1}$$

and simplifying, we have

$$v^3 + 3v^2 + 2v + x^{\frac{1}{2}} = 0; \tag{2}$$

whence $v = -x^{\frac{1}{2}}/2 + \cdots$, and therefore $y = x^{\frac{1}{2}}(1 - x^{\frac{1}{2}}/2 + \cdots) = x^{\frac{1}{2}} - x/2 + \cdots$. To find the next term, set $v = x^{\frac{1}{2}}(-1/2 + v')$ in (2), and so on.

* For a fuller discussion of the methods of this section and the use in connection with them of Newton's parallelogram see Chrystal's *Algebra*, II, pp. 349–371; also Frost's *Curve Tracing* and Johnson's *Curve Tracing*.

Taylor's theorem. If $f(x) = a_0 + a_1x + a_2x^2 + \cdots$ when $|x| < \lambda$, and we replace x by $x + h$, we obtain **984**

$$f(x+h) = a_0 + a_1(x+h) + a_2(x+h)^2 + \cdots.$$

It follows from § 976 that when $|x| + |h| < \lambda$ we may transform this series into a power series in h by expanding $(a+h)^2$, $(a+h)^3, \cdots$ by the binomial theorem, and then collecting terms which involve like powers of h. By the method employed in § 848 it may be shown that the result will be

$$f(x+h) = f(x) + f'(x)h + f''(x)\cdot\frac{h^2}{2!} + \cdots + f^{(n)}(x)\cdot\frac{h^n}{n!} + \cdots,$$

where $f'(x)$, $f''(x)$, $\cdots$ denote the sums of the series whose terms are the first, second, $\cdots$ derivatives of the terms of the given series $a_0 + a_1x + a_2x^2 + \cdots$, namely,

$$f'(x) = a_1 + 2\,a_2x + 3\,a_3x^2 + \cdots,$$
$$f''(x) = 2\,a_2 + 3\cdot 2\,a_3x + \cdots, \text{ and so on.}$$

If in the preceding identity we replace x by a and h by $x - a$, where $|a| + |x - a| < \lambda$, we obtain the expansion of $f(x)$ in powers of $x - a$, namely, **985**

$$f(x) = f(a) + f'(a)(x-a) + \cdots + f^n(a)\frac{(x-a)^n}{n!} + \cdots.$$

From this last expansion and § 971 it follows that if $f(x) = a_0 + a_1x + \cdots$ when $|x| < \lambda$, and if $|a| < \lambda$, then **986**

$$\lim_{x \doteq a} f(x) = f(a).$$

EXERCISE LXXXIX

1. Show that $(1 + x + x^2 + \cdots)^2 = 1 + 2x + 3x^2 + 4x^3 + \cdots$.

2. Show that $(1 + x + x^2 + \cdots)^3 = 1 + 3x + 6x^2 + 10x^3 + \cdots$.

3. Show that $(1 + x^2 + x^4 + \cdots)/(1 + x + x^2 + \cdots) = 1 - x + x^2 - \cdots$.

4. Assuming that $(1 - x + 2x^2)^{\frac{1}{2}} = 1 + a_1x + a_2x^2 + \cdots$, find a_1, a_2, a_3, a_4 by squaring the given equation and applying § 973.

5. By a similar method find the first four terms of the expansions of

(1) $(8 - 3x)^{\frac{1}{3}}$. (2) $(1 + x - x^2)^{\frac{3}{2}}$.

6. Expand each of the following fractions in ascending powers of x to the fourth term by the method of the example in § 980.

(1) $\dfrac{2 + x - 3x^2 + 5x^3}{1 + 2x + 3x^2}$. (2) $\dfrac{x + 5x^2 - x^3}{1 - x + x^2 - x^3}$.

7. Expand each of the following fractions in ascending powers of x to the fourth term by the method of undetermined coefficients.

(1) $\dfrac{3x^2 + x^3}{1 + x + x^2}$. (2) $\dfrac{x + 5x^4}{x^3 + 2x^4 + 3x^5}$.

8. Expand each of the following fractions to the fifth term by the method of the first example in § 981 and indicate the limits of convergence of the expansions.

(1) $\dfrac{9x - 22}{(x^2 - 4)(x - 3)}$. (2) $\dfrac{5x + 6}{(2x + 3)(x + 1)^2}$.

9. Expand each of the following fractions to the fourth term in descending powers of x. For what values of x will the first of these expansions converge?

(1) $\dfrac{2x + 3}{2x^2 + x - 15}$. (2) $\dfrac{x^4 + 1}{x^4 + x^3 + x^2 + x + 1}$.

10. Revert each of the following series to the fourth term.

(1) $y = x + x^2 + x^3 + x^4 + \cdots$. (2) $y = x - \dfrac{x^2}{2} + \dfrac{x^3}{3} - \dfrac{x^4}{4} + \cdots$.

11. From $y = 1 + x + x^2/2 + x^3/3 + \cdots$ derive to the fourth term a series for x in powers of $y - 1$.

12. From $y = x^2 + 3x^3$ derive to the fourth term a series for x in powers of $y^{\frac{1}{2}}$.

13. By the method of § 983 find the first three terms of the expansions of the values of y in terms of x which satisfy the following equations and vanish when $x = 0$.

(1) $x^2 + y^2 + y - 3x = 0$. (2) $x^3 + y^3 - xy = 0$.

14. By aid of the theorem of § 976 show that

$$\frac{x}{1 - x} + \frac{2x^2}{1 - x^2} + \frac{3x^3}{1 - x^3} + \cdots = \frac{x}{(1 - x)^2} + \frac{x^2}{(1 - x^2)^2} + \frac{x^3}{(1 - x^3)^2} + \cdots.$$

XXXIV. THE BINOMIAL, EXPONENTIAL, AND LOGARITHMIC SERIES

The binomial series. When m is a positive integer, **987**

$$1+\frac{m}{1}x+\frac{m(m-1)}{1\cdot 2}x^2+\frac{m(m-1)(m-2)}{1\cdot 2\cdot 3}x^3+\cdots \quad (1)$$

is a finite or terminating series and its sum is $(1+x)^m$.

When m is not a positive integer, (1) is an infinite series, but one which converges, that is, has a sum, when $|x|<1$, § 968. We proceed to demonstrate that if m has any rational value whatsoever, this sum is $(1+x)^m$.

The series (1) is a function of both x and m, but since we are now concerned mainly with its relation to m we shall represent it by $\phi(m)$.

For convenience let m_r denote the coefficient of x^r in (1), so that $m_r = m(m-1)\cdots(m-r+1)/r!$.

Then if m and n denote any two numbers, we have

$$\phi(m) = 1 + m_1 x + m_2 x^2 + m_3 x^3 + \cdots, \quad (2)$$

$$\phi(n) = 1 + n_1 x + n_2 x^2 + n_3 x^3 + \cdots, \quad (3)$$

$$\phi(m+n) = 1 + (m+n)_1 x + (m+n)_2 x^2 + \cdots. \quad (4)$$

We can prove that $\phi(m)\cdot\phi(n) = \phi(m+n)$.

For when $|x|<1$, so that (2) and (3) converge, we have

$$\phi(m)\cdot\phi(n) = 1 + \left.\begin{matrix} m_1 \\ +n_1 \end{matrix}\right| x + \left.\begin{matrix} m_2 \\ +m_1 n_1 \\ +n_2 \end{matrix}\right| x^2 + \left.\begin{matrix} m_3 \\ +m_2 n_1 \\ +m_1 n_2 \\ +n_3 \end{matrix}\right| x^3 + \cdots. \quad (5)$$

But in §§ 773, 774 it is shown that

$$m_1 + n_1 = (m+n)_1,\ m_2 + m_1 n_1 + n_2 = (m+n)_2, \cdots,$$
$$m_r + m_{r-1} n_1 + \cdots + m_1 n_{r-1} + n_r = (m+n)_r.$$

Hence $\phi(m)\cdot\phi(n)$

$$= 1 + (m+n)_1 x + (m+n)_2 x^2 + (m+n)_3 x^3 + \cdots$$
$$= \phi(m+n). \quad (6)$$

By repeated applications of (6), we have

$$\phi(m)\cdot\phi(n)\cdot\phi(p)=\phi(m+n)\cdot\phi(p)=\phi(m+n+p),$$

and so on, for any finite number of factors of the form $\phi(m)$, $\phi(n)$, $\phi(p)$, $\phi(q)$, $\cdots$.

We are now prepared to prove the binomial theorem for all rational values of the exponent m, namely:

988 **Theorem.** *If* m *be any rational number whatsoever, the sum of the series*

$$\phi(\mathrm{m})=1+\frac{\mathrm{m}}{1}\mathrm{x}+\frac{\mathrm{m}(\mathrm{m}-1)}{1\cdot 2}\mathrm{x}^2+\frac{\mathrm{m}(\mathrm{m}-1)(\mathrm{m}-2)}{1\cdot 2\cdot 3}\mathrm{x}^3+\cdots,$$

when $|\mathrm{x}|<1$, *is* $(1+\mathrm{x})^{\mathrm{m}}$.

Notice first of all that when $m=0$ the series reduces to 1, and that when $m=1$ the series reduces to $1+x$.

Hence $$\phi(0)=1 \text{ and } \phi(1)=1+x. \qquad (1)$$

1. Let m be a positive integer.

Then
$$\begin{aligned}\phi(m)&=\phi(1+1+\cdots \text{ to } m \text{ terms})\\ &=\phi(1)\cdot\phi(1)\cdots \text{ to } m \text{ factors}\\ &=[\phi(1)]^m=(1+x)^m, \text{ by (1)}, \qquad (2)\end{aligned}$$

which proves the theorem for a positive integral exponent.

2. Let m be any positive rational fraction p/q.

Then
$$\begin{aligned}[\phi(p/q)]^q&=\phi(p/q)\cdot\phi(p/q)\cdots \text{ to } q \text{ factors}\\ &=\phi(p/q+p/q+\cdots \text{ to } q \text{ terms})\\ &=\phi(p)=(1+x)^p, \text{ by (2)}.\end{aligned}$$

Therefore $$\phi(p/q)=(1+x)^{\frac{p}{q}}. \qquad (3)$$

For it follows from the equation $[\phi(p/q)]^q=(1+x)^p$ and § 986 that the values which $\phi(p/q)$ takes for all values of x such that $|x|<1$ must be the corresponding values of one and the same qth root of $(1+x)^p$.

Moreover this root must be the *principal* qth root, namely, $(1+x)^{\frac{p}{q}}$; for this is the only one of the qth roots of $(1+x)^p$ which has the same value as $\phi(p/q)$ when $x=0$.

3. Let m be any negative rational number $-s$.

Since $\phi(-s)\cdot\phi(s)=\phi(-s+s)=\phi(0)=1,$ by (1)

we have $\phi(-s)=1/\phi(s)=1/(1+x)^s$ by (3)

$$=(1+x)^{-s}, \quad (4)$$

which proves the theorem for any rational exponent.

It is not difficult to extend the theorem to irrational values of the exponent.

Example. Expand $(1+2x+3x^2)^{\frac{1}{3}}$ in ascending powers of x.

We have $(1+2x+3x^2)^{\frac{1}{3}}=[1+(2x+3x^2)]^{\frac{1}{3}}$

$$=1+\tfrac{1}{3}\cdot(2x+3x^2)+\frac{\frac{1}{3}(-\frac{2}{3})}{2}(2x+3x^2)^2$$

$$+\frac{\frac{1}{3}(-\frac{2}{3})(-\frac{5}{3})}{2\cdot 3}(2x+3x^2)^3+\cdots$$

$$=1+\frac{2x}{3}+\frac{5x^2}{9}-\frac{68x^3}{81}\cdots.$$

The expansion converges when $2|x|+3|x^2|<1$;
therefore when $9|x^2|+6|x|+1<4$;
therefore when $3|x|+1<2$;
therefore when $|x|<1/3$.

Corollary. *If* m *is rational and* $|x|<|a|$, *we have* **989**

$$(a+x)^m=a^m+ma^{m-1}x+\frac{m(m-1)}{1\cdot 2}a^{m-2}x^2+\cdots$$

For $(a+x)^m=a^m\left(1+\dfrac{x}{a}\right)^m$

$$=a^m\left[1+m\frac{x}{a}+\frac{m(m-1)}{1\cdot 2}\frac{x^2}{a^2}+\cdots\right] \quad (1)$$

$$=a^m+ma^{m-1}x+\frac{m(m-1)}{1\cdot 2}a^{m-2}x^2+\cdots, \quad (2)$$

where (1) and therefore (2) converge if $|x/a|<1$, or $|x|<|a|$.

990 The exponential series. We have already shown that the series

$$1 + x/1 + x^2/2! + x^3/3! + \cdots + x^n/n! + \cdots \qquad (1)$$

converges for all finite values of x, § 968.

Let e denote its sum when $x = 1$, so that

$$e = 1 + 1 + 1/2! + 1/3! + \cdots = 2.71828\cdots.$$

We are to prove that the sum of (1) for any real value of x is e^x.

For let $f(x)$ denote the sum of (1), so that

$$f(x) = 1 + x/1 + x^2/2! + x^3/3! + \cdots + x^n/n! + \cdots,$$

$$f(y) = 1 + y/1 + y^2/2! + y^3/3! + \cdots + y^n/n! + \cdots.$$

Then by the rule for multiplying infinite series, § 978,

$$f(x) \cdot f(y) = 1 + (x + y) + \left(\frac{x^2}{2!} + xy + \frac{y^2}{2!}\right)$$

$$+ \left(\frac{x^3}{3!} + \frac{x^2}{2!}\frac{y}{1} + \frac{x}{1}\frac{y^2}{2!} + \frac{y^3}{3!}\right) + \cdots$$

$$= 1 + (x + y) + \frac{(x + y)^2}{2!} + \frac{(x + y)^3}{3!} + \cdots$$

$$= f(x + y).$$

From this result it follows that

$f(x) \cdot f(y) \cdot f(z) = f(x + y) \cdot f(z) = f(x + y + z)$, and so on.

Hence, observing that $f(0) = 1$ and $f(1) = e$, we may prove successively, precisely as in § 988, that

1. When x is a positive integer m,

$$f(m) = [f(1)]^m = e^m.$$

2. When x is a positive fraction p/q,

$$f(p/q) = \sqrt[q]{[f(1)]^p} = \sqrt[q]{e^p} = e^{\frac{p}{q}}.$$

3. When x is a negative rational $-s$,

$$f(-s) = 1/f(s) = 1/e^s = e^{-s}.$$

Therefore when x is rational we have $f(x) = e^x$, that is,

$$e^x = 1 + x + x^2/2! + x^3/3! + \cdots + x^n/n! + \cdots. \qquad (2)$$

Moreover (2) is also true for irrational values of the exponent x. For if b denote any given irrational number and x be made to approach b as limit through a sequence of rational values, for all these rational values of x we have $f(x) = e^x$ and therefore $\lim_{x \doteq b} f(x) = \lim_{x \doteq b} e^x$.

But $\lim_{x \doteq b} f(x) = f(b)$, § 986, and $\lim_{x \doteq b} e^x = e^b$, § 728. Therefore $f(b) = e^b$, that is, $1 + b + b^2/2! + \cdots = e^b$.

The second member of (2) is also a convergent series, that is, has a sum, when x is imaginary. Hence (2) may be used to *define* e^x for imaginary values of the exponent x. Thus, by *definition*,

$$e^i = 1 + i + i^2/2! + i^3/3! + \cdots + i^n/n! + \cdots.$$

Series for a^x. Let a denote any positive number and x any real number. **991**

Since $a = e^{\log_e a}$, § 732, we have $a^x = e^{x \log_e a}$, § 730.

Therefore, substituting $x \log_e a$ for x in the series, § 990, (2), we have

$$a^x = 1 + x \log_e a + x^2 (\log_e a)^2/2! + \cdots + x^n (\log_e a)^n/n! + \cdots.$$

It can be proved, as in § 968, 1, that this series converges for all finite values of x.

The logarithmic series. If in the series just obtained for a^x we replace a by $1 + x$ and x by y, we have **992**

$$(1 + x)^y = 1 + \log_e(1 + x) \cdot y + [\log_e(1 + x)]^2 y^2/2! + \cdots. \quad (1)$$

But by the binomial theorem, § 988, when $|x| < 1$,

$$(1 + x)^y = 1 + yx + \frac{y(y-1)}{1 \cdot 2} x^2 + \frac{y(y-1)(y-2)}{1 \cdot 2 \cdot 3} x^3 + \cdots. \quad (2)$$

By carrying out the indicated multiplications and collecting terms we can transform (2) into a power series in y.

The coefficient of y in this series will be

$$x + \frac{(-1)x^2}{1 \cdot 2} + \frac{(-1)(-2)x^3}{1 \cdot 2 \cdot 3} + \cdots, \text{ or } x - \frac{x^2}{2} + \frac{x^3}{3} - \cdots.$$

Equate this to the coefficient of y in (1). We obtain, if $|x| < 1$,

$$\log_e(1+x) = x - x^2/2 + x^3/3 - x^4/4 + \cdots. \quad (3)$$

This series is called the *logarithmic series.* In § 968 we proved that it converges when $|x| < 1$.

In the proof just given we have assumed that the series (2) remains equal to $(1+x)^y$ after it has been transformed into a power series in y. But this follows from § 976 when $|x| < 1$. For if x' and y' denote $|x|$ and $|y|$ respectively, the series $U_1' + U_2' + \cdots$ of § 976, corresponding to (2), is

$$1 + y'x' + \frac{y'(y'+1)}{1 \cdot 2} x'^2 + \frac{y'(y'+1)(y'+2)}{1 \cdot 2 \cdot 3} x'^3 + \cdots,$$

and this series is convergent when $x' < 1$, its sum being $(1-x')^{-y'}$, § 988.

As we have proved the truth of the binomial theorem only for rational values of the exponent, it may be observed that (3) follows from (1) and (2) even when y is restricted to rational values (see the remark at the end of § 972).

Example. Show that $\lim_{n \doteq \infty}\left(1+\frac{1}{n}\right)^n = e$.

We have

$$\log_e\left(1+\frac{1}{n}\right)^n = n\log_e\left(1+\frac{1}{n}\right) = n\left(\frac{1}{n} - \frac{1}{2n^2} + \cdots\right) = 1 - \frac{1}{2n} + \cdots.$$

Hence $\lim_{n \doteq \infty} \log_e\left(1+\frac{1}{n}\right)^n = 1$, and therefore $\lim_{n \doteq \infty}\left(1+\frac{1}{n}\right)^n = e^1 = e$.

For $\lim u = \lim e^{\log_e u} = e^{\lim(\log_e u)}$, §§ 726–729, 731.

993 **Computation of natural logarithms.** The logarithms of numbers to the base e are called their *natural logarithms.* A table of natural logarithms may be computed as follows:

We have $\log_e(1+x) = x - x^2/2 + x^3/3 - x^4/4 + \cdots$, (1)

and therefore $\log_e(1-x) = -x - x^2/2 - x^3/3 - x^4/4 - \cdots$. (2)

Subtract (2) from (1). Since

$$\log_e(1+x) - \log_e(1-x) = \log_e\frac{1+x}{1-x},$$

we obtain $\log_e\dfrac{1+x}{1-x} = 2\left(x + \dfrac{x^3}{3} + \dfrac{x^5}{5} + \cdots\right)$. (3)

In (3) set $\frac{1+x}{1-x}=\frac{n+1}{n}$ and therefore $x=\frac{1}{2n+1}$.
We obtain

$$\log_e\frac{n+1}{n}=2\left(\frac{1}{2n+1}+\frac{1}{3(2n+1)^3}+\frac{1}{5(2n+1)^5}+\cdots\right),$$

or $\log_e(n+1)$

$$=\log_e n+2\left(\frac{1}{2n+1}+\frac{1}{3(2n+1)^3}+\frac{1}{5(2n+1)^5}+\cdots\right). \quad (4)$$

Setting $n=1$ in (4),

$$\log_e 2=2(1/3+1/3^4+1/5\cdot 3^5+\cdots)=.6931\cdots.$$

Setting $n=2$ in (4),

$$\log_e 3=\log_e 2+2(1/5+1/3\cdot 5^3+\cdots)=1.0986\cdots,$$

and so on to any integral value of n.

Modulus. By § 755, $\log_a n=\log_e n/\log_e a$. Hence the logarithms of numbers to any base a may be obtained by multiplying their natural logarithms by $1/\log_e a$. We call $1/\log_e a$ or its equivalent, § 756, $\log_a e$, the *modulus* of the system of logarithms to the base a. In particular, the modulus of the system of *common* logarithms is $\log_{10}e=.43429\cdots$. **994**

EXERCISE XC

1. Compute $\log_e 4$ and $\log_e 5$ each to the fourth decimal figure.

2. Show that $e^{-1}=2/3!+4/5!+6/7!+\cdots$.

3. Show by multiplication that

$$\left(1+\frac{x}{1}+\frac{x^2}{2!}+\frac{x^3}{3!}+\cdots\right)\left(1-\frac{x}{1}+\frac{x^2}{2!}-\frac{x^3}{3!}+\cdots\right)=1.$$

4. Show that $(e^{ix}+e^{-ix})/2=1-x^2/2!+x^4/4!-x^6/6!+\cdots$.

5. Show that $(e^{ix}-e^{-ix})/2i=x-x^3/3!+x^5/5!-x^7/7!+\cdots$.

6. Show that the $(r+1)$th term in the expansion of $(1-x)^{-n}$ by the binomial theorem is $\frac{n(n+1)\cdots(n+r-1)}{r!}x^r$.

7. Find the term in the expansion of $(8+x)^{\frac{2}{3}}$ which involves x^4.

8. Find the term in the expansion of $(1-x^{\frac{1}{2}})^{-\frac{1}{2}}$ which involves x^3.

9. For what values of x will the expansion of $(9-4x^2)^{\frac{1}{2}}$ and of $(12+x+x^2)^{\frac{2}{3}}$ in ascending powers of x converge?

10. Expand $(1-x+2x^2)^{\frac{3}{4}}$ in powers of x to the term involving x^4.

11. Find the first three terms of the expansion of $(8+3x)^{\frac{2}{3}}(9-2x)^{-\frac{1}{2}}$ in powers of x. For what values of x will the expansion converge?

12. Find the limiting values of the following expressions.

(1) $\lim\limits_{x \doteq 0} \dfrac{e^{2x}-e^{-2x}}{3x}$. (2) $\lim\limits_{x \doteq 0} \dfrac{(1+x^2)^{\frac{1}{2}}-(1-x^2)^{\frac{1}{3}}}{(1+3x)^{\frac{1}{3}}-(1+4x)^{\frac{1}{4}}}$.

13. Prove that $\lim\limits_{n \doteq \infty}\left(1+\dfrac{1}{n}\right)^{nx} = e^x$.

14. Prove that $\lim\limits_{x \doteq 0}(e^x+x)^{\frac{1}{x}} = e^2$.

15. Expand $\log_e(1+x+x^2)$ in powers of x to the term involving x^4. For what values of x will this expansion converge?

16. Show that $\log_e \dfrac{m}{n} = \dfrac{m-n}{n} - \dfrac{1}{2}\left(\dfrac{m-n}{n}\right)^2 + \dfrac{1}{3}\left(\dfrac{m-n}{n}\right)^3 - \cdots$.

17. Show that $\log_e \dfrac{n^2}{n^2-1} = \dfrac{1}{n^2} + \dfrac{1}{2n^4} + \dfrac{1}{3n^6} + \cdots$.

18. Show that

$$\frac{1}{n+1} + \frac{1}{2(n+1)^2} + \frac{1}{3(n+1)^3} + \cdots = \frac{1}{n} - \frac{1}{2n^2} + \frac{1}{3n^3} - \cdots.$$

XXXV. RECURRING SERIES

995 **Recurring series.** A series $a_0 + a_1x + a_2x^2 + \cdots$ in which every $r+1$ consecutive coefficients are connected by an identity of the form

$$a_n + p_1a_{n-1} + p_2a_{n-2} + \cdots + p_ra_{n-r} = 0,$$

where $p_1, p_2, \cdots, p_r$ are constant for all values of n, is called a *recurring series of the rth order*, and the identity is called its *scale of relation*.

Thus, in $\quad 1 + 3x + 5x^2 + 7x^3 + \cdots + (2n+1)x^n + \cdots \quad$ (1)

every three consecutive coefficients are connected by the formula

$$a_n - 2a_{n-1} + a_{n-2} = 0; \qquad (2)$$

for $5 - 2\cdot 3 + 1 = 0$, $7 - 2\cdot 5 + 3 = 0$, $2n+1-2(2n-1)+2n-3=0$.

Hence (1) is a recurring series of the second order whose scale of relation is (2).

A geometric series is a recurring series of the first order.

996 Every power series which is the expansion of a proper fraction whose denominator is of the rth degree is a recurring series of the rth order.

Thus, if $\dfrac{2+x}{1+2x+3x^2} = a_0 + a_1 x + a_2 x^2 + \cdots + a_n x^n + \cdots$, (1)

we have

$$2 + x = a_0 + \left.\begin{matrix} a_1 \\ +2a_0 \end{matrix}\right| x + \left.\begin{matrix} a_2 \\ +2a_1 \\ +3a_0 \end{matrix}\right| x^2 + \cdots + \left.\begin{matrix} a_n \\ +2a_{n-1} \\ +3a_{n-2} \end{matrix}\right| x^n + \cdots$$

Therefore $a_0 = 2, \quad a_1 + 2a_0 = 1, \quad a_2 + 2a_1 + 3a_0 = 0,$

and, in general, $a_n + 2a_{n-1} + 3a_{n-2} = 0.$ (2)

Hence (1) is a recurring series of the second order whose scale of relation is (2).

997 If a few terms at the beginning of a power series be given, it is always possible to find a scale of relation which these terms will satisfy. By aid of this scale the series may be continued, as a recurring series, to any required term.

Example. Given $1 + 4x + 7x^2 + 10x^3 + 13x^4 + \cdots$. Find the scale of relation which these terms satisfy, and then find two additional terms.

As $1 + 4x + 7x^2 + \cdots$ is not a geometric series, we begin by testing the scale of the second order, $a_n + pa_{n-1} + qa_{n-2} = 0$.

If all the given terms are to satisfy this scale, we must have

$$7 + 4p + q = 0, \quad 10 + 7p + 4q = 0, \quad 13 + 10p + 7q = 0.$$

Solving the first two of these equations, $p = -2$, $q = 1$.

And these values satisfy the third equation, since $13 - 10 \cdot 2 + 7 = 0$.

Hence the required scale is $a_n - 2a_{n-1} + a_{n-2} = 0$.

We may therefore find the required coefficients a_5, a_6 as follows :

$$a_5 - 2 \cdot 13 + 10 = 0, \ \therefore a_5 = 16; \quad a_6 - 2 \cdot 16 + 13 = 0, \ \therefore a_6 = 19.$$

If the given terms had been $1 + 4x + 7x^2 + 10x^3 + 14x^4$, they would not have satisfied a scale of the form $a_n + pa_{n-1} + qa_{n-2} = 0$. But we might then have assigned a sixth term ax^5 *arbitrarily* and have found a scale of the *third* order, $a_n + pa_{n-1} + qa_{n-2} + ra_{n-3} = 0$, which the *six* terms would satisfy, namely, by solving for p, q, r the equations

$$10 + 7p + 4q + r = 0, \quad 14 + 10p + 7q + 4r = 0, \quad a + 14p + 10q + 7r = 0.$$

From the example it will be seen that ordinarily when $2r$ terms are given the series may be continued in one way as a recurring series of the rth order, and that when $2r+1$ terms are given it may be continued in infinitely many ways as a recurring series of the $(r+1)$th order.

998 The generating function of a recurring power series. Every recurring power series of the rth order is the expansion of a proper fraction whose denominator is of the rth degree (compare § 996). This fraction is called the *generating function of the series.* It is the sum of the series when the series is convergent.

Thus, let $$a_0 + a_1x + a_2x^2 + \cdots + a_nx^n + \cdots \quad (1)$$
be a recurring series of the second order whose scale of relation is
$$a_n + pa_{n-1} + qa_{n-2} = 0. \quad (2)$$

$$\begin{aligned}
\text{Set}\quad S_n &= a_0 + a_1x + a_2x^2 + \cdots + a_{n-1}x^{n-1}\\
\therefore px\,S_n &= \qquad pa_0x + pa_1x^2 + \cdots + pa_{n-2}x^{n-1} + pa_{n-1}x^n\\
\therefore qx^2S_n &= \qquad\qquad qa_0x^2 + \cdots + qa_{n-3}x^{n-1} + qa_{n-2}x^n + qa_{n-1}x^{n+1},\\
\hline
\therefore (1+px+qx^2)S_n &= a_0 + (a_1+pa_0)x \qquad + (pa_{n-1}+qa_{n-2})x^n \qquad + qa_{n-1}x^{n+1}
\end{aligned}$$

the remaining terms on the right disappearing because of (2).

When (1) is convergent, as n increases S_n will approach S, the sum of (1), as limit, and x^n will approach 0.

Therefore $$(1+px+qx^2)S = a_0 + (a_1+pa_0)x,$$

that is, $$S = \frac{a_0 + (a_1+pa_0)x}{1+px+qx^2}. \quad (3)$$

999 The general term of a recurring power series. This may be obtained, when the generating function is known, by the method illustrated in the following example.

Example. Find the generating function and the general term of the recurring series whose scale is $a_n - a_{n-1} - 2a_{n-2} = 0$ and whose first two terms are $5 + 4x$.

Here $$p = -1,\quad q = -2,\quad a_0 = 5,\quad a_1 = 4.$$

Therefore, by § 998, (3), $$S = \frac{5-x}{1-x-2x^2} = \frac{5-x}{(1+x)(1-2x)}. \quad (1)$$

Separating (1) into its partial fractions, § 537,

$$\frac{5-x}{(1+x)(1-2x)} = \frac{2}{1+x} + \frac{3}{1-2x} = 2(1+x)^{-1} + 3(1-2x)^{-1}.$$

But if $|x|<1$, $\quad 2(1+x)^{-1} = 2[1 - x + x^2 - \cdots + (-1)^n x^n + \cdots]$.

And if $|x| < 1/2$, $3(1-2x)^{-1} = 3[1 + 2x + 4x^2 + \cdots + 2^n x^n + \cdots]$.

Therefore the general term is $[(-1)^n 2 + 3 \cdot 2^n] x^n$.

EXERCISE XCI

1. If the first three terms of a recurring series of the third order are $2 - 3x + 5x^2$ and the scale of relation is $a_n + 2a_{n-1} - a_{n-2} + 3a_{n-3} = 0$, find the fourth and fifth terms.

2. Find the scale of relation and two additional terms in each of the following:

 (1) $1 + 3x + 2x^2 - x^3 - 3x^4 + \cdots$.

 (2) $2 - 5x + 4x^2 + 7x^3 - 26x^4 + \cdots$.

 (3) $1 - 3x + 6x^2 - 10x^3 + 15x^4 - 21x^5 + \cdots$.

3. Find the generating function and the general term of each of the following:

 (1) $2 + x + 5x^2 + 7x^3 + 17x^4 + \cdots$.

 (2) $3 + 7x + 17x^2 + 43x^3 + 113x^4 + \cdots$.

4. Prove that in a recurring series $a_0 + a_1 x + \cdots$ of the third order, whose scale of relation is $a_n + pa_{n-1} + qa_{n-2} + ra_{n-3} = 0$, the generating function is $\dfrac{a_0 + (a_1 + pa_0)x + (a_2 + pa_1 + qa_0)x^2}{1 + px + qx^2 + rx^3}$.

5. By aid of the preceding formula find the generating function and the general term of the series

$$1 + 2x + 11x^2 + 24x^3 + 85x^4 + 238x^5 + \cdots.$$

6. Show that $a + (a+d)x + (a+2d)x^2 + (a+3d)x^3 + \cdots$ is a recurring series of the second order, and find its generating function.

7. Show that $1^2 + 2^2x + 3^2x^2 + 4^2x^3 + \cdots$ is a recurring series of the third order whose generating function is $(1+x)/(1-x)^3$.

8. Show that $1 \cdot 2 + 2 \cdot 3x + 3 \cdot 4x^2 + 4 \cdot 5x^3 + \cdots$ is a recurring series of the third order, and find its sum when convergent.

XXXVI. INFINITE PRODUCTS

1000 **Infinite products.** This name is given to expressions of the form

$$\Pi(1+a_r)=(1+a_1)(1+a_2)\cdots(1+a_r)\cdots,$$

in which the number of the factors is supposed to be infinite.

Such a product is said to be *convergent* or *divergent* according as $(1+a_1)(1+a_2)\cdots(1+a_n)$ approaches or does not approach a finite limit as n is indefinitely increased.

1001 **Theorem.** *If all the numbers* a_r *are positive, the infinite product* $\Pi(1+a_r)$ *is convergent or divergent according as the infinite series* Σa_r *is convergent or divergent.*

First, suppose that Σa_r is convergent and has the sum S.

Then since $1+x<e^x$ when x is positive, § 990,

we have $(1+a_1)(1+a_2)\cdots(1+a_n)<e^{a_1}\cdot e^{a_2}\cdots e^{a_n}$,

that is, $\qquad <e^{a_1+a_2+\cdots+a_n}<e^S$.

Hence, as n increases, $(1+a_1)(1+a_2)\cdots(1+a_n)$ increases but remains less than the finite number e^S. It therefore approaches a limit, § 192, that is, $\Pi(1+a_r)$ is convergent.

Second, suppose that Σa_r is divergent.

In this case $\lim(a_1+a_2+\cdots+a_n)=\infty$.

But $(1+a_1)(1+a_2)\cdots(1+a_n)>1+(a_1+a_2+\cdots+a_n)$.

Therefore $\lim\,(1+a_1)(1+a_2)\cdots(1+a_n)=\infty$, that is, $\Pi(1+a_r)$ is divergent.

Thus, $\Pi(1+1/n^p)$ is convergent when $p>1$, divergent when $p \overline{\gtrless} 1$.

Example 1. If Σa_r is a divergent positive series whose terms are all less than 1, show that $\Pi(1-a_r)=0$.

Since $a_r<1$ and $1-a_r^2<1$, we have $1-a_r<1/(1+a_r)$ numerically.

Hence $(1-a_1)(1-a_2)\cdots(1-a_n)<1/(1+a_1)(1+a_2)\cdots(1+a_n)$.

But $\qquad \lim(1+a_1)(1+a_2)\cdots(1+a_n)=\infty$.

Therefore $\Pi(1-a_r)=\lim(1-a_1)(1-a_2)\cdots(1-a_n)=0$.

Example 2. Show that when $x = 1$ the binomial series, § 987, converges if $m + 1 > 0$, but diverges if $m + 1 < 0$.

When $x = 1$, the binomial series becomes

$$1 + m + \frac{m(m-1)}{1 \cdot 2} + \cdots + \frac{m(m-1)\cdots(m-n+1)}{1 \cdot 2 \cdots n} + \cdots. \quad (1)$$

In this series we have

$$\frac{u_{n+1}}{u_n} = \frac{m-n+1}{n} = -1 + \frac{m+1}{n}. \quad (2)$$

Hence, if $m + 1 < 0$, we have $|u_{n+1}/u_n| > 1$, and (1) is divergent, § 951.

But if $m + 1 > 0$, after a certain term the series will be of the kind described in § 958 and will therefore converge.

For if r denote the first integer greater than $m + 1$, it follows from (2) that when $n > r$, u_{n+1}/u_n is negative and numerically less than 1. Hence the terms of (1) after u_r are alternately positive and negative and decrease numerically. Therefore (1) converges if $\lim u_n = 0$.

But $$u_{n+1} = \frac{m}{1} \cdot \frac{m-1}{2} \cdots \frac{m-n+1}{n}$$

$$= (-1)^n \left(1 - \frac{m+1}{1}\right)\left(1 - \frac{m+1}{2}\right) \cdots \left(1 - \frac{m+1}{n}\right),$$

and it follows from Ex. 1 that as n increases the product on the right approaches 0 as limit.

EXERCISE XCII

1. Show that $\frac{3}{2} \cdot \frac{5}{4} \cdot \frac{9}{8} \cdot \frac{17}{16} \cdots$ and $\frac{5}{4} \cdot \frac{10}{9} \cdot \frac{17}{16} \cdot \frac{26}{25} \cdots$ are convergent.

2. For what positive values of x are the following infinite products convergent?

(1) $$\Pi\left(1 + \frac{x^n}{n^2}\right) = \left(1 + \frac{x}{1^2}\right)\left(1 + \frac{x^2}{2^2}\right)\left(1 + \frac{x^3}{3^2}\right) \cdots$$

(2) $$\Pi\left(1 + \frac{x^n}{n!}\right) = \left(1 + \frac{x}{1!}\right)\left(1 + \frac{x^2}{2!}\right)\left(1 + \frac{x^3}{3!}\right) \cdots.$$

(3) $$\Pi\left(1 + \frac{x^n}{3^n}\right) = \left(1 + \frac{x}{3}\right)\left(1 + \frac{x^2}{9}\right)\left(1 + \frac{x^3}{27}\right) \cdots.$$

3. Show that $\lim \frac{a(a+1)(a+2)\cdots(a+n)}{b(b+1)(b+2)\cdots(b+n)} = \infty$ or 0 according as $a > b$ or $a < b$.

XXXVII. CONTINUED FRACTIONS

1002 Continued fractions. This name is given to expressions of the form $a + \cfrac{b}{c + \cfrac{d}{e + \cdots}}$, or $a + \frac{b}{c +}\frac{d}{e +}\cdots$ as they are usually written.

We shall consider *simple* continued fractions only. These have the form $a_1 + \frac{1}{a_2 +}\frac{1}{a_3 +}\cdots$, where a_1 is a positive integer or 0 and $a_2, a_3, \cdots$ are positive integers.

The numbers $a_1, a_2, \cdots$ are called the first, second, $\cdots$ *partial quotients* of the continued fraction.

According as the number of these quotients is finite or infinite, the fraction is called *terminating* or *nonterminating*.

1003 Terminating fractions. Evidently every terminating simple continued fraction has a positive rational value, for it may be reduced to a simple fraction.

Thus, $2 + \frac{1}{3 +}\frac{1}{4} = 2 + \frac{4}{13} = \frac{30}{13}$; $\frac{1}{2 +}\frac{1}{3 +}\frac{1}{4} = \frac{13}{30}$.

Conversely, every positive rational number may be converted into a terminating simple continued fraction. This will be evident from the following example.

Example. Convert 67/29 into a continued fraction.

Applying the method for finding the greatest common divisor of two integers to 67 and 29, we have

$$\begin{array}{l} 29)\,67\,(2 = a_1 \\ \underline{58} \\ 9)\,29\,(3 = a_2 \\ \underline{27} \\ 2)\,9\,(4 = a_3 \\ \underline{8} \\ 1)\,2\,(2 = a_4 \\ \underline{2} \\ 0 \end{array}$$

$$\therefore \frac{67}{29} = 2 + \frac{9}{29} = 2 + \frac{1}{29/9}. \quad (1)$$

$$\therefore \frac{29}{9} = 3 + \frac{2}{9} = 3 + \frac{1}{9/2}. \quad (2)$$

$$\therefore \frac{9}{2} = 4 + \frac{1}{2}. \quad (3)$$

Substituting (2) in (1), and (3) in the result, we have

$$\frac{67}{29} = 2 + \cfrac{1}{3 + \cfrac{1}{4 + \cfrac{1}{2}}} = 2 + \frac{1}{3 +}\,\frac{1}{4 +}\,\frac{1}{2}, \text{ as required.}$$

Since $29/67 = 1 \div 67/29$, we also have

$$\frac{29}{67} = \frac{1}{2 +}\,\frac{1}{3 +}\,\frac{1}{4 +}\,\frac{1}{2}.$$

Convergents. The fractions $\frac{a_1}{1}$, $a_1 + \frac{1}{a_2}$, $a_1 + \frac{1}{a_2 +}\,\frac{1}{a_3}$, $\cdots$, **1004** are called the first, second, third, $\cdots$ *convergents* of the fraction $a_1 + \frac{1}{a_2 +}\,\frac{1}{a_3 + \cdots}$.

When a_1 is 0, the first convergent is written $\frac{0}{1}$.

Theorem 1. *Each odd convergent is less and each even convergent is greater than every subsequent convergent.* **1005**

This follows from the fact that a fraction decreases when its denominator increases.

Thus,

1. $a_1 < a_1 + \cdots$.

2. $a_1 + \frac{1}{a_2} > a_1 + \frac{1}{a_2 + \cdots}$, since $\frac{1}{a_2} > \frac{1}{a_2 + \cdots}$.

3. $a_1 + \cfrac{1}{a_2 + \cfrac{1}{a_3}} < a_1 + \cfrac{1}{a_2 + \cfrac{1}{a_3 + \cdots}}$, since $a_2 + \frac{1}{a_3} > a_2 + \frac{1}{a_3 + \cdots}$,

by 2; and so on.

Reduction of convergents. On reducing the first, second, third, **1006** $\cdots$ convergents of $a_1 + \frac{1}{a_2 +}\,\frac{1}{a_3 + \cdots}$ to the form of simple fractions, we obtain

$$\frac{a_1}{1}, \quad \frac{a_1a_2 + 1}{a_2}, \quad \frac{a_1a_2a_3 + a_1 + a_3}{a_2a_3 + 1}, \quad \cdots. \tag{1}$$

Let $p_1, p_2, p_3, \cdots$ denote the numerators, and $q_1, q_2, q_3, \cdots$ the denominators of the convergents as thus reduced, so that

$$p_1 = a_1,\ p_2 = a_1a_2 + 1,\ p_3 = a_1a_2a_3 + a_1 + a_3, \cdots \quad (2)$$

$$q_1 = 1,\quad q_2 = a_2, \qquad q_3 = a_2a_3 + 1, \cdots. \quad (3)$$

Since $a_1, a_2, a_3, \cdots$ are positive integers, it follows from (2), (3) that as n increases p_n and q_n continually increase, and that they approach ∞ if the given fraction does not terminate.

By examining (2) and (3) it will be found that

$$p_3 = a_3p_2 + p_1 \text{ and } q_3 = a_3q_2 + q_1. \quad (4)$$

This is an illustration of the following theorem.

1007 **Theorem 2.** *The numerator and denominator of any convergent are connected with those of the two preceding convergents by the formulas*

$$\mathrm{p}_n = \mathrm{a}_n\mathrm{p}_{n-1} + \mathrm{p}_{n-2},\quad \mathrm{q}_n = \mathrm{a}_n\mathrm{q}_{n-1} + \mathrm{q}_{n-2}.$$

For suppose that these formulas have been proved to hold good for the kth convergent, so that

$$p_k = a_kp_{k-1} + p_{k-2},\quad q_k = a_kq_{k-1} + q_{k-2}, \quad (1)$$

$$\frac{p_k}{q_k} = \frac{a_kp_{k-1} + p_{k-2}}{a_kq_{k-1} + q_{k-2}}. \quad (2)$$

The $(k+1)$th convergent may be derived from the kth by merely replacing a_k by $a_k + 1/a_{k+1}$, § 1004. Therefore, since $p_{k-1}, p_{k-2}, q_{k-1}, q_{k-2}$ do not involve a_k, it follows from (2) that

$$\frac{p_{k+1}}{q_{k+1}} = \frac{(a_k + 1/a_{k+1})p_{k-1} + p_{k-2}}{(a_k + 1/a_{k+1})q_{k-1} + q_{k-2}}$$

$$= \frac{a_{k+1}(a_kp_{k-1} + p_{k-2}) + p_{k-1}}{a_{k+1}(a_kq_{k-1} + q_{k-2}) + q_{k-1}} = \frac{a_{k+1}p_k + p_{k-1}}{a_{k+1}q_k + q_{k-1}}, \text{ by (1)};$$

that is,

$$p_{k+1} = a_{k+1}p_k + p_{k-1},\quad q_{k+1} = a_{k+1}q_k + q_{k-1}.$$

We have thus proved that if the formulas $p_n = a_np_{n-1} + p_{n-2}$, $q_n = a_nq_{n-1} + q_{n-2}$ hold good for any particular convergent, they hold good for the next convergent also. But we have already shown that they hold good for the third convergent. Hence they hold good for the fourth, hence for the fifth, and so on to every convergent after the third (compare § 791).

Example. Compute the convergents of $3 + \frac{1}{2} + \frac{1}{3} + \frac{1}{4} + \frac{1}{5}$.

Since $3 = 3/1$ and $3 + 1/2 = 7/2$, we have $p_1 = 3$, $p_2 = 7$, $q_1 = 1$, $q_2 = 2$.

Hence $p_3 = 3 \cdot 7 + 3 = 24$, $p_4 = 4 \cdot 24 + 7 = 103$, $p_5 = 5 \cdot 103 + 24 = 539$,
and $q_3 = 3 \cdot 2 + 1 = 7$, $q_4 = 4 \cdot 7 + 2 = 30$, $q_5 = 5 \cdot 30 + 7 = 157$.

Therefore the convergents are $\frac{3}{1}, \frac{7}{2}, \frac{24}{7}, \frac{103}{30}, \frac{539}{157}$.

Theorem 3. *The numerators and denominators of every two consecutive convergents are connected by the formula* **1008**

$$p_n q_{n-1} - p_{n-1} q_n = (-1)^n.$$

The formula holds good when $n = 2$. For, by § 1006, we have $p_2 q_1 - p_1 q_2 = (a_1 a_2 + 1) - a_1 a_2 = 1 = (-1)^2$.

Moreover we can prove that if the formula holds good when $n = k$, it also holds good when $n = k + 1$.

$$\text{For } p_{k+1} q_k - p_k q_{k+1} = (a_{k+1} p_k + p_{k-1}) q_k - p_k (a_{k+1} q_k + q_{k-1})$$
$$= -(p_k q_{k-1} - p_{k-1} q_k). \qquad \S\ 1007$$

Hence, if $$p_k q_{k-1} - p_{k-1} q_k = (-1)^k,$$

then $$p_{k+1} q_k - p_k q_{k+1} = (-1)^{k+1}.$$

Therefore, since the formula is true for $n = 2$, it is true for $n = 2 + 1$ or 3, therefore for $n = 3 + 1$ or 4, and so on to any positive integral value of n.

Corollary 1. *Every convergent* p_n/q_n *is an irreducible fraction.* **1009**

For if p_n and q_n had a common factor, it would follow from the relation $p_n q_{n-1} - p_{n-1} q_n = (-1)^n$ that this factor is a divisor of $(-1)^n$, which is impossible.

1010 **Corollary 2.** *For the differences between convergents we have the formulas*

1. $\frac{p_n}{q_n} - \frac{p_{n-1}}{q_{n-1}} = \frac{(-1)^n}{q_n q_{n-1}}$. 2. $\frac{p_n}{q_n} - \frac{p_{n-2}}{q_{n-2}} = \frac{(-1)^{n-1}a_n}{q_n q_{n-2}}$.

The first formula is an immediate consequence of the relation $p_n q_{n-1} - p_{n-1} q_n = (-1)^n$.

The second follows from the fact, §§ 1007, 1008, that

$$\begin{aligned} p_n q_{n-2} - p_{n-2} q_n &= (a_n p_{n-1} + p_{n-2}) q_{n-2} - p_{n-2}(a_n q_{n-1} + q_{n-2}) \\ &= a_n (p_{n-1} q_{n-2} - p_{n-2} q_{n-1}) = (-1)^{n-1} a_n. \end{aligned}$$

The theorem of § 1005 may be derived from these formulas.

1011 **Theorem 4.** *The* n*th convergent of a nonterminating simple continued fraction approaches a definite limit as* n *is indefinitely increased.*

For, by § 1005, the odd convergents p_1/q_1, p_3/q_3, $\cdots$ form a never-ending increasing sequence, every term of which is less than the finite number p_2/q_2. Hence, § 192, a variable which runs through this sequence will increase toward some number λ as limit.

Similarly a variable which runs through the sequence of even convergents p_2/q_2, p_4/q_4, $\cdots$ will decrease toward some number μ as limit.

But $\mu = \lambda$, since $\mu - \lambda = \lim_{m \doteq \infty} \left[\frac{p_{2m}}{q_{2m}} - \frac{p_{2m-1}}{q_{2m-1}}\right] = \lim_{m \doteq \infty} \frac{(-1)^{2m}}{q_{2m} q_{2m-1}} = 0.$

Therefore a variable which runs through the complete sequence of convergents $p_1/q_1, p_2/q_2, p_3/q_3, \cdots$ will approach λ as limit.

1012 By the *value* of a nonterminating simple continued fraction is meant the number $\lim_{n \doteq \infty} (p_n/q_n)$. It follows from § 1003 that this number is always irrational.

The value of a terminating fraction is that of its last convergent, § 1004.

In the statements of the following theorems, §§ 1013, 1014, the understanding is that when the fraction terminates, "convergent" means any convergent except the last one.

Corollary 1. *The value of a simple continued fraction lies between the values of every two consecutive convergents.* **1013**

Corollary 2. *The difference between the value of a continued fraction and that of its* nth *convergent is numerically less than* $1/q_nq_{n+1}$ *and greater than* a_{n+2}/q_nq_{n+2}. **1014**

For let λ denote the value of the fraction, and to fix the ideas suppose that n is odd.

We then have $\frac{p_n}{q_n} < \frac{p_{n+2}}{q_{n+2}} < \lambda < \frac{p_{n+1}}{q_{n+1}}$. §§ 1005, 1013

Hence $\lambda - \frac{p_n}{q_n} < \frac{p_{n+1}}{q_{n+1}} - \frac{p_n}{q_n}, \therefore < \frac{1}{q_nq_{n+1}}$, § 1010, 1

and $\lambda - \frac{p_n}{q_n} > \frac{p_{n+2}}{q_{n+2}} - \frac{p_n}{q_n}, \therefore > \frac{a_{n+2}}{q_nq_{n+2}}$. § 1010, 2

Evidently $1/q_nq_{n+1} < 1/q_n^2$, and by making use of the relation $q_{n+2} = a_{n+2}q_{n+1} + q_n$, § 1007, it may readily be shown that $a_{n+2}/q_nq_{n+2} \geqq 1/q_n(q_n + q_{n+1})$. Hence the difference between λ and p_n/q_n is less than $1/q_n^2$ and greater than $1/q_n(q_n + q_{n+1})$.

Corollary 3. *Each convergent is a closer approximation to the value of the fraction than is any preceding convergent.* **1015**

For, by § 1014, if λ denote the value of the fraction, the difference between λ and p_n/q_n is numerically less than $1/q_nq_{n+1}$, while the difference between λ and p_{n-1}/q_{n-1} is numerically greater than $a_{n+1}/q_{n-1}q_{n+1}$; and $1/q_nq_{n+1} < a_{n+1}/q_{n-1}q_{n+1}$, since $q_{n-1} < a_{n+1}q_n$, § 1006.

Corollary 4. *The convergent* p_n/q_n *is a closer approximation to the value of the fraction than is any other rational fraction whose denominator does not exceed* q_n. **1016**

For if a/b is a closer approximation to the value of the fraction than p_n/q_n is, it must also, § 1015, be a closer approximation than p_{n-1}/q_{n-1} is, and must therefore, § 1013, lie between p_n/q_n and p_{n-1}/q_{n-1}.

To fix the ideas, suppose that n is even.

We then have $\dfrac{p_{n-1}}{q_{n-1}} < \dfrac{a}{b} < \dfrac{p_n}{q_n}$.

Hence $\dfrac{p_n}{q_n} - \dfrac{p_{n-1}}{q_{n-1}} > \dfrac{a}{b} - \dfrac{p_{n-1}}{q_{n-1}}$;

that is, $\dfrac{1}{q_n q_{n-1}} > \dfrac{aq_{n-1} - bp_{n-1}}{bq_{n-1}}$,

or $b > q_n(aq_{n-1} - bp_{n-1})$.

But $aq_{n-1} - bp_{n-1}$ is positive, since $a/b > p_{n-1}/q_{n-1}$. Hence $b > q_n$; that is, if a/b is a closer approximation to the value of the continued fraction than p_n/q_n is, its denominator b must be *greater* than q_n.

1017 **Recurring fractions.** A nonterminating continued fraction in which a single partial quotient or a group of consecutive partial quotients continually recurs is called a *recurring* fraction. And such a fraction is called *pure* or *mixed* according as it begins or does not begin with these recurring partial quotients.

The value of a recurring fraction may be found as follows.

Example 1. Find the value of $\dot{2} + \dfrac{\dot{1}}{3+\cdots} = 2 + \dfrac{1}{3+}\dfrac{1}{2+}\dfrac{1}{3+\cdots}$.

This is a pure recurring fraction with the *period* $2 + \dfrac{1}{3}$. Hence, if x denote its value, we have

$$x = 2 + \frac{1}{3+}\frac{1}{x}, \quad \therefore x = \frac{7x+2}{3x+1}, \quad \therefore 3x^2 - 6x - 2 = 0, \quad \therefore x = \frac{3+\sqrt{15}}{3}.$$

Example 2. Find the value of $4 + \dfrac{1}{5+}\dfrac{\dot{1}}{\dot{2}+}\dfrac{\dot{1}}{3+\cdots}$.

This is a mixed recurring fraction with the period $2 + 1/3$. Hence, if x denote the value of the recurring part $\dot{2} + \dfrac{\dot{1}}{3+\cdots}$, and y the value of the entire fraction, we have, by Ex. 1,

$$y = 4 + \frac{1}{5+}\frac{1}{x} = \frac{21x+4}{5x+1} = \frac{21(3+\sqrt{15})/3+4}{5(3+\sqrt{15})/3+1} = \frac{75+21\sqrt{15}}{18+5\sqrt{15}}.$$

In general, if x denote the value of a pure recurring fraction with the period $a_1 + \cdots \frac{1}{+ a_k}$, we have, § 1007,

$$x = a_1 + \cdots \frac{1}{+ a_k} \frac{1}{+ x} = \frac{p_k x + p_{k-1}}{q_k x + q_{k-1}},$$

and therefore $\quad q_k x^2 + (q_{k-1} - p_k) x - p_{k-1} = 0.$

Since the absolute term $-p_{k-1}$ of this quadratic is negative, it has one and but one positive root, and this root is the value of the fraction.

Again, if y denote the value of the mixed recurring fraction

$$a_1 + \cdots \frac{1}{+ a_r} \frac{1}{+ a_{r+1}} + \cdots \frac{1}{+ a_{r+k}} + \cdots,$$

we find the value x of the recurring part as above, and then have, § 1007,

$$y = a_1 + \cdots \frac{1}{+ a_r} \frac{1}{+ x} = \frac{p_r x + p_{r-1}}{q_r x + q_{r-1}}.$$

On converting irrational numbers into continued fractions. **1018** Every positive irrational number is the value of a definite nonterminating simple continued fraction which may be obtained to any required partial quotient by the following process.

If b denote the number in question, first find a_1, the greatest integer less than b. Then $b = a_1 + 1/b_1$, where b_1 is some irrational number greater than 1. Next find a_2, the greatest integer less than b_1. Then $b_1 = a_2 + 1/b_2$, where b_2 is some irrational number greater than 1. Continuing thus, we have

$$b = a_1 + \frac{1}{b_1} = a_1 + \frac{1}{a_2} \frac{1}{+ b_2} = \cdots = a_1 + \frac{1}{a_2} \frac{1}{+ a_3} + \cdots.$$

It can be proved that when b is a quadratic surd the continued fraction thus obtained is a recurring fraction.

Example. Convert $\sqrt{11}$ into a continued fraction.

The greatest integer less than $\sqrt{11}$ is 3, and, § 603,

$$\sqrt{11} = 3 + (\sqrt{11} - 3) = 3 + \frac{2}{\sqrt{11} + 3} = 3 + \frac{1}{(\sqrt{11} + 3)/2}. \quad (1)$$

The greatest integer less than $(\sqrt{11} + 3)/2$ is 3, and

$$\frac{\sqrt{11} + 3}{2} = 3 + \frac{\sqrt{11} - 3}{2} = 3 + \frac{2}{2(\sqrt{11} + 3)} = 3 + \frac{1}{\sqrt{11} + 3}. \quad (2)$$

The greatest integer less than $\sqrt{11} + 3$ is 6, and

$$\sqrt{11} + 3 = 6 + (\sqrt{11} - 3) = 6 + \frac{2}{\sqrt{11} + 3} = 6 + \frac{1}{(\sqrt{11} + 3)/2}. \quad (3)$$

The last fraction in (3) is the same as the last in (1). Hence the steps from (3) on will be (2), (3) repeated indefinitely; that is, the partial quotients 3 and 6 will recur. Hence, substituting (2) in (1), and (3) in the result, and so on, we have $\sqrt{11} = 3 + \frac{1}{3+}\frac{1}{6+\cdots}$.

1019 A given irrational number can be expressed in *only one way* as a simple continued fraction. This follows from the fact that two nonterminating simple continued fractions cannot be equal unless their corresponding partial quotients are equal.

For if $a + \alpha = c + \gamma$, where a and c denote positive integers and α and γ denote positive numbers which are less than 1, then $a = c$, since otherwise it would follow from $a - c = \gamma - \alpha$ that an integer, not 0, is numerically less than 1.

Hence, if $a_1 + \frac{1}{a_2+}\frac{1}{a_3+\cdots} = c_1 + \frac{1}{c_2+}\frac{1}{c_3+\cdots}$, where $a_1, a_2, a_3, \cdots$, $c_1, c_2, c_3, \cdots$ denote positive integers, we have $a_1 = c_1$, $\therefore \frac{1}{a_2+}\frac{1}{a_3+\cdots}$ $= \frac{1}{c_2+}\frac{1}{c_3+\cdots}$, $\therefore a_2 + \frac{1}{a_3+\cdots} = c_2 + \frac{1}{c_3+\cdots}$, $\therefore a_2 = c_2$, and so on.

1020 If we compute the continued fraction to which a given irrational number b is equal as far as the nth partial quotient, we can find its nth convergent p_n/q_n, and this rational fraction p_n/q_n will express b approximately with an error less than $1/q_n^2$, § 1014. Moreover p_n/q_n will be a closer approximation to b than is any other rational fraction whose denominator does not exceed q_n, § 1016.

Thus, the first four convergents of $\sqrt{11} = 3 + \frac{1}{3+}\frac{1}{6+}\frac{1}{3+}\cdots$ are $\frac{3}{1}, \frac{10}{3}, \frac{63}{19}, \frac{199}{60}$, and $\frac{199}{60}$ expresses $\sqrt{11}$ with an error less than $\frac{1}{60^2}$.

Solution of indeterminate equations of the first degree. **1021** Given any equation of the form $ax + by = c$, where a, b, c denote integers of which a and b have no common factor, § 672. If we convert a/b into a continued fraction, the last convergent of this fraction will be a/b itself, and if the convergent next to the last be p/q, we have $aq - bp = \pm 1$, § 1008. This fact makes it possible always to find a pair of integral values of x and y which satisfy $ax + by = c$. The method is illustrated in the following example.

Example. Find an integral solution of $205x + 93y = 7$.

As in § 1003, Ex., we find $\frac{205}{93} = 2 + \frac{1}{4+}\frac{1}{1+}\frac{1}{8+}\frac{1}{2}$.

The convergents, found as in § 1007, Ex., are $\frac{2}{1}, \frac{9}{4}, \frac{11}{5}, \frac{97}{44}, \frac{205}{93}$.

Hence $$205 \cdot 44 - 93 \cdot 97 = -1,$$

or, multiplying by -7, $205(-44 \cdot 7) + 93(97 \cdot 7) = 7$.

Therefore $x = -308$, $y = 679$ is a solution of $205x + 93y = 7$.

The general solution is $x = -308 + 93t$, $y = 679 - 205t$, § 674.

Similarly we may show that $205x - 93y = 7$ has the solution $x = -308$, $y = -679$.

EXERCISE XCIII

Compute the convergents of the following:

1. $3 + \frac{1}{4+}\frac{1}{1+}\frac{1}{5}$.

2. $\frac{1}{1+}\frac{1}{1+}\frac{1}{3+}\frac{1}{10+}\frac{1}{12}$.

Convert each of the following into a continued fraction. For each of the last three compute the fourth convergent and estimate the error made in taking this convergent as the value of the fraction.

3. $\frac{10}{12}$.

4. $\frac{457}{56}$.

5. $\frac{142}{513}$.

6. 3.54.

7. .1457.

8. $\frac{233}{177}$.

9. $\frac{421}{972}$.

10. $\frac{23456}{31827}$.

Convert each of the following into a recurring continued fraction and compute the fifth convergents and the corresponding errors for the first four of them.

11. $\sqrt{17}$. **12.** $\sqrt{26}$. **13.** $\sqrt{6}$. **14.** $\sqrt{38}$.

15. $\sqrt{105}$. **16.** $1/\sqrt{23}$. **17.** $\sqrt{19}$. **18.** $\sqrt{71}$.

19. $3\sqrt{3}$. **20.** $(\sqrt{10}-2)/2$. **21.** $(\sqrt{2}+1)/(\sqrt{2}-1)$.

Find the values of the following recurring fractions.

22. $\frac{\dot{1}}{1+}\frac{1}{2+}\frac{\dot{1}}{3+}\cdots$. **23.** $\frac{\dot{1}}{2+}\frac{1}{1+}\frac{\dot{1}}{3+}\cdots$. **24.** $3+\frac{1}{4+}\frac{\dot{1}}{5+}\frac{\dot{1}}{2+}\cdots$.

25. $\dot{2}+\frac{1}{3+}\frac{1}{4+}\frac{\dot{1}}{5+}\cdots$. **26.** $\frac{1}{2+}\frac{1}{7+}\frac{\dot{1}}{1+}\frac{1}{2+}\frac{\dot{1}}{1+}\cdots$.

27. Show that $\sqrt{a^2+1}=a+\frac{1}{2a+}\frac{1}{2a+}\cdots$.

28. Show that $\sqrt{a^2+2}=a+\frac{1}{a+}\frac{1}{2a+}\frac{1}{a+}\frac{1}{2a+}\cdots$.

29. Show that

$$\frac{\dot{1}}{a+}\frac{1}{b+}\frac{\dot{1}}{c+}\cdots=\frac{-(abc+a-b+c)+\sqrt{(abc+a+b+c)^2+4}}{2(ab+1)}.$$

30. Convert the positive root of $x^2+x-1=0$ into a continued fraction.

31. Show that $\frac{p_n}{q_n}=\frac{p_1}{q_1}+\frac{1}{q_1q_2}-\frac{1}{q_2q_3}+\cdots+\frac{(-1)^n}{q_{n-1}q_n}$.

32. Show that $\frac{1}{a_2+}\frac{1}{a_3+}\cdots=\frac{1}{q_1q_2}-\frac{1}{q_2q_3}+\frac{1}{q_3q_4}-\cdots$.

33. What rational fraction having a denominator less than 1000 will most nearly express the ratio of the diagonal of a square to its side? Estimate the error made in taking this fraction as the value of the ratio.

34. Find the simplest fraction which will express $\pi=3.14159265\cdots$ with an error which is less than .000001.

35. Compute the sixth convergent of $e=2.71828\cdots$ and estimate the error made in taking it as the value of e.

36. Find an integral solution of $127x-214y=6$.

37. Find an integral solution of $235x+412y=10$.

38. Find the general integral solution of $517x-323y=31$.

XXXVIII. PROPERTIES OF CONTINUOUS FUNCTIONS

FUNCTIONS OF A SINGLE VARIABLE

Functions. If the variable y depends on the variable x in such a manner that to each value of x there corresponds a definite value or set of values of y, we call y a *function* of x. **1022**

In what follows when we say that y is a function of x and write $y = f(x)$, we shall mean that it is a *one-valued* function; in other words, that to each value of x there corresponds but *one* value of y. And $f(a)$ will denote the value of y which corresponds to the value a of x.

Evidently y is a function of x if it be equal to an algebraic expression in x, as when $y = x^2 + 1$. But a relation which defines y as a function of x may be one which cannot be expressed by an equation. Thus, y is a function of x if y is 1 for all rational values of x and -1 for all other values of x. But this relation between y and x cannot be expressed by an equation.

We call y a function of x even when there are *exceptional* values of x for which the given relation between y and x fails to determine y, § 1024.

Sometimes y is defined as a function of x only for a certain class of values of x or only for values of x which lie between certain limits. Thus, the equation $y = x + 2x^2 + 3x^3 + \cdots$, by itself considered, determines y for those values only of x which are numerically less than 1.

Continuity of a function. Let $f(x)$ denote a given function of x. We say that $f(x)$ is *continuous* at a, that is, when $x = a$, if $f(a)$ has a definite finite value, and if $\lim_{x=a} f(x) = f(a)$. **1023**

In the contrary case we say that $f(x)$ is *discontinuous* at a.

Here and in what follows the notation $\lim_{x=a} f(x) = f(a)$ means that $f(x)$ will approach $f(a)$ as limit whenever x approaches a as limit, that is, no matter what the sequence of values may be through which x runs in approaching a as limit.

In the case of a function y defined by a given equation $y = f(x)$ it may happen that the expression $f(x)$ assumes an **1024** indeterminate form when $x = a$, §§ 513–518. The equation $y = f(x)$ by itself considered then fails to define y when $x = a$. But if $\lim_{x \doteq a} f(x)$ has a definite finite value, we assign this as the value of $f(a)$, § 519, which makes $f(x)$ *continuous* at a. If $\lim_{x \doteq a} f(x) = \infty$, we assign to $f(a)$ the value ∞, § 515; $f(x)$ is then *discontinuous* at a. Finally, if $\lim_{x \doteq a} f(x)$ is indeterminate, we have no reason for assigning any single value to $f(a)$. Evidently we can assign none for which $\lim_{x \doteq a} f(x) = f(a)$. In this case also $f(x)$ is *discontinuous* at a.

1. Thus, every rational function $f(x)$ is continuous except perhaps when the denominator of some fraction occurring in $f(x)$ vanishes.

For example, consider the function $f(x) = (x-1)/(x^2-1)$.

This function is continuous except when $x^2 - 1 = 0$, that is, when $x = 1$ or -1. For if a is not 1 or -1, $f(a) = (a-1)/(a^2-1)$ has a definite finite value and $\lim_{x \doteq a} f(x) = f(a)$, § 509.

When $x = 1$, the expression $(x-1)/(x^2-1)$ assumes the indeterminate form 0/0. But $\lim_{x \doteq 1} f(x) = \lim_{x \doteq 1} [(x-1)/(x^2-1)] = \lim_{x \doteq 1} [1/(x+1)] = 1/2$, and by assigning to $f(1)$ the value 1/2 we make $f(x)$ continuous when $x = 1$.

When $x = -1$, $f(x)$ is discontinuous; for $\lim_{x \doteq -1} f(x) = \infty$.

2. Consider the following function:

$$f(x) = \frac{2^{\frac{1}{x}} + 3}{2^{\frac{1}{x}} + 1} = \frac{1 + 3/2^{\frac{1}{x}}}{1 + 1/2^{\frac{1}{x}}} = \frac{1/2^{-\frac{1}{x}} + 3}{1/2^{-\frac{1}{x}} + 1}.$$

Here $f(0)$ has the indeterminate form ∞/∞, § 517.

But if we write $f(x)$ in the second form and then make x approach 0 through *positive* values, we have $\lim 2^{\frac{1}{x}} = \infty$, and therefore $\lim f(x) = 1$.

If we write $f(x)$ in the third form and then make x approach 0 through *negative* values, we have $\lim 2^{-\frac{1}{x}} = \infty$, and therefore $\lim f(x) = 3$.

Finally, if we make x approach 0 through values which are alternately positive and negative, $f(x)$ will not approach any limit.

Hence $f(x)$ is discontinuous at 0. No value can be assigned to $f(0)$ for which $\lim_{x \doteq 0} f(x) = f(0)$.

1025 From the definition of continuity in § 1023 it immediately follows, § 189, that

The sufficient and necessary condition that $f(x)$ *be continuous at* a *is that* $f(a)$ *have a definite finite value, and that for every positive number* δ *which can be assigned it shall be possible to find a corresponding positive number* ϵ *such that*

$$|f(x) - f(a)| < \delta \text{ whenever } |x - a| < \epsilon.$$

Thus in the neighborhood of a value of x, as a, at which $f(x)$ is continuous, very small changes in the value of x are accompanied by very small changes in the value of $f(x)$, and the change in the value of x can be taken small enough to make the corresponding change in the value of $f(x)$ as small as we please. This is not true of $f(x)$ in an interval containing a value of x at which $f(x)$ is discontinuous. See the examples in § 1024.

Theorem 1. *If both of the functions* $f(x)$ *and* $\phi(x)$ *are continuous at* a, *the same is true of* $f(x) \pm \phi(x)$ *and* $f(x) \cdot \phi(x)$, *also of* $f(x)/\phi(x)$ *unless* $\phi(a) = 0$. **1026**

If $f(x)$ *is continuous at* a, *the same is true of* $\sqrt[n]{f(x)}$.

This follows immediately from the definition of continuity at a, § 1023, and the theorems of §§ 203–205, according to which $\lim [f(x) + \phi(x)] = \lim f(x) + \lim \phi(x)$, and so on.

Real functions. In what follows x will denote a *real variable*, that is, one which takes real values only, and $f(x)$ will denote a *real function* of x, that is, one which has real values when x is real. **1027**

Number intervals. The practice of picturing real numbers by points on a straight line, §§ 134, 209, suggests the following convenient nomenclature. **1028**

Let us call the assemblage of all real numbers between a and b, a and b themselves included, the *number interval* a, b, and represent it by the symbol (a, b).

Moreover, it being understood that $a < b$, let us call a and b the *left* and *right extremities* of the interval (a, b). Also, if

$c = (a + b)/2$, let us say that (a, b) is divided at c into the two *equal intervals* (a, c) and (c, b); and so on.

Thus, (1, 7) is divided at 4 into the two equal intervals (1, 4), (4, 7); and at 3 and 5 into the three equal intervals (1, 3), (3, 5), (5, 7).

1029 We say that the function $f(x)$ is *continuous throughout the interval* (a, b) if it is continuous for every value of x in this interval.

1030 **Theorem 2.** *If* f (x) *is continuous throughout the interval* (a, b), *and* f (a) *and* f (b) *have contrary signs, there is in* (a, b) *a number* x_0 *such that* $f(x_0) = 0$.

To fix the ideas, suppose that $f(a)$ is + and that $f(b)$ is −. Divide (a, b) into any number of equal intervals, say into the two equal intervals (a, c) and (c, b).

If $f(c) = 0$, our theorem is proved, c being x_0. But if $f(c) \neq 0$, it must be true of one of the intervals (a, c) or (c, b) that $f(x)$ is + at its left extremity and − at its right. Thus, if $f(c)$ is −, this is true of (a, c), and if $f(c)$ is +, it is true of (c, b). Select this interval and for convenience call it (a_1, b_1). Then $f(a_1)$ is + and $f(b_1)$ is −.

Deal with this interval (a_1, b_1) as we have just dealt with (a, b), and so on indefinitely. We shall either ultimately come upon an interval extremity for which $f(x) = 0$, which is then the x_0 sought, or we shall define a never-ending sequence of intervals within intervals,

$$(a, b),\ (a_1, b_1),\ (a_2, b_2),\ \cdots,\ (a_n, b_n),\ \cdots,$$

such that $\quad f(a),\ f(a_1),\ f(a_2),\ \cdots,\ f(a_n),\ \cdots$ are +

and $\quad f(b),\ f(b_1),\ f(b_2),\ \cdots,\ f(b_n),\ \cdots$ are −.

It follows from §§ 192, 193 that as n increases a_n and b_n approach the same number as limit. For a_n remains less than b and never decreases, and b_n remains greater than a and never increases, and $\lim (b_n - a_n) = \lim (b - a)/2^n = 0$.

Call this limit x_0. Then $f(x_0) = 0$.

For since $f(x)$ is continuous at x_0, $\lim f(a_n) = \lim f(b_n) = f(x_0)$. But since $f(a_n)$ is always positive, its limit $f(x_0)$ cannot be negative, and since $f(b_n)$ is always negative, its limit $f(x_0)$ cannot be positive. Therefore $f(x_0)$ is 0.

Thus, if $f(x) = 1 - x^2/2! + x^4/4! - x^6/6! + \cdots$, it may readily be shown that $f(1)$ is positive and $f(2)$ negative. Hence this $f(x)$ will vanish for some value of x between 1 and 2.

Simpler illustrations of the theorem will be found in §§ 833, 836.

Maximum and minimum values. Superior and inferior limits. **1031** Consider the following infinite assemblages of numbers:

$$2,\ 1\tfrac{1}{2},\ 1\tfrac{1}{4},\ 1\tfrac{1}{8},\ \cdots\ \text{(A)}, \qquad 2,\ 2\tfrac{1}{2},\ 2\tfrac{3}{4},\ 2\tfrac{7}{8},\ \cdots\ \text{(B)}.$$

In (A) there is a greatest number, namely 2, but no least number; and in (B) there is a least number, namely 2, but no greatest number.

On the other hand, while there is no least number in (A), among the numbers which are less than those in (A) there is a greatest, namely 1. Similarly among the numbers which are greater than those in (B) there is a least, namely 3.

The like is true of all infinite assemblages of finite numbers, that is, of numbers which lie between two given finite numbers a and b. In other words,

Theorem 3. *Let* $a_1, a_2, \cdots, a_n, \cdots$ (A) *denote any infinite assemblage of finite numbers. Then* **1032**

1. *Either among the different numbers in* (A) *there is a greatest or among the numbers greater than those in* (A) *there is a least.*

2. *Either among the different numbers in* (A) *there is a least or among the numbers less than those in* (A) *there is a greatest.*

To prove 1 assign all numbers greater than those in (A) to a class R_2, and all other real numbers, including those in (A), to a class R_1. Since each number in R_1 will then be less than every number in R_2, there will be, § 159, either a greatest number in R_1 or a least in R_2, — which means either a greatest among the different numbers in (A) or a least among the numbers which are greater than those in (A).

By similar reasoning 2 may be proved.

1033 If among the different numbers of an assemblage there is a greatest, we call that number the *maximum* number of the assemblage; if a least, its *minimum* number.

The *superior limit* of an assemblage is the maximum number, if there be one. If not, it is the least number which is greater than every number in the assemblage.

The *inferior limit* of an assemblage is the minimum number, if there be one. If not, it is the greatest number which is less than every number in the assemblage.

An assemblage like 1, 2, 3, 4, $\cdots$ which contains numbers greater than every assignable number is said to have the superior limit ∞. Similarly an assemblage like -1, -2, -3, -4, $\cdots$ is said to have the inferior limit $-\infty$.

Evidently, if an assemblage has a finite superior limit λ, either λ is its maximum number or we can find in the assemblage numbers which differ from λ as little as we please.

1034 By the "values of $f(x)$ in (a, b)" we shall mean those which correspond to values of x in (a, b). And if this assemblage has a maximum or a minimum value, we shall call it the *absolute maximum* or *minimum value* of $f(x)$ in (a, b). The maximum and minimum values defined in § 639 may or may not be the *absolute* maximum and minimum values.

1035 **Theorem 4.** *If* f(x) *is continuous throughout the interval* (a, b), *it has an absolute maximum and an absolute minimum value in* (a, b).

For since the values of $f(x)$ in (a, b) are finite, § 1023, they have finite superior and inferior limits. Call these limits λ and μ respectively.

We are to demonstrate that in (a, b) there is a number x_0 such that $f(x_0) = \lambda$, and a number x_1 such that $f(x_1) = \mu$.

As the proofs of these two theorems are essentially the same, we shall give only the first of them.

Divide (a, b) into any number of equal intervals, say into two such intervals. Evidently λ will be the superior limit of the values of $f(x)$ in at least one of these half intervals. For convenience call this half interval (a_1, b_1).

Deal with the interval (a_1, b_1) as we have just dealt with (a, b), and so on indefinitely.

We thus obtain a never-ending sequence of intervals within intervals,

$$(a, b),\ (a_1, b_1),\ (a_2, b_2),\ \cdots,\ (a_n, b_n),\ \cdots,$$

in each of which λ is the superior limit of the values of $f(x)$.

As n is indefinitely increased, a_n and b_n approach the same number as limit (see § 1030).

If we call this limit x_0, then $f(x_0) = \lambda$.

For if not, since both $f(x_0)$ and λ denote constants, their difference must be some constant, as α, different from 0, so that

$$\lambda - f(x_0) = \alpha. \tag{1}$$

Since $f(x)$ is continuous at x_0, we can make the interval (a_n, b_n) so small that for every value of x in (a_n, b_n) we have, § 1025,

$$|f(x) - f(x_0)| < \alpha/2. \tag{2}$$

And since λ is the superior limit of the values of $f(x)$ in (a_n, b_n), we can choose x in (a_n, b_n) and (2) so that, § 1033,

$$\lambda - f(x) < \alpha/2. \tag{3}$$

But it will then follow from (2) and (3) that

$$\lambda - f(x_0) < \alpha. \tag{4}$$

Therefore, since (4) contradicts (1), (1) is false; that is, $\lambda - f(x_0) = 0$, or $\lambda = f(x_0)$, as was to be proved.

1036 **Corollary.** *If* f(x) *is continuous throughout the interval* (a, b), *it will have in* (a, b) *every value intermediate to its maximum and minimum values in this interval.*

For let c denote the value in question and consider the function $f(x) - c$, which is continuous in (a, b), § 1026.

If $f(x_0)$ and $f(x_1)$ denote the absolute maximum and minimum values of $f(x)$ in (a, b), $f(x_0) - c$ is + and $f(x_1) - c$ is −. Hence, § 1030, between x_0 and x_1 there is a number, call it x_2, such that $f(x_2) - c = 0$, or $f(x_2) = c$, as was to be proved.

1037 **Oscillation of a function.** By the *oscillation* of $f(x)$ in (a, b) is meant the difference between the superior and inferior limits of the values of $f(x)$ in (a, b).

1038 **Theorem 5.** *Let* f(x) *be continuous throughout* (a, b). *If any positive number* α *be assigned, however small, it is possible to divide* (a, b) *into a finite number of equal intervals in each of which the oscillation of* f(x) *is less than* α.

For divide (a, b) into any number of equal intervals, say into two such intervals, each of these in turn into two equal intervals, and so on. The process must ultimately yield intervals in each of which the oscillation of $f(x)$ is less than α.

For if not, there must be in (a, b) at least one half interval in which the oscillation of $f(x)$ is not less than α; in this, in turn, a half interval in which the oscillation of $f(x)$ is not less than α; and so on without end.

Let this never-ending sequence of intervals within intervals be

$$(a, b),\ (a_1, b_1),\ (a_2, b_2),\ \cdots,\ (a_n, b_n),\ \cdots,$$

and, as in § 1030, let $\lim a_n = \lim b_n = x_0$.

Since $f(x)$ is continuous throughout (a, b), it has an absolute maximum and an absolute minimum value in each of the intervals (a, b), (a_1, b_1), $\cdots$, (a_n, b_n), $\cdots$, § 1035.

Let $f(\alpha_n)$ denote the absolute maximum and $f(\beta_n)$ the absolute minimum value of $f(x)$ in (a_n, b_n).

Then, by hypothesis, $f(\alpha_n) - f(\beta_n) \overline{\gtrless} \alpha$,

and therefore $\lim f(\alpha_n) - \lim f(\beta_n) \overline{\gtrless} \alpha$.

But this is impossible. For since α_n and β_n are in (a_n, b_n), and $\lim a_n = \lim b_n = x_0$, we have $\lim \alpha_n = \lim \beta_n = x_0$.

Therefore, since $f(x)$ is continuous at x_0, $\lim f(\alpha_n) = \lim f(\beta_n)$; that is, $\lim f(\alpha_n) - \lim f(\beta_n) = 0$, $\therefore$ not $\overline{\gtrless} \alpha$.

FUNCTIONS OF TWO INDEPENDENT VARIABLES

Functions of two variables. We say that the variable u is a *function* of the variables x and y when to each pair of values of x and y there corresponds a definite value or set of values of u. **1039**

We shall confine ourselves to the case in which to each pair of values of x, y there corresponds a *single* value of u.

The notation $u = f(x, y)$ will mean that u is a function of x and y, and $f(a, b)$ will mean the value which u has when $x = a$ and $y = b$.

Thus, u is a function of x and y if $u = f(x, y) = x^2 - 2y + 1$. Here, when $x = 1$, $y = 2$, we have $u = f(1, 2) = 1 - 4 + 1 = -2$.

The note at the end of § 1022 applies, *mutatis mutandis*, here also.

Continuity of such a function. Let $f(x, y)$ denote a given function of x and y. We say that $f(x, y)$ is *continuous* at a, b, that is, when $x = a$ and $y = b$, if $f(a, b)$ has a definite finite value and if $f(x, y)$ will always approach $f(a, b)$ as limit when x and y are made to approach a and b respectively as limits. **1040**

In the contrary case we say that $f(x, y)$ is *discontinuous* at a, b, that is, when $x = a$ and $y = b$.

From this definition and § 189, it immediately follows that

The sufficient and necessary condition that f(x, y) *be continuous at* a, b *is that* f(a, b) *have a definite finite value, and that for every positive number* δ *which can be assigned it shall be possible to find a corresponding positive number* ϵ *such that* **1041**

$$|f(x, y) - f(a, b)| < \delta \text{ whenever } |x - a| < \epsilon \text{ and } |y - b| < \epsilon.$$

1042 **Theorem 1.** *If both of the functions* $f(x, y)$ *and* $\phi(x, y)$ *are continuous at* a, b, *the same is true of* $f(x, y) \pm \phi(x, y)$ *and* $f(x, y) \cdot \phi(x, y)$, *also of* $f(x, y)/\phi(x, y)$, *unless* $\phi(a, b) = 0$.

If $f(x, y)$ *is continuous at* a, b, *the same is true of* $\sqrt[n]{f(x, y)}$.

This follows immediately from § 1040 and §§ 203–205.

1043 **Number regions.** In what follows it is to be understood that x and y denote *real variables*, and $f(x, y)$ a *real function* of these variables (compare § 1027).

As is shown in § 382, pairs of values of x and y may be pictured by points in a plane. Evidently, if employing this method we draw the lines which are the graphs of the equations $x = a$, $x = b$, $y = c$, $y = d$, § 384, the rectangle bounded by these lines will contain the graphs of all pairs of values of x, y such that $a \leqq x \leqq b$, $c \leqq y \leqq d$. With this rectangle in mind, we shall call the assemblage of all such pairs of values of x, y the *number region* $(a, b;\ c, d)$.

1044 We say that $f(x, y)$ is *continuous throughout the region* $(a, b;\ c, d)$ if it is continuous for every pair of values of x, y in this region.

1045 **Theorem 2.** *If* $f(x, y)$ *be continuous throughout the region* $(a, b;\ c, d)$, *it has a maximum and a minimum value in this region.*

Since $f(x, y)$ is continuous throughout the given region, its values within this region have finite superior and inferior limits, §§ 1032, 1040. Call these limits λ and μ.

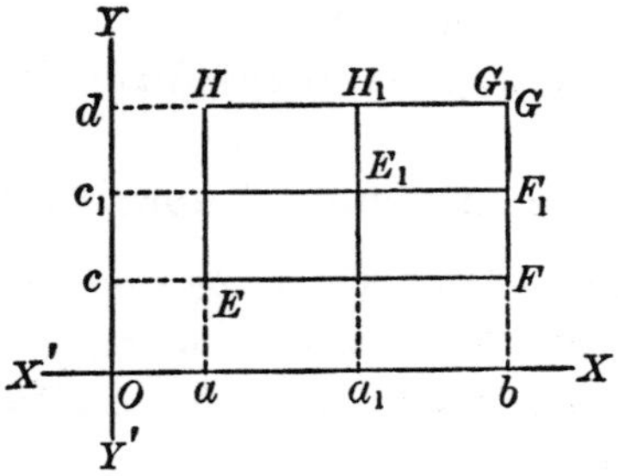

We are to prove that in $(a, b;\ c, d)$ there is a value pair x_0, y_0 such that $f(x_0, y_0) = \lambda$; and similar reasoning will show that there is also a value pair x_1, y_1 such that $f(x_1, y_1) = \mu$.

For construct the rectangle $EFGH$ which pictures the number region $(a, b;\ c, d)$, § 1043.

By the "values of $f(x, y)$ in $EFGH$" we shall mean the values of $f(x, y)$ corresponding to all pairs of values of x, y in $(a, b;\ c, d)$.

Divide $EFGH$ into four equal rectangles as in the figure. Evidently λ will be the superior limit of the values of $f(x, y)$ in at least one of these quarter rectangles. Call this quarter rectangle $E_1F_1G_1H_1$.

Deal with the rectangle $E_1F_1G_1H_1$ as we have just dealt with $EFGH$, and so on indefinitely. We thus obtain a never-ending sequence of rectangles within rectangles,

$$EFGH,\ E_1F_1G_1H_1,\ \cdots,\ E_nF_nG_nH_n,\ \cdots, \tag{1}$$

in each of which λ is the superior limit of the values of $f(x, y)$.

Let a_n denote the abscissa of E_n, and c_n its ordinate. As is proved in § 1030, when n is indefinitely increased a_n and c_n approach definite limits.

If $\lim a_n = x_0$ and $\lim c_n = y_0$, then $f(x_0, y_0) = \lambda$.

For if not, let $$\lambda - f(x_0, y_0) = \alpha. \tag{2}$$

Since $f(x, y)$ is continuous at x_0, y_0, we can so choose $E_nF_nG_nH_n$ that for every pair of values of x, y in this rectangle we have, § 1041,

$$|f(x, y) - f(x_0, y_0)| < \alpha/2. \tag{3}$$

And since λ is the superior limit of the values of $f(x, y)$ in $E_nF_nG_nH_n$, we can so choose x, y in $E_nF_nG_nH_n$ and in (3) that

$$\lambda - f(x, y) < \alpha/2. \tag{4}$$

From (3) and (4) it then follows that

$$\lambda - f(x_0, y_0) < \alpha. \tag{5}$$

But (5) contradicts (2). Hence (2) is false and therefore $f(x_0, y_0) = \lambda$, as was to be demonstrated.

THE FUNDAMENTAL THEOREM OF ALGEBRA

We are now in a position to prove that *every rational integral equation has a root*, § 797. We proceed as follows.

1046 **Theorem.** *Given* $\phi(\mathrm{z}) = 1 + \mathrm{bz^m} + \mathrm{cz^{m+1}} + \cdots + \mathrm{kz^n}$, *where* b, c, $\cdots$, k *denote constants, real or complex, and* z *a complex variable; it is always possible so to choose* z *that* $|\phi(\mathrm{z})| < 1$.

For let the expressions for z and b in terms of absolute value and amplitude, § 877, be

$$z = \rho(\cos\theta + i\sin\theta),\quad b = |b|\cdot(\cos\beta + i\sin\beta).$$

Then $bz^m = \rho^m|b|\cdot[\cos(m\theta + \beta) + i\sin(m\theta + \beta)]$. §§ 879, 881

First choose θ so that $\qquad m\theta + \beta = \pi. \qquad (1)$

Then $\quad bz^m = \rho^m|b|\cdot(\cos\pi + i\sin\pi) = -\rho^m|b|,$

since $\qquad \cos\pi = -1$ and $\sin\pi = 0$. §§ 877, 878

Next choose ρ so that, § 854,

$$|c|\rho^{m+1} + \cdots + |k|\rho^n < |b|\rho^m < 1. \qquad (2)$$

If z_0 denote the value of z which corresponds to the values of θ and ρ thus chosen, then $|\phi(z_0)| < 1$.

For since $\quad \phi(z_0) = (1 - \rho^m|b|) + cz_0^{m+1} + \cdots + kz_0^n,$

we have, § 235,

$$|\phi(z_0)| \leqq 1 - \rho^m|b| + |c|\rho^{m+1} + \cdots + |k|\rho^n,\ \therefore < 1, \text{ by (2)}.$$

1047 **Corollary.** *Given the function* $\mathrm{f(z)} = \mathrm{a_0z^n} + \mathrm{a_1z^{n-1}} + \cdots + \mathrm{a_n}$; *if* f(z) *does not vanish when* z = b, *we can always choose* z *so that* $|\mathrm{f(z)}| < |\mathrm{f(b)}|$.

For in $f(z)$ set $z = b + h$ and develop by Taylor's theorem, § 848. It may happen that certain of the derivatives $f'(z)$, $f''(z)$, and so on vanish when $z = b$; but they cannot all vanish since $f^{(n)}(z) = n!\,a_0$. Let $f^m(z)$ denote the first one which does not vanish.

Then $$f(b+h) = f(b) + f^m(b)\frac{h^m}{m!} + \cdots + f^{(n)}(b)\frac{h^n}{n!},$$

and therefore $$\frac{f(b+h)}{f(b)} = 1 + \frac{f^m(b)}{f(b)}\cdot\frac{h^m}{m!} + \cdots + \frac{f^{(n)}(b)}{f(b)}\cdot\frac{h^n}{n!}.$$

The second member of the last equation is a polynomial in h of the form considered in § 1046. Hence we can so choose h that $|f(b+h)/f(b)| < 1$ and therefore $|f(b+h)| < |f(b)|$.

1048 **Theorem.** *Given* $f(z) = a_0z^n + a_1z^{n-1} + \cdots + a_n$; *a value of* z *exists for which* f(z) *vanishes.*

For in $f(z)$ set $z = x + iy$, where x and y are real, and having expanded $a_0(x+iy)^n$, $a_1(x+iy)^{n-1}$, $\cdots$ by aid of the binomial theorem, collect all the real terms in the result, and likewise all the imaginary terms. We may thus reduce $f(z)$ to the form $f(z) = \phi(x, y) + i\psi(x, y)$, where $\phi(x, y)$ and $\psi(x, y)$ denote real polynomials in x, y, and therefore have, § 232,

$$|f(z)| = [\phi(x, y)^2 + \psi(x, y)^2]^{\frac{1}{2}}.$$

By § 855, we can find a positive number, as c, such that the roots of $f(z) = 0$, if there be any, are all of them numerically less than c; and if $c' = c/\sqrt{2}$, evidently $|z|$, or $(x^2 + y^2)^{\frac{1}{2}}$, is less than c for all values of x, y such that $-c' < x < c'$, $-c' < y < c'$.

But in this number region $(-c', c';\ -c', c')$ the expression $[\phi(x, y)^2 + \psi(x, y)^2]^{\frac{1}{2}}$ is a continuous function of x and y, § 1042. It therefore has a minimum value in this region, § 1045, say when $x = x_0$, $y = y_0$.

If $z_0 = x_0 + iy_0$, then $|f(z_0)| = [\phi(x_0, y_0)^2 + \psi(x_0, y_0)^2]^{\frac{1}{2}} = 0$.

For since $|f(z_0)|$ is the minimum value of $|f(z)|$, we cannot make $|f(z)| < |f(z_0)|$. Therefore $|f(z_0)| = 0$, since otherwise, § 1047, we could so choose z that $|f(z)| < |f(z_0)|$.

Hence $|f(z)|$, and therefore $f(z)$, vanishes when $z = z_0$; that is, z_0 is a root of the equation $f(z) = 0$.

ANSWERS

I. Page 89

1. Seven, four, eight; ten; seventeen. **2.** Seven.
3. $n = 7,\ a_0 = 3,\ a_1 = 1,\ a_2 = 0,\ a_3 = -4,\ a_4 = 1,\ a_5 = a_6 = 0,\ a_7 = -12.$
4. $f(0) = 3,\ f(-1) = 0,\ f(3) = 48,\ f(8) = 963.$
5. $f(0) = 2/5,\ f(-2) = 12,\ f(6) = 20/17.$
6. $f(1) = 5,\ f(4) = 9,\ f(5) = 8 + \sqrt{5}.$
7. $f(x-2) = 2x - 1,\ f(x^2+1) = 2x^2 + 5.$
8. $f(0, 0) = 8,\ f(1, 0) = 10,\ f(0, 1) = 7,\ f(1, 1) = 9,\ f(-2, -3) = 1$

II. Page 97

1. $2x^2y(b-a)$. **2.** $a^2 + a + b^2$. **3.** -9. **4.** $-4a^3 - 4a^2b - 21ab^2 + 11b^3$.
5. $-a + 3b - 7c$. **6.** $x^3 + 4x^2 + 5x - 2$. **7.** $b^3 - 5a^2b$.
8. $y^3 + 3y - 1$. **9.** $-10a + 24b$. **10.** $8x + 11$.
11. $-5a + 5b + 9c$. **12.** $2y - 4x - 2z$. **13.** $x^3 - x^2 - 8x - 12$.
14. $-x^4 + 9x^2 + y^2 + x - 3y - 7$.

§ 316. Page 105

1. $a^2 + b^2 + 4c^2 + 9d^2 - 2ab + 4ac - 6ad - 4bc + 6bd - 12cd$.
2. $1 + 4x + 10x^2 + 12x^3 + 9x^4$.
3. $x^6 - 2x^5y + 3x^4y^2 - 4x^3y^3 + 3x^2y^4 - 2xy^5 + y^6$.

III. Page 106

1. $6x^7 - 13x^6 + 7x^5 + 15x^4 - 34x^3 + 35x^2 - 21x + 5$.
2. $15x^5 - 14x^4a - x^3a^2 + 7x^2a^3 - 5xa^4 - 2a^5$.
3. $x^6 - y^6$. **4.** $6x^6 - 4x^5 - 9x^4 + 35x^3 - 10x^2 - 21x + 35$.
5. $28x^2 - 43xy + 10y^2$. **6.** $a^2b + (a^2 - ab + b^2)x - ax^2 - x^3$.
7. $x^6 - 2x^5 + 9x^4 - 10x^3 + 17x^2 - 6x$.
8. $2x^{2n-2} - 2x^{2n-3} - 3x^{2n-4} + 8x^{2n-5} - 5x^{2n-6}$.
9. $a^4 - a^2b^2 + 6ab^3 - 9b^4$. **10.** $x^2 - 9y^2 - 4z^2 + 12yz$
11. $x^8 - y^8 - 3xy - 1$. **12.** $a^3 + b^3 - c^3 + 3abc$

13. $3x^2 - 14xy + 8y^2 + 23x - 32y + 30$.

14. $2x^2 + 7y^2 + 24z^2 + 15xy - 59yz - 14zx$.

15. $b^4 - x^4$.

16. $x^8 + x^4 + 1$.

17. $2x^2y^2 + 2y^2z^2 + 2z^2x^2 - x^4 - y^4 - z^4$.

18. $1 + 1 + 1,\ 1 + 2 + 3 + 2 + 1,\ 1 + 3 + 6 + 7 + 6 + 3 + 1,$
$1 + 4 + 10 + 16 + 19 + 16 + 10 + 4 + 1$.

19. $1 + 5 + 10 + 10 + 5 + 1$
$1 + 6 + 15 + 20 + 15 + 6 + 1$
$1 + 7 + 21 + 35 + 35 + 21 + 7 + 1$
$1 + 8 + 28 + 56 + 70 + 56 + 28 + 8 + 1$
$1 + 9 + 36 + 84 + 126 + 126 + 84 + 36 + 9 + 1$
$1 + 10 + 45 + 120 + 210 + 252 + 210 + 120 + 45 + 10 + 1$.

20. $16x^2 - 24xy + 9y^2,\ 64x^3 - 144x^2y + 108xy^2 - 27y^3$.

21. $x^2 + 4y^2 + 9z^2 + 16u^2 + 4xy + 6xz - 8xu + 12yz - 16yu - 24zu$.

22. $x^3 + 8y^3 + 27z^3 + 6x^2y + 12xy^2 + 9x^2z + 27xz^2 + 36y^2z + 54yz^2$
$+ 36xyz,\ x^3 + 8y^3 - 27z^3 + 6x^2y + 12xy^2 - 9x^2z + 27xz^2 - 36y^2z$
$+ 54yz^2 - 36xyz$.

23. $a^4 - 8a^2b^2 + 16b^4$.

24. $a_0b_{17} + a_1b_{16} + \cdots + a_{17}b_0,\ a_{12}b_{19} + a_{13}b_{18} + \cdots + a_{27}b_4$.

25. $-4,\ -14,\ -55$.

27. $32a^{10}x^{15}y^{35},\ -x^{35}y^{56}z^{63},\ a^{4n}b^{2mn}c^{6n},\ a^{mn}b^{n^2}c^{2n^2}$.

28. $a^{12}b^4c^{18},\ -8a^2x^{16}y^{34}$.

IV. Page 110

1. $3a^2/2b$.
2. $-3y^4z/4ax^5$.
3. $-5x^m/4y^m$.
4. $3a^7b^8/c$.
5. $xy/(x + y)$.
6. $(x^2 + xy + y^2)(x + y)$.
7. $(a - b)(b - c)(c - a)$.
8. $-6abc + 5a^2c^2 - 4a^3b^2c^4$.
9. $3(x - y)^2 - 2(x - y) + 5$.
10. $6a^7b^3c$.
11. a^3.
12. $-2ax^2/y^2$.

V. Page 119

1. $-1/2$.
2. -2.
3. $-5/2$.
4. 16.
5. 6.
6. 1.
7. 5.
8. $-13/4$.
9. $-39/5$.
10. 779/1439.
11. $(4a + 5b)/3(b + c)$.
12. 1.
13. a.
14. $(a + b)/a$.
15. $(m - n)/(m + n)$.
16. $1/2,\ 1/3,\ -1/4,\ -2/5$.
17. 0, 1, 14.
18. 0, 8/3.
19. c/b, 0.
20. 1, 1, 2.

VI. Page 124

1. 68.
2. 14.
3. 26 and 324.
4. 84.
5. 5.
6. 40 and 10; 5 years hence.
7. 2 hours and 24 minutes.
8. A, 25 days; B, $16\frac{2}{3}$ days.

9. In the same direction at $43\frac{7}{11}$ minutes after 8; in opposite directions at $10\frac{10}{11}$ minutes after 8.
10. In $5\frac{5}{11}$ minutes.
11. 30 seconds per hour.
12. A, \$540; B, \$360; C, \$240; D, \$160.
13. \$42,000.
14. 576 square feet.
15. 80 feet.
16. 5 dollar pieces, 10 half-dollar pieces, 15 dimes.
17. \$3750 at 6%, \$1250 at 4%.
18. In the ratio 2 : 3.
19. $\frac{1}{3}$ of a pound.
20. One gallon the first time, two gallons the second time.
The original liquid was 60% alcohol.
21. The trains pass at 11 : $12\frac{12}{19}$ o'clock, $35\frac{10}{19}$ miles from Jersey City.
22. Their rates are 60 and 36 miles per hour. They pass at 11 : 15 A.M.
23. 250.
24. 57 of gold, 330 of silver.
25. \$28.
26. 75 square inches.
27. Units digit is $(9a + b)/18$, tens digit is $(9a - b)/18$.
28. $(bc - a)/(1 - c)$ years from the present time.

VII. Page 134

1. $x = 37,\ y = 25.$
2. $x = 10,\ y = 7.$
3. $x = 3,\ y = -2.$
4. $x = 5,\ y = -7.$
5. $x = 1/3,\ y = 1/2.$
6. $x = 4,\ y = 6.$
7. $x = -3/20,\ y = 4/5.$
8. $x = 5/2,\ y = 1.$
9. $x = -2,\ y = 1.$
10. $x = (a^2 + b^2)/(a + b) + 2,\ y = (a^2 + b^2)/(a + b).$
11. $x = cq/(aq + bp),\ y = cp/(aq + bp).$
12. $x = a + b,\ y = a - b.$
13. $x = 10,\ y = 8.$
14. $x = -1,\ y = 1/2.$
15. $x = (bc' + b'c)aa'/(ab' + a'b)cc',\ y = (ac' - a'c)bb'/(ab' + a'b)cc'.$
16. $x = y = ab/(a + b - ab).$

VIII. Page 136

1. $x = 48,\ y = -32/7.$
2. $x = 1/5,\ y = 2.$
3. $x = 2,\ y = 3.$
4. $x,\ y = 0,\ 1/2;\ 0,\ 2;\ 1,\ 0;\ -2/3,\ 0$
5. $x,\ y = 1,\ 1;\ -5/3,\ 0.$
6. $x,\ y = 0,\ 5/2;\ 5/3,\ 5/3;\ -5,\ 5.$
7. $x,\ y = 1,\ 3;\ 2,\ 2.$
8. $x,\ y = 2,\ 1;\ -4,\ 5.$
9. $x,\ y = -1/3,\ 4/3;\ -3,\ 2;\ -3/5,\ 8/5;\ -7/5,\ 2/5.$
10. $x,\ y = -3,\ 1;\ 2,\ 2;\ -2,\ -1;\ 4,\ -3.$

X. Page 147

1. $x = 5,\ y = 6,\ z = 7.$
2. $x = 1,\ y = -3,\ z = 3.$
3. $x = 5,\ y = 2,\ z = 2.$
4. $x = -7,\ y = -1,\ z = -3.$
5. $x = 33/5,\ y = 22/5,\ z = 11/10.$
6. $x = 1,\ y = -1,\ z = 7.$
7. $x = 1/8,\ y = 1/7,\ z = 1.$
8. $x = 2,\ y = 3,\ z = 5,\ u = -4.$
9. $x = 0,\ y = 4,\ z = -3,\ u = 0.$
10. $x = \dfrac{l - m + n}{2c},\ y = \dfrac{l + m - n}{2b},\ z = \dfrac{-l + m + n}{2a}.$

11. $x = \dfrac{mnd}{amn + bnl + clm}$, $y = \dfrac{nld}{amn + bnl + clm}$, $z = \dfrac{lmd}{amn + bnl + clm}$.

12. $x = -12$, $y = -8$, $z = -4$.

13. $(x - y - 3) + (y - z + 5) + (z - x - 2) \equiv 0$.

14. $2(3x - 8y + 7z - 10) + 5(2x + 5y - 3z - 12) - (16x + 9y - z - 80) \equiv 0$.

XI. Page 150

1. 5, 6, and 9. **2.** 33, 14, and 4. **3.** 87. **4.** A, \$4600 ; B, \$600.

5. A, $(p - q + r)/2$; B, $(p + q - r)/2$; C, $(-p + q + r)/2$.

6. \$2345, $4\frac{1}{2}$%. **7.** \$15,500. **8.** 12 square inches.

9. A, \$27 ; B, \$13. **10.** A, 12 days ; B, 9 days ; C, 8 days.

11. 18 and 12 feet per second. **12.** 20 and 16 miles per hour.

13. A's, 15 miles per hour ; B's, 10 miles per hour. **14.** $10\frac{10}{43}$ yards.

15. A's, $7\frac{17}{29}$ yards per second ; B's, 7 yards per second. **16.** 100 pounds.

17. A, 1/3 ; B, 11/3 ; C, 5. **18.** In A, 64% ; in B, 32%.

19. Velocity of sound 1100 feet per second; of bullet, $1447\frac{7}{19}$ feet per second.

20. Capacity of tank, 18,000 gallons. 300 gallons per minute pass through A, 200 through B, 600 through C.

XII. Page 154

1. $3(x-2)^3 + 17(x-2)^2 + 34(x-2) + 19$.

2. $(2x+3)^2 - 2(2x+3) + 4$. **3.** $f(x) = 43x^2/24 - 8x + 29/24$.

4. $f(x) = 3x^3 + x + 5$. **5.** $f(x, y) = 2x - 3y + 4$.

6. $2x + y - 7 = 0$. **7.** No. **8.** $x + 3y - 11 = 0$.

9. $c = 3$. **10.** $2x - 5y + 11 = 0$, and $4x - y - 5 = 0$.

11. $x^3 - 2x^2 - 5x + 6 = 0$. **12.** $x^2 + xy - x - 4y = 0$.

13. $3x + 2y - 3 \equiv -(x + y - 1) + (2x - y + 2) + 2(x + 2y - 3)$.

§ 401. Page 158

2. $Q = x^2 + 3$, $R = -4x + 4$.

§ 404. Page 161

2. $Q = 3x^3 + 5x^2 - 11x/2 + 9/4$, $R = -9x/4 + 5/2$.

§ 405. Page 161

2. $l = -11$, $m = -6$.

§ 408. Page 164

2. $2 - x + 6x^2 - 7x^3 + 13x^4$.

§ 409. Page 164

2. $2x + 3y - 1$.

XIII. Page 165

1. $Q = 2x^2 - 3x + 4$, $R = 2x + 4$. **2.** $Q = 3x^2 - 8x + 5$, $R = 0$.
3. $Q = 2x^3 - x^2 + 3x - 5$, $R = 20$.
4. $Q = 4x^4 + x^2 - x + 3$, $R = -2x^2 + 4x - 8$.
5. $Q = 2x + 3$, $R = 6x - 11$.
6. $Q = 2x^2 - 3x + 6$, $R = -12x^2 + 24x - 17$.
7. $3x^3 - 5x^2 - 7x + 12 \equiv (x - 2)(3x^2 + x - 5) + 2$.
8. $a = -3/2$, $b = -3/2$. **9.** $a = -2$, $b = 5$.
10. $Q = x^4 + x^2 + 1$, $R = 0$. **11.** $Q = 2x - y + 3$, $R = 0$.
12. $Q = a - b + 2c$, $R = 0$. **13.** $Q = a + b - c$, $R = 0$.
14. $Q = x^2 + ax + 2a$, $R = 0$. **15.** $Q = 4x^2 + 6xy + 9y^2$, $R = 0$.
16. $Q = x^3 + x^2y + xy^2 - 3y^3$, $R = 0$. **17.** $Q = 3a^3 + 2a^2b - ab^2 + 4b^3$, $R = 0$.
18. $Q = x^2 - yx/2 + 3y^2/4$, $R = y^3/4$; $Q' = y^2 - xy + 2x^2$, $R' = -2x^3$.
19. $Q = 1 - 4x + 6x^2$, $R = 6x^3 - 18x^4$.
20. $Q = 1 + 4x + 10x^2$, $R = 10x^3 - 50x^4$.
21. $1 + 2x + 4x^2 + 8x^3$. **22.** $2 + 5x + 5x^2 - 5x^3$.

§ 412. Page 169

3. $Q = 5x^4 + 15x^3 + 44x^2 + 132x + 397$, $R = 1193$.
4. $Q = x^2 + 3x + 2$, $R = 0$. **5.** $Q = x^2 + 4$, $R = 0$.

§ 414. Page 170

2. $f(2) = 48$, $f(-2) = 96$, $f(4) = 756$, $f(-2/3) = 112/9$.

XIV. Page 173

1. $Q = x^3 + x^2 + 3x + 1$, $R = 0$.
2. $Q = 5x^4 + 9x^3 + 19x^2 + 64x + 198$, $R = 597$.
3. $Q = 3x^3 - 6x^2 + 13x - 26$, $R = 51$.
4. $Q = x^2 + 5x - 6$, $R = 0$. **5.** $Q = x^2 - 5x/3 - 2/9$, $R = 16/9$.
6. $Q = x^2 - (b + c)x + bc$, $R = 0$. **7.** $Q = 2x^3 + 3x^2y - xy^2 + 5y^3$, $R = 0$.
8. $f(1) = 0$, $f(2) = 9$, $f(5) = 228$, $f(-1) = 6$, $f(-3) = -36$, $f(-6) = -399$.
9. $m = 3$. **10.** $l = -22$, $m = -24$. **13.** $2x^3 - 6x^2 - 12x + 16$.
14. Misprint. For $x = -1$ read $x = 1$.
The answer is then $5x^3 - 24x^2 + 25x + 6$.

XV. Page 176

1. $(x^2+1)^2+(x-2)(x^2+1)-x$.
2. $(2x^2+1)^2+x(2x^2+1)+5$.
3. $(2x+1)(x^3-x^2+x+3)^2-(2x^2+8x)(x^3-x^2+x+3)+(6x^2-3)$.
4. $(x-2y)(x^2+xy+y^2)^2+(2x+3y)y^2(x^2+xy+y^2)-2xy^4$.
5. $2(x-3)^3+10(x-3)^2+7(x-3)-9$.
6. $(x+2)^5-7(x+2)^4+10(x+2)^3+30(x+2)^2-99(x+2)+85$.
7. $(x+3)^3-27$.
8. $(x+1)^3-2(x+1)$.

XVI. Page 180

1. $2x^2y^3(3x^2z^2-6yz+4)$.
2. $3n(n-1)$.
3. $(a+1)(b-1)$.
4. $(m-n)(x-n)$.
5. $(3y-2)(x-4)$.
6. $(2x+y)(5y+3)$.
7. $xy(x-y)(xy+2)$.
8. $x(x+1)(x^2+1)$.
9. $(a-b)(c-d)$.
10. $a(c+d)(a-b)$.
11. $(d+e)(a+b+c)$.
12. $(a+d)(a-b+c)$.

§ 435. Page 181. Ex. 3

1. $(x+7)^2$.
2. $(3-a)^2$.
3. $(3xy+5)^2$.
4. $(x-2y+3)^2$.
5. $(8a^4-3)^2$.
6. $(a-b+c)^2$.

§ 436. Page 181

1. $(x^2+y^3)(x^2-y^3)$.
2. $6a(a+b)(a-b)$.
3. $3ax(2ax+5y)(2ax-5y)$.
4. $(5x^n+7x^m)(5x^n-7x^m)$.
5. $(6x^2+1)(\sqrt{6}x+1)(\sqrt{6}x-1)$.
6. $(x^2+xy-y^2)(x^2-xy-y^2)$.

§ 438. Page 182

1. $(4x-5y)(16x^2+20xy+25y^2)$.
2. $(3x+1)(9x^2-3x+1)$.
3. $(4x^2+9y^2)(2x+3y)(2x-3y)$.

§ 439. Page 184

1. $(x^2+\sqrt{2}xy+y^2)(x^2-\sqrt{2}xy+y^2)$.
2. $(x^4+y^4)(x^2+y^2)(x+y)(x-y)$.
3. $(x+y)(x^2-xy+y^2)(x^6-x^3y^3+y^6)$.

XVII. Page 184

1. $xy(2x-5y)^2$.
2. $7t(2x+3y)(2x-3y)$.
3. $(x-2y+3z)^2$.
4. $45(a+b)(a-b)(a^2+b^2)$.
5. $48x(x+1)^2(x-1)$.
6. $(2+a+2b)(2-a-2b)$.
7. $(x^2+x+1)(x^2-x+1)$.
8. $(a^2+2ab-b^2)(a^2-2ab-b^2)$.
9. $(a^2+2a+4)(a^2-2a+4)$.
10. $(3x^2+3xy+4y^2)(3x^2-3xy+4y^2)$.

11. $(-a+b+c+d)(a-b+c+d)(a+b-c+d)(a+b+c-d)$.
12. $9y^3(2x+y^2)(2x-y^2)(4x^2+2xy^2+y^4)(4x^2-2xy+y^4)$.
13. $(x-y)(x^2+xy+y^2)(x^6+x^3y^3+y^6)$.
14. $(x+y)(x-y)(x^2+y^2)(x^2+xy+y^2)(x^2-xy+y^2)(x^4-x^2y^2+y^4)$.
15. $(x^2+y^2)(x^8-x^6y^2+x^4y^4-x^2y^6+y^8)$.
16. $(x-2)(x^4+2x^3+4x^2+8x+16)$.
17. $(x+y^2)(x^6-x^5y^2+x^4y^4-x^3y^6+x^2y^8-xy^{10}+y^{12})$.

XVIII. Page 185

1. $(x+1)(x-1)(x^2-x+1)$.
2. $(x+1)(x-1)(x-2)(x^2+2x+4)$.
3. $(x+1)(x-1)^3$.
4. $(x+2)(x-2)(x-7)$.
5. $(x+y)^2(x-y)^2(x^2+y^2)$.
6. $(x+1)(x^2+x+2)$.
7. $(x+1)(x^4+x^3+2x^2+x+1)$.
8. $(x^2+2x+3)^2$.

§ 442. Page 186. Ex. 4

1. $(x+1)(x+2)$.
2. $(x-1)(x-15)$.
3. $(x+2)(x-6)$.
4. $(x+6)(x-5)$.
5. $(x+8)(x+12)$.
6. $(x-5)(x-16)$.

§ 443. Page 187. Ex. 4

1. $(3x-2)(2x-3)$.
2. $(x+3)(5x-1)$.
3. $(2x+1)(7x-3)$.
4. $(3x+1)(6x+5)$.
5. $(7x+4)(7x+11)$.
6. Misprint. $abx^2-(ac-b^2)x-bc=(ax+b)(bx-c)$.

§ 444. Page 188. Ex. 4

1. $(x+5+\sqrt{2})(x+5-\sqrt{2})$.
2. $(x-4)(x-6)$.
3. $(x-6+3i)(x-6-3i)$.
4. $[x+(1+\sqrt{3}i)/2][x+(1-\sqrt{3}i)/2]$.
5. $2[x+(3+\sqrt{7}i)/4][x+(3-\sqrt{7}i)/4]$.
6. $(x-2b)(x-4a+2b)$.

§ 445. Page 189. Ex. 2

1. $(x+y)(x+4y)$.
2. $\left(x-\frac{1+\sqrt{3}i}{2}y\right)\left(x-\frac{1-\sqrt{3}i}{2}y\right)$.

§ 446. Page 190

2. $(2x-y+z)(x-3y+2z)$.

XIX. Page 190

1. $(x-6)(x-8)$. **2.** $[x-(21+\sqrt{921})/2][x-(21-\sqrt{921})/2]$.
3. $(x-11)(5x+2)$. **4.** $(4x+7)(4x+9)$.
5. $(6x-1)(9x-2)$. **6.** $4(x+2y)(3x-y)$.
7. $(x+2)(x+3)(x-2)(x-3)$. **8.** $xy(x+3y)(x-6y)$.
9. $[x-(3+\sqrt{3}i)/2][x-(3-\sqrt{3}i)/2]$.
10. $3[x+(1+\sqrt{10})/3][x+(1-\sqrt{10})/3]$.
11. $[x-(2+\sqrt{6})y][x-(2-\sqrt{6})y]$. **12.** $(x+3b)(x-6a-3b)$.
13. $(ax+a-b)(bx-a-b)$. **14.** $[(1+c)x+b][x+d]$.
15. $(x-3y-1)(x-5y+3)$. **16.** $(x+y+2z)(x+2y+z)$.

XX. Page 194

1. $(x-1)(x-2)(x+3)$. **2.** $(x+1)(x+2)(x+3)$.
3. $(x-1)(x-2)(x-3)(x-4)$. **4.** $(x-1)(x+2)(x^2-x+1)$.
5. $(x-3)(2x+1)(3x+1)$. **6.** $[x-(1+\sqrt{3})y][x-(1-\sqrt{3})y](2x-y)$.
7. $(x-1)^3(2x+5)$.
8. $(x+1)(x-1)(x+3)(x-3)(2x+1)(2x-1)$.
9. $(x+2)(2x+1)(3x+2)(x^2+1)$.
10. $(x+1)(x-1)(x-2)(5x-2)(x^2+x+1)$.
11. $x=-2,\ 6$. **12.** $x=1/2,\ 2/3$. **13.** $x=7,\ -2$.
14. $x=-3\pm\sqrt{11}$. **15.** $x=2,\ 3,\ 4$. **16.** $x=1,\ 1,\ -1,\ -3$.
17. $x=1,\ (-1\pm\sqrt{3}i)/2$. **18.** $x=1,\ 2/5,\ -1/2$.

XXI. Page 195

1. $(3x-2)(2y+5)$. **2.** $(ac-bd)(ab-cd)$.
3. $(a-b)^2(a^2+ab+b^2)$. **4.** $a(a+3b)(a-3b)(a^2+9b^2)$.
5. $ab(a-b)(a+b)^2$. **6.** $(3ax+y)(bx-2y)$.
7. $3(x+2y)(x-2y)(x^2+2xy+4y^2)(x^2-2xy+4y^2)$.
8. $(x-1)(x+2)(x^4+2x^3+3x^2+2x+4)$.
9. $y^3(2x+y^2)(2x-y^2)(4x^2+2xy^2+y^4)(4x^2-2xy^2+y^4)$.
10. $(x-a)(x+b)$. **11.** $(x^n+3)(x^n-6)$.
12. $-(x+6)(x-7)$. **13.** $3(x+1)(x-2)(x^2+2x+4)$.
14. $(x+2)(x+3)(x-3)(x^2+2x+4)$. **15.** $(x+a+b+c)(x-a-b+c)$.
16. $(x+2)(x+4)(x-2)(x-4)$. **17.** $(a-b-2)(a-b-3)$.
18. $(x+y)(x+3y)(x-y)(x-3y)$. **19.** $(2x+y)(3x-5y-2)$.
20. $(x+a)(x-a)(x+b)(x-b)$.
21. $(-x+y+z+u)(x-y+z+u)(x+y-z+u)(x+y+z-u)$.
22. $(2x+3)(7x-1)$. **23.** $(1+22y)(1-3y)$.
24. $xy(y+4x)(y+51x)$. **25.** $(a+3bc)^2(a-3bc)^2$.

26. $x(x-1)(x-6)(x-7)$.
27. $(5y-x)(7y^2-10xy+19x^2)$.
28. $(x+y)(x-y)^3$.
29. $(x+a)(x+a-b)$.
30. $5xy(x-y)(x^2-xy+y^2)$.
31. $(x+1)^2(x-1)^3$.
32. $(b^2+b+1)(b^2-b+1)$.
33. $(2x+y-1)(x+3y+5)$.
34. $(a^2+2a+2)(a^2-2a+2)$.
35. $(x+y+3z)(x-2y+z)$.
36. $(2a^2+3ab+3b^2)(2a^2-3ab+3b^2)$.
37. $(x+5b)(x-8a-5b)$.
38. $(x^2+x+1)(x^2-x+1)(x^4-x^2+1)$.
39. $4x^2(2x-1)$.
40. $(a+b)(a-b)(x+y)(x-y)$.
41. $(x-a)(x+b)(x-b)$.
42. $(x-a)(x+b)(x^2+ax+a^2)$.
43. $(a+3b-2c)(a-3b+2c)$.
44. $(2a+1)^3$.
45. $(x^2-x+1)^2$.
46. $(a^2+b^2)(x^2+y^2)$.
47. $(x+1)(x+3)(x-3)(2x+1)(2x-1)$.
48. $(x-1)^2(x^2+2x+3)$.
49. $(x+3a+b)(x+2a-b)$.
50. $(x+2)(3x-2)(5x+3)$.
51. $(cx+ab)(abx+c)$.
52. $(x+a)(x-2a)(2x+a)$.
53. $(x+1)[(a-b)x+(a+b)]$.
54. $(x-y)(x^2+xy+y^2)(x^4+x^3y+x^2y^2+xy^3+y^4)$
$(x^8-x^7y+x^5y^3-x^4y^4+x^3y^5-xy^7+y^8)$.
55. $(x+1)(x-1)(x-2)(x-4)$.
56. $(x-1)(2x+1)^2$.
57. $(x-1)(x-4)(3x+2)(x^2+x+1)$.
58. $(x-2)(x+3)(x+4)(5x-1)$.
59. $(ab+c+d)(ac-d)$.
60. $(x+y+z)(x-y-z)(-x+y-z)(-x-y+z)$.

§ 459. Page 197. Ex. 2

1. x^3y^2. 2. $x+y$. 3. $x+2$. 4. $(x-2)(x-3)$.

§ 460. Page 198. Ex. 3

1. $x+3$. 2. $(x+2)(x+3)$. 3. $(x-1)(x-3)^2$.

§ 462. Page 199. Ex. 3

1. $x-2$. 2. $x+4$.

§ 468. Page 203

2. $2x^2+x+1$.

XXII. Page 204

1. x^3yz^3.
2. $(a+b)(a-b)$.
3. y^2-y+1.
4. $a+1$.
5. $x-1$.
6. x^2+y^2.
7. $x+2$.
8. $(x-1)(x-2)$.
9. x^2+x+1.

10. $(x+1)^3$. **11.** $(x-1)(x-2)^2$. **12.** $x+2$.
13. x^2+x+1. **14.** x^2+x-6. **15.** x^3+x^2-x-1.
16. $3x-7$. **17.** $3x+5$. **18.** $x(6x^2+7x-3)$
19. $2x^2-x+3$. **20.** $2x^3-4x^2+x-1$. **21.** x^2+x+1.
22. $x+2$. **23.** x^2-3. **24.** $y(x-2y)$.
25. $(x-1)(x+3)$. **26.** $(2x-3)^2(x+1)$.

XXIII. Page 207

1. $(9x^2+1)(9x^2-1)$. **2.** $(a^5+b^5)(a^5-b^5)$. **3.** $a^3(a+1)(a^3-1)$.
4. $(x-y)^4(x+y)^2(x^2+xy+y^2)(x^2+y^2)$. **5.** $(x-1)(x-2)(x-3)$.
6. $(x+y+z)(x-y-z)(-x+y-z)(-x-y+z)$.
7. $(2x-3y)(x+3y)(3x-y)$. **8.** $(x+1)(x-1)(x^2+1)$.
9. $(a^2+xy)(2x+3y)(2x-3y)$.
10. $x(2x+3y)(2x-3y)(2x-y)(4x+5y)$. **11.** $(x^3+y^3)(x^3-y^3)$.
12. $(x^2-1)(x^4+x^2+1)(3x-5)(x^2+1)$.
13. $(2x+3)(4x^2-6x+9)(4x^2+6x+9)(3x-2)$.
14. $(x^2+4a^2)(x+2a)(x-2a)$. **15.** $x(x+2)(x+b)(x-b)(x+a)$
16. $(x+1)(x+2)(x+3)(x+4)(2x-5)$.
17. $(x-2)(x^2+2x+4)(3x+1)(9x^2-3x+1)(2x+1)(x^2+x+1)$.
18. $(x-1)(x-2)(x-3)(x+3)(2x-1)$.
19. $(2x^4-x^3+10x^2+4x+5)(x^2+x+1)(x^2+3)$.
20. $(2x^4+3x^3-4x^2+13x-6)(x+1)^2$.

§ 491. Page 212

Ex. $m=5,\ n=-14$.

XXIV. Page 215

1. $xy(x+2y)$. **2.** $\dfrac{x^2+xy+y^2}{x^2+y^2}$. **3.** $\dfrac{x+3}{x+9}$.

4. $\dfrac{x-3}{x+2}$. **5.** $\dfrac{3(x-3b)}{2(x+3b)}$. **6.** $\dfrac{5x+a}{5x-3a}$.

7. $\dfrac{(x+5)(x-5)}{(x+3)(x-2)}$. **8.** $\dfrac{3x-5}{2x-3}$. **9.** $\dfrac{1}{x^2-y^2}$.

10. $\dfrac{x-y+z}{x+y-z}$. **11.** $1-y^2$. **12.** $\dfrac{m-6n}{3m-n}$.

13. $\dfrac{2x^2+5x-12}{2x^2+x-15}$. **14.** $\dfrac{x-1}{2x+1}$. **15.** $\dfrac{x^2+4}{2(x^2+6)}$.

16. $\dfrac{x-3}{x^2+2x+4}$. **17.** $\dfrac{x^2-2bx+c^2}{x^2+2bx-c^2}$. **18.** 3.

XXV. Page 222

1. $\dfrac{2}{2a+3b}$.
2. $\dfrac{x^3-x^2+2x-1}{(x^3+1)(x-1)}$.
3. $\dfrac{3}{(x-1)(x-3)}$.
4. $\dfrac{3x^2-11}{(x-1)(x-2)(x-3)}$.
5. $\dfrac{2x(b^2-c^2)}{(x^2-b^2)(x^2-c^2)}$.
6. 0.
7. a.
8. $-\dfrac{8x^2+6x-9}{8x^3-27}$.
9. 4.
10. $3a^2+3b^2+3c^2+4ab+4ac+4bc$.
11. $\dfrac{7}{x^2-x-3}$.
12. $\dfrac{2}{x^4-4x^2-x+2}$.
13. $\dfrac{a^6+a^4+a^2+1}{a^3}$.
14. a^3+1.
15. $\dfrac{x+1}{x-1}$.
16. $\dfrac{2x-3}{x(x+1)}$.
17. $\dfrac{x^3}{x+a}$.
18. $x^2-y^2+z^2-2xz$.
19. $\dfrac{4ab}{(a-b)^2}$.
20. $\dfrac{-x+y+z}{x-y+z}$
21. $-\dfrac{a^2b^2}{(a-b)^2}$.
22. $\dfrac{x^2-3x+1}{x^2-4x+1}$.
23. $\dfrac{x^4+3x^2+1}{x^3+2x}$.

XXVI. Page 230

1. 1/2.
2. -3.
3. ∞.
4. 0.
5. $-5/4$.
6. ∞.
7. $3/2, \infty, 0, 2$.
8. $-1/9$.
9. ∞.
10. 3.
11. 2/3.

XXVII. Page 235

1. $x=1$.
2. $x=15/31$.
3. $x=6$.
4. $x=4$.
5. $x=5/6$.
6. $x=-5/3$.
7. $x=0$.
8. $x=-(a+b)$.
9. $x=-10$.
10. $x=-2$.
11. $x=-2/3$.
12. $x=-7$.
13. $x=3/4$.
14. $x=3$.
15. $x=-4$.
16. $x=\dfrac{cq+dp+(a+b)pq}{c+d+ap+bq}$.
17. $x=-71/33$.
18. $x=2bc/(b+c)$.
19. $x=(a+b+c)/3$.
20. $x=-5$.
21. No root.
22. $x=3,\ y=-2$.
23. $x=1,\ y=3$
24. $x=\dfrac{2abc}{ab+bc-ca},\ y=\dfrac{2abc}{-ab+bc+ca},\ z=\dfrac{2abc}{ab-bc+ca}$.
25. $x=2,\ y=-2,\ z=4$.

§ 534. Page 239

2. $-\dfrac{4}{x}+\dfrac{4x^2+4x+9}{x^3+x^2+x-1}$.

XXVIII. Page 244

1. $\dfrac{3}{x-2} - \dfrac{1}{x+3}$.

2. $\dfrac{8}{5(2x+1)} + \dfrac{3}{5(3x-1)}$.

3. $-\dfrac{2}{x+1} + \dfrac{8}{x+2} - \dfrac{6}{x+3}$.

4. $-\dfrac{1}{x-1} + \dfrac{11}{2(x-2)} - \dfrac{9}{x-3} + \dfrac{9}{2(x-4)}$

5. $\dfrac{1}{x+1} + \dfrac{1}{x^2-x+1}$.

6. $\dfrac{2}{x} + \dfrac{3}{1+x} + \dfrac{5}{1-x}$.

7. $x + 2 + \dfrac{1}{x-1} - \dfrac{2}{x-2}$.

8. $\dfrac{2}{x-2} + \dfrac{11}{(x-2)^2} + \dfrac{20}{(x-2)^3} + \dfrac{13}{(x-2)^4}$.

9. $\dfrac{1}{12x} + \dfrac{3}{44(x-4)} - \dfrac{10}{33(2x+3)}$.

10. $-\dfrac{4}{2x^2+1} - \dfrac{1}{x+1} + \dfrac{1}{x-1}$.

11. $\dfrac{2}{(x+3)^2} - \dfrac{21}{(x+3)^3} + \dfrac{76}{(x+3)^4} - \dfrac{98}{(x+3)^5}$.

12. $\dfrac{x}{x^2+1} - \dfrac{x-1}{x^2+2}$.

13. $\dfrac{3}{2(x-1)} - \dfrac{1}{2(x-3)} + \dfrac{13}{(x-3)^2}$.

14. $\dfrac{1}{x-2} - \dfrac{x-1}{x^2+1}$.

15. $\dfrac{2x-4}{x^2+x+1} + \dfrac{2x+6}{(x^2+x+1)^2} - \dfrac{3x+1}{(x^2+x+1)^3}$.

16. $\dfrac{1}{x} + \dfrac{1}{x^2} - \dfrac{1}{x-1} + \dfrac{2}{(x-1)^2}$.

17. $\dfrac{1}{x^2+2} + \dfrac{2}{3(x-2)} - \dfrac{2}{3(x+1)}$.

18. $\dfrac{a^2+pa+q}{(a-b)(a-c)} \cdot \dfrac{1}{x-a} + \dfrac{b^2+pb+q}{(b-a)(b-c)} \cdot \dfrac{1}{x-b} + \dfrac{c^2+pc+q}{(c-a)(c-b)} \cdot \dfrac{1}{x-c}$.

19. $-\dfrac{2}{27x} + \dfrac{25}{256(x-1)} - \dfrac{3}{64(x-1)^2} - \dfrac{163}{6912(x+3)} - \dfrac{35}{288(x+3)^2} - \dfrac{25}{48(x+3)^3}$.

20. $\dfrac{4x+3}{2(x^2+x+1)} - \dfrac{2x-3}{2(x^2-x+1)}$.

XXIX. Page 249

1. x and z.

2. $\Sigma a^2b^2 = a^2b^2 + b^2c^2 + c^2a^2$.

$\Sigma a^3b^4 = a^3b^4 + b^3a^4 + b^3c^4 + c^3b^4 + c^3a^4 + a^3c^4$

$\Sigma a^2/b = a^2/b + b^2/a + b^2/c + c^2/b + c^2/a + a^2/c$.

$\Sigma a^2b^3c^5 = a^2b^3c^5 + a^2c^3b^5 + b^2c^3a^5 + b^2a^3c^5 + c^2a^3b^5 + c^2b^3a^5$.

$\Sigma a^2b^2c^4 = a^2b^2c^4 + a^2c^2b^4 + b^2c^2a^4$.

$\Sigma (a+b)c = (a+b)c + (b+c)a + (c+a)b$.

$\Sigma (a+b^2)c^3 = (a+b^2)c^3 + (b+a^2)c^3 + (b+c^2)a^3 + (c+b^2)^3$
$\qquad + (c+a^2)b^3 + (a+c^2)b^3$.

$\Sigma (a+2b+3c) = (a+2b+3c) + (a+2c+3b) + (b+2c+3a)$
$\qquad + (b+2a+3c) + (c+2a+3b) + (c+2b+3a)$.

4. No.

5. $y^2 - x^2,\ x^2 - z^2,\ z^2 - y^2;\ a^2bc,\ b^2cd,\ c^2da,\ d^2ab;\ (a-c)(b-a)$,
$(b-a)(c-b),\ (c-b)(a-c)$.

6. $ab^3c^2 + bc^3d^2 + cd^3a^2 + da^3b^2$, $a(b-c) + b(c-d) + c(d-a) + d(a-b)$, $(b+2c)(a+d) + (c+2d)(b+a) + (d+2a)(c+b) + (a+2b)(d+c)$, $\frac{a^2}{(a-b)(a-c)} + \frac{b^2}{(b-c)(b-d)} + \frac{c^2}{(c-d)(c-a)} + \frac{d^2}{(d-a)(d-b)}$.

XXX. Page 251

1. $-(x-y)(y-z)(z-x)$.
2. $-(x-y)(y-z)(z-x)$.
3. $3(x-y)(y-z)(z-x)$.
4. $(x-y)(y-z)(z-x)(x+y+z)$.
5. $(x-y)(y-z)(z-x)(xy+yz+zx)$.
6. $-(x+y)(y+z)(z+x)(x-y)(y-z)(z-x)$.
7. $3(x+y)(y+z)(z+x)$.
8. $5(x-y)(y-z)(z-x)(x^2+y^2+z^2-xy-yz-zx)$.
9. $80xyz(x^2+y^2+z^2)$.
10. $-2(x-y)(y-z)(z-x)(x+y+z)$.
11. $(x+y)(y+z)(z+x)$.
12. $-(x-y)(y-z)(z-x)(x^3+y^3+z^3+x^2y+y^2x+x^2z+z^2x+y^2z+z^2y+xyz)$.
13. $a^2+b^2+c^2+ab+bc+ca$.
14. 0.
15. 0.
16. 1.
17. $\frac{x^2}{(x-a)(x-b)(x-c)}$.

XXXI. Page 259

1. $27x^3 + 54x^2y + 36xy^2 + 8y^3$.
2. $a^8 - 8a^7b + 28a^6b^2 - 56a^5b^3 + 70a^4b^4 - 56a^3b^5 + 28a^2b^6 - 8ab^7 + b^8$.
3. $1 + 14x^2 + 84x^4 + 280x^6 + 560x^8 + 672x^{10} + 448x^{12} + 128x^{14}$.
4. $16 + \frac{32}{x} + \frac{24}{x^2} + \frac{8}{x^3} + \frac{1}{x^4}$.
5. $x^6 - 18x^4 + 135x^2 - 540 + \frac{1215}{x^2} - \frac{1458}{x^4} + \frac{729}{x^6}$.
6. $\frac{x^5}{y^5} - 5\frac{x^3}{y^3} + 10\frac{x}{y} - 10\frac{y}{x} + 5\frac{y^3}{x^3} - \frac{y^5}{x^5}$.
7. $1 - 4x + 14x^2 - 28x^3 + 49x^4 - 56x^5 + 56x^6 - 32x^7 + 16x^8$.
8. $a^6 + 3a^5x - 5a^3x^3 + 3ax^5 - x^6$.
9. $231x^5/16$.
10. $-3153199104a^5b^7$.
11. $-8064a^{10}b^5c^5$.
12. $126x^4$, $-126x^5$.
13. 56.
14. 15120.
15. 15.
16. -648.
17. 924.
18. 2,795,520.
19. $x^3 - 6x^2y - xy^2 + 30y^3$.
20. $x^4 - 4x^3 - 19x^2 + 46x + 120$.
21. 96.
22. 64, 400.
23. 16, 16, 24.

§ 568. Page 260

1. $\frac{8a^2b^3}{5c^4d^6}$.
2. $3xy^2z^3$.
3. $\frac{2x^2y^6}{az^5}$.

10

§ 569. Page 262

3. $2x^2 - x + 1$.

§ 570. Page 264

$5x^2 - 4x + 3.$

§ 572. Page 264

2. $2 - \frac{x}{4} + \frac{15x^2}{64}.$

XXXII. Page 269

1. $-\frac{3x^2y^5}{5a^3z^4}.$ **2.** $\frac{23a^2b^3}{25cd^4}.$ **3.** $xy(x-y).$

4. $x^2 - x + 1.$ **5.** $x^3 - x + 3.$ **6.** $2x^3 + 3x^2y - y^3.$

7. $2x - 5 - 3/x.$ **8.** $7 - 6x - 5x^2.$ **9.** $x^4 + x^3 - x^2 + x - 2.$

10. $x^2 - 2x - 1.$ **11.** $2x^2 - 3xy + 4.$ **12.** $\frac{x}{y} + xy + \frac{y}{x}.$

13. $1 - x - x^2/2 - x^3/2.$ **14.** $2 - x/4 + 47x^2/64 + 47x^3/512.$

15. $x^2 + x + 1.$ **16.** $3x^4 + x^2 - 1.$ **17.** $2x^2 - 3ax + 3a^2.$

18. $\frac{x}{y} + 1 + \frac{y}{x}.$ **19.** $1 - \frac{1}{3}x + \frac{2}{9}x^2.$ **20.** $x^2 - x + 1.$

21. $x^2 + x + 1.$ **22.** $a = 6,\ b = 1.$ **23.** 167.

24. Misprint. The number should be 2313.61. Its square root is 48.1.

25. 24.15. **26.** 2037. **27.** .0566. **28.** 3.004. **29.** 1.414.

30. 7.449. **31.** 15.315. **32.** 123. **33.** 55.1. **34.** 10.12.

XXXIII. Page 274

1. $3\sqrt{2}.$ **2.** $14\sqrt{3}.$ **3.** $-9.$ **4.** $-\sqrt[3]{10}.$

5. $\sqrt{6}/2.$ **6.** $\sqrt[3]{12}/2.$ **7.** $\sqrt[4]{6}/2.$ **8.** $\sqrt[5]{6}/2.$

9. $ab^2c^3d\sqrt[5]{25d}.$ **10.** $2c\sqrt[6]{2a^2b^4c^2}.$ **11.** $z\sqrt[4]{2x^2y^3z}.$ **12.** $\sqrt[n]{5ab^2c^3}.$

13. $\sqrt[3]{ab^2c^3}.$ **14.** $a^2b^3c^4\sqrt[n]{ab^2}.$ **15.** $x\sqrt{y^2 - z^2}.$ **16.** $(x+y)\sqrt{(x-y)}.$

17. $x\sqrt[3]{x^3 - y^3}.$ **18.** $b\sqrt{a(a-b)}.$ **19.** $\frac{1}{4ab}\sqrt[3]{2a^2b(a^3+b^3)}.$

20 $\frac{1}{a-b}\sqrt{a^2 - b^2}.$ **21.** $\frac{1}{3(x+1)}\sqrt[3]{3(x^3+1)}.$ **22.** $\sqrt[3]{b^3 - a^3}/b.$

23. $\frac{c}{a^nb^{n+1}}\sqrt[3]{bc^n}.$ **24.** $\frac{ax-b}{b^2}\sqrt{b}.$ **25.** $\sqrt{27a^3}.$ **26.** $\sqrt{\frac{a+b}{a-b}}$

27. $\sqrt[4]{3ax}.$ **28.** $3\sqrt{2},\ 5\sqrt{2},\ \sqrt{2}/4.$

29. Misprint. Replace $\sqrt[6]{192}$ by $\sqrt[3]{192}$. The three radicals then reduce to $2\sqrt[3]{3},\ 4\sqrt[3]{3},\ 2\sqrt[3]{3}/3.$

30. $(x-y)\sqrt{x^2 + xy + y^2},\ xy\sqrt{x^2 + xy + y^2}.$

XXXIV. Page 277

1. $\sqrt[30]{243}, \sqrt[30]{27}, \sqrt[30]{9}$. **2.** $\sqrt[12]{a^8}, \sqrt[12]{8a^9b^6}, \sqrt[12]{49b^{10}}$.

3. $3\sqrt{2} = \sqrt[6]{5832},\ 2\sqrt[3]{3} = \sqrt[6]{576}.\ \therefore\ 3\sqrt{2} > 2\sqrt[3]{3}$.

4. $\sqrt{3} = \sqrt[12]{729},\ \sqrt[3]{4} = \sqrt[12]{256},\ \sqrt[4]{5} = \sqrt[12]{125}.\ \therefore\ \sqrt{3} > \sqrt[3]{4} > \sqrt[4]{5}$.

5. 5. **6.** $2\sqrt{5}$. **7.** $2\sqrt[3]{4}$. **8.** 30.

9. $30\sqrt[3]{3}$. **10.** $\sqrt{6}/3$. **11.** $2\sqrt[12]{2}$. **12.** $\sqrt[12]{17578125}/5$.

13. 10. **14.** $a^2b^3c^6\sqrt[6]{ab^5c}$. **15.** $\sqrt[6n]{a^5}$. **16.** $\sqrt[3]{a^2b^2}$.

17. abc^2. **18.** $\sqrt[6n]{a^{2n+1}}$. **19.** $\sqrt[18]{ab^{17}}/b$. **20.** $\frac{1}{ab}\sqrt[3]{b^2}$.

21. $24\sqrt{3}$. **22.** a^4. **23.** $64xy^3z^4\sqrt{xz}$. **24.** $\sqrt[6]{a}$.

25. $\sqrt{2}$. **26.** $\sqrt[10]{ab^2c^7}/c$. **27.** $\sqrt[3]{4}$. **28.** $\sqrt[4]{8}$.

29. $\sqrt[3]{4}$. **30.** $\sqrt[12]{32}$. **31.** $\sqrt[2n]{a}$. **32** $\sqrt[4]{a^p}$.

33. $10\sqrt{3}$. **34.** $\frac{51}{10}\sqrt{5} + 3\sqrt{7}$. **35.** $\frac{5}{2}\sqrt[3]{4}$. **36.** $\frac{a+b+c}{abc}\sqrt{abc}$.

37. $\frac{7}{2}\sqrt{2} - \frac{2}{3}\sqrt[3]{3}$. **38.** 0. **39.** $(3+2a)\sqrt{ax}$. **40.** $\frac{1}{x^2-y^2}\sqrt{x^2-y^2}$

41. $2\sqrt{3} + 3\sqrt{2} + 6$. **42.** $\sqrt{3} + \sqrt{5} + \sqrt{7}$. **43.** $7\sqrt{3} + 4\sqrt{10}$.

44. $\sqrt{17}$. **45.** $10 + 6\sqrt{3}$. **46.** $a + \sqrt{a} + 1$.

XXXV. Page 282

1. $a^{\frac{3}{3}}$. **2.** $c^{\frac{2}{3}}$. **3.** $a^{\frac{9}{8}}$. **4.** $b^{\frac{19}{8}}$.

5. $\sqrt[3]{a^2}$. **6.** $\frac{1}{c^2}\sqrt{c}$. **7.** $\frac{1}{d^4}$. **8.** $\sqrt{e}$.

9. $\frac{b^3c^2}{a}$. **10.** $\frac{1}{x^{\frac{1}{2}}y^{\frac{3}{2}}}$. **11.** $\frac{1}{x^{10}}$. **12.** $\frac{y^{\frac{1}{2}}}{x^{\frac{1}{2}}}$.

13. $bc - c^2b^{-1}$. **14.** $125.2^{\frac{1}{2}}/32$. **15.** 27. **16.** 9.

17. $\frac{2^{\frac{1}{2}}}{256}$. **18.** $a^{\frac{9}{4}}$. **19.** 1. **20.** $(ab)^{\frac{5}{4}}$.

21. a^4b^{-3}. **22.** $a^{\frac{1}{6}}$. **23.** $a^2b^4c^{-6}$. **24.** $-8a^6$.

25. b^6a^{-4}. **26.** $b^{\frac{3}{4}}$. **27.** $a^{-1}b^{\frac{1}{3}}c^{\frac{5}{3}}$. **28.** $5a^{11}/4$.

29. $a^{\frac{1}{3}}b^{-1}c$. **30.** $a^{-\frac{1}{12}}$. **31.** a^2. **32.** x^{x^3}.

33. $x^{x^2y}y^{xy^{-}}$. **34.** $x^{\frac{1}{2}}y^{\frac{1}{4}} - x^{\frac{1}{4}}y^{\frac{1}{2}}$. **35.** $x^{\frac{1}{2}} + x^{\frac{1}{4}}y^{\frac{1}{4}} + y^{\frac{1}{2}}$

36. $a^{\frac{5}{3}} + a^{\frac{4}{3}}b^{\frac{1}{2}} + ab + a^{\frac{2}{3}}b^{\frac{3}{2}} + a^{\frac{1}{3}}b^2 + b^{\frac{5}{2}}$.

37. $x^2 - 4x^{\frac{3}{2}}y^{\frac{1}{4}}z^{\frac{1}{4}} + 6xy^{\frac{1}{2}}z^{\frac{1}{2}} - 4x^{\frac{1}{2}}y^{\frac{3}{4}}z^{\frac{3}{4}} + yz$.

38. $e^x - e^{-x}$. **39.** $x + 2x^{\frac{1}{2}}y^{\frac{1}{2}} + 3x^{-\frac{1}{2}}y^{\frac{1}{2}}$. **40.** $x + 1 + x^{-1}$.

XXXVI. Page 284

1. $1+\frac{x}{3}-\frac{x^2}{9}+\frac{5x^3}{81}$.

2. $a^{-\frac{1}{3}}-\frac{a^{-1}x^{-\frac{2}{3}}}{2}+\frac{3a^{-\frac{5}{3}}x^{-\frac{4}{3}}}{8}-\frac{5a^{-\frac{7}{3}}x^{-2}}{16}$

3. $9-\frac{4x}{3^2}-\frac{4x^2}{3^6}-\frac{32x^3}{3^{11}}$.

4. $a+\frac{1}{m}a^{1-m}x+\frac{1-m}{m^2 2!}a^{1-2m}x^2+\frac{(1-m)(1-2m)}{m^3 3!}a^{1-3m}x^3$.

5. $a^4+4a^5b^{-\frac{1}{2}}+10a^6b^{-1}+20a^7b^{-\frac{3}{2}}$.

6. $x^{-3}-6x^{-\frac{7}{2}}y^{\frac{1}{3}}+21x^{-4}y^{\frac{2}{3}}-56x^{-\frac{9}{2}}y$.

7. $\frac{1}{2}-\frac{3x}{4}+\frac{9x^2}{8}-\frac{27x^3}{16}$.

8. $1-\frac{2x}{5}+\frac{7x^2}{25}-\frac{28x^3}{125}$.

9. $1-\frac{9x^{\frac{1}{2}}}{2}+\frac{135x}{8}-\frac{945x^{\frac{3}{2}}}{16}$.

10. $-55x^9$.

11. $\frac{-663}{2^{10}}x^{\frac{21}{2}}y^2$.

12. $-\frac{19\cdot 23\cdot 31\cdot 33}{2^{25}}x^{\frac{3}{2}}$.

13. $-\frac{5}{2^5}x^{-2}$.

14. 1. 9.9498. 2. 3.9578. 3. 1.9873.

XXXVII. Page 287

1. $a^{\frac{2}{7}}$.

2. $a^{\frac{1}{3}}b^{\frac{1}{2}}$.

3. $x^{\frac{1}{2}}$.

4. $\sqrt{a}-\sqrt{bc}$.

5. $(\sqrt{x}+\sqrt{y}-\sqrt{z})(x+y-z-2\sqrt{xy})$.

6. $(\sqrt{xy}+\sqrt{yz}-\sqrt{zx})(xy+yz-zx-2y\sqrt{xz})$.

7. $[\sqrt{x}+\sqrt{y}+\sqrt{z}+\sqrt{u}][x+y-u-z-2(\sqrt{xy}-\sqrt{uz})]$
$[x^2+y^2+z^2+u^2-2xy-2xz-2xu-2yz-2yu-2zu-8\sqrt{xyzu}]$.

8. $(\sqrt{x}-\sqrt[4]{x}+1)(x-\sqrt{x}+1)$.

9. $x^{\frac{2}{3}}-x^{\frac{1}{3}}y^{\frac{1}{3}}+y^{\frac{2}{3}}$.

10. $a^{\frac{4}{3}}+a^{\frac{2}{3}}b^{\frac{2}{3}}+b^{\frac{4}{3}}$.

11. $(x^{\frac{1}{4}}+y^{\frac{1}{4}})(x^{\frac{1}{2}}+y^{\frac{1}{2}})$.

12. $(x^9)^{\frac{11}{12}}-(x^9)^{\frac{10}{12}}(y^8)^{\frac{1}{12}}+(x^9)^{\frac{9}{12}}(y^8)^{\frac{2}{12}}-\cdots+(y^8)^{\frac{11}{12}}$.

13. $(1-x^{\frac{1}{2}}y^{\frac{1}{3}})(1+xy^{\frac{2}{3}}+x^2y^{\frac{4}{3}})$.

14. $(x^{\frac{1}{3}}-1)$.

15. $3+\sqrt{5}$.

16. $(1+\sqrt{2}-\sqrt{3})\sqrt{2}$.

17. $(1-\sqrt[3]{2}+\sqrt[3]{4})$.

18. $(1-\sqrt[3]{3})$.

19. $\sqrt[3]{3^2}(1-\sqrt[3]{2})$.

20. $\frac{\sqrt{a}\cdot\sqrt[5]{b^3}}{ab}$.

21. $\frac{a^2+2a\sqrt{b}+b}{a^2-b}$.

22. $\frac{3-\sqrt{6}}{15}$.

23. $\frac{b-\sqrt{b^2-a^2}}{a^2}$.

24. $\frac{x+\sqrt{x^2-y^2}}{y}$.

25. $\sqrt{2}+\sqrt{3}$.

26. $\frac{1-\sqrt{2}-\sqrt{3}+\sqrt{6}}{2}$.

27. $\frac{(\sqrt{x}+\sqrt{y})(\sqrt{x}+\sqrt{y}-\sqrt{x+y})}{2}$.

28. $\frac{\sqrt[3]{3}(\sqrt[3]{3^4}+\sqrt[3]{3^2}+1)}{4}$.

29. .447.

30. 2.756

31. 1.732.

XXXVIII. Page 290

1. 256. **2.** 1/9. **3.** $\pm 16\sqrt{2}$. **4.** 14. **5.** 1.
6. $\left(\frac{d}{\sqrt{a}+\sqrt{b}+\sqrt{c}}\right)^2$. **7.** 3. **8.** 5. **9.** -1.
10. 34/15. **11.** 2. **12.** 2. **13.** 6. **14.** 5/4.
15. $\begin{cases} x=-1, \ y=3, \\ x=3, \ y=(73-10\sqrt{10})/9. \end{cases}$ **16.** $x=10$, $y=3$.

XXXIX. Page 293

1. $\sqrt{7}+\sqrt{2}$. **2.** $2(\sqrt{2}+\sqrt{3})$. **3.** $5-\sqrt{7}$. **4.** $(\sqrt{10}+\sqrt{15})/5$.
5. $(3\sqrt{2}-\sqrt{10})/2$. **6.** $\sqrt[4]{2}(\sqrt{5}+\sqrt{3})$. **7.** $\sqrt{a+b}+\sqrt{a-b}$.
8. $\sqrt{a}-\sqrt{b-a}$. **9.** $1+\sqrt{2}$. **10.** $1+2\sqrt{3}$.

XL. Page 297

1. $7i$. **2.** $3\sqrt{2}i$. **3.** $-4\sqrt{6}$. **4.** $2i$. **5.** -2.
6. 1 **7.** i. **8.** $-i$. **9.** $(x-y)i$.
10. $2-\sqrt{6}+(2\sqrt{2}+\sqrt{3})i$. **11.** $648\sqrt{6}$. **12.** -22. **13.** 0.
14. 10. **15.** 16. **16.** $-i$. **17.** $\frac{a^2-b^2}{a^2+b^2}+\frac{2ab}{a^2+b^2}i$.
18. $1-\sqrt{2}i$. **19.** $1-\sqrt{3}i$. **20.** $\sqrt{2}(1+i)$. **23.** $x=5/3$, $y=-4/3$.
24. $3+2i$. **25.** $1+i$. **26.** $(a+b)+(a-b)i$.

§ 630. Page 299. Exs. 1, 2

1. 2, -4. 2. 3, 1/2. 3. 0, 5. 4. $\pm\sqrt{6}/3$.
1. $6x^2+13x+6=0$ 2. $x^2-a^2=0$. 3. $4x^2-x=0$.

XLI. Page 301

1. 5, -7. **2.** 3/2, $-1/2$. **3.** $5\pm\sqrt{7}$.
4. $(-1\pm 2i)/3$. **5.** $(-3\pm\sqrt{41})/4$. **6.** 9/2, 1/2.
7. 12, -21. **8.** 17/6, $-15/2$. **9.** 23/4, 9/2.
10. 32/5, $-2/3$. **11.** $1+\sqrt{3}$, $2-\sqrt{3}$. **12.** 2, $4+i$.
13. 4, 4. **14.** 2, 2. **15.** $(1\pm\sqrt{5})/2$.
16. 3, -5. **17.** 3/2. **18.** 5, 16/7.
19. 1/2. **20.** 5, 5/2. **21.** 1, $-58/91$.
22. -2, 1/4. **23.** 1/3, $-3a$. **24.** $a+b$, $a-b$.
25. b/c, $-a/c$. **26.** $2a+b$, $2a-b$. **27.** $a(3c\pm 2b)$.
28. $\frac{a+b}{a-b}$, $\frac{a-b}{a+b}$. **29.** $\frac{a+b+c\pm\sqrt{a^2+b^2+c^2-ab-ac-bc}}{3}$.
30. $\frac{a^2+b^2}{2a}$, $\frac{a^2+b^2}{2b}$.

XLII. Page 302

1. 22, 23. **2.** 15, 16. **3.** 5, 6.
4. 13, 14, 15. **5.** 86. **6.** 7/5.
7. 42. **8.** \$40. **9.** 5%.
10. 4%. **11.** 100. **12.** 6, $6\frac{2}{3}$ ft.
13. 144, 112 sq. in. **14.** $2(\sqrt{2}-1)$. **15.** 21 gal.
16. 40, 45 mi. per hour. **17.** $1\frac{1}{2}$ hr., 4 mi. per hour. **18.** 4 hr. 20 min.
19. 6. **20.** 2 hr.
21. (1) 1/4 sec.; 1 sec.; no. (2) In 1/2 sec. (3) In 3/2 sec.

XLIII. Page 308

1. 2 and -1. **2.** ∞, $-2/3$; ∞, ∞. **3.** $(x+2y-2)(3x-y+1)$. **4.** ± 1.
5. p^2-4q, $p^4-4p^2q+2q^2$, $(p^2-2q)/q$. **6.** $-45/32$, $-7/2$.
7. $x^2-x+2=0$, $2x^2+x+1=0$, $x^2+2x+8=0$, $x^2-x+2=0$.
8. 1. min. $=-13$. 2. min. $=31/8$. 3. max. $=5$.
4. min. $=-1/2$, max. $=1/2$. 5. min. $=4$.
6. max. $=-(2+\sqrt{2})/4$, min. $=-(2-\sqrt{2})/4$.
9. A square.
10. Toward a point $2\frac{2}{3}$ miles from the nearest point.
11. 36 ft., 3/2 sec.

§ 643. Page 310. Ex. 3

1. $1/2$, $(2\pm i\sqrt{2})/3$. 2. -1, 4, $1\pm\sqrt{2}$.

§ 644. Page 311. Ex. 7

1. ± 3, $\pm\sqrt{6}/3$. 2. $(3\pm\sqrt{5})/2$, $(3\pm\sqrt{17})/2$.
3. 0, $-a$, $-a(1\pm\sqrt{57})/2$. 4. -3, 2, $-(1\pm\sqrt{19})/3$.

§ 645. Page 313. Ex. 4

1. 1, $(1\pm i\sqrt{3})/2$. 2. $(3\pm\sqrt{5})/2$, $(1\pm i\sqrt{3})/2$.
3. -1, $(1\pm i\sqrt{3})/2$, $(-1\pm i\sqrt{3})/2$.

§ 646. Page 313. Ex. 3

1. -2. $1\pm i\sqrt{3}$. 2. $\sqrt{2}(1\pm i)/2$, $-\sqrt{2}(1\pm i)/2$.
3. $\pm i$, $(\sqrt{3}\pm i)/2$, $(-\sqrt{3}\pm i)/2$.

§ 648. Pages 314, 315

2. $1/4, -3/4$. **4.** $1, 6$. **7.** $3, 7$. **9.** $(6+\sqrt{6})/3$.

XLIV. Page 316

1. $\pm 3/2, \pm\sqrt{2}$.
2. $7\sqrt[3]{63}/9$.
3. $\pm 2, \pm 3\sqrt{2}$.
4. $-1, -1, -1, 2/3, 2/3, 3/2$.
5. $1, -2, \pm\sqrt{3}\,i$.
6. $1, -1, 1 \pm i$.
7. $1, -1, 5/3, -1/3$.
8. $7, -1, 3 \pm \sqrt{5}$.
9. $(-1 \pm \sqrt{17})/4, (-1 \pm \sqrt{7}\,i)/4$.
10. $1, 1, \pm i$.
11. $\pm i, (-1 \pm i\sqrt{3})/2$.
12. $1, 2 \pm \sqrt{3}, 3 \pm 2\sqrt{2}$.
13. $3, 3(-1 \pm \sqrt{5} + i\sqrt{10 \pm 2\sqrt{5}})/4, 3(-1 \pm \sqrt{5} - i\sqrt{10 \pm 2\sqrt{5}})/4$.
14. $1, (1 \pm i\sqrt{3})/4$.
15. $0, \pm\sqrt{3}\,i$.
16. $5, -1, 2 \pm 3i$.
17. $-a, -b, -(a+b)/2$.
18. $(a+b)/2, [a+b \pm \sqrt{2-(a-b)^2}]/2$.
19. $0, -9/5, -(15 \pm \sqrt{401})/22$.
20. $\pm a, \pm 1/a$.
21. $1, -1/3, (1 \pm \sqrt{19})/3$.
22. $1, -1/2$. **23.** $2, 3$. **24.** $3, -2$. **25.** $3/2$.
26. $16/25$. **27.** $1, -3$. **28.** $0, 16, 81$. **29.** 3.
30. $(6+\sqrt{6})/3$. **31.** $1/5$. **32.** $4/5$. **33.** 1.
34. $\dfrac{27abc-(a+b+c)^3}{9(a^2+b^2+c^2-ab-bc-ca)}$.
35. $6, -3, (3 \pm i\sqrt{71})/2$.
36. $0, -2a$.
37. $\pm\sqrt{182}/14, \pm 3\sqrt{7}/7$.

XLV. Page 320

1. $2/3, -11/9; -4, -13/3$.
2. $-1, -1; 5/3, 3/5$.
3. $1/15, 2/15$; one infinite solution.
4. $3, 1; -1/15, -41/5$.
5. $0, 0; 51/14, -17/14$.
6. $1, 1/2; 12/5, 32/15$.
7. No finite solution.
8. $5, 9; 333/28, 185/42$.
9. $1, 1/4; -12/13, -4/9$.
10. $1, -3; -1, 3; -1, 3; 2, -3$.
11. $2, -2; -2, 2; -1 \pm \sqrt{3}, 1 \pm \sqrt{3}$.
12. $0, 0; 15/8, 9/8$; two infinite solutions.
13. $m = 1$ or -1.
14. $m=1, c=1/3; m=-1/2, c=-1/3$.
15. $(2x-y+4)(x+y+3)$.

§ 655. Page 321. Ex. 2

1. $1, 0; -1, 0$.
2. $3, 1/5; 3, -1/5; -3, 1/5; -3, -1/5$.

§ 656. Page 321. Ex. 3

1. $2, 1; 3, -1$; two infinite solutions.
2. $3i\sqrt{14}/14, i\sqrt{14}/7; 3i\sqrt{14}/14, -i\sqrt{14}/7;$
$-3i\sqrt{14}/14, i\sqrt{14}/7; -3i\sqrt{14}/14, -i\sqrt{14}/7$.

§ 657. Page 323. Ex. 5

1. 0, 0; 0, 0; 5, -5; -3, -9.
2. -2, 3; -3, 2; $(3 \pm \sqrt{17})/2$, $(-3 \pm \sqrt{17})/2$.

§ 658. Page 323. Ex. 2

1, 2; -1, 1; $(1 \pm i\sqrt{23})/4$, $(1 \pm i\sqrt{23})/4$.

XLVI. Page 324

1. 2, 3; 2, -3; -2, 3; -2, -3.
2. 2, 1/3; 2, $-1/3$; -2, 1/3; -2, $-1/3$.
3. 7, -1; -5, 2; two infinite solutions.
4. -5, -1; 15/2, 3/2; $2\sqrt{3}$, $3\sqrt{3}$; $-2\sqrt{3}$, $-3\sqrt{3}$.
5. 2, -3; $-2/3$, 7/3; $(-6 \pm 2\sqrt{6})/3$, $(3 \mp 4\sqrt{6})/3$.
6. 1, 0; -1, 2; two infinite solutions.
7. 5, -1; 5, -21; -7, $19 \pm 2\sqrt{85}$.
8. 5, 5; -5, -5; $\sqrt{5}\,i$, $-\sqrt{5}\,i$; $-\sqrt{5}\,i$, $\sqrt{5}\,i$.
9. 3, 1; -3, -1; $3\sqrt{5}\,i$, $-7\sqrt{5}\,i/5$; $-3\sqrt{5}\,i$, $7\sqrt{5}\,i/5$.
10. 1, 1/2; -1, $-1/2$; $\sqrt{91}/5$, $\sqrt{91}/16$; $-\sqrt{91}/5$, $-\sqrt{91}/16$; 0, 0; 0, 0; 0, 0; 0, 0; eight infinite solutions.
11. $3\sqrt{2}$, $2\sqrt{2}$; $-3\sqrt{2}$, $-2\sqrt{2}$; $2\sqrt{2}$, $3\sqrt{2}$; $-2\sqrt{2}$, $-3\sqrt{2}$.
12. 5, 4; -4, -5; $10 \pm 2\sqrt{30}$, $-10 \pm 2\sqrt{30}$.
13. 2, 4; 3, -1; $(-5 \pm \sqrt{13})/2$, $(-3 \pm 5\sqrt{13})/2$.

XLVII. Page 325

1. 4, 1; -1, -4; one infinite solution.
2. 125, -27; -27, 125; one infinite solution.
3. 5, 3; -5, -3; 3, 5; -3, -5; four infinite solutions.
4. 2, 3; $-1+\sqrt{3}\,i$, $3(-1+\sqrt{3}\,i)/2$; $-1-\sqrt{3}\,i$, $-3(1+\sqrt{3}\,i)/2$; six infinite solutions.
5. -2, 2; -2, -2; $(2 \pm 2\sqrt{7})/3$, $(1 \pm \sqrt{7})/3$; two infinite solutions.
6. 0, 0; 6, 9; 50/21, $-20/21$; one infinite solution.

XLVIII. Page 328

1. 9, -4; -4, 9.
2. 14, -2; -2, 14
3. 17, 2; -17, -2; 2, 17; -2, -17.
4. 9, 2; -2, -9.
5. 8, 1; 1, 8; one infinite solution.
6. 7, 5; 5, 7; 7ω, 5ω; 5ω, 7ω; $7\omega^2$, $5\omega^2$; $5\omega^2$, $7\omega^2$, where $\omega = (-1+\sqrt{3}\,i)/2$; three infinite solutions.

7. $4/3, 1/4$; $1/3, 1$; one infinite solution.
8. $3, 1$; $-1, -3$; $1 \pm i\sqrt{10}, -1 \pm i\sqrt{10}$.
9. $2, 0$; $0, 2$; $1 \pm i\sqrt{3}, 1 \mp i\sqrt{3}$; one infinite solution.
10. $4, -7/2$; $-7/2, 4$.
11. $5, -4$; $-4, 5$; $-10 + 3\sqrt{11}, -10 - 3\sqrt{11}$; $-10 - 3\sqrt{11}, -10 + 3\sqrt{11}$; two infinite solutions.
12. $3, 1$; $1, 3$; $-3, -1$; $-1, -3$; $\pm\sqrt{5}\,i, \pm\sqrt{6}\,i$; $\pm\sqrt{6}\,i, \pm\sqrt{5}\,i$.
13. $2, -5$; $-5, 2$; $(3 \pm \sqrt{31}\,i)/2, (3 \mp \sqrt{31}\,i)/2$.
14. $0, 2$; $2, 0$; $(-23 \pm \sqrt{389})/12, (-23 \mp \sqrt{389})/12$.
15. $0, 0$; $\pm\sqrt{5}, \pm\sqrt{5}$; $\pm\sqrt{5}\,i, \mp\sqrt{5}\,i$; $\pm\sqrt{10}(1+i)/2, \mp\sqrt{10}(1-i)/2$; $\pm\sqrt{10}(1-i)/2, \mp\sqrt{10}(1+i)/2$.

XLIX. Page 329

1. $4, -1, 3$; $-2, 5, -3$.
2. Misprint. Second equation should be $y(z + x) = 6$. The solutions are then $4, 1, 2$; $-4, -1, -2$.
3. $-(a^2 \pm bc)/a, -(b^2 \pm ca)/b, -(c^2 \pm ab)/c$.

L. Page 330

1. $2/3, -11/9$; $-4, -13/3$.
2. Misprint. Second equation should be $x - y = 1$. Solutions are then $4, 3$; $-3, -4$.
3. $(b + a)/2, (b - a)/2$; $(a - b)/2, -(a + b)/2$.
4. $\sqrt{\dfrac{a-b}{a^2+b^2}}, \pm\sqrt{\dfrac{b-a}{a^2+b^2}}$; $-\sqrt{\dfrac{a-b}{a^2+b^2}}, \pm\sqrt{\dfrac{b-a}{a^2+b^2}}$.
5. $1, 1/4$; $-12/13, -4/9$.
6. a, b; $2a - b, 2b - a$.
7. $5, 1/2$; $1/2, 5$.
8. $a, b/2$; $-a, -b/2$; $a/2, b$; $-a/2, -b$.
9. $4, 1$; $-1, -4$; $0, 0$; $0, 0$.
10. $\dfrac{2ab}{b-a}, \dfrac{2ab}{b+a}$; $0, 0$.
11. $3, -7$; $35/2, 15/2$; $0, 0$; $0, 0$.
12. $5, 2$; $-5, 2$; $2, -4/5$; $-2, -4/5$.
13. $3, 1$; $-3, -1$; $6, 3$; $-6, -3$.
14. $6, 1$; $-27, -10$; $(7 \pm 3\sqrt{65})/2, (3 \pm \sqrt{65})/2$.
15. $7/2, 2$; $-2, -7/2$; $136/65, 238/65$; $-238/65, -136/65$.
16. α, β; $\alpha\omega, \beta\omega$; $\alpha\omega^2, \beta\omega^2$; where $\alpha = \dfrac{\sqrt[3]{a^2b^2}}{b}$, $\beta = \dfrac{\sqrt[3]{a^2b^2}}{a}$, and $\omega = (-1 + i\sqrt{3})/2$; six infinite solutions.
17. α, β; $\alpha\omega, \beta\omega$; $\alpha\omega^2, \beta\omega^2$; where $\alpha = \dfrac{\sqrt[3]{4(a^2-b^2)^2}}{2(a-b)}$, $\beta = \dfrac{\sqrt[3]{4(a^2-b^2)^2}}{2(a+b)}$, and $\omega = (-1 + i\sqrt{3})/2$; six infinite solutions.

18. $a/(a^2+b^2)$, $b/(a^2+b^2)$; 0, 0; two infinite solutions.
19. $4a$, a; $-4a$, $-a$.
20. 3/2, 1/2; $-1/2$, $-3/2$; one infinite solution.
21. a, 0; 0, a; $a(1 \pm \sqrt{7}\,i)/2$, $a(1 \mp \sqrt{7}\,i)/2$.
22. 7, 3; 3, 7; $(-17 \pm \sqrt{646})/5$, $(-17 \mp \sqrt{646})/5$.
23. $(3 \pm \sqrt{3}\,i)/2$, $(3 \mp \sqrt{3}\,i)/2$; 0, 0; 0, 0.
24. $\pm a\sqrt{3}$, 0; $2a$, a; $-2a$, $-a$.
25. 1, 4; 4, 1; an infinite solution.
26. 3, 0; 0, 4; 0, 0; 0, 0.
27. 5, 0; 0, 5; $\sqrt{10}$, $-\sqrt{10}$; $-\sqrt{10}$, $\sqrt{10}$.
28. 2, 4; -9, 15; -1, -1; 0, 0.
29. 3/2, 1/2; 1/2, 3/2; $1 \pm i\sqrt{1155}/35$, $1 \mp i\sqrt{1155}/35$.
30. $\dfrac{\sqrt{a^2+1}}{a}$, $\sqrt{a^2+1}$; $\dfrac{-\sqrt{a^2+1}}{a}$, $-\sqrt{a^2+1}$.
31. 2, 0; -3, 5; one infinite solution.
32. 1, 5/3; 2, $-4/3$; $(-9 \pm \sqrt{21})/6$, $(-1 \pm 3\sqrt{21})/6$.
33. 1, 3, 1; 149/5, $-1/5$, -15.
34. 1, 2, -3, 0; -1, -2, 3, 0.
35. 3, 1, 1; -3, -1, -1.
36. 1, -1, 2; $-14/13$, 23/13, 17/13; $(-29 \pm 3\sqrt{105})/26$, $(-11 \mp 2\sqrt{105})/13$, $(-1 \pm \sqrt{105})/26$.

LI. Page 331

1. 7 and 5. **2.** 6 and 3. **3.** 5/6.
4. 8, 9, and 20. **5.** 12 and 5 ft. **6.** 15, 9, and 12 in.
7. 39, 36, and 15 in. **8.** 16, 14, 9 ft. **9.** 14 and 12 in.
10. A, \$4.50; B, \$5. **11.** \$1200, 4%.
12. Two children, each of whom receives \$10,000; five grandchildren, each of whom receives \$8000.
13. 4 and 2 mi. per hour. **14.** A, 3 hr.; B, 4 hr.; C, 6 hr.
15. A, 24 in. per second; B, 20 in.
16. A, 8 in. per second; B, 2 in. **17.** 18 mi. **18.** 96 mi.

LII. Page 339

17. Meets the x-axis at the points $3 \pm \sqrt{8}$, 0; touches the y-axis at the point 0, 1.
20. $m = \pm 1$. **21.** $c = 1$ or 4/3. **23.** $y = 2x + 1$ and $x + 2y = 0$.
24. When $|\lambda| = 2$, parabola; when $|\lambda| < 2$, ellipse; when $|\lambda| > 2$, hyperbola.

LIII. Page 341

6. $x < 30$. 7. $x < -2$ or > 4. 8. $-1 < x < 3$, or $x > 6$.

LIV. Page 346

1. $x = 20 - 17t$, $y = 6 - 6t$. The solutions corresponding to $t = 0, -1, -2, \cdots$ are positive.
2. $x = 2 - 12t$, $y = -6 - 43t$. The solutions corresponding to $t = -1, -2, \cdots$ are positive.
3. $x = -17 + 39t$, $y = 7 - 16t$. No positive solution.
4. $x = -2 + 23t$, $y = 43 - 72t$. No positive solution.
5. $x = 16 - 27t$, $y = 28 - 49t$. The solutions corresponding to $t = 0, -1, -2, \cdots$ are positive.
6. $x = 54 - 97t$, $y = 21 - 47t$. The solutions corresponding to $t = 0, -1, -2, -3, \cdots$ are positive.
7. $x = 2 + 21t$, $y = 3 - 26t$, $z = -1 - 11t$. No positive solution.
8. $x = 10 - 14t$, $y = -4 + 35t$, $z = -2 + 15t$. No positive solution.
9. $x = u$, $y = 2v + 1$, $z = 2u + 3v$. The solutions corresponding to positive values of u and v are positive.
10. $x = 7 - 3u - 2v$, $y = 1 + 2u$, $z = v$. The positive solutions are 7, 1, 0; 5, 1, 1; 3, 1, 2; 1, 1, 3; 4, 3, 0; 2, 3, 1; 0, 3, 2; 1, 5, 0.
11. Fifty positive solutions. 12. $\frac{41}{35} = \frac{3}{5} + \frac{4}{7}$.
13. 2 calves, 16 lambs; or 8 calves, 9 lambs; or 14 calves, 2 lambs.
14. 15, 2, 6; 12, 4, 7; 9, 6, 8; 3, 10, 10. 15. 314.
16. The 167th and 169th divisions measured from either end.

LV. Page 350

1. 32; 192/5; $10a^2b^2$; $\sqrt{30}$. 2. $-3 : 2$; $1 : 5$.
3. $x : y = -1 : 2$ or $3 : 1$. $y : x = -2 : 1$ or $1 : 3$.
13. (1) 0, 0, 3/2. (2) 0, 0, 5, 8/7.
14. 104, 156, 260. 15. 8 gallons from A, 10 gallons from B.

LVI. Page 353

1. $-14/5$. 2. $\pm 2\sqrt{3}/3$. 3. $x^2 + 3y - 4 = 0$. 4. -72.
7. 3. 8. 15/8. 9. $\frac{a\sqrt{2}}{2}, \frac{a\sqrt[3]{4}}{2}$.

LVII. Page 356

1. 60, 630; $25\frac{1}{2}$, 225. **2.** $n(n+1)/2$; n^2; $n(n+1)$.
3. $n(3n-2)$. **4.** $-1/3, 0, 1/3, 2/3, 1, 4/3, 5/3, 2, 7/3, 8/3$.
5. $-1/2, 0, 1/2, 1, 3/2$. **6.** $l = 20$, $S = 160$. **7.** $a = 3$, $S = -14$.
8. $d = 3$, $S = 178$. **9.** $n = 52$, $d = -1/2$. **10.** $d = 2$, $l = 12\frac{3}{7}$.
11. $n = 28$, $S = 187\frac{3}{5}$. **12.** $a = 31$, $l = -1$. **13.** $a = 25$, $d = -3$.
14. $n = 6$, $a = -72$. **15.** $n = 5$, $l = 18$. **18.** $1, 1\frac{1}{6}, 1\frac{1}{3}, 1\frac{1}{2}, 1\frac{2}{3}, 1\frac{5}{6}, 2, \cdots$.
19. 3, 5, 7. **20.** 55,350. **21.** $1716.
22. In 9 hours, midway between their starting points.

LVIII. Page 360

1. 162, 122. **2.** $13\frac{1}{2}$, $32\frac{1}{2}$. **3.** 8; 2/3; 25/12.
4. $\frac{341}{999}$, $\frac{78}{1375}$, $8\frac{4516}{9999}$. **5.** $l = -3000$, $S = -3333.33$. **6.** $r = 1/2$, $S = 95\frac{1}{4}$.
7. $n = 8$, $S = 255/16$. **8.** $a = 1$, $S = 61$. **9.** $n = 5$, $l = 2/3$.
10. $r = 5$, $l = -375$. **11.** $n = 5$, $r = 4/3$. **12.** $a = 45/8$, $l = -20/27$.
13. Misprint. Should be $83\frac{1}{8}$. Then $n = 6$, $a = 4$. **14.** $a = 10$, $r = 3/5$.
15. $a = 3$, $r = 2$. **16.** ab. **17.** 15, 45, 135.
18. 3/16. **19.** 8, 20, 50, 125. **20.** 1, -3, 9.
21. 3, 12, 21 or 63, 12 -39. **22.** -2, 2, 6, 18. **23.** 75 feet.

LIX. Page 363

1. 3/11, 3/13. **2.** 30/23. **3.** 75/7, 150/13, 75/6, 150/11.
4. 2. **5.** 3 and 5. **6.** 2 and 8.

LX. Page 369

1. 191 and 1350. **2.** 6560 and 180,360.
3. (1) Second order; eighteenth term is 224.
(2) Third order; twentieth term is 10,118.
(3) Third order; twelfth term is -1.
(4) Fourth order; tenth term is 10,100.
4. Third order; nth term is $n(n+1)(n+2)$. Sum of first n terms is $n(n+1)(n+2)(n+3)/4$.
Fourth order; nth term is $2n(n+1)^3$.
5. 560, 105. **6.** 1149. **7.** 962.
8. 2024. **9.** 220, 385. **10.** 1274.
13. Second order; sum of first n terms is $n(n^2+2)/3$.
Third order; sum of first n terms is $n(n+1)(n+2)(n+3)/4!$.

LXI. Page 374

1. $y = 4 + 23x + 26x^2 + 7x^3$; when $x = -5/2$, then $y = -3/8$; when $x = -1/2$, then $y = -15/8$.
2. $f(x) = -90 + 73x - 16x^2 + x^3$; $f(12) = 210$.
3. 704.3716. **4.** 110.592. **5.** .04237. **6.** 20.8734.
7. $(2520 - 806x - 9x^2 - 13x^3)/840$.

LXII. Page 378

1. 2, 1/2, 6, 4, 6, − 3, − 6, − 3, − 2/3, 7/3, 3/2.
2. 1.0791, .6532, .1505, .2594. **3.** $(3 \log_a 2 + \log_a 3 + 2 \log_a 5)6$.
4. (1) $\frac{2}{3} \log_a b - \frac{1}{2} \log_a c - \frac{4}{5} \log_a d$. (2) $\frac{1}{12} - \frac{1}{3} \log_a b$.

LXIII. Page 389

The answers given below are obtained by the use of the four-place table, pp. 384, 385, and by following in every case the rule at the end of § 751.

1. 36,460. **2.** 82.28. **3.** 1.210×10^{13}. **4.** -1.744×10^{-5}.
5. 22.13. **6.** .1151. **7.** -4.558×10^{7}. **8.** 16.38.
9. − .4255. **10.** 77. **11.** 9.472. **12.** 137.
13. 1.413. **14.** 8.218. **15.** − .676. **16.** 13.46.
17. .01. **18.** 5.461. **19.** 1.101. **20.** 5.34.
21. 9108. **22.** 632.8. **23.** − 50.22. **24.** 2.453×10^{-6}.
25. − 73.6. **26.** 2.647×10^{13}. **27.** .5381. **28.** 96.56.
29. 15.77. **30.** 2.652×10^{5}.

LXIV. Page 392

The formulas at the top of p. 391 for interest compounded semiannually, quarterly, and so on should be $A = P(1 + r/2)^{2n}$, $A = P(1 + r/4)^{4n}$, and so on.

1. 3.925, − 1.578, .8361.
2. (1) $x = 6$. (2) $x = 1$ or 2. (3) $x = 2.152$.
3. (1) $x = 2, -5$. (2) $x = 4.642$. (3) $x = 3.051$. (4) $x = 3.537$.
4. $41,410. **5.** $10,010. **7.** $694.80. **8.** $8030.
9. $20,730, $30,000. **10.** $b = 496.4$, area = 77,500. **11.** 5179.
12. Area of surface, 6998; volume, 55050.

§ 763. Page 396

4. 65. **5.** 380 single tickets. **6.** 144.
7. 1800. **9.** 325. **10.** 3600

§ 764. Page 397

3. 144.

§ 765. Page 397

1. 5, 25, 125. **2.** 343.

§ 766. Page 398

4. 630,630. **5.** 715.

§ 769. Page 401

6. 136, 252, 8855. **7.** 15. **8.** 8. **9.** 220.
10. 792, 330, 462. **11.** 120, 420, 252. **12.** 3,953,520.
13. 10010 · 7!. **14.** 52!/(13!)⁴, 52!/(13!)⁴ · 4!. **15.** 2250.

§ 770. Page 402

1. 15. **3.** 41.

§ 771. Page 402

$C_6^{12} = 924, \ C_7^{15} = 6435.$

LXV. Page 405

1. 24. **2.** 720. **3.** 336. **4.** 210, 5040. **5.** 161,700.
6. 30. **7.** 420. **8.** (1) 90,720. (2) 322,560, (3) 4320.
9. 60,480, 504, 84. **10.** 60. **11.** 1680.
12. 6. **13.** (1) 399. (2) 259. **14.** 360.
15. 1232, 224. **16.** 31. **17.** 3,783,780.
18. 462. **19.** 1446. **20.** 5096520 · 15!.
21. 825. **22.** (1) 60. (2) 325. **23.** 2970.
25. 63,063. **26.** 145,152. **27.** 300.
28. 5^{15}, $15!/(3!)^5$. **29.** 5760. **30.** 37,740.
32. 126. **33.** 252. **34.** 5040, 840, 7560.

LXVI. Page 409

1. $\Sigma a^3 + 3\,\Sigma a^2b + 6\,\Sigma abc$.
2. $\Sigma a^5 + 5\,\Sigma a^4b + 10\,\Sigma a^3b^2 + 20\,\Sigma a^3bc + 30\,\Sigma a^2b^2c + 60\,\Sigma a^2bcd$.
3. 83160, 34,650, 16,632. **4.** 12,600. **5.** 15,120. **6.** 4455. **7.** 26,396.

LXVII. Page 414

1. 5 to 3 against. 5/8. **2.** 10/19, 9/19. **3.** \$37.50.
5. (1) 7/16, 3/8, 3/16. (2) 1/8, 7/40. (3) 1/560, 143/280, 9/40.
(4) 21/65, 27/65. (5) 45/286.
6. 1/6, 4/9. **7.** 1/6. **8.** 11/36, 5/18. **9.** 1/12.
10. 13/18, 5/18. **11.** 5/42, 5/14, 1/21. **12.** 256/270725, 4/270725.
13. $10(143)^4 13!\ 39!/52!$ **14.** 1/36, 1/54. **15.** 2/7.

§ 785. Page 417

4. 255/256. **5.** 104/105. **6.** 35/81. **7.** 5/8, 1/8, 5/24, 1/24.

§ 787. Page 420

5. 607/2970. **6.** 5/6. **7.** 6/11, 5/11.

§ 790. Page 421

3. 105/512, 193/512. **4.** 5/324, 7/432. **5.** 112/243. **6.** 16/81

LXVIII. Page 422

1. (1) 7/225. (2) 14/75. **2.** 66/455. **3.** \$3.58.
4. 11/16. **5.** 1/24, 1/4, 11/24, 1/4. **6.** 7 to 2.
7. 7 to 1, 103 to 5. **8.** 7/165, 16/33. **9.** 13 to 11.
10. 30 to 31. **11.** 7/15, 53/165, 7/33; 9/19, 6/19, 4/19.
12. \$11. **13.** 3/5. **14.** No.
15. 59049/100000. **16.** 64/81. **17.** $191 \cdot 3^5/5^7$.
18. \$1. **19.** 5103/32768, 1012581/1048576.
20. 81/128. **21.** \$48, \$16.

LXX. Page 431

1. (1) $x^3 - 2ax^2 + (a^2 - ab - b^2)x + a^2b + ab^2 = 0$.
(2) $6x^5 - 43x^4 + 78x^3 - 5x^2 - 12x = 0$.
4. 6 and -2. **6.** 2, 3, -4.
7. 2, 5, -5. **8.** $-1/3$, $(1 \pm \sqrt{3}\,i)/2$.
9. 0, 1/2, $-2 \pm \sqrt{3}$. **10.** -1, -1, $-1 \pm \sqrt{2}\,i$.
11. 1, -1, -2, $-3/2$. **12.** -3, $-2/3$, $\pm i$.
13. -2, 5, 7, $(-1 \pm \sqrt{3}\,i)/2$. **14.** -1, 2, 2, $(1 \pm i)/2$.
15. -1, 2, 3, -3, -4. **16.** 2, 1/2, $-1/2$, 2/3.

17. $-1, -1, 3, 3, 3.$

18. $-2, 4, 5, 3/2.$

19. $-2, 1/2, 3/2, \pm i.$

20. $2, 2, -3, -4, -5.$

21. $-2, -2, -1/2, (-1 \pm \sqrt{3}\, i)/2.$

22. $6, 8, 1/2, 1/3.$

23. $-1/2, -3/5, (-3 \pm \sqrt{5})/2.$

24. $1/2, 1/3, 2/3, 3/2.$

25. $-1/2, -2/3, -3/2, (1 \pm \sqrt{11}\, i)/2.$

26. $1, -1, -2, -3/2, \pm \sqrt{2}\, i.$

27. $1, -1, -2, -3/2, -1 \pm \sqrt{2}\, i.$

28. $1, -1, 2, 2/5, (-1 \pm \sqrt{3}\, i)/2.$

LXXI. Page 435

1. $3/2.$

2. (1) $1, -3/2, 9/4.$ (2) $1 + 2\sqrt{2}\, i, -3, 1 - 2\sqrt{2}\, i.$

3. (1) $-2 - \sqrt{5}, -2, -2 + \sqrt{5}.$ (2) $1, 3, 5.$

5. $r^2 - rp + q - 1 = 0.$

6. $-1 \pm \sqrt{2}\, i, -1 \pm \sqrt{2}\, i.$

7. $-1, 3/2, 3/7.$

8. $-4, -6, 6, 5.$

9. (1) $x^3 - px^2 + qx - r = 0.$

(2) $x^3 + kpx^2 + k^2qx + k^3r = 0.$

(3) $rx^3 + qx^2 + px + 1 = 0.$

(4) $x^3 + (p - 3k)x^2 + (q - 2kp + 3k^2)x + (r - kq + pk^2 - k^3) = 0.$

(5) $x^3 + (2q - p^2)x^2 + (q^2 - 2pr)x - r^2 = 0.$

(6) $r^2x^3 + (q^2 - 2pr)x^2 - (2q - p^2)x + 1 = 0.$

10. (1) $17/4.$ (2) $-37/8.$ (3) $1.$ (4) $5/2.$

11. (1) $2/3.$ (2) $-11/3.$ (3) $-1.$ (4) $-31.$ (5) $-7/3.$

LXXII. Page 443

1. $x^7 - 3x^4 + 2x^2 + 6x - 7 = 0.$

2. (1) $x^4 - x^3 - 8x^2 + 24x + 64 = 0.$

(2) $162x^4 + 27x^3 - 36x^2 - 18x + 8 = 0.$

3. $10x^6 + 9x^5 + 3x^3 - x^2 + 5 = 0.$

4. (1) $2x^5 + 21x^4 + 88x^3 + 181x^2 + 180x + 74 = 0.$

(2) $2x^5 - 9x^4 + 16x^3 - 17x^2 + 12x + 2 = 0.$

5. $x^4 - 10x^3 + 225x^2 + 1080x - 16{,}875 = 0.$

6. $3x^4 - 162x^2 - 647x - 733 = 0.$

7. (1) $x^3 - 3x^2 + 10 = 0$, or $x^3 + 3x^2 + 6 = 0.$

(2) $x^3 + 2x^2 - 4 = 0$, or $27x^3 - 54x^2 - 76 = 0.$

8. $x^4 - x^3 + 6x^2 - x + 4 = 0.$

9. $x^4 + 9x^3 + 29x^2 + 39x + 17 = 0.$

10. (1) $rx^3 + (q^2 - 2pr)x^2 + r(p^2 - 2q)x + r^2 = 0.$

(2) $(r - pq)x^3 + (p^3 - 2pq + 3r)x^2 + (3r - pq)x + r = 0.$

11. (1) $x^3 + 4x^2 - 3x + 2 = 0.$ (2) $x^3 - 6x^2 + 17x - 8 = 0.$

(3) $4x^3 - 9x^2 + 18x - 27 = 0.$ (4) $8x^3 + 16x^2 + 9x + 2 = 0.$

(5) $16x^3 + 24x^2 + 27x - 8 = 0.$

12. (1) 2 and -6. (2) 4 and -1. (3) 4 and -7.

(4) 2 and -5. (5) 9 and -2. (6) 2 and -1.

LXXIII. Page 449

1. $-1,\ -1/2,\ 1 \pm 2i$.
2. $1,\ 1/2,\ 2 \pm \sqrt{2}$.
3. $x^4 + 12x^3 + 44x^2 + 18x - 116 = 0$.
4. $x^4 - 2x^2 + 9 = 0$.
5. (1) Four imaginary roots.
   (2) At least *two* imaginary roots. Of the real roots not more than *one* can be positive nor more than *one* negative.
   (3) No positive roots.
   (4) No negative roots.
   (5) At least *four* imaginary roots. Of the real roots not more than *two* can be positive nor more than *one* negative.
   (6) At least *four* imaginary roots. Of the real roots not more than *one* can be positive nor more than *two* negative.
   (7) At least *two* imaginary roots. Of the real roots not more than *two* can be positive nor more than *one* negative.
   (8) When n is odd, at least $3n - 3$ imaginary roots. Of the real roots not more than *two* can be positive nor more than *one* negative.
   When n is even, at least $3n - 6$ imaginary roots. Of the real roots not more than *two* can be positive nor more than *four* negative.
7. Two positive and three negative.
8. That $x^{2n+1} + 1 = 0$ has one negative and $2n$ imaginary roots; that $x^{2n} - 1 = 0$ has at least $2n - 2$ imaginary roots, not more than one positive nor more than one negative root; that $x^{2n+1} - 1 = 0$ has one positive and $2n$ imaginary roots.

LXXIV. Page 453

1. Between 0 and 1, 2 and 3, -1 and -2.
2. Between 1 and 2, 0 and -1, -2 and -3.
3. Between 1 and 2, 3 and 4, -1 and -2.
4. Between 2 and 3, -1 and -2, -2 and -3.
5. Between 1 and 2, 4 and 5, -1 and -2.
6. Between -2 and -3, -4 and -5, -6 and -7.
7. Between -2 and -3.
8. Between 3 and 4, -3 and -4.
9. Between 0 and 1, 2 and 3, 5 and 6, 0 and -1.
10. Between 1 and 2, 0 and -1, -1 and -2, -4 and -5.
11. Between 1 and 2, 4 and 5, 5 and 6, 0 and -1.
12. Between 1 and 2, 3 and 4, 0 and -1, -2 and -3, -3 and -4.

LXXV. Page 459

1. 1.213411. **2.** 2.469545. **3.** .179989. **4.** 2.137811.
5. 2.768345. **6.** -1.945341. **7.** 1.903211. **8.** -5.134578.
9. 3.236067. **10.** -2.157451. **11.** 2.356895 and 2.692021.
12. 1.602, 3.292, and -1.895. **13.** 1.246, $-.445$, and -1.802.
14. .347, 1.532, and -1.879. **15.** 1.558, $-.578$, -1.904, -4.075.
16. 2.5712. **17.** 2.884 and 3.054. **18.** 13.24.
19. Two roots between 0 and 1, one between 0 and -1.
20. 2/3, -1, .254, 1.860, -2.114.

LXXVI. Page 464

1. $10x^4 - 16x^3 + 2x - 20$, $40x^3 - 48x^2 + 2$, $120x^2 - 96x$, $240x - 96$, 240.
2. $(x^4 - 2x^3 + 1) + 2(2x^3 - 3x^2)h + 6(x^2 - x)h^2 + 2(2x - 1)h^3 + h^4$.
3. (1) $3 - 6(x+1) + 7(x+1)^2 - 4(x+1)^3 + (x+1)^4$.
(2) $80(x-2) + 80(x-2)^2 + 40(x-2)^3 + 10(x-2)^4 + (x-2)^5$.
(3) $\dfrac{2 + 3(x-1) + 3(x-1)^2 + (x-1)^3}{2 + 2(x-1) + (x-1)^2}$.
4. (1) $-1, -1, 2$. (2) $1/3, 1/3, -2$.
(3) $\pm\sqrt{6}\,i/2, \pm\sqrt{6}\,i/2$. (4) $1 \pm \sqrt{3}, 1 \pm \sqrt{3}$.
(5) $3, 3, \pm\sqrt{2}\,i/2$, (6) $(-1 \pm \sqrt{3}\,i)/2, (-1 \pm \sqrt{3}\,i)/2, 2$.
(7) $2, 2, -1 \pm \sqrt{2}\,i$. (8) $1, 1, 1, -1, -1$.
(9) $\pm i, \pm i, 2/3$.
6. $a = \pm 16$.
7. $a = 3, b = 1/9, x = -1/3$; or $a = -3, b = -1/9, x = 1/3$.
9. $108p^5 = 3125r^3$. **10.** $(ax + b)^n$.
11. $(-1 \pm \sqrt{3}\,i)/2, \pm i$; $(-1 \pm \sqrt{3}\,i)/2, \pm 1$.
12. $2 \pm 2\sqrt{2}, -4$; $1 \pm \sqrt{2}, -1$.

LXXVII. Page 471

1. Maximum value corresponds to $x = 0$ and is 4; minimum value corresponds to $x = 2$ and is 0.
2. (1) Minimum when $x = 1/4$.
(2) Maximum when $x = (1 - \sqrt{3})/2$, minimum when $x = (1 + \sqrt{3})/2$.
(3) Maximum when $x = -2$, minimum when $x = 2$.
(4) Maximum when $x = 1/3$, minimum when $x = 3$.
(5) Maximum when $x = 0$, minimum when $x = 2$.
(6) Maximum when $x = 1/2$, minimum when $x = -1$ or 2.
(7) Has neither a maximum nor a minimum.
(8) Maximum when $x = -1$, minimum when $x = -(2 \pm \sqrt{10})/2$.

LXXVIII. Page 477

1. Two between 3 and 4, one between -1 and -2.
2. Two between 3 and 4, one between -3 and -4.
3. One between 0 and -1, two imaginary.
4. One between -1 and -2, two imaginary.
5. Two between 2 and 3, one between -4 and -5.
6. Two between 3 and 4, two between -1 and -2.
7. One between 0 and 1, one between -1 and -2, two imaginary.
8. Two between 0 and 1, two between 3 and 4.
9. Two between 0 and 1, one between 2 and 3, one between -3 and -4.
10. One between 0 and 1, one between 1 and 2, two between -2 and -3.
11. One real root.
12. No real root.
13. When n is even, no real root; when n is odd, one real root.
14. Two real roots.

LXXIX. Page 482

1. $s_3 = -(a_1^3 - 3a_0a_1a_2 + 3a_0^2a_3)/a_0^3$.
   $s_4 = \quad (a_1^4 - 4a_0a_1^2a_2 + 4a_0^2a_1a_3 + 2a_0^2a_2^2)/a_0^4$.
2. $\Sigma 1/\alpha^2 = (q^2 - 2pr)/r^2$. $\Sigma 1/\alpha^3 = (-q^3 + 3pqr - 3r^2)/r^3$. $\Sigma\alpha\beta^2 = 3r - pq$.
3. $x^3 + 7x^2 + 12x - 1 = 0$.
4. (1) $s_1 = 1$, $s_2 = -5$, $s_3 = -20$, $s_4 = -9$.
   (2) 29. (3) 20. (4) -60. (5) 449/256. (6) $-111/4$.

§ 871. Page 485

2. $-1 + \sqrt[3]{A} + \sqrt[3]{B}$, $-1 + \omega\sqrt[3]{A} + \omega^2\sqrt[3]{B}$, $-1 + \omega^2\sqrt[3]{A} + \omega\sqrt[3]{B}$,
   where $A = (-1 + \sqrt{5})/2$ and $B = (-1 - \sqrt{5})/2$.

§ 874. Page 486

2. $(1 \pm \sqrt{5})/2$, $(3 \pm \sqrt{13})/2$.

§ 875. Page 488

2. $(1 \pm \sqrt{3}\,i)/2$, $\sqrt{2}\,(1 \pm i)/2$, $\sqrt{2}\,(-1 \pm i)/2$.

LXXX. Page 491

1. $4, -2 \pm i\sqrt{3}$.
2. $8, (1 \pm i\sqrt{3})/2$.
3. $\sqrt[3]{A} + \sqrt[3]{B},\ \omega\sqrt[3]{A} + \omega^2\sqrt[3]{B},\ \omega^2\sqrt[3]{A} + \omega\sqrt[3]{A}$,
   where $A = 2 + \sqrt{3},\ B = 2 - \sqrt{3}$.
4. $A = 3/4 + \sqrt{1887}/72,\ B = 3/4 - \sqrt{1887}/72$.
5. $-1 + \sqrt[3]{A} + \sqrt[3]{B},\ -1 + \omega\sqrt[3]{A} + \omega^2\sqrt[3]{B},\ -1 + \omega^2\sqrt[3]{A} + \omega\sqrt[3]{B}$,
   where $A = 4 + 2\sqrt{6},\ B = 4 - 2\sqrt{6}$.
6. $1 + \sqrt[3]{A} + \sqrt[3]{B},\ 1 + \omega\sqrt[3]{A} + \omega^2\sqrt[3]{B},\ 1 + \omega^2\sqrt[3]{A} + \omega\sqrt[3]{B}$,
   where $A = -5/2 + 5\sqrt{749}/54,\ B = -5/2 - 5\sqrt{749}/54$.
7. $(-\sqrt{3} \pm \sqrt{4\sqrt{3} - 5})/2,\ (\sqrt{3} \pm i\sqrt{5 + 4\sqrt{3}})/2$.
8. $(1 \pm \sqrt{5})/2,\ (3 \pm \sqrt{13})/2$.
9. $-1 \pm \sqrt{2},\ 1 \pm 2i$.
10. $(-5 \pm \sqrt{33})/2,\ (-3 \pm \sqrt{13})/2$.
11. $(1 \pm \sqrt{3}i)/2,\ (-1 \pm \sqrt{3}i)/2,\ (1 \pm 2\sqrt{2}i)/3$.
12. $\pm i,\ (1 \pm \sqrt{3}i)/2,\ (3 \pm \sqrt{5})/2,\ (1 \pm \sqrt{15}i)/4$.
13. $1, 1, -1, (-1 \pm \sqrt{15}i)/4, (-1 \pm \sqrt{35}i)/6$.
14. $z^3 + z^2 - 2z - 1 = 0$.
15. $(2a^2 - 3ab + c)^2/4 + (b - a^2)^3 \gtreqless 0$.
16. $\cos\frac{2k\pi}{5} + i\sin\frac{2k\pi}{5},\ k = 0, \cdots, 4;\ \cos\frac{(2k+1)\pi}{6} + i\sin\frac{(2k+1)\pi}{6}$,
    $k = 0, \cdots, 5$.
17. (1) $2\cos 20°, 2\cos 140°, 2\cos 260°$.
    (2) $2\sqrt{2}\cos 15°, 2\sqrt{2}\cos 135°, 2\sqrt{2}\cos 255°$.
18. 3.
19. Radius of base is $2\frac{1}{2}$, altitude, 8.
20. The problem as it stands has no solution. This is indicated by the fact that if we attempt to solve for the altitude, we obtain a negative result. It can be shown that the volume of the greatest cylinder which can be inscribed in a right circular cone is four ninths that of the cone. Let the student prove that in the case of the cone described in the example the altitude of this greatest cylinder is 2.

LXXXI. Page 497

1. $p^3 - p(q^2 + r^2 + s^2) + 2qrs$.
2. $(b - c)x + (c - a)y + (a - b)z$
3. $p(p^2 + q^2 + r^2 + s^2)$.
4. 0.
5. 18.
6. 74.
7. -30.
10. (1) $a_1b_2c_3d_4 - a_2b_1c_3d_4$. (2) $a_1b_2c_3d_4 - a_1b_3c_2d_4$. (3) $-a_2b_3c_4d_1$.
    (4) $a_1b_2c_3d_4 + a_1b_3c_4d_2 + a_1b_4c_2d_3 - a_1b_4c_3d_2 - a_1b_3c_2d_4 - a_1b_2c_4d_3$.
    (5) $a_1b_2c_3d_4 + a_2b_4c_3d_1 + a_4b_1c_3d_2 - a_4b_2c_3d_1 - a_2b_1c_3d_4 - a_1b_4c_3d_2$.
11. $a_2b_4c_3d_1e_5,\ -a_4b_2c_1d_5e_3,\ a_5b_4c_3d_2e_1,\ -c_1d_2a_3e_4b_5,\ -c_1b_2e_3a_4d_5,\ -d_3a_2e_4b_1c_5$.

LXXXII. Page 501

1. (1) $-22{,}680$. (2) 0. (3) $4\,abcdef$.

LXXXIII. Page 507

1. 0. 2. -4. 3. 0. 4. $-357{,}840$.

5. $\begin{vmatrix} 0 & bc-a^2 & b^2-ac \\ b^2-ac & 0 & bc-a^2 \\ bc-a^2 & b^2-ac & 0 \end{vmatrix}$. 6. $\begin{vmatrix} ap+cr & ap & cr \\ ap & ap+bq & bq \\ cr & bq & bq+cr \end{vmatrix}$.

7. $\begin{vmatrix} a & -a & a^2+ab & ac+ad \\ -b & b & ab+b^2 & bc+bd \\ c & c & -ac+bc & -c^2+cd \\ d & d & ad-bd & cd-d^2 \end{vmatrix}$.

8. $\begin{vmatrix} l^2+m^2+n^2 & lm+mn+nl & ln+ml+nm \\ ml+nm+ln & m^2+n^2+l^2 & mn+nl+lm \\ nl+lm+mn & nm+ln+ml & n^2+l^2+m^2 \end{vmatrix}$.

10. This exercise should read: "Prove that a determinant reduces to its leading term when all the elements at either side of the leading diagonal are zero."

LXXXIV. Page 511

1. $x=10/7,\ y=1,\ z=4/7$. 2. $x=1,\ y=1/2,\ z=1/3$.

3. $x=\dfrac{d(d-b)(d-c)}{a(a-b)(a-c)},\ y=\dfrac{d(d-c)(d-a)}{b(b-c)(b-a)},\ z=\dfrac{d(d-a)(d-b)}{c(c-a)(c-b)}$.

4. $x=1,\ y=1/2,\ z=1/3,\ t=-1$. 5. $x:y:z=-1:1:1$.

6. $x:y:z=k:l:-1$, if $a_1b_2-a_2b_1\neq 0$.

7. $\lambda=0$ or $-3\pm 2\sqrt{21}$.

LXXXV. Page 519

1. Common root is $-3/2$.

2. $\begin{vmatrix} a_0 & a_1 & a_2 & 0 \\ 0 & a_0 & a_1 & a_2 \\ b_0 & b_1 & b_2 & 0 \\ 0 & b_0 & b_1 & b_2 \end{vmatrix}$.

3. $(a+d)^3+b^3+c^3-3\,bc(a+d)$.

4. (1) $4\,p^3+27\,q^2$. (2) $c(4\,b^3+27\,a^2c)$. 5. The double root is -2.

6. $x,\ y=0,\ 0;\ 3,\ -1;\ -2,\ 2$; and one infinite solution.

§ 950. Page 527. Ex. 3

(1) $\frac{1}{2\cdot 3}+\frac{3}{3\cdot 4}+\frac{5}{4\cdot 5}+\frac{7}{5\cdot 6}+\cdots$; divergent.

(2) $\frac{1}{2}+\frac{\sqrt{2}}{5}+\frac{\sqrt{3}}{10}+\frac{2}{17}+\cdots$; convergent.

(3) $\frac{1}{9}+\frac{3}{35}+\frac{5}{91}+\frac{7}{189}+\cdots$; convergent.

§ 952. Page 528

5. When $x<1$.

6. When $x>1$.

LXXXVI. Page 530

1. Convergent.
2. Convergent.
3. Divergent.
4. Convergent.
5. Divergent.
6. Divergent.
7. Convergent.
8. Convergent.
9. Divergent.

10. $\frac{2}{1\cdot 3}+\frac{3}{2\cdot 4}+\frac{4}{3\cdot 5}+\frac{5}{4\cdot 6}$; divergent.

11. $\frac{1}{\sqrt[3]{2}}+\frac{\sqrt[3]{2}}{\sqrt[3]{9}}+\frac{\sqrt[3]{3}}{\sqrt[3]{28}}+\frac{\sqrt[3]{4}}{\sqrt[3]{65}}$; divergent.

12. $(\sqrt{2}-1)+(\sqrt{5}-2)+(\sqrt{10}-3)+(\sqrt{17}-4)$; divergent.

13. Convergent.
14. Divergent.
15. When $x<1$.
16. When $x<1$.

LXXXVII. Page 534

1. (1) Convergent. (2) Divergent. (3) Convergent.

2. (1) Convergent for all real values of x except $x=0,\ 1,\ -1/2,\ 1/3,\ \cdots,\ (-1)^{n-1}/n,\ \cdots$.

(2) Convergent when x is less than 1, otherwise it is divergent.

3. Should read: "If the series $u_1+u_2+u_3+\cdots$ is *absolutely* convergent, etc."

LXXXVIII. Page 538

1. ∞.
2. 1/2.
3. 1/2.

4. For $x=-1$ and for all values of x between -1 and 2.

5. For all real values of x.

6. For values of x which are greater than 1/3 and less than 1/2.

LXXXIX. Page 551

4. $a_1 = -1/2$, $a_2 = 7/8$, $a_3 = 7/16$, $a_4 = -21/64$.

5. (1) $2 - x/4 - x^2/32 - 5x^3/768$. (2) $1 + 3x/2 - 9x^2/8 - 13x^3/16$.

6. (1) $2 - 3x - 3x^2 + 20x^3 + \cdots$. (2) $x + 6x^2 + 4x^3 - x^4 + \cdots$.

7. (1) $3x^2 - 2x^3 - x^4 + 3x^5 + \cdots$. (2) $x^{-2} - 2x^{-1} + 1 + 9x + \cdots$.

8. (1) $-11/6 + 5x/36 - 89x^2/216 + 65x^3/1296 - 761x^4/7776 + \cdots$; converges when $|x| < 2$.

(2) $2 - 11x/3 + 46x^2/9 - 173x^3/27 + 616x^4/81 + \cdots$; converges when $|x| < 1$.

9. (1) $x^{-1} + x^{-2} + 7x^{-3} + 4x^{-4} + \cdots$; converges when $|x| > 3$.

(2) $1 - x^{-1} + x^{-4} - x^{-6} + \cdots$.

10. (1) $x = y - y^2 + y^3 - y^4 + \cdots$. (2) $x = y + y^2/2 + y^3/6 + y^4/24 + \cdots$.

11. $x = (y-1) - (y-1)^2/2 + (y-1)^3/6 - (y-1)^4/24 + \cdots$.

12. $x = y^{\frac{1}{2}} - 3y/2 + 45y^{\frac{3}{2}}/8 - 27y^2 + \cdots$.

13. (1) $y = 3x - 10x^2 + 60x^3 + \cdots$.

(2) $y = x^2 + x^5 + 3x^8 + \cdots$. $y = x^{\frac{1}{2}} - x^2/2 - 3x^{\frac{7}{2}}/8 + \cdots$.

$y = -x^{\frac{1}{2}} - x^2/2 + 3x^{\frac{7}{2}}/8 + \cdots$.

XC. Page 559

1. $\log_e 4 = 1.3862$. $\log_e 5 = 1.6093$. **7.** $-7x^4/3^5 \cdot 2^{10}$. **8.** $231x^3/2^{10}$

9. The first series converges when $|x| < 3/2$, the second when $|x| < 3$.

10. $1 - 3x/4 + 45x^2/32 + 43x^3/128 - 333x^4/2048$.

11. $4/3 + 13x/27 + 53x^2/1296$; converges when $|x| < 8/3$.

12. (1) $4/3$. (2) ∞.

15. $x + x^2/2 - 2x^3/3 + x^4/4 \cdots$; converges when $|x| < (\sqrt{5} - 1)/2$.

XCI. Page 563

1. $-19x^3 + 52x^4$.

2. (1) Scale is $a_n - a_{n-1} + a_{n-2} = 0$; terms are $-2x^5 + x^6$.

(2) Scale is $a_n + 2a_{n-1} + 3a_{n-2} = 0$; terms are $31x^5 + 16x^6$.

(3) Scale is $a_n + 3a_{n-1} + 3a_{n-2} + a_{n-3} = 0$; terms are $28x^6 - 36x^7$.

3. (1) Generating function is $(2-x)/(1-x-2x^2)$; general term is $[2^n + (-1)^n]x^n$.

(2) Generating function is $(3-8x)/(1-5x+6x^2)$; general term is $(2^{n+1} + 3^n)x^n$.

5. Generating function is $(1 + x + 4x^2)/(1 - x - 5x^2 - 3x^3)$; general term is $[3^n + (-1)^n n]x^n$.

6. Generating function is $[a - (a-d)x]/(1 - 2x + x^2)$.

8. Sum is $2/(1 - 3x + 3x^2 - x^3)$.

XCII. Page 565

(1) $x < 1$. (2) $x < \infty$. (3) $x < 3$.

XCIII. Page 575

1. $\frac{3}{1}, \frac{13}{4}, \frac{16}{5}, \frac{93}{29}.$

2. $\frac{0}{1}, \frac{1}{1}, \frac{1}{2}, \frac{4}{7}, \frac{41}{72}, \frac{496}{871}.$

3. $\frac{1}{1+}\frac{1}{5}.$

4. $8+\frac{1}{6+}\frac{1}{4+}\frac{1}{2}.$

5. $\frac{1}{3+}\frac{1}{1+}\frac{1}{1+}\frac{1}{1+}\frac{1}{1+}\frac{1}{2+}\frac{1}{1+}\frac{1}{1+}\frac{1}{4}.$

6. $3+\frac{1}{1+}\frac{1}{1+}\frac{1}{5+}\frac{1}{1+}\frac{1}{3}.$

7. $\frac{1}{6+}\frac{1}{1+}\frac{1}{6+}\frac{1}{3+}\frac{1}{9+}\frac{1}{7}.$

8. $1+\frac{1}{3+}\frac{1}{6+}\frac{1}{4+}\frac{1}{2}$; fourth convergent is $\frac{104}{79}$; error $< \frac{1}{79^2}$.
(Exact error is 1/79.177.)

9. $\frac{1}{2+}\frac{1}{3+}\frac{1}{4+}\frac{1}{5+}\frac{1}{6}$; fourth convergent is $\frac{13}{30}$; error $< \frac{1}{30^2}$.

10. $\frac{1}{1+}\frac{1}{2+}\frac{1}{1+}\frac{1}{4+}\frac{1}{19+}\frac{1}{3+}\frac{1}{1+}\frac{1}{2+}\frac{1}{1+}\frac{1}{5}$; fourth convergent is $\frac{3}{4}$; error $< 1/4 \cdot 19$.

11. $4+\frac{1}{8+}\frac{1}{8+}\cdots$; fifth convergent is $\frac{17684}{4289}$; error $< \frac{1}{4289^2}$.

12. $5+\frac{1}{10+}\frac{1}{10+}\cdots$; fifth convergent is $\frac{52525}{10301}$; error $< \frac{1}{10301^2}$.

13. $2+\frac{1}{2+}\frac{1}{4+}\cdots$; fifth convergent is $\frac{218}{89}$; error $< \frac{1}{89^2}$.

14. $6+\frac{1}{6+}\frac{1}{12+}\cdots$; fifth convergent is $\frac{33294}{5401}$; error $< \frac{1}{5401^2}$.

15. $10+\frac{1}{4+}\frac{1}{20+}\cdots.$

16. $\frac{1}{4+}\frac{1}{1+}\frac{1}{3+}\frac{1}{1+}\frac{1}{8+}\cdots.$

17. $4+\frac{1}{2+}\frac{1}{1+}\frac{1}{3+}\frac{1}{1+}\frac{1}{2+}\frac{1}{8+}\cdots.$

18. $8+\frac{1}{2+}\frac{1}{2+}\frac{1}{1+}\frac{1}{7+}\frac{1}{1+}\frac{1}{2+}\frac{1}{2+}\frac{1}{16+}\cdots.$

19. $5+\frac{1}{5+}\frac{1}{10+}\cdots.$

20. $\frac{1}{1+}\frac{1}{1+}\frac{1}{2+}\frac{1}{1+}\cdots.$

21. $5+\frac{1}{1+}\frac{1}{4+}\cdots.$

22. $(\sqrt{37}-4)/3.$

23. $(\sqrt{37}-5)/3.$

24. $\frac{3\sqrt{35}+50}{\sqrt{35}+15}.$

25. $\frac{\sqrt{1806}+36}{34}.$

26. $\frac{20+\sqrt{10}}{43+2\sqrt{10}}.$

30. $\frac{1}{1+}\frac{1}{1+}\frac{1}{1+}\cdots.$

33. 577/408; error $< 1/408^2$.

34. 355/113.

35. 87/32; error $< 1/32^2$.

36. $x = 546$, $y = 324$.

37. $x = 1350$, $y = -770$.

38. $x = 155 + 323t$, $y = 248 + 517t$.

INDEX

Numbers refer to pages

CHEL/354.H

$$x^3+y^3=(x+y)(x^2-xy+y^2)$$

$$x^3-y^3=(x-y)(x^2+xy+y^2)$$

$$(x+y)^3=x^3+3x^2y+3xy^2+y^3$$

$$(x-y)^3=x^3-3x^2y+3xy^2-y^3$$